Handbook of Biosensors

Handbook of Biosensors: Nanotechnology for Biomedical Diagnostics provides a comprehensive overview of the transformative role played by nanotechnology in the development of biosensors for biomedical applications. Highlighting interdisciplinary collaboration and technological innovation, this book demonstrates how to accelerate the translation of nanotechnology-enabled biosensors from the laboratory to the clinic, ultimately improving patient outcomes and quality of life. The inclusion of case studies highlighting successful applications of biosensors in healthcare, personalized medicine, and point-of-care diagnostics further enhances the book's relevance and utility for researchers, educators, and industry professionals alike. The book serves as a comprehensive guide for researchers, educators, and industry professionals interested in harnessing the power of nanotechnology to advance healthcare through innovative biosensor technologies.

Key Features:

- Offers an in-depth exploration of the role of nanotechnology in biomedical diagnostics, focusing specifically on biosensors.
- Features contributions from leading experts and researchers in the field of nanotechnology and biomedical diagnostics. These expert insights provide readers with valuable perspectives, practical advice, and cutting-edge research findings, enhancing the credibility and relevance of the content.
- Readers will benefit from hands-on guidance on biosensor design, fabrication, characterization, and optimization, facilitating the translation of research findings into practical solutions.

Series in Sensors

Series Editors: Barry Jones and Haiying Huang
Other recent books in the series:

Radiation Sensors with 3D Electrodes
Cinzia Da Vià, Gian-Franco Dalla Betta, Sherwood Parker

Semiconductor Radiation Detectors
Alan Owens

Advanced Chromatic Monitoring
Gordon R. Jones, Joseph W. Spencer

Nanosensors: Physical, Chemical, and Biological, ***2nd Edition***
Vinod Kumar Khana

Multi-Component Torque Sensing Systems
Qiaokang Liang

Ultra Fast Silicon Detectors: Design, Tests, and Performances
Marco Ferrero, Roberta Arcidiacono, Marco Mandurrino, Valentina Sola, Nicolò Cartiglia

A Hands-On Course in Sensors Using the Arduino and Raspberry Pi, Second Edition
Volker Ziemann

Multifunctional Sensors: Design, Construction, Methodology and Uses
Bansari Deb Majumder and Joyanta Kumar Roy

IoT Sensors: An Exploration of Sensors for Internet of Things
Vinod Kumar Khanna

Handbook of Carbon Sensors: Understanding and Applications
J. G. Manjunatha

Fabrication and Advanced Applications of Nanomaterial Based Electrochemical Sensors
Shashanka Rajendrachari, Vinayak Adimule

Handbook of Biosensors: Nanotechnology for Biomedical Diagnostics
J.G. Manjunatha

Handbook of Biosensors

Nanotechnology for Biomedical Diagnostics

Edited by
J.G. Manjunatha

CRC Press
Taylor & Francis Group
Boca Raton London New York

CRC Press is an imprint of the
Taylor & Francis Group, an **informa** business

Designed cover image: Shutterstock

First edition published 2026
by CRC Press
2385 NW Executive Center Drive, Suite 320, Boca Raton FL 33431

and by CRC Press
4 Park Square, Milton Park, Abingdon, Oxon, OX14 4RN

CRC Press is an imprint of Taylor & Francis Group, LLC

ISBN: 9781032873411 (hbk)
ISBN: 9781032881164 (pbk)
ISBN: 9781003536239 (ebk)

DOI: 10.1201/9781003536239

Typeset in Times
by Deanta Global Publishing Services, Chennai, India

Contents

Editor

J.G. Manjunatha is Assistant Professor at the Department of Chemistry, FMKMCC Madikeri, Mangalore University.

Contributors

Maliana El Aamri
Laboratory of Process Engineering & Environment, Faculty of Sciences and Techniques
Hassan II University of Casablanca
Casablanca, Morocco

Aziz Amine
Laboratory of Process Engineering & Environment, Faculty of Sciences and Techniques
Hassan II University of Casablanca
Casablanca, Morocco

Luciano Benedini
INQUISUR-CONICET, Department of Chemistry
Universidad Nacional del Sur Bahía Blanca
Argentina

Rishitha Gayathri Botlapally
Centre for Biotechnology, UCESTH
Jawaharlal Nehru Technological University
Hyderabad, India

Kalyani Chepuri
Centre for Biotechnology, UCESTH
Jawaharlal Nehru Technological University
Hyderabad, India

Maya Devi
Functional Materials Lab, Department of Physics, School of Applied Sciences
Kalinga Institute of Industrial Technology (KIIT)
Odisha, India

Radha Doddamani
Department of Chemistry
Karnatak University
Dharwad, India

Amar K. Durgannavar
Department of Chemistry
Karnatak University
Dharwad, India

Dalia S. Fathy
Petrochemical Department, Polymer Lab
Egyptian Petroleum Research Institute
Cairo, Egypt

Vadakavila Geethikalal
Centre for Biotechnology, UCESTH
Jawaharlal Nehru Technological University
Hyderabad, India

Ishita Ghosh
Department of Physics
Netaji Subhas University
Jharkhand, India

Kathuroju Harikrishna
Centre for Biotechnology, UCESTH
Jawaharlal Nehru Technological University
Hyderabad, India

Ramesha Hanumanthappa
Neurochemistry Lab, Department of Biochemistry
Karnatak University
Dharwad, India

Geetha B. Heggannavar
Department of Chemistry
Karnatak University
Dharwad, India

Nimisha Jadon
School of Studies in Environmental Chemistry
Jiwaji University
Gwalior, India

Ramavath Bharath Kumar
Centre for Biotechnology, UCESTH
Jawaharlal Nehru Technological University
Hyderabad, India

Ranabothu Madhurima
Centre for Biotechnology, UCESTH
Jawaharlal Nehru Technological University
Hyderabad, India

Dipak Maity
Department of Chemistry and Chemical Biology and Integrated Nanosystems Development Institute
Indiana University
Indianapolis, IN

Nahla A. Mansour
Petrochemical Department, Polymer Lab
Egyptian Petroleum Research Institute
Cairo, Egypt

Azza M. Mazrouaa
Petrochemical Department, Polymer Lab
Egyptian Petroleum Research Institute
Cairo, Egypt

Manal G. Mohamed
Petrochemical Department, Polymer Lab
Egyptian Petroleum Research Institute
Cairo, Egypt

Hasna Mohammadi
Laboratory of Process Engineering & Environment, Faculty of Sciences and Techniques
Hassan II University of Casablanca
Casablanca, Morocco

Pratiksha Pandey
Centre for Biotechnology, UCESTH
Jawaharlal Nehru Technological University
Hyderabad, India

Swetapadma Praharaj
Functional Materials Lab, Department of Physics, School of Applied Sciences
Kalinga Institute of Industrial Technology (KIIT)
Odisha, India

Chittepu Pranitha
Centre for Biotechnology, UCESTH
Jawaharlal Nehru Technological University
Hyderabad, India

Nutan Rani
Department of Chemistry
University of Delhi
New Delhi, India

Swati Rani
Department of Chemistry
University of Delhi
New Delhi, India

Ramon Rial
Soft Matter and Molecular Biophysics Group, Department of Applied Physics and Institute of Materials (iMATUS)
University of Santiago de Compostela
Santiago de Compostela, Spain

Dibyaranjan Rout
Functional Materials Lab, Department of Physics, School of Applied Sciences
Kalinga Institute of Industrial Technology (KIIT)
Odisha, India

Kalawati Saini
Department of Chemistry
University of Delhi
New Delhi, India

Swati Shrivastava
School of Studies in Environmental Chemistry
Jiwaji University
Gwalior, India

Sumitha M. S.
Centre for Advanced Materials Research, Department of Physics, Government College for Women
University of Kerala
Thiruvananthapuram, India

Sanchaita Tewary
Centre for Biotechnology, UCESTH
Jawaharlal Nehru Technological University
Hyderabad, India

Priyanka Thakral
Division of School of Interdisciplinary Arts and Sciences
University of Washington
Tacoma, WA

Puja Tomar
School of Studies in Environmental Chemistry
Jiwaji University
Gwalior, India

Volukula Vaishnavi
Centre for Biotechnology, UCESTH
Jawaharlal Nehru Technological University
Hyderabad, India

Xavier T. S.
Centre for Advanced Materials Research, Department of Physics, Government College for Women
University of Kerala
Thiruvananthapuram, India

Sapna Yadav
Department of Chemistry
University of Delhi
New Delhi, India

Ghita Yammouri
Laboratory of Process Engineering & Environment, Faculty of Sciences and Techniques
Hassan II University of Casablanca
Casablanca, Morocco

Preface

In recent decades, the convergence of nanotechnology and biosensing has revolutionized the landscape of biomedical diagnostics. This book, *Handbook of Biosensors: Nanotechnology for Biomedical Diagnostics*, aims to provide a comprehensive overview of the current progress, innovations, and future directions in this rapidly evolving interdisciplinary field. With increasing demands for rapid, accurate, and cost-effective diagnostic tools, biosensors based on nanomaterials have emerged as powerful platforms for disease detection, personalized medicine, and real-time health monitoring. Nanomaterials such as nanoparticles, nanotubes, quantum dots, and nanowires have brought about a paradigm shift in biosensor technology. Their unique physicochemical properties enhance sensitivity, selectivity, and miniaturization, enabling the detection of minute concentrations of biomarkers and pathogens. By integrating these materials with cutting-edge techniques in chemistry, biology, and electronics, researchers are pushing the boundaries of early diagnostics and point-of-care testing. This book brings together contributions from leading researchers across academia and industry, highlighting the design principles, fabrication strategies, and real-world applications of nanobiosensors. It covers a broad spectrum of diagnostic areas including cancer, infectious diseases, metabolic disorders, and neurological conditions. Particular emphasis is placed on the role of emerging nanostructures, hybrid materials, and smart sensing interfaces in improving analytical performance and biocompatibility. The objective of this volume is not only to inform but also to inspire innovation by offering insights into current challenges and future trends. It is our hope that this compilation will serve as a valuable resource for students, researchers, clinicians, and technologists working at the interface of nanotechnology and biosensing. We express our sincere gratitude to all the contributors for their scholarly efforts and to the reviewers for their valuable feedback. We also thank the editorial team for their unwavering support throughout the preparation of this book. As we move forward into an era where healthcare is becoming increasingly personalized and digital, we believe that nanobiosensors will play a pivotal role in shaping the future of diagnostics. May this book serve as both a reference and a catalyst for further research and collaboration.

Dr J. G. Manjunatha
Editor

1 Introduction to Nanotechnology in Biomedical Diagnostics

Sapna Yadav, Nutan Rani, Kalawati Saini and Dipak Maity

1.1 INTRODUCTION

Nanotechnology, often known as nanoscience, is the investigation and development of applied sciences at the atomic or molecular level, which generally ranges from 1 to 100 nanometers (nm). The term "nano" is claimed to be derived from the Greek word meaning "dwarf." The nanoscale, despite its small size, offers enormous promise. Nanotechnology is modifying matter on an extremely tiny scale, where new physical, chemical, and biological properties can emerge that vary from those at larger scales [1, 2]. A nanometer is one-billionth of a meter. Nanomaterials possess unique optical, magnetic, electrical, and physical properties, including reactivity, strength, surface area, sensitivity, and stability. This field encompasses medical (drug delivery systems, sensors), electronics (nanoscale transistors), energy (solar cells, supercapacitors, batteries), and materials science. In essence, nanotechnology is the creation and use of systems, materials, or devices having traits and capabilities resulting from their extremely tiny scale [3–6].

Nanomaterials vary in shape, size, and chemical content. According to their dimensions, they may be divided into four primary categories: zero-dimensional, one-dimensional, two-dimensional, or three-dimensional (0D, 1D, 2D, or 3D, respectively) [7].

Zero-dimensional (0D) nanomaterials: A material is classified as having a zero-dimensional nanostructure if all three of its dimensions are within the nanometer scale (less than 100 nm in all directions) [8, 9]. They include nanoparticles, quantum dots, and so on [9].

One-dimensional (1D) nanomaterials: A material is classified as having a one-dimensional nanostructure if any two of its dimensions are within the nanometer scale and one dimension is outside the nanoscale range [8, 9]. They include nanotubes, nanorods, nanowires, and so on [9].

Two-dimensional (2D) nanomaterials: These materials, which frequently form sheet-like structures, have two dimensions within the nanoscale, and one dimension

DOI: 10.1201/9781003536239-1

is outside the nanoscale [8, 9]. They include nanofilms, nanosheets, nanolayers, and so on [9].

Three-dimensional (3D) nanomaterials: These materials contain nanoscale structures in three dimensions. This indicates they are larger than 100 nm in all three dimensions. These include nanocomposites, core shells, multi-nanolayers, nanowire bundles, and nanotube bundles [9].

Nanostructures vary greatly in manufacture, characteristics, and applications. These three nanomaterials have distinct electrical, optical, magnetic, and surface characteristics, making them suitable for varied applications [8]. The classification of nanomaterials is also given in Figure 1.1.

There are specific categories of nanomaterials, such as:

- Carbon-based nanomaterials: These include carbon nanofiber, carbon black, graphene, carbon nanotubes, and fullerenes. They have unique mechanical, thermal, and electrical properties [10].
- Metal-based nanomaterials: Metal nanoparticles often involve zinc, cobalt, aluminum, copper, cadmium, iron, gold, lead, and silver. Nanoparticles range in size from 10 to 100 nm, with unique surface features including high surface-to-volume ratio, pore size, surface charge, crystalline structures, color, reactivity, morphologies, and sensitivity [10].
- Metal oxide-based nanomaterials: These include ZnO, NiO, CrO_2, MoO_3, MnO_2, TiO_2, Fe_2O_3, Co_3O_4, CeO_2, Bi_2O_3, and so on. These nanoparticles have unique properties that set them apart from their metal counterparts [3–10, 11].
- Polymeric nanomaterials: These are made from nanoscale polymer particles, useful in tissue engineering and drug delivery [10].
- Ceramic nanomaterials: Ceramic nanoparticles are inorganic solids composed of carbonates, carbides, oxides, carbonates, carbides, and phosphates. They are created by heating and then cooling them down. They have high strength and stability and often used in several applications such as the photodegradation of dyes, drug delivery, imaging applications, and catalysts [3].

Nanotechnology facilitates the development and utilization of nano-sized materials to develop previously unimaginable instruments, diagnostics, and treatments. In the biomedical field, nanotechnology enables scientists and engineers to create systems and devices that can interact with biological processes at the cellular and molecular levels, improving the precision, efficacy, and scope of medical interventions [12].

Nanotechnology has had a substantial influence on drug delivery systems, diagnostic and health monitoring, and regenerative medicine, among other disciplines. Traditional medication delivery techniques frequently have disadvantages, such as limited selectivity, poor bioavailability, and unwanted side effects. However, nanoparticles can be designed to deliver medications directly to particular cells or tissues, decreasing adverse effects and boosting therapeutic results [13]. Nanotechnology can also be used to alter diagnostic procedures, making them quicker, more precise,

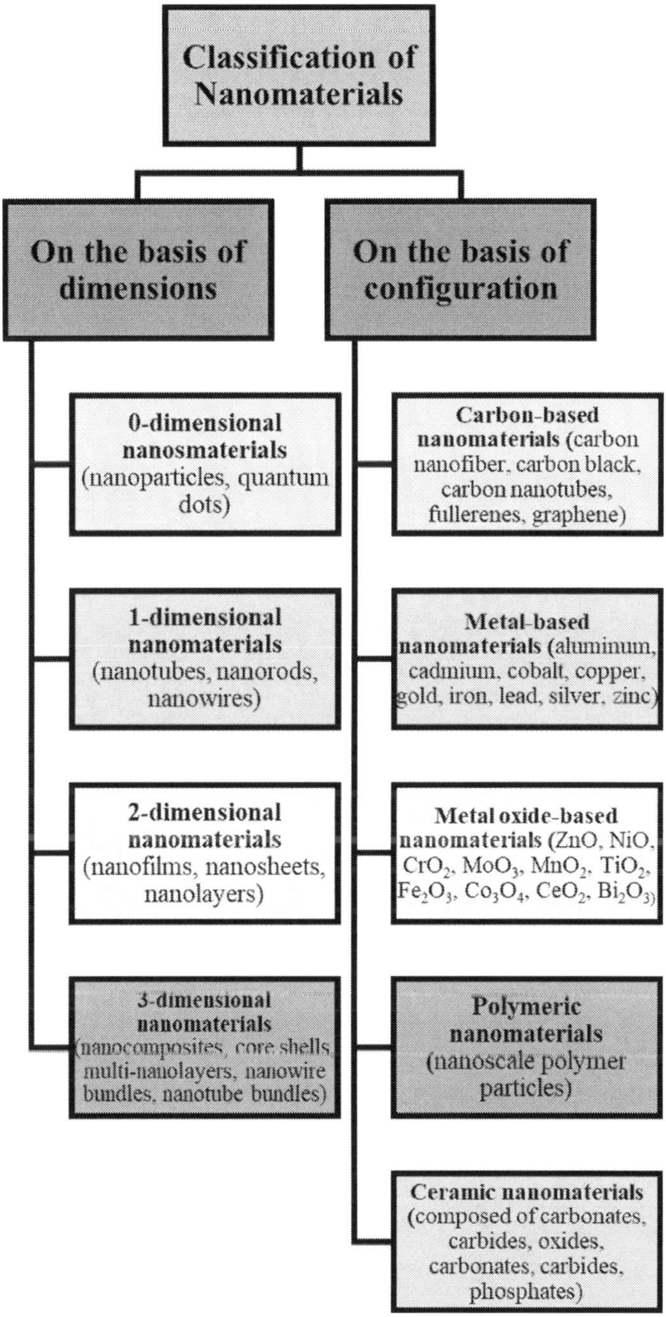

FIGURE 1.1 The classification of nanomaterials.

and less intrusive [14, 15]. Nanotechnology's promise goes even further, including the development of improved biosensors and wearable health monitoring systems. Nanosensors can help to fight various health issues by detecting and analyzing certain biomolecules [16–18]. Nowadays, the use of nanotechnology in the field of wound healing and tissue engineering is opening up new possibilities. Nanomaterials can be developed in a way that they resemble the extracellular matrix of human tissues. These nanomaterials can support tissue regeneration, cell proliferation, and differentiation. Therefore, nanomaterials are promising materials for providing alternatives for tissue engineering, such as bone, skin, cartilage, and so on [19, 20].

Conventionally used diagnostic methods have several drawbacks, such as the requirement of huge sample amounts/volumes, incorrect results, time-consuming processing, the need for expert professionals, and very expensive equipment. On the other hand, nanosensors and lab-on-a-chip devices require minimal sample amount/volume and can detect diseases at an early stage [21]. Researchers are developing specific biomarkers functionalized nanoparticles for accurate and fast diagnostic results. These biomarkers are associated with specific diseases such as cancer, cardiovascular disease, and infections. Another important application of nanomaterials is in point-of-care systems. Point-of-care nanodevices are very effective for on-the-spot monitoring of patients' health status. This on-time health monitoring facilitates timely medical intervention [22–24].

Nanotechnology based nanosensors have the capacity to be fine-tuned for recognition and interaction with specific biomolecules linked to certain illnesses. Also, nanosensors are able to reduce the likelihood of false positive diagnostic results and increase the accuracy of diagnostic results. Additionally, nanotechnology may detect many biomarkers in a single test, which gives a more extensive image of a patient's health condition by increasing overall diagnostic capability. Nanotechnology-based diagnostic techniques give earlier and more precise therapies. Therefore, the results from these diagnostic techniques can be used for the development of personalized medicine for patients [21–25].

Nanotechnology has promising applications as well as drawbacks in the field of medical diagnostics. Researchers must overcome these drawbacks (possible toxicity, environmental impact, and ethical considerations) before its commercial use. Currently, the use of nanotechnologies in the field of biomedical applications is growing very fast. Researchers across the globe are working on the development of new nanodevices for better diagnostic and therapeutic approaches for personalized patient care. Finally, we can say that nanotechnology can improve the field of biomedical applications by improving understanding, diagnosis, and treatment of diseases [26–28].

Another interesting field is the utilization of nanotechnology and nanomaterials in drug molecules analysis. Drugs are chemicals which bind with body proteins and cells and change their original biological structure and physiological functions. Drugs do not just bind with specific proteins or cells; they also interact with other sites. This is the main reason for their toxicity and side effects. Nanotechnology plays a vital role in drug delivery and drug discovery [29, 30]. Pushpanjali et al. have designed a poly(L- leucine) layered carbon nanotube paste electrode for the sensitive and selective investigation of cetirizine in presence of paracetamol [31]. Monnappa et al. have developed a voltammetric sensor based on a cetyl trimethylammonium bromide-immobilized carbon paste electrode for the investigation of alloxan [32]. Raril et al. have designed a poly-threonine modified graphite-carbon nanotube paste electrode for the determination of tyrosine [33]. Tigari et al. have prepared a poly(glutamine) film-modified carbon nanotube paste electrode for the determination of curcumin [34]. Hareesha et al. have designed a poly-thiazole yellow-G layered multi-walled carbon nanotube paste electrode for the analysis of formoterol fumarate in the presence of uric acid [35]. Manjunatha et al. have developed a poly(maleic acid) modified carbon paste electrode for the determination of dopamine in the presence of uric acid [36].

Herein, we are aiming to provide an extensive study of the role of nanotechnology in biomedical diagnostics. In this chapter we have explored the role of nanotechnology in various areas of biomedical applications such as cancer diagnostics, disease biomarker detection, infectious disease detection, point-of-care diagnostics, contrast agents, and targeted imaging. Also, we have given an overview about the advantages and disadvantages of nanomaterials in the healthcare industry. Hopefully this chapter will offer a comprehensive insight into the role of nanotechnology in the healthcare and monitoring sector and motivate more and more researchers for innovation in this field.

1.2 NANOMATERIALS IN BIOMEDICAL DIAGNOSTICS

Nowadays, nano-sized materials have become very popular for their magnificent performance in biomedical diagnostics. Nanomaterials can interact with biological systems at molecular and cellular levels. As we know that the nanomaterials with very small sizes are responsible for their special qualities. Besides their very small size, nanomaterials have high surface area and adjustable electric, magnetic, and optical properties. Also, nanomaterials can be functionalized according to biological interactions required. Herein, we will discuss the significant role of nano-sized materials in the field of biomedical diagnostic approaches. A summary of nanomaterials in biomedical diagnostics is given in Figure 1.2.

1.2.1 Nanoparticles in Imaging Modalities

A key component of clinical practice is early illness detection and diagnosis, particularly for cancer. For instance, it has been found that patients with gastrointestinal cancer who benefited from early identification had a much greater two-year survival rate than those who did not. Furthermore, individuals with breast cancer who benefited from early identification had a 17%–28% lower death rate [37]. Molecular imaging is the non-invasive evaluation of biological and metabolic processes in live beings, particularly in humans and experimental animals. Molecular imaging plays a very significant role in disease diagnosis and therapy. The imaging methods have been broadly divided into two groups: functional or molecular imaging methods and morphological or anatomical imaging methods. Magnetic resonance imaging

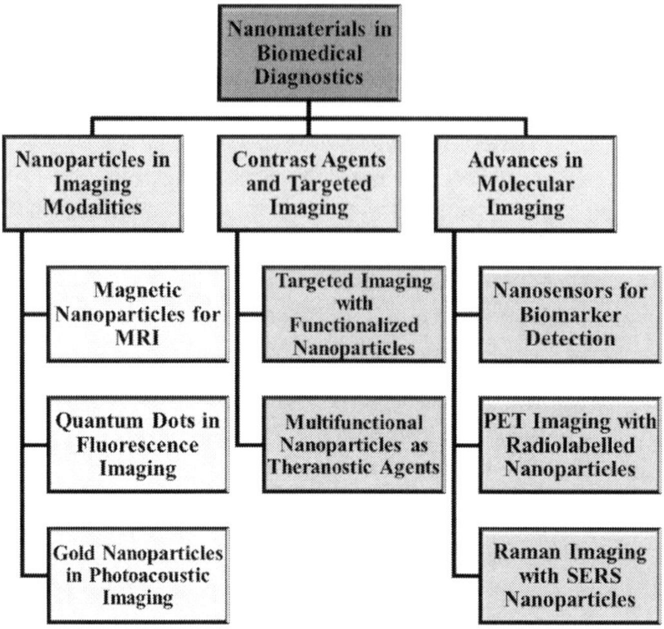

FIGURE 1.2 Summary of nanomaterials in biomedical diagnostics.

(MRI), computed tomography (CT), and ultrasound (US) are the most crucial techniques for clinical translation and high spatial resolution. Nowadays, optical imaging (OI), positron emission tomography (PET), and single photon emission computed tomography (SPECT) are being used to characterize molecular/functional imaging modalities [38].

Nanomedicine and nanotechnology have proven to offer a new promising field in the health care industry. As we have already discussed that the nanomaterials in size from 1 nm to 100 nm have a significant impact on diagnostics and imaging. Due to their tiny size, nanoparticles can be used more efficiently in imaging techniques as compared to traditional materials. Conventional imaging modalities like computed tomography (CT), magnetic resonance imaging (MRI), and ultrasound have several shortcomings in resolution, sensitivity, and specificity. Nanomaterials provide solutions to these shortcomings by offering insights into biological processes and higher-resolution images [38, 39].

MRI is a non-invasive, radiation-free technology used for medical diagnosis. MRI instruments apply an external magnetic field to align magnetic nuclei, allowing for resonance with a radiofrequency pulse [40]. Since ^1H is the most prevalent magnetic nucleus in the human body and has the greatest gyromagnetic ratio in nature, it is the most often investigated subject in MRI [40]. MRI contrast agents are classified into T1 and T2 categories based on their capacity to interact with magnetic fields. T1 contrast agents are primarily transition metal ion complexes with various chelating agents, particularly Gd(III). T2 contrast agents consist of magnetic nanoparticles (MNPs), primarily γ-Fe_2O_3 and Fe_3O_4. Iron oxide nanoparticles with magnetic nuclei smaller than 20 nm exhibit strong magnetization and are classified as superparamagnetic [41]. Because of their intrinsic ultrafine size, biocompatibility, and superparamagnetic qualities, magnetic nanoparticles (MNPs) are being explored more and more for a variety of biological applications. The size, functionality, stability, dispensability, and interfacial surfaces of iron oxide nanoparticles all affect their usefulness. Polymer-stabilized magnetic nanoparticles are employed for MRI, cell movement tracking, and *in vivo* cell differentiation state monitoring due to their excellent spatial resolution [42].

Magnetic nanoparticles, specifically iron oxide nanoparticles, have gained popularity in MRI as contrast agents. These nanoparticles can modify the relaxation durations of adjacent hydrogen atoms, resulting in higher contrast in MRI pictures. Iron oxide nanoparticles, which are commonly coated with biocompatible polymers, are capable of being functionalized with targeting ligands that allow them to attach to certain cell types, such as cancer cells. This method not only improves contrast but also offers information about specific molecular targets, which aids in tumor identification and characterization [42, 43].

In recent years, a thorough evaluation of fluorescence microscopy staining and labeling methods has been conducted. Fluorescence microscopy is the primary method used in biomedical optical imaging. Confocal microscopes may provide sectional views if fluorescent dyes are used to stain tissue, cells, or cellular compartments. Fluorescent dyes can also be used to selectively label biomolecules to track biomolecular interactions at transmembrane-bound receptors or inside cells [44]. Perhaps more than any other nanomaterial, semiconductor nanocrystals, also known as quantum dots (QDs), have influenced bioimaging research.

When the link between semiconductor materials' size and band-gap was explained, the variety of QD-based applications seemed unimaginable [45]. Nanomaterials provide a great deal of potential for clinical diagnostics and bioimaging.

Less than 10 nm in size, semiconductor quantum dots (QDs) are molecular-sized nanocrystals. Strong quantum confinement effects provide QDs with several special photoelectric characteristics, including reduced photo-bleaching and luminescence that vary with wavelength and size. These characteristics are combined in QDs-based imaging to enhance fluorescence imaging, which has the advantages of real-time, precise, and in vivo observations with high sensitivity, quick reactions, cheap costs, and no radiation [46].

The photo acoustic (PA) effect, which was initially noted by Alexander G. Bell in 1880, is the basis of the swiftly evolving biomedical imaging technique known as photo acoustic imaging (PAI). In the absence of ionizing radiation, PAI provides real-time, high-spatial-resolution information on the morphological, functional, and molecular content of sick tissues. Because acoustic signals scatter far less than light in tissue, PAI provides a deeper imaging depth and a higher spatial resolution (down to 5 μm) than fluorescence optical imaging. Because PAI is based on the optical characteristics of various tissues, it provides superior tissue contrast to ultrasonic imaging, which is constrained by the mechanical characteristics of biological tissues [47].

The uses of PAI are significantly expanded by the introduction of contrast agents based nanoparticles. These enable PAI to image more clearly and deeply inside tissue. Nanoparticles are often constructed with maximal absorption in the near-infrared range, where tissue has minimal optical attenuation, allowing for deep light penetration. Nanoparticles can be coated with targeted peptides and antibodies, allowing for cell-specific contrast and imaging [48]. Since metal nanoparticles are biocompatible and have a wide absorption cross-section, they first served as photo-acoustic contrast agents. Gold nanoparticles are desirable because they allow for multiplexing in photo-acoustic imaging [49]. Gold nanoparticles may absorb and scatter light at certain wavelengths, known as localized surface plasmon resonance (LSPR) [50]. Gold nanoparticles have been extensively used in photo acoustic tomography imaging applications due to their optical characteristics. Gold particles are excitable in the near-infrared (NIR) region (650 nm–900 nm), allowing for deeper light penetration, lesser auto fluorescence, and reduced light scattering. This enables decreased light scattering, lower auto fluorescence, and deeper light penetration. Unlike small molecule fluorophores, which can also be activated in the NIR region using single- or two-photon excitation, gold particles have a significant advantage in that they are resistant to photo-bleaching [51]. To go into clinical trials with Au nanoparticles, it's necessary to have a thorough understanding of their interactions with biological samples. To eliminate the need for hazardous components in synthetic procedures, it is crucial to use non-toxic green chemicals during synthesis [52].

1.2.2 Contrast Agents and Targeted Imaging

Contrast agents are chemicals that increase the visibility of particular tissue abnormalities or structures, making them simpler to identify and examine. By facilitating targeted imaging and boosting contrast for increased diagnostic precision, nanomaterials have progressed the field of contrast agents. Nanomaterials provide specificity and functionality that conventional agents cannot match. Nanotechnology developments might result in a tumor-targeting contrast agent based on NPs that would improve tumor imaging's sensitivity and specificity and allow for the early detection of metastases [53].

By altering their surface, functionalization improves the qualities and features of nanoparticles and makes it possible for them to be used extensively in medicine. Labeling particular locations or cells is crucial for quantitative analysis of illness or tissue damage [54]. Targeted imaging is achieved by functionalizing nanoparticles with biomolecules, such as antibodies, peptides, or small molecules, that bind selectively to specific biomarkers. In cancer diagnostics, where early identification of cancerous cells may greatly enhance the survival of patients, this targeting power is very helpful [55]. For many years, nanoplatforms for cancer imaging and treatment have been developed, including superparamagnetic iron oxides, gold nanoparticles, liposomes, protein-bound particles, polymers, and quantum dots. Gold nanoparticles enveloped by targeting moieties, therapeutic payloads, and hydrophilic polymer coatings such as polyethylene glycol (PEG), polysaccharides, poloxamines, or poloxamers generate multifunctionalized gold nanoparticles. These platforms have been developed as contrast agents for computed tomography (CT) imaging and as enhancement agents for radiotherapy since they are compatible with X-rays. They have special qualities such as low inherent toxicity, biologically relevant size, ease of functionalization with biomolecules, and high surface area that contribute to better therapeutic delivery, imaging, and non-invasive disease diagnostics. They also have improved optical properties [56, 57].

Similarly, iron oxide nanoparticles can be coated with antibodies or ligands that target-specific tumor markers. The accuracy of MRI scans is increased by this functionalization, which permits the selective accumulation of nanoparticles in tumor tissues. These contrast chemicals reduce false positives and increase diagnostic confidence by attaching exclusively to the targeted cells, making it easier to distinguish between healthy and sick tissue [58, 59].

Theranostic agents combine therapeutic and diagnostic functionalities in a single nanoparticle platform, enabling simultaneous disease detection, treatment, and monitoring. This subject is being pushed by advances in nanoparticle systems that can provide the required functionality. Nanomedicines enable personalized treatment solutions for individuals [60]. For example, mesoporous silica nanoparticles have been used in drug delivery systems due to their great drug-carrying capacity. The unique morphology, mono-distribution, tuneable particle sizes, uniform pore volume, excellent biocompatibility, thermal stability, exceptional physicochemical properties (e.g., colloidal stability/water dispersibility), versatile surface chemistry, high surface area, and high loading capacity contribute to their success [61]. Iron Oxide nanoparticles can be used as carriers to deliver medications directly to cancer cells as well as contrast agents for imaging. Theranostics eliminates the need for distinct diagnostic and therapeutic processes by employing a single chemical to both find and treat tumors, enabling real-time evaluation of treatment efficacy [62, 63].

Theranostic agents can also be programmed to release therapeutic chemicals in response to particular stimuli like pH changes or enzyme activity in the tumor microenvironment. This targeted release mechanism ensures that medications are given precisely where they are required, reducing side effects to healthy tissues and improving therapeutic success [64, 65].

1.2.3 Advances in Molecular Imaging

Molecular imaging is a medical imaging technology that uses sensitive instruments and contrast to identify, localize, and monitor molecular processes in cells, tissues, and organisms. This field of medical imaging has grown in parallel with the development of molecular medicine, which considers each patient's genetic composition when developing a treatment plan [66].

The rapid advancement of biocompatible nanotechnologies currently holds up the possibility of broadening the scope of molecular imaging and treatment using a variety of innovative substances. Nanotechnology has propelled significant advancements in molecular imaging, particularly through the development of nanoparticles that can detect and report on specific molecular events. These developments have improved the ability to detect diseases at an early stage, track disease progression, and monitor responses to treatment [67].

Nanosensors are nanoparticles that detect specific biological molecules, such as proteins or nucleic acids, associated with diseases. These sensors are highly sensitive and can detect biomarkers at very low concentrations, making them ideal for early disease detection [67]. For example, carbon (graphene, graphene oxide, and carbon nanotube nanomaterials) and non-carbon (gold, silver, copper, platinum, inorganic-organic hybrid nanoparticles, etc.) nanosensors have been developed to detect cancer biomarkers such as prostate-specific antigen (PSA) in blood samples. These nanosensors offer a non-invasive diagnostic method with high sensitivity, potentially enabling early cancer detection when treatment is most effective [68].

Nanosensors can also be designed for *in vivo* applications, where they continuously monitor biomarker levels and transmit data in real time. Such sensors are valuable for chronic disease management, where tracking the progression of disease markers can inform treatment adjustments and improve patient outcomes. By detecting biomarkers before symptoms appear, nanosensors enable a shift toward preventive healthcare, reducing the burden of disease on healthcare systems [69, 70].

Numerous radiolabeled nanoparticles have been created and assessed in recent years for their prospective application as probes in the imaging of a wide range of malignancies using positron emission tomography (PET). PET imaging technology, one of the many imaging modalities employed, offers extremely sensitive and quantitative data and is anticipated to become more and more significant in the early diagnosis of diseases and better treatment decision-making. Using positron emission tomography (PET) for non-invasive imaging during the development phase is a great way to examine pharmacological profiles and choose the best nanodimensional design for the targeted drug delivery system. Moreover, molecular imaging techniques are crucial for patient screening, selection, and monitoring in a personalized therapeutic strategy [71, 72].

PET imaging can track the distribution and concentration of positron-emitting radionuclides when imaging probes labeled with them are injected. The positron from the nucleus finally collides with a neighboring negatively charged electron. Two 511 keV γ-rays that are around 180° apart are created during the annihilation and picked up by a PET scanner. The distribution and concentration of isotopes can be reflected in the γ-ray pictures produced by the PET scanner [73].

With half-lives ranging from 122 s to 109.7 min, the most often employed radionuclides with short half-lives are gallium-68, fluorine-18, carbon-11, nitrogen-13, and oxygen-15. The longer-lived positron emitters have half-lives of 100.22 h for iodine-124, 78.4 h for zirconium-89, and 12.701 h for copper-64. The most often employed chelators in indirect labeling are diethylenetriamine-*N*,*N*,*N*′ ,*N*,*N*-pentaacetic acid, pentetic acid, (carboxymethyl)imino]*bis*(ethylenenitrilo)-tetra-acetic acid (DTPA) and deferoxamine (DFO), 2,2′,2′ ′,2′ ′ ′-(1,4,7,10-tetraazacyc lododecane-1,4,7,10-tetrayl)tetraacetic acid (DOTA), and 1,4,7-triazacyclononane-*N*,*N*′,*N*′ ′-triacetic acid (NOTA). Most of the research published in the literature is on using NPs radiolabeled with positron-emitting isotopes for tumor imaging, although they have also been utilized to detect neurological, cardiovascular, and rheumatic conditions [74].

Surface-enhanced Raman scattering (SERS) is an advanced imaging technique that enhances the weak Raman signals typically produced by biomolecules. SERS nanoparticles are coated with a metallic surface, usually gold or silver, which amplifies Raman signals by several orders of magnitude, making it possible to detect minute amounts of biomolecules with high sensitivity. SERS nanoparticles can be functionalized to target-specific molecules, allowing for the detection of disease markers at extremely low concentrations [75–77].

In cancer diagnostics, SERS nanoparticles are often conjugated with targeting ligands that bind to tumor-specific antigens. When exposed to a laser, these nanoparticles generate a unique Raman signal, making it possible to identify and quantify the presence of cancer cells. This technique provides high-resolution molecular information and has the potential for use in intraoperative imaging, where real-time detection of tumor margins is essential for successful surgical outcomes [75–77].

1.3 APPLICATIONS OF NANOTECHNOLOGY IN DISEASE DIAGNOSIS

1.3.1 Nanotechnology in Cancer Diagnostics

Cancer is a category of illnesses characterized by the uncontrolled development and spread of abnormal cells. Death may ensue if the spread of cancer cells is not stopped at metastasis [78]. Throughout human history, cancer has been a leading cause of death. As humans developed industrially and technologically, cancer's hold shifted. Nevertheless, this century has seen a sharp decrease in the risk of certain cancer kinds in wealthy nations. However, the prevalence of the disease's most notable types has gone up. In nations where risk factors, including cigarette smoking, unhealthy eating patterns, and exposure to hazardous chemicals at work or in the environment, are increasingly prevalent, the incidence of lung, breast, prostate, colon, and rectal cancers has increased [79].

One of the key components of broad cancer care is biomedical imaging. Real-time monitoring, accessibility without causing tissue damage, low or no invasiveness, and the ability to operate over a broad range of time and size scales involved in biological and pathological processes are just a few of the numerous benefits of biomedical imaging [80].

Around the world, oncologists treat cancer patients using a range of methods, such as chemotherapy, radiation therapy, and surgery. However, several restrictions must be addressed to treat tumor tissue, which drives the growing interest in using nanomaterials. A growing understanding of the tumor microenvironment over the past ten years has spurred our attempts to use nanoparticles in a cutting-edge cancer treatment and diagnostic approach [81].

Improving treatment results requires early cancer detection before the disease has spread or caused permanent harm. But when it comes to sensitivity, precision, and the capacity to identify cancer in its early stages, conventional diagnostic techniques frequently fall short. Nanotechnology has demonstrated enormous promise in this area. Researchers have created nanoparticles with special qualities that have the potential to completely change how we detect and track cancer [82]. The role of nanotechnology in cancer theranostics is summarized in Figure 1.3.

Enhancing early detection is one of the main benefits of nanotechnology in cancer diagnosis. Conventional procedures like imaging (CT, MRI, and mammography) and biopsy frequently don't identify cancer until the tumor has grown to a size that can be felt with a physical examination or seen on imaging. These techniques are frequently intrusive and expensive, and do not always provide a comprehensive view of the degree of malignancy. To overcome these constraints, nanotechnology has made it possible to create extremely sensitive and precise diagnostic instruments that can identify cancer in its early stages, often even before the tumor is big enough to be seen using traditional imaging methods. Nanoparticles can be customized to interact with cancer cells and their molecular constituents [83, 84].

It is possible to create nanoparticles that will selectively bind to biological markers or receptors on the surface of cancer cells. Biomarkers and antigens expressed by cancer cells are frequently different from those expressed by healthy cells. Nanoparticles such as gold nanoparticles, carbon nanotubes, and quantum dots can be engineered to bind to these biomarkers, allowing cancer cells to be detected at extremely low quantities [85, 86]. For example, gold nanoparticles can be functionalized with antibodies that recognize cancer cell surface receptors, allowing them to attach to tumor cells. Once these nanoparticles have attached to tumor cells, a variety of methods, including optical, magnetic, and fluorescence-based methods, may be used to detect and analyze the cancer cells. The sensitivity of these techniques

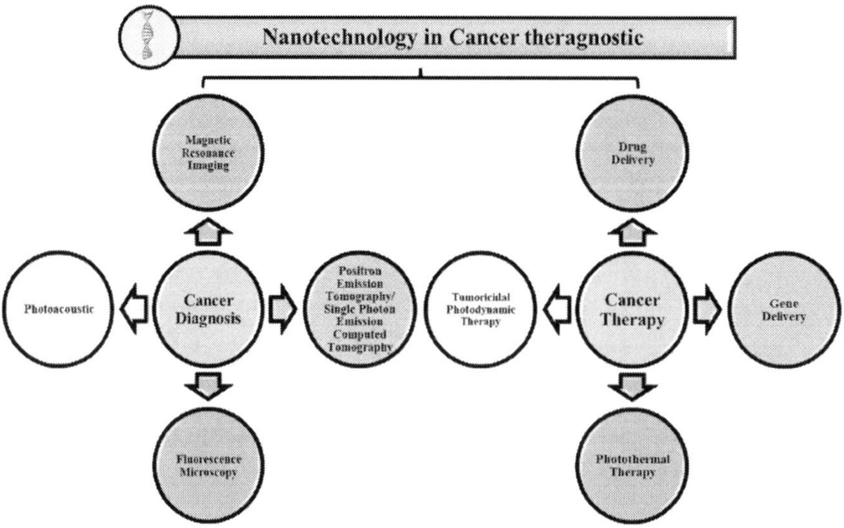

FIGURE 1.3 Role of nanotechnology in cancer theragnostic.

is effectively higher for the identification of cancer-related biomarkers in various bodily fluids like blood, urine, and so on. Consequently, they make the diagnosis of cancer cells less invasive and more efficient [87–89].

As already indicated, nanomaterials have a tendency to improve accuracy, resolution, and early detection of functional groups of imaging techniques. MRI, CT, and PET scanning are important imaging techniques used in the detection, staging, and analysis of cancer. However, there are some disadvantages such as a lack of the precision required to identify tiny tumors and to distinguish between benign and malignant growths. Nanomaterials can be used as contrast agents for these imaging methods. These nano-contrast agents can increase the sensitivity and provide a variety of imaging methods, and give a more detailed image of the tumor. Magnetic nanomaterials can work as MRI contrast agents. First, these nanomaterials are injected into the patient's body. Then they build up in tumor tissues to make tumors more visible on MRI scans. Therefore, these nanomaterials make it possible to see tumors more clearly, even the smallest ones that could not be seen using traditional techniques.

Quantum dots, also known as semiconductor nanocrystals, are nanoparticles with special optical characteristics controlled by the quantum confinement effect. They are made up of elements from Groups II–VI, II–V, III–V, IV–VI, I–VI, I–III–VI, IV, or II. The particle size of quantum dots is equal to or less than the exciton Bohr radius of the comparable bulk material, which is responsible for these characteristics. By adjusting their size and elemental composition, one may be able to achieve a variety of fluorescent characteristics. In the presence of light, quantum dot nanoparticles (linked to certain molecules or antibodies that bind to cancer cells) release fluorescence that makes it possible to identify cancer cells even at extremely low concentrations [90]. Quantum dots will complement existing fluorophores and fluorescent protein-fusion technologies for the need for enhanced photostability, near-infrared region emission, or single-molecule sensitivity over long periods. The resolution of optical imaging methods is greatly increased by this method, making it possible to see tumors at the cellular and subcellular levels. Another benefit of quantum dots is their stability and ability to emit several colors, which can help in identifying different kinds of cancer or stages of the illness in a single test [91].

Another important application of nanotechnology in cancer diagnostics is in the field of liquid biopsy. A liquid biopsy is a non-invasive procedure for identifying cancer-related genetic material or tumor cells in the blood circulation. Liquid biopsies can offer real-time data on cancer development and therapy response. Traditional biopsy techniques can be intrusive and uncomfortable, and they may not always give a complete picture of the malignancy because in traditional biopsy techniques, a tissue sample is taken from the tumor. However, circulating tumor cells (CTCs), extracellular vesicles (EVs) released into the blood by tumors, and genetic alterations linked to cancer can all be found using liquid biopsy. These elusive biomarkers may be extracted from the blood via nanoparticle engineering, enabling a less intrusive and more thorough diagnostic. This approach provides valuable insights into the genetic information of a patient's cancer and can help identify mutations that may be driving tumor growth [92, 93].

Based on the information above, we may conclude that nanotechnology is revolutionizing cancer diagnostics by improving imaging methods, facilitating early detection, and offering real-time tracking of cancer growth. The ability to identify the disease early on might greatly increase the survival rates and quality of life of cancer patients. Furthermore, nanotechnology enables more accurate, more individualized, and less invasive diagnostic methods that can assist direct treatment decisions and reduce the need for repeated biopsies. Nanotechnology holds great promise for the detection of cancer, and as further research is conducted, it is likely to become an important part of cancer therapy in the future.

1.3.2 Nanotechnology in Infectious Disease Detection

Numerous viruses and other diseases have emerged in recent years. Pandemics have been caused by the global spread of several of these infectious illnesses. Even though several kinds of biosensors have been used to detect viruses, their low sensitivity is still a problem. Because of this, it is very important to develop and modify various diagnostic methods for the sake of effective diagnosis of viruses and other infectious diseases. Nanotechnology has gained attention nowadays as an evolutionary tool in the field of infectious diagnostics. Nanotechnology gives unparalleled sensitivity, specificity, and speed in infection monitoring. Nanotechnology has the ability to functionalize materials at atomic or molecular levels. Therefore, it can be used to revolutionize infection diagnostics by identification of biomarkers, pathogens, and other diseases. Nanotechnology helps in the early detection of infection, which is the most crucial weapon for fighting against infections and diseases by reducing transmission and improving treatment outcomes [94]. Nanosensors have several advantages, such as their mobility, specificity, and ease of use by non-experts. Nanosensors are one of the most exciting applications of nanotechnology in the detection of infectious diseases. Nanomaterials including gold nanoparticles, carbon nanotubes, and quantum dots are used notably for accurate pathogen detection [94, 95].

The COVID-19 virus is a single-stranded, positive-sense, enclosed virus. There are five phases in the COVID-19 cycle inside the host's body. After attaching itself to host receptors, the coronavirus and the plasma membrane undergo endocytosis or direct membrane fusion (penetration). After that, the RNA genome of the virus is discharged into the cytoplasm and replicates inside the nucleus. The RNA genome starts transcription, translation, and protein synthesis as soon as it enters the nucleus. The lungs are the main organ immediately impacted by COVID-19. Fever, dry cough, and dyspnea are common symptoms of COVID-19. Nevertheless, the virus can enter the circulation by way of the respiratory tract's epithelial cells, affecting circulatory systems and organs like the liver, kidney, heart, gastrointestinal tract, and brain [96]. Molecular testing and chest computerized tomography (CT) scans have been used to screen for and diagnose COVID-19. Both protein and nucleic acid diagnostic techniques rely heavily on data like (1) the pathogen's genomic and proteomic makeup or (2) changes in the host's protein/gene expression after infection. Nowadays, a variety of nanomaterials are used for viral detection, including silica nanoparticles (NPs), polymeric nanoparticles, metallic nanoparticles, carbon nanotubes, and quantum dots [97].

Particularly at extremely low concentrations, nucleic acids are among the most important biomarkers that play a vital role in both fundamental research and clinical diagnosis. Researchers have created DNA nanobiosensors, which present a promising opportunity for the detection of viral genomes through the integration of DNA structures with nanotechnology. The sequence-specific hybridization detection of HIV-1 DNA, as well as its crucial aspects, such as the detection limit/range of specified bioassays, are highlighted by using the types of sensors that have been covered here [98]. Metal nanoclusters, metal oxide nanoparticles, metal nanoparticles, quantum dots, and carbon-based nanoparticles are widely used in HIV diagnostics. Using nanoparticles effectively improved performance, resulting in higher sensitivity and reduced detection limits by orders of magnitude [99]. Nanotechnology's key principles provide new ways to monitor various biomarkers. Modern fabrication techniques have led to the development of more sensitive optical, mechanical, and electrical biosensor platforms for cell and biomolecule sensing. These platforms can replace or eliminate costly and bulky components found in clinical diagnostic instruments. Microfluidic technologies allow for precise manipulation of tiny amounts of biological fluids, resulting in reduced blood sample size and less intrusive collecting approaches. Smaller instrument footprints offer great mobility in sample-to-answer systems [100].

The infectious disease known as tuberculosis (TB) is caused by the *Mycobacterium tuberculosis* (*M. tuberculosis*/MTB) bacteria. When TB sufferers cough, sneeze, or spit, it usually targets the lungs that have been disseminated via the air. This infectious illness is the most infectious killer disease, killing around 1.5 million people annually. Although *M. tuberculosis* has been identified for millennia, diagnosing TB remains difficult even with advances in technology. Its limited expansion, which restricts its on-site patient care method, is the primary requirement. Second, pulmonary TB patients do not show symptoms at the beginning of the infection, which further delays detection [101]. Numerous biosensors for MTB detection have been created thanks to significant advancements in nanotechnology and microfluidics. A biosensor platform is made up of a biological sensor that reacts to physicochemical changes in the detecting region and analytical equipment. As will be covered below, TB biosensors can be classified as mass/piezoelectric, biochemical, electrical, or optical sensors, depending on the signal-producing mechanism used. These sensing platforms rely on the detection of nucleic acid hybridization, entire mycobacteria, or antibody–antigen interactions [102].

Another notable benefit of nanotechnology is its capacity to magnify detection signals, overcoming the limits imposed by low pathogen concentrations in clinical samples. Nanomaterials are used in techniques such as nanoparticle-enhanced PCR (polymerase chain reaction) and isothermal amplification to improve nucleic acid amplification efficiency and speed [103]. Additionally, nanoscale markers, such as fluorescent or magnetic nanoparticles, might improve the visualization of diagnostic data, making detection easier and more reliable [104]. These characteristics are critical in identifying illnesses such as the Zika virus or Ebola when early containment and diagnosis are critical for managing epidemics [105].

Consequently, nanotechnology gives innovative resolutions to the obstacles of infectious disease diagnosis. Nanosensors are a great invention in the field of modern diagnostics. They have enhanced sensitivity, specificity, and speed. In the future of research and development, nanotechnology will embrace the potential of reshaping global health by early, accurate, and accessible detection of infectious diseases.

1.4 NANOSENSORS FOR DISEASE BIOMARKER DETECTION

Nanotechnology and nanomaterials have extremely unique qualities that make them ideal for use in nanosensors. Various nanomaterials such as nanowires, nanoparticles, and quantum dots have been employed for detection and analysis of target analytes with sensitivity and specificity [106]. Because of their small size, nanosensors have the ability to interact at atomic and molecular levels. This feature allows them to detect the extremely small amounts of biomarkers [106].

Nanosensors work on different principles such as optical, electrical, mechanical, and magnetic transductions. Surface plasmon resonance (SPR), fluorescence, and colorimetric changes are some of the various optical properties of materials that change when they interact with various target molecules. These optical changes should be observed for the detection of a patient's illness [107–109]. For example, when quantum dots bond with a biomarker, they can release light at particular wavelengths. These wavelengths give a quantifiable signal [110]. The functioning of electrical nanosensors depends on variations of electrical conductivity and resistance when they interact with various target molecules. Silicon nanowires or carbon nanotubes are examples of electrical nanosensors. The attachment of different biomarkers to the surface of these nanosensors alters the electrical properties [111, 112].

The basic operating principles for mechanical nanosensors are changes in vibrational frequencies and mass. Microcantilevers have tiny beam-like structures. Various biomolecules bind to the surface of microcantilevers and lead to changes in their mass. This change in mass help in the detection of biomarkers. In the case of magnetic nanosensors, magnetic nanoparticles are used as source material. Magnetic nanoparticles respond to external magnetic fields, which can be changed by the presence of biomarkers. The resourcefulness of these working procedures permits nanosensors to target an extensive range of analytes with high precision [113–115].

In nanosensors, nanomaterials are often bound with biorecognition components such as enzymes, aptamers, antibodies, and so on, on their surface. These biorecognition components connect to specific biomarkers favorably to improve their effectiveness. This lowers the possibility of false positive and negative results by increasing the specificity. Utilization of a precise biomarker is crucial for early diagnosis, analysis, and the enhancement of specificity. [116–119].

Early detection and analysis of particular stages of cancer is one of the most auspicious applications of nanosensors. Carcinoembryonic antigen (CEA), alpha-fetoprotein (AFP), and prostate-specific antigen (PSA) are the most widely used cancer biomarkers. They can be detected at extremely low concentrations using nanosensors [120, 121]. The use of gold nanoparticles functionalized with antibodies represents one example, where they can bind to specific cancer biomarkers and produce a measurable optical signal [122]. Similarly, graphene-based nanosensors can detect microRNAs associated with cancer. They provide an understanding of cancer progression and possible medicinal targets [123].

Additionally, nanosensors are essential for identifying biomarkers of viral diseases. Nanosensors have been used for the detection of viral proteins or RNA from infections such as SARS-CoV-2. SARS-CoV-2 is the virus that causes COVID-19. Early and fast identification of these biomarkers is important for diagnosis and timely control of the spread of infectious diseases [124, 125]. Nanosensors also have great uses in the diagnosis of bacterial infections; by recognizing certain proteins or nucleic acids linked to harmful bacteria, nanosensors can help control antibiotic resistance [126, 127].

Alzheimer's disease, Parkinson's disease, and multiple sclerosis are some of the most serious neurological disorders. Biomarkers of these neurological disorders are generally present at low concentrations in cerebrospinal fluid or blood [128]. Nanosensors provide the sensitivity essential for identification of these biomarkers, including amyloid-beta peptides and tau proteins. Nanosensors help in early diagnosis of these disorders, which is crucial for managing neurodegenerative diseases and improving patient outcomes [129, 130].

Nanosensors attract a wide range of attention in the field of cardiovascular disease diagnostics. Biomarkers like troponin, C-reactive protein (CRP), and B-type natriuretic peptide (BNP) can indicate myocardial infarction, heart failure, or other cardiovascular conditions. Nanosensors can identify these biomarkers rapidly and accurately, enabling timely intervention and decreasing death rates [131–133].

Nanosensors have achieved significant advancement in the field of diabetic diagnosis. Nanosensors have been used to detect glucose levels, insulin, and glycated hemoglobin. Nanomaterials-based real-time glucose monitors can be used to properly manage the condition. To further increase their usefulness in the treatment of chronic diseases, nanosensors are also being developed to identify indicators for metabolic syndrome and obesity [134, 135].

Nanosensors may function at the molecular and atomic level, but they offer unmatched sensitivity and specificity, making them extremely promising for the detection of disease biomarkers. However, they have a lot of obstacles that prevent them from being widely used. Their scalability and repeatability are a significant problem since it is still difficult and expensive to produce nanosensors with reliable performance on a wide scale. Additionally, to guarantee dependable long-term operation, their stability and endurance in intricate biological contexts need to be improved. Because new technologies frequently need thorough evaluation and adaptation to clinical workflows, integration with current diagnostic systems and regulatory approval procedures can present challenges [136–138]. Despite these obstacles, potential avenues are provided by developments in lab-on-a-chip technologies, machine learning for data processing, and nanomaterial fabrication. Nanosensors are a key component in improving precision healthcare and early illness detection, with future possibilities including real-time health monitoring, non-invasive diagnostics, and personalized medical applications. Progress depends on cooperation between regulators, industry, and academics [138, 139].

1.5 NANOTECHNOLOGY IN POINT-OF-CARE DIAGNOSTICS

In the healthcare system, point-of-care (POC) diagnostic tools are essential, especially for illness monitoring and diagnosis. POC testing has several benefits, such as quick and precise results, simplicity of use, affordability, and minimal equipment requirements. One of the objectives of point-of-care testing is to create a chip-based, compact, portable, and self-contained system that enables the assessment of several analytes in complicated samples. POC testing represents a paradigm change from conventional diagnostic tests conducted in clinical laboratories to near-patient settings, giving doctors access to rapid diagnostic data that helps them make better decisions about diagnosis and treatment. Small handheld devices and larger bench-top devices are the two types of POC devices. Modern methods of microfabrication are used to create small handheld devices that automate sample preparation, assay procedures, analysis, and signal detection. The larger bench-top machines are smaller, simpler versions of the mainframe central lab equipment [140].

Nanoparticles' distinct physicochemical characteristics have facilitated the creation of biosensors (nanosensors) for point-of-care disease detection. Their tiny size often increases electron transfer rates and shortens enzyme-to-electrode lengths, which enhances the performance of techniques like electrochemical and enzymatic biosensors. Additionally, localized surface plasmon resonance (SPR) can be improved by noble metal nanoparticles, which in turn can boost optical biosensors [141].

Accurate measurements for colorimetric or fluorescent, microscopic, or surface plasmon resonance (SPR) and electrochemical detections are made possible by the combination of portable sensors and smartphone apps. By interacting with target-specific molecules including aptamers, dyes, enzymes, and antibodies, these sensors may be used to describe and determine the concentration of chemical and biological samples. Global notice and attention have been drawn to the adaptation of POC smartphone-based biosensing devices for rural areas or poor countries [142]. Nanotechnology has revolutionized the field of point-of-care (POC) diagnostics by enabling highly sensitive, portable, and rapid diagnostic tools. This advancement is reshaping healthcare delivery, particularly in resource-limited settings, through enhanced disease detection, management, and accessibility [142]. Portable nanodevices utilize nanomaterials such as nanoparticles, nanosensors, and nanostructured surfaces to detect biomarkers at low concentrations. These devices are compact, lightweight, and capable of providing real-time results, making them ideal for use in remote or underserved areas [143, 144]. For instance, nanotechnology-enabled lab-on-a-chip (LOC) platforms integrate multiple laboratory functions into a single microfluidic chip, allowing healthcare providers to perform complex diagnostic tests at the bedside or in the field. The high surface-to-volume ratio of nanomaterials enhances the sensitivity and specificity of these devices, enabling the detection of diseases at early stages [145]. Such innovations have been applied in diagnosing infectious diseases like malaria, HIV, and tuberculosis, as well as chronic conditions like diabetes.

Nanotechnology-enhanced rapid diagnostic tests (RDTs) are revolutionizing the speed and precision of illness detection. In RDTs, nanomaterials such as quantum dots and gold nanoparticles are used as labels or probes to enhance signals, enabling very sensitive and quick detection. Because these tests may produce answers in a matter of minutes, they eliminate the need for sophisticated lab equipment and lengthy processing times [146–148]. Gold nanoparticles can be used in in lateral flow tests to find antigens or antibodies. When it comes to controlling outbreaks, where prompt detection can assist stop the spread of illness, these nanotechnology-based RDTs are very helpful. The effectiveness of medical reactions is further increased by the use of multiplexed nanodiagnostic tests, which can identify many infections at once [149, 150].

Nanotechnology is a cornerstone in advancing POC diagnostics, offering portable, rapid, and accessible solutions that bridge gaps in healthcare delivery. This paradigm shift holds the potential to improve public health, particularly in regions where traditional healthcare infrastructure is lacking.

1.6 CHALLENGES AND FUTURE PERSPECTIVES

Safety, biocompatibility, and regulatory approval are major obstacles to the use of nanotechnology in biomedical diagnostics [151]. Concerns regarding toxicity and other adverse consequences are raised by the tiny size and strong surface reactivity of nanoparticles. These particles can cross biological barriers, gather in different tissues, and interact with cells in unpredictable ways, occasionally leading to toxicity, inflammation, or oxidative stress [152]. Extensive research is necessary to establish safe levels and appropriate doses for therapeutic applications since long-term consequences are still poorly understood.

Another major problem is assuring biocompatibility. Nanoparticles must be designed to minimize immune responses and to be compatible with the body's biological processes. Even small variations in size, shape, or surface coating can affect how nanoparticles interact with cells, tissues, and immune components, potentially producing undesirable reactions. Researchers must ensure that nanoparticles degrade or are appropriately eliminated from the body after being employed for diagnostic reasons [153, 154].

Nanotechnology in biomedical diagnostics has the potential to improve healthcare by enabling better accuracy, faster diagnosis, and highly personalized patient treatment. Emerging developments in nanotechnology indicate the development of multifunctional nanoparticles capable of diagnosing and treating illnesses, a notion known as theranostics [62, 63]. Particularly in cancer and infectious disorders, such nanoparticles might administer modified medicines and identify early disease signs, enabling real-time monitoring of therapy success. Developments in nanosensors and biomarker identification will also help with early detection, increasing the sensitivity and specificity of diagnostics [67, 116]. It is anticipated that nanosensors that can identify trace amounts of disease biomarkers in human fluids like blood or saliva would streamline diagnostic testing and open the door to less intrusive, point-of-care testing [87–89].

As these technologies advance, diagnostics will likely shift towards more personalized, preventative, and networked care approaches. Reaching this goal will require advances in nanomaterial biocompatibility, safety standards, and regulatory frameworks, all of which are critical to the effective application of nanotechnology-enhanced diagnostics in mainstream healthcare.

1.7 CONCLUSION

In conclusion, the incorporation of nanotechnology into biomedical diagnostics represents a paradigm change in disease detection, monitoring, and treatment. This growing subject takes advantage of the unique features of nanoscale materials, which display behavior and interactions that differ from those at the macro scale. The exploration in this study reveals that nanotechnology has the potential to take the diagnostic to the next level.

Nanotechnology offers various advantages for biomedical diagnostic devices such as sensitivity and specificity. Various nanomaterials such as metal and metal oxide nanoparticles, carbon-based nanomaterials, and quantum dots are synthesized and modified to interact with biomolecules such as proteins, nucleic acids, and cell receptors. These interactions between nanomaterials and biomolecules permit early and precise detection of disease biomarkers. This helps in analyzing and providing suitable treatment to the patient. Nanosensors provide a higher diagnostic accuracy, which helps to diminish false positive and negative results. This improves diagnostic test reliability and enables fast and effective medical responses.

Imaging is another promising area of interest for diagnostic techniques. The use of nanomaterials as contrast agents in various imaging techniques opens a huge window of possibilities. The utilization of functionalized nanoparticles in MRI, CT, PET, and optical imaging offers real-time monitoring of disease at the cellular and molecular levels. These advancements help healthcare teams to develop personalized medicine and targeted approach therapy.

Theranostics include the diagnosis and treatment of patients under the supervision of healthcare experts. Nanomaterials show multifunctional properties because of their tiny size. These multifunctional properties of nanomaterials make them good candidates for theragnostic applications. Multifunctional nanomaterials can be used as diagnostic as well as therapeutic agents. In other words, a single nanoparticle can detect the illness of a patient as well as customize the treatment of the patient. This theranostic behavior of nanomaterials is helpful in early, accurate, and real-time monitoring of patient health conditions along with effective and personalized treatment. This integrated approach permitted by nanomaterials certainly simplifies processes by reducing the overall time and cost required during the treatment.

Overall, nanotechnology has the potential to completely transform the future of diagnostic methods. Nanotechnology has advantages, such as early detection, accurate results, specificity, and improved resolution in imaging. Because of these advantages, nanotechnology has attracted the attention of researchers across the globe in the field of biomedical diagnosis and patient-centered care. Along with these advantages, nanotechnology has some limitations such as biocompatibility and safety. More thorough research is required to overcome these problems. Finally, we can say nanotechnology is the future of biomedical diagnosis and theranostics. It has the potential to improve patients quality of life and change the healthcare industry.

REFERENCES

1. Abiodun-Solanke, I. M. F.; Ajayi, D. M.; Arigbede, A. O. Nanotechnology and its Application in Dentistry. *Ann. Med. Health Sci. Res.* 2014, 4(3), 171–177.
2. Nasr-Eldahan, S.; Nabil-Adam, A.; Shreadah, M. A.; Maher, A. M.; Tamer El-Sayed Ali, T. E.-S. A review article on nanotechnology in aquaculture sustainability as a novel tool in fish disease control. *Aquac. Int.* 2021, 29, 1459–1480.
3. Mekuye, B.; Abera, B. Nanomaterials: An overview of synthesis, classification, characterization, and applications. *Nano Select* 2023, 4, 486–501.
4. Hulla, J. E.; Sahu, S. C.; Hayes, A.W. Nanotechnology: History and future. *Hum. Exp. Toxicol.* 2015, 34(12), 1318–1321.
5. Silva, G. A. Introduction to nanotechnology and its applications to medicine. *Surg. Neurol.* 2004, 61, 216–220.
6. Yadav, S.; Rani, N.; Saini, K. A review on transition metal oxides based nanocomposites, their synthesis techniques, different morphologies and potential applications. *IOP Conf. Ser.: Mater. Sci. Eng.* 2022, 1225, 012004.
7. Baig, N.; Kammakakam, I.; Falath, W. Nanomaterials: A review of synthesis methods, properties, recent progress, and challenges. *Mater. Adv.* 2021, 2, 1821.
8. Adul-Rasool, A. A.; Athair, D. M.; Zaidan, H. K.; Rheima, A. M.; Al-Sharify, Z. T.; Mohammed, S. H.; Kianfar, E. 0,1,2,3D nanostructures, types of bulk nanostructured materials, and drug nanocrystals: An overview. *Cancer Treat. Res. Commun.* 2024, 40, 100834.
9. Kolahalam, L. A.; Viswanath, I. V. K.; Diwakar, B. S.; Govindh, B.; Reddy, V.; Murthy, Y. L. N. Review on nanomaterials: Synthesis and applications. *Mater. Today: Proc.* 2019, 18, 2182–2190.
10. Salem, S. S.; Hammad, E. N.; Mohamed, A. A.; El-Dougdoug, W. A Comprehensive review of nanomaterials: Types, synthesis, characterization, and applications. *Biointerface Res. Appl. Chem.* 2023, 13 (1), 41.
11. Amin, M. T.; Alazba, A. A.; Manzoor, U. A review of removal of pollutants from water/wastewater using different types of nanomaterials. *Adv. Mater. Sci. Eng.* 2014, 825910.
12. Harish, V.; Tewari, D.; Gaur, M.; Yadav, A.B.; Swaroop, S.; Bechelany, M.; Barhoum, A. Review on nanoparticles and nanostructured materials: Bioimaging, biosensing, drug delivery, tissue engineering, antimicrobial, and agro-food applications. *Nanomaterials* 2022, 12, 457.
13. Suri, S. S.; Fenniri, H.; Singh, B. Nanotechnology-based drug delivery systems. *J. Occup. Med. Toxicol.* 2007, 2, 16.
14. Jackson, T. C.; Patani, B. O.; Ekpa, D. E. Nanotechnology in diagnosis: A review. *Adv. Nanopart.* 2017, 6, 93–102.
15. Qasim, M.; Lim, D-J.; Park, H.; Na, D. Nanotechnology for diagnosis and treatment of infectious diseases. *J. Nanosci. Nanotechnol.* 2014, 14(10), 7374–7387.
16. Ramesh, M.; Janani, R.; Deepa, C.; Rajeshkumar, L. Nanotechnology-enabled biosensors: A review of fundamentals, design principles, materials, and applications. *Biosensors* 2023, 13, 40.

17. Barbosa, A. I.; Rebelo, R.; Reis, R. L.; Bhattacharya, M.; Correlo, V. M. Current nanotechnology advances in diagnostic biosensors. *Med. Devices Sens.* 2021, 4, e10156.
18. Verma, D.; Singh, K. R. B.; Yadav, A. K.; Nayak, V.; Singh, J.; Solanki, P. R.; Singh, R. P. Internet of things (IoT) in nano-integrated wearable biosensor devices for healthcare applications. *Biosens. Bioelectron. X* 2022, 11, 100153.
19. Wong, K. K. Y.; Liu, X. Nanotechnology meets regenerative medicine: A new frontier. *Nanotechnol. Rev.* 2013, 2(1), 59–71.
20. Khalilov, R. A comprehensive review of advanced nano-biomaterials in regenerative medicine and drug delivery. *ABES* 2023, 8(1), 5–18.
21. Thwala, L.N.; Ndlovu, S.C.; Mpofu, K.T.; Lugongolo, M.Y.; Mthunzi-Kufa, P. Nanotechnology-based diagnostics for diseases prevalent in developing countries: Current advances in point-of-care tests. *Nanomaterials* 2023, 13, 1247.
22. Chakraborty, M.; Jain, S.; Rani, V. Nanotechnology: Emerging tool for diagnostics and therapeutics. *Appl. Biochem. Biotechnol.* 2011, 165, 1178–1187.
23. Laroui, H.; Rakhya, P.; Xiao, B.; Viennois, E.; Merlin, D. Nanotechnology in diagnostics and therapeutics for gastrointestinal disorders. *Dig. Liver Dis.* 2013, 45, 995–1002.
24. Hauck, T. S.; Giri, S.; Gao, Y.; Chan, W. C. W. Nanotechnology diagnostics for infectious diseases prevalent in developing countries. *Adv. Drug Deliv. Rev.* 2010, 62, 438–448.
25. Jain, K. K. Nanotechnology in clinical laboratory diagnostics. *Clin. Chim. Acta* 2005, 358, 37–54.
26. Patel, S.; Nanda, R.; Sahoo, S. Nanotechnology in healthcare: Applications and challenges. *Med. Chem.* 2015, 5, 12.
27. Abbasi, R.; Shineh, G.; Mobaraki, M.; Doughty, S.; Tayebi, L. Structural parameters of nanoparticles affecting their toxicity for biomedical applications: A review. *J. Nanopart. Res.* 2023, 25, 43.
28. Egbuna, C.; Parmar, V. K.; Jeevanandam, J.; Ezzat, S. M.; Patrick-Iwuanyanwu,K. C.; Adetunji, C. O.; Khan, J.; Onyeike, E. N.; Uche, C. Z.; Akram, M.; Ibrahim, M. S.; Mahdy, N. M. E.; Awuchi, C. G.; Saravanan, K.; Tijjani, H.; Odoh, U. E.; Messaoudi, M.; Ifemeje, J. C.; Olisah, M. C.; Ezeofor, N. J.; Chikwendu, C. J.; Ibeabuchi, C. G. Toxicity of nanoparticles in biomedical application: Nanotoxicology. *J. Toxicol.* 2021, 2021, 9954443.
29. Pushpanjali, P. A.; Manjunatha, J. G.; Hareesha, N. An overview of recent developments of carbon-based sensors for the analysis of drug molecules. *J. Electrochem. Sci. Eng.* 2021, 11(3), 161–177.
30. Jain, K. K. The role of nanobiotechnology in drug discovery. *DDT* 2005, 10(21), 1435–1442.
31. Pushpanjali, P. A.; Manjunatha, J. G.; Hareesha, N.; D'Souza, E. S.; Charithra, M. M.; Prinith, N. S. Voltammetric analysis of antihistamine drug cetirizine and paracetamol at poly(L-leucine) layered carbon nanotube paste electrode. *Surf. Interfaces* 2021, 24, 101154.
32. Monnappa, A. B.; Manjunatha, J. G.; Bhatt, A. S. Design of a sensitive and selective voltammetric sensor based on a cationic surfactant-modified carbon paste electrode for the determination of alloxan. *ACS Omega* 2020, 5, 23481–23490.
33. Raril, C.; Manjunatha, J. G.; Ravishankar, D. K.; Fattepur, S.; Siddaraju, G.; Nanjundaswamy, L. Validated electrochemical method for simultaneous resolution of tyrosine, uric acid, and ascorbic acid at polymer modified nano-composite paste electrode. *Surf. Eng. Appl. Electrochem.* 2020, 56(4), 415–426.
34. Tigari, G.; Manjunatha, J. G. Poly(glutamine) flm-coated carbon nanotube paste electrode for the determination of curcumin with vanillin: An electroanalytical approach. *Monatsh. Chem.* 2020, 151, 1681–1688.

35. Hareesha, N.; Manjunatha, J. G. Electro-oxidation of formoterol fumarate on the surface of novel poly(thiazole yellow-G) layered multi-walled carbon nanotube paste electrode. *Sci. Rep.* 2021, 11, 12797.
36. Manjunatha, J. G.; Swamy, K.; Mamatha, G. P.; Gilbert, O.; Sherigara, B. S. Poly(maleic acid) modified carbon paste electrode for simultaneous detection of dopamine in the presence of uric acid: A cyclic voltammetric study. *Anal. Bioanal. Electrochem.* 2011, 3(2), 146–159.
37. Han, X.; Xu, K.; Taratula, O.; Farsad, K. Applications of nanoparticles in biomedical imaging. *Nanoscale* 2019, 11, 799.
38. Padmanabhan, P.; Kumar, A.; Kumar, S.; Chaudhary, R. K.; Gulyás, B. Nanoparticles in practice for molecular-imaging applications: An overview. *Acta Biomater.* 2016, 41, 1–16.
39. Key, J.; James F Leary, J. F. Nanoparticles for multimodal in vivo imaging in nanomedicine. *Int. J. Nanomedicine* 2014, 9, 711–726.
40. Zhou, Z.; Yang, L.; Gao, J.; Chen, X. Structure–relaxivity relationships of magnetic nanoparticles for magnetic resonance imaging. *Adv. Mater.* 2019, 31, 1804567.
41. Abakumov, M. A.; Nukolova, N. V.; Sokolsky-Papkov, M.; Shein, S. A.; Sandalova, T. O.; Vishwasrao, H. M.; Grinenko, N. F.; Gubsky, I. L.; Abakumov, A. M.; Kabanov, A. V.; Chekhonin, V. P. VEGF-targeted magnetic nanoparticles for MRI visualization of brain tumor. *Nanomedicine* 2015, 11, 825–833.
42. Yallapu, M. M.; Othman, S. F.; Curtis, E. T.; Gupta, B. K.; Jaggi, M.; Chauhan, S. C. Multi-functional magnetic nanoparticles for magnetic resonance imaging and cancer therapy. *Biomaterials* 2011, 32, 1890–1905.
43. Rümenapp, C.; Gleich, B.; Haase, A. Magnetic nanoparticles in magnetic resonance imaging and diagnostics. *Pharm. Res.* 2012, 29, 1165–1179.
44. Schäferling, M. The art of fluorescence imaging with chemical sensors. *Angew. Chem. Int. Ed.* 2012, 51, 3532–3554.
45. Wegner, K. D.; Hildebrandt, N. Quantum dots: Bright and versatile *in vitro* and *in vivo* fluorescence imaging biosensors. *Chem. Soc. Rev.* 2015, 44, 4792–4834.
46. Kang, Y.-F.; Li, Y.-H.; Fang, Y-W.; Xu, Y.; Wei, X.-M.; Yin, X.-B. Carbon quantum dots for zebrafish fluorescence imaging. *Sci. Rep.* 2015, 5, 11835.
47. Li, W.; Chen, X. Gold nanoparticles for photoacoustic imaging. *Nanomedicine* (Lond.) 2015, 10(2), 299–320.
48. Yang, X.; Stein, E. W.; Ashkenazi, S.; Wang, L. V. Nanoparticles for photoacoustic imaging. *Wires Nanomed. Nanobior.* 2009, 1, 360–368.
49. Lemaster, J. E.; Jokerst, J. V. What is new in nanoparticle-based photoacoustic imaging. *Wiley Interdiscip Rev Nanomed Nanobiotechnol.* 2017, 9(1), e1404.
50. Li, W.; Brown, P. K.; Wang, L. V.; Xia, Y. Gold nanocages as contrast agents for photoacoustic imaging. *Contrast Media Mol. Imaging* 2011, 6, 370–377.
51. Pan, D.; Pramanik, M.; Wickline, S. A.; Wang, L. V.; Lanza, G. M. Recent advances in colloidal gold nanobeacons for molecular photoacoustic imaging. *Contrast Media Mol. Imaging* 2011, 6, 378–388.
52. Manohar, S.; Ungureanu, C.; Van Leeuwen, T. G. Gold nanorods as molecular contrast agents in photoacoustic imaging: The promises and the caveats. *Contrast Media Mol. Imaging* 2011, 6, 389–400.
53. Naseri, N.; Ajorlou, E.; Asghari, F.; Pilehvar-Soltanahmadi, Y. An update on nanoparticle-based contrast agents in medical imaging. *Artif. Cells Nanomed. Biotechnol.* 2018, 46(6), 1111–1121.
54. Thiruppathi, R.; Mishra, S.; Ganapathy, M.; Padmanabhan, P.; Gulyás, B. Nanoparticle functionalization and its potentials for molecular imaging. *Adv. Sci.* 2017, 4, 1600279.

55. McCarthy, J. R.; Weissleder, R. Multifunctional magnetic nanoparticles for targeted imaging and therapy. *Adv. Drug Deliv. Rev.* 2008, 60, 1241–1251.
56. Jiao, P. F.; Zhou, H. Y.; Chen, L. X.; Yan, B. Cancer-targeting multifunctionalized gold nanoparticles in imaging and therapy. *Curr. Med. Chem.* 2011, 18, 2086–2102.
57. Kong, F.-Y.; Zhang, J.-W.; Li, R.-F.; Wang, Z.-X.; Wang, W.-J.; Wang, W. Unique roles of gold nanoparticles in drug delivery, targeting and imaging applications. *Molecules* 2017, 22, 1445.
58. Rosen, J. E.; Chan, L.; Shieh, D.-B.; Gu, F. X. Iron oxide nanoparticles for targeted cancer imaging and diagnostics. *Nanomedicine* 2012, 8, 275–290.
59. Peng, X.-H.; Qian, X.; Mao, H.; Wang, A. Y.; Chen, Z.; Nie, S.; Shin, D. M. (2008) Targeted magnetic iron oxide nanoparticles for tumor imaging and therapy. *Int. J. Nanomedicine* 2008, 3(3), 311–321.
60. Cole, J. T.; Holland, N. B. Multifunctional nanoparticles for use in theranostic applications. *Drug Deliv. and Transl. Res.* 2015, 5, 295–309.
61. Kandasamy, G.; Maity, D. Multifunctional theranostic nanoparticles for biomedical cancer treatments—A comprehensive review. *Mater. Sci. Eng. C* 2021, 127, 112199.
62. Santhosh, P. B.; Ulrih, N. P. Multifunctional superparamagnetic iron oxide nanoparticles: Promising tools in cancer theranostics. *Cancer Lett.* 2013, 336, 8–17.
63. Yen, S. K.; Padmanabhan, P.; Selvan, S. T. Multifunctional iron oxide nanoparticles for diagnostics, therapy and macromolecule delivery. *Theranostics* 2013, 3(12), 986–1003.
64. Jia, F.; Liu, X.; Li, L.; Mallapragada, S.; Narasimhan, B.; Wang, Q. Multifunctional nanoparticles for targeted delivery of immune activating and cancer therapeutic agents. *J. Control. Release* 2013, 172, 1020–1034.
65. Dykman, L. A.; Khlebtsov, N. G. Multifunctional gold-based nanocomposites for theranostics. *Biomaterials* 2016, 108, 13–34.
66. Achilefu, S. Introduction to concepts and strategies for molecular imaging. *Chem. Rev.* 2010, 110, 2575–2578.
67. Wickline, S. A.; Lanza, G. M. Nanotechnology for molecular imaging and targeted therapy. *Circulation* 2003, 107(8), 1092–1095.
68. Numan, A.; Singh, S.; Zhan, Y.; Li, L.; Khalid, M.; Rilla, K.; Ranjan, S.; Cinti, S. Advanced nanoengineered-customized point-of-care tools for prostate-specifc antigen. *Microchim. Acta* 2022, 189, 27.
69. Eckert, M. A.; Vu, P. Q.; Zhang, K.; Kang, D.; Ali, M. M.; Xu, C.; Zhao, W. Novel molecular and nanosensors for in vivo sensing. *Theranostics* 2013, 3(8), 583–594.
70. Salvati, E.; Stellacci, F.; Krol, S. Nanosensors for early cancer detection and for therapeutic drug monitoring. *Nanomedicine* (Lond.) 2015, 10(23), 3495–512.
71. Chakravarty, R.; Goel, S.; Dash, A.; Cai, W. Radiolabeled inorganic nanoparticles for positron emission tomography imaging of cancer: An overview. *Q. J. Nucl. Med. Mol. Imaging* 2017, 61(2), 181–204.
72. Stockhofe, K.; Postema, J. M.; Schieferstein, H.; Ross, T. L. Radiolabeling of nanoparticles and polymers for PET imaging. *Pharmaceuticals* 2014, 7, 392–418.
73. Sun, X.; Cai, W.; Chen, X. Positron emission tomography imaging using radiolabeled inorganic nanomaterials. *Acc. Chem. Res.* 2015, 48, 286–294.
74. Bentivoglio, V.; Varani, M.; Lauri, C.; Ranieri, D.; Signore, A. Methods for radiolabelling nanoparticles: PET Use (Part 2). *Biomolecules* 2022, 12, 1517.
75. Cialla, D.; März, A.; Böhme, R.; Theil, F.; Weber, K.; Schmitt, M.; Popp. J. Surface-enhanced Raman spectroscopy (SERS): Progress and trends. *Anal. Bioanal. Chem.* 2012, 403, 27–54.
76. Du, Z.; Qi, Y.; He, J.; Zhong, D.; Zhou, M. Recent advances in applications of nanoparticles in SERS in vivo imaging. *WIREs Nanomed. Nanobiotechnol.* 2020, 13(2), e1672.

77. Zhang, Y.; Hong, H.; Myklejord, D. V.; Cai, W. Molecular imaging with SERS-active nanoparticles. *Small* 2011, 7(23), 3261–3269.
78. Mathur, G.; Nain, S.; Sharma, P. K. Cancer: An overview. *Academic J. Cancer Res.* 2015, 8 (1), 01–09.
79. Trichopoulos, D.; Li, F. P.; Hunter, D. J. What causes cancer. *Sci. Am.* 1996, 275(3), 80–87.
80. Fass, L. Imaging and cancer: A review. *Mol. Oncol.* 2008, 2, 115–152.
81. Mosleh-Shirazi, S.; Abbasi, M.; Moaddeli, M. R.; Vaez, A.; Shafiee, Kasaee, M. S. R.; Amani, A. M.; Hatam, S. Nanotechnology advances in the detection and treatment of cancer: An overview. *Nanotheranostics* 2022, 6, 400–423.
82. Choi, Y.-E.; Kwak, J.-W.; Park, J. W. Nanotechnology for early cancer detection. *Sensors* 2010, 10, 428–455.
83. Liu, C. H.; Grodzinski, P. Nanotechnology for cancer imaging: Advances, challenges, and clinical opportunities. *Radiol. Imaging Cancer* 2021, 3(3), e200052.
84. Gunasekera, U. A.; Pankhurst, Q. A.; Douek, M. Imaging applications of nanotechnology in cancer. *Targ. Oncol.* 2009, 4, 169–181.
85. Kemp, J. A.; Kwon, Y. J. Cancer nanotechnology: Current status and perspectives. *Nano Converg.* 2021, 8, 34.
86. Hartman, K. B.; Wilson, L. J.; Rosenblum, M. G. Detecting and treating cancer with nanotechnology. *Mol. Diag. Ther.* 2008, 12(1), 1–14.
87. Goddard, Z. R.; Marin, M. J.; Russell, D. A.; Searcey, M. Active targeting of gold nanoparticles as cancer therapeutics. *Chem. Soc. Rev.* 2020, 49, 8774.
88. Dykman, L. A.; Khlebtsov, N. G. Uptake of engineered gold nanoparticles into mammalian cells. *Chem. Rev.* 2014, 114, 1258–1288.
89. Chacko, N.; Ankri, R. Non-invasive early-stage cancer detection: Current methods and future perspectives. *Clin. Exp. Med.* 2025, 25, 17.
90. McHugh, K. J.; Jing, L.; Behrens, A. M.; Jayawardena, S.; Tang, W.; Gao, M.; Langer, R.; Jaklenec, A. Biocompatible semiconductor quantum dots as cancer imaging agents. *Adv. Mater.* 2018, 30(18), 1706356.
91. Michalet, X.; Pinaud, F. F.; Bentolila, L. A.; Tsay, J. M.; Doose, S.; Li, J. J.; Sundaresan, G.; Wu, A. M.; Gambhir, S. S.; Weiss, S. Quantum dots for live cells, in Vivo imaging, and diagnostics. *Science* 2005, 307, 538–544.
92. Huang, Q.; Wang, Y.; Chen, X.; Wang, Y.; Li, Z.; Du, S.; Wang, L.; Chen, S. Nanotechnology-based strategies for early cancer diagnosis using circulating tumor cells as a liquid biopsy. *Nanotheranostics* 2018, 2, 21–41.
93. Kalogianni, D. P. Nanotechnology in emerging liquid biopsy applications. *Kalogianni Nano Converg.* 2021, 8, 13.
94. Ramalingam, M.; Jaisankar, A.; Cheng, L.; Krishnan, S.; Lan, L.; Hassan, A.; Sasmazel, H. T.; Kaji, H.; Deigner, H.-P.; Pedraz, J. L.; Kim, H.-W.; Shi, Z.; Marrazza, G. Impact of nanotechnology on conventional and artificial intelligence-based biosensing strategies for the detection of viruses. *Discov. Nano* 2023, 18, 58.
95. Khan, H.; Jan, Z.; Ullah, I.; Alwabli, A.; Alharbi, F.; Habib, S.; Islam, M.; Shin, B.-J.; Young Lee, M.; Koo, J. A deep dive into AI integration and advanced nanobiosensor technologies for enhanced bacterial infection monitoring. *Nanotechnol. Rev.* 2024, 13, 20240056.
96. Vahedifard, F.; Chakravarthy, K. Nanomedicine for COVID-19: The role of nanotechnology in the treatment and diagnosis of COVID-19. *Emergent Mater.* 2021, 4, 75–99.
97. Rasmi, Y.; Saloua, K.S.; Nemati, M.; Choi, J.R. Recent progress in nanotechnology for COVID-19 prevention, diagnostics and treatment. *Nanomaterials* 2021, 11, 1788.

98. Mozhgani, S.-H.; Kermani, H. A.; Norouzi, M.; Arabi, M.; Soltani, S. Nanotechnology based strategies for HIV-1 and HTLV-1 retroviruses gene detection. *Heliyon* 2020, 6, e04048.
99. Farzin, L.; Shamsipur, M.; Samandari, L.; Sheibani, S. HIV biosensors for early diagnosis of infection: The intertwine of nanotechnology with sensing strategies. *Talanta* 2020, 206, 120201.
100. Damhorst, G. L.; Watkins, N. N.; Bashir, R. Micro- and nanotechnology for HIV/AIDS diagnostics in resource-limited settings. *IEEE. Trans. Biomed. Eng.* 2013, 60(3), 715–726.
101. Muthukrishnan, L. Multidrug resistant tuberculosis-Diagnostic challenges and its conquering by nanotechnology approach-An overview. *Chem. Biol. Interact.* 2021, 337, 109397.
102. Wang, S.; Inci, F.; Libero, G. D.; Singhal, A.; Demirci, U. Point-of-care assays for tuberculosis: Role of nanotechnology/microfluidics. *Biotechnol. Adv.* 2013, 31, 438–449.
103. Su, Y.; Chu, H.; Tian, J; Du, Z.; Xu, W. Insight into the nanomaterials enhancement mechanism of nucleic acid amplification reactions. *Trends Anal. Chem.* 2021, 137, 116221.
104. Yao, J.; Yang, M.; Duan, Y. Chemistry, biology, and medicine of fluorescent nanomaterials and related systems: New insights into biosensing, bioimaging, genomics, diagnostics, and therapy. *Chem. Rev.* 2014, 114, 6130–6178.
105. Ramakrishnan, S. G.; Robert, B.; Salim, A.; Ananthan, P.; Sivaramakrishnan, M.; Subramaniam, S.; Natesan, S.; Suresh, R.; Rajeshkumar, G.; Maran, J. P.; Al-Dhabi, N. A.; Karuppiah, P.; Arasu, M. V. Nanotechnology based solutions to combat zoonotic viruses with special attention to SARS, MERS, and COVID 19: Detection, protection and medication. *Microb. Pathog.* 2021, 159, 105133.
106. Darwish, M. A.; Abd-Elaziem, W.; Elsheikh, A.; Zayed, A. A. Advancements in nanomaterials for nanosensors: A comprehensive review. *Nanoscale Adv.* 2024, 6, 4015.
107. Ghasemi, F.; Fahimi-Kashani, N.; Bigdeli, A.; Alshatteri, A. H.; Abbasi-Moayed, S.; Al-Jaf, S. H.; Merry, M. Y.; Omer, K. M.; Hormozi-Nezhad, M. R. Paper-based optical nanosensors -A review. *Anal. Chim. Acta* 2023, 1238, 340640.
108. Sargazi, S.; Fatima, I.; Kiani, M. H.; Mohammadzadeh, V.; Arshad, R.; Bilal, M.; Rahdar, A.; Díez-Pascual, A. M.; Behzadmehr, R. Fluorescent-based nanosensors for selective detection of a wide range of biological macromolecules: A comprehensive review. *Int. J. Biol. Macromol.* 2022, 206, 115–147.
109. Amirjani, A.; Rahbarimehr, E. Recent advances in functionalization of plasmonic nanostructures for optical sensing. *Microchim. Acta* 2021, 188, 57.
110. Costa-Fernańdez, J. M; Pereiro, R.; Sanz-Medel, A. The use of luminescent quantum dots for optical sensing. *Trends Anal. Chem.* 2006, 25(3), 207–218.
111. Munawar, A.; Ong, Y.; Schirhagl, R.; Tahir, M. A.; Khan, W. S.; Bajwa, S. Z. Nanosensors for diagnosis with optical, electric and mechanical transducers. *RSC Adv.* 2019, 9, 6793.
112. Hahm, J.-I.; Lieber, C. M. Direct ultrasensitive electrical detection of DNA and DNA sequence variations using nanowire nanosensors. *Nano Lett.* 2004, 4(1), 51–54.
113. Hang, X.; He, S.; Dong, Z.; Minnick, G.; Rosenbohm, J.; Chen, Z.; Yang, R.; Chang, L. Nanosensors for single cell mechanical interrogation. *Biosens. Bioelectron.* 2021, 179, 113086.
114. Bruns, N.; Pustelny, K.; Bergeron, L. M.; Whitehead, T. A.; Clark, D. S. Mechanical nanosensor based on FRET within a thermosome: Damage-reporting polymeric materials. *Angew. Chem.* 2009, 121, 5776–5779.
115. Javaid, M.; Haleem, A.; Singh, R. P.; Rab, S.; Suman, R. Exploring the potential of nanosensors: A brief overview. *Sens. Int.* 2021, 2, 100130.

116. Xu, L.; Shoaie, N.; Jahanpeyma, F.; Zhao, J.; Mostafa Azimzadeh, M.; Al, K. T. Optical, electrochemical and electrical (nano)biosensors for detection of exosomes: A comprehensive overview. *Biosens. Bioelectron.* 2020, 161, 112222.
117. Welch, E. C.; Powell, J. M.; Clevinger, T. B.; Fairman, A. E.; Shukla, A. Advances in biosensors and diagnostic technologies using nanostructures and nanomaterials. *Adv. Funct. Mater.* 2021, 31(44), 2104126.
118. Fitzgerald, R. C.; Antoniou, A. C.; Fruk, L.; Rosenfeld, N. The future of early cancer detection. *Nat. Med.* 2022, 28, 666–677.
119. Califf, R. M. Biomarker definitions and their applications. *Exp. Biol. Med.* 2018, 243, 213–221.
120. Laraib, U.; Sargazi, S.; Rahdar, A.; Khatami, M.; Pandey, S. Nanotechnology-based approaches for effective detection of tumor markers: A comprehensive state-of-the-art review. *Int. J. Biol. Macromol.* 2022, 195, 356–383.
121. Ye, F.; Zhao, Y.; El-Sayed, R.; Muhammed, M.; Hassan, M. Advances in nanotechnology for cancer biomarkers. *Nano Today* 2018, 18, 103–123.
122. Devi, R. V.; Doble, M.; Verma, R. S. Nanomaterials for early detection of cancer biomarker with special emphasis on gold nanoparticles in immunoassays/sensors. *Biosens. Bioelectron.* 2015, 68, 688–698.
123. Zhang, C.; Miao, P.; Sun, M.; Yan, M.; Liu, H. Progress in miRNA detection using graphene material–based biosensors. *Small* 2019, 15(38), 1901867.
124. Bardhan, N. M.; Jansen, P.; Belcher, A. M. Graphene, carbon nanotube and plasmonic nanosensors for detection of viral pathogens: Opportunities for rapid testing in pandemics like COVID-19. *Front. Nanotechnol.* 2021, 3, 733126.
125. Rabiee, N.; Bagherzadeh, M.; Ghasemi, A.; Zare, H.; Ahmadi, S.; Fatahi, Y.; Dinarvand, R.; Rabiee, M.; Ramakrishna, S.; Shokouhimehr, M.; Varma, R. S. Point-of-use rapid detection of SARS-CoV-2: Nanotechnology-enabled solutions for the COVID-19 pandemic. *Int. J. Mol. Sci.* 2020, 21, 5126.
126. Bruno, A.; Tripodi, F.; Armanni, A.; Barbieri, L.; Colombo, A.; Fumagalli, S.; Moukham, H.; Tomaino, G.; Kukushkina, E.; Lorenzi, R.; Marchesi, L.; Monguzzi, A.; Paleari, A.; Ronchi, A.; Secchi, V.; Sironicd, L.; Colombo, M. Advancements in nanosensors for detecting pathogens in healthcare environments. *Environ. Sci. Nano* 2024, 11, 4449.
127. Saxena, S.; Punjabi, K.; Ahamad, N.; Singh, S.; Bendale, P.; Banerjee, R. Nanotechnology approaches for rapid detection and theranostics of antimicrobial resistant bacterial infections. *ACS Biomater. Sci. Eng.* 2022, 8, 2232–2257.
128. Pathak, N.; Vimal, S. K.; Tandon, I.; Agrawal, L.; Hongyi, C.; Bhattacharyya, S. Neurodegenerative disorders of Alzheimer, Parkinsonism, amyotrophic lateral sclerosis and multiple sclerosis: An early diagnostic approach for precision treatment. *Metab. Brain Dis.* 2022, 37, 67–104.
129. Phan, L. M. T.; Hoang, T. X.; Vo, T. A. T.; Pham, H. L.; Le, H. T. N.; Chinnadayyala, S. R.; Kim, J. Y.; Lee, S.-M.; Cho, W. W.; Kim, Y. H.; Choi, S. H.; Cho, S. Nanomaterial-based optical and electrochemical biosensors for amyloid beta and Tau: Potential for early diagnosis of Alzheimer's disease. *Expert Rev. Mol. Diagn.* 2021, 21(2), 175–193.
130. Kaushik, A.; Jayant, R. D.; Tiwari, S.; Vashist, A.; Nair, M. Nano-biosensors to detect beta-amyloid for Alzheimer's disease management. *Biosens. Bioelectron.* 2016, 80, 273–287.
131. Wang, X.-Y.; Zhang, F.; Zhang, C.; Zheng, L.-R.; Yang, J. The biomarkers for acute myocardial infarction and heart failure. *BioMed Res. Int.* 2020, 2020, 2018035.
132. Tang, X.; Zhu, Y.; Guan, W.; Zhou, W.; Wei, P. Advances in nanosensors for cardiovascular disease detection. *Life Sci.* 2022, 305, 120733.
133. Sabir, F.; Barani, M.; Mukhtar, M.; Rahdar, A.; Cucchiarini, M.; Zafar, M.N.; Behl, T.; Bungau, S. Nanodiagnosis and nanotreatment of cardiovascular diseases: An overview. *Chemosensors* 2021, 9, 67.

134. Yadav, K. Nanotechnology in diabetes management: Revolutionizing treatment and diagnostics. *J. Mol. Liq.* 2024, 414, 126117.
135. Liu, Y.; Zeng, S.; Ji, W.; Yao, H.; Lin, L.; Cui, H.; Santos, H. A.; Pan, G. Emerging theranostic nanomaterials in diabetes and its complications. *Adv. Sci.* 2022, 9, 2102466.
136. Yang, T.; Duncan, T. V. Challenges and potential solutions for nanosensors intended for use with foods. *Nat. Nanotechnol.* 2021, 16, 251–265.
137. Leong, Y. X.; Tan, E. X.; Leong, S. X.; Koh, C. S. L.; Nguyen, L. B. T.; Chen, J. R. T.; Xia, K.; Ling, X. Y. Where nanosensors meet machine learning: Prospects and challenges in detecting disease X. *ACS Nano* 2022, 16, 13279–13293.
138. Kim, S.-J.; Choi, S.-J.; Jang, J.-S.; Cho, H.-J.; Kim, I.-D. Innovative nanosensor for disease diagnosis. *Acc. Chem. Res.* 2017, 50, 1587–1596.
139. Adam, T.; Gopinath, S. C. B. Nanosensors: Recent perspectives on attainments and future promise of downstream applications. *Process Biochem.* 2022, 117, 153–173.
140. Syedmoradi, L.; Daneshpour, M.; Alvandipour, M.; Gomez, F. A.; Hajghassem, H.; Omidfar, K. Point of care testing: The impact of nanotechnology. *Biosens. Bioelectron.* 2017, 87, 373–387.
141. Noah, N. M.; Ndangili, P. M. Current trends of nanobiosensors for point-of-care diagnostics. *J. Anal. Methods Chem.* 2019, 2019, 2179718.
142. Zarei, M. Portable biosensing devices for point-of-care diagnostics: Recent developments and applications. *Trends Anal. Chem.* 2017, 91, 26–41.
143. Kim, C.; Kang, M. S.; Raja, I. S.; Joung, Y. K.; Han, D.-W. Advancements in nanobiosensor technologies for in-vitro diagnostics to point of care testing. *Heliyon* 2024, 10, e40306.
144. Eissa, M. E. Nanosensors for early detection and diagnosis of cancer: A review of recent advances. *J. Cancer Res. Rev.* 2024, 1(1), 1–13.
145. Chen, X.; Xu, J.; Ji, B.; Fang, X.; Jin, K.; Qian, J. The role of nanotechnology-based approaches for clinical infectious diseases and public health. *Front. Bioeng. Biotechnol.* 2023, 11, 1146252.
146. Castillo-León, J.; Trebbien, R.; Castillo, J. J.; Svendsena, W. E. Commercially available rapid diagnostic tests for the detection of high priority pathogens: Status and challenges. *Analyst* 2021, 146, 3750.
147. Teles, F. S. R. R.; Pires de Tavira, L. A.; Pina da Fonseca, L. J. Biosensors as rapid diagnostic tests for tropical diseases. *Crit. Rev. Clin. Lab. Sci.* 2010, 47(3), 139–169.
148. Ha, Y. Exploiting the potential of magnetic nanoparticles for rapid diagnosis tests (RDTs): Nanoparticle-antibody conjugates and color development strategies. *Diagnostics* 2023, 13, 3033.
149. Baker, A. N.; Hawker-Bond, G. W.; Georgiou, P. G.; Dedol, S.; Field, R. A.; Gibson, M. I. Glycosylated gold nanoparticles in point of care diagnostics: From aggregation to lateral flow. *Chem. Soc. Rev.* 2022, 51, 7238.
150. Giri, S.; Sykes, E. A.; Jennings, T. L.; Chan, W. C. W. Rapid screening of genetic biomarkers of infectious agents using quantum dot barcodes. *ACS Nano* 2011, 5(3), 1580–1587.
151. Ma, X.; Tian, Y.; Yang, R.; Wang, H.; Allahou, L. W.; Chang, J.; Williams, G.; Knowles, J. C.; Poma, A. Nanotechnology in healthcare, and its safety and environmental risks. *J. Nanobiotechnol.* 2024, 22, 715.
152. Zoroddu, M. A.; Medici, S.; Ledda, A.; Nurchi, V. M.; Lachowicz, J. I.; Peana, M. Toxicity of nanoparticles. *Curr. Med. Chem.* 2014, 21, 3837–3853.
153. Li, X.; Wang, L.; Fan, Y.; Feng, Q.; Cui, F.-Z. Biocompatibility and toxicity of nanoparticles and nanotubes. *J. Nanomater.* 2012, 2012, 548389.
154. Gamucci, O.; Bertero, A.; Gagliardi, M.; Bardi, G. Biomedical nanoparticles: Overview of their surface immune-compatibility. *Coatings* 2014, 4, 139–159.

2 Surface Functionalization Techniques for Biosensors

Azza M. Mazrouaa, Dalia S. Fathy, Nahla A. Mansour and Manal G. Mohamed

2.1 INTRODUCTION

Biosensors' innovative methods have created new opportunities and are widely employed in forensic medicine and industry, alongside their established biomedical uses such as pregnancy testing, bacterial infection detection, cholesterol monitoring, and rapid troponin T testing. Biosensors are highly adaptable and can be used in a wide range of disciplines, providing fast and dependable findings for different substances being analyzed. Biosensors play a crucial role in forensic medicine by detecting hazardous chemicals, illicit narcotics, and other forensic markers in biological samples, aiding in investigations and court proceedings. Biosensors are employed in the industrial domain to ensure quality control, monitor environmental conditions, and ensure food safety. They facilitate the precise and effective identification of contaminants or specific analytes (Chen et al. 2020; Rohaizad et al. 2021). The application of biosensors in sectors beyond traditional biomedicine showcases their extensive potential and emphasizes their importance in other disciplines. Natural nanosystems possess considerable potential for the advancement of innovative biosensors and bioanalytical systems through diverse biochemical modifications. A physicochemical transducer and biological components work together in a biosensor to produce an electrical signal that indicates the results of the bioanalysis. The differences between biosensor analysis and bioanalysis/bioassay methods pertain to the procedural steps and the use of reagents (Verma and Yadav 2024). The biosensor gadget employs biologically relevant components to identify a certain species. This device comprises a bioreceptor, which is a biological recognition system that interacts with a transducer and the analyte that turns this biochemical interaction into a measurable and manageable signal (Alvarado-Meza 2018). Biosensors can be categorized into nucleic acid antibody-based, enzyme-based, affinity-based, and whole-cell biosensors, based on the biorecognition element type employed (Zahra et al. 2024). The biosensor performance is significantly influenced by the choice of materials for surface modification as well as the surface chemistry and functionalization approach employed for immobilizing the bioreceptors. There are various

DOI: 10.1201/9781003536239-2

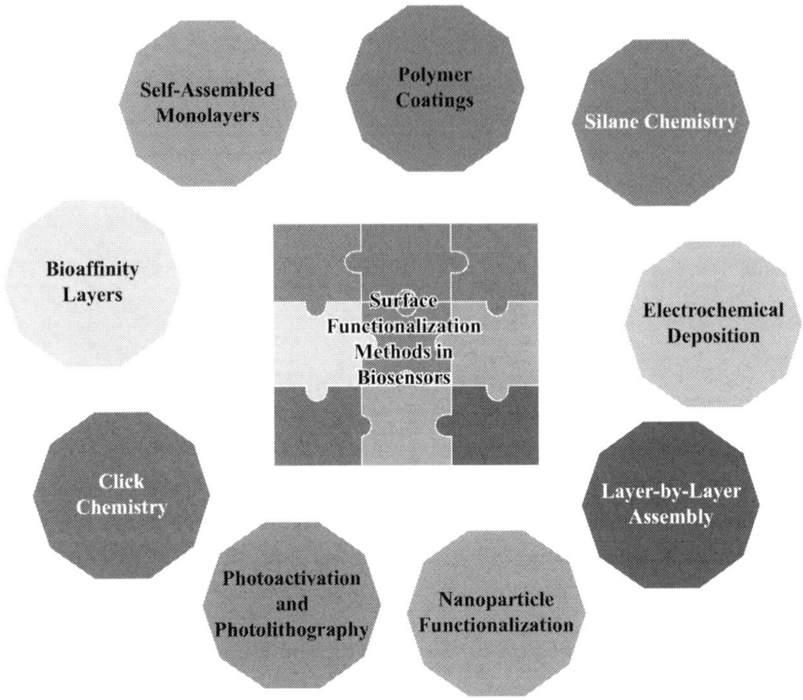

FIGURE 2.1 Different surface functionalization methods in biosensors.

surface modification methodologies described in the literature, and each type of modification produces varied results in terms of biosensor performance. The specific substrate to which one applies an immobilization technique determines its selection (Sandhyarani 2019). Surface functionalization is a crucial element in the design of biosensors, as it allows for the precise and effective interaction between the surface of the sensor and the substances being analyzed. Here are the many commonly used surface functionalization methods in biosensors: self-assembled monolayers (SAMs), polymer coatings, silane chemistry, electrochemical deposition, layer-by-layer assembly, nanoparticle functionalization, photoactivation and photolithography, click chemistry, and bioaffinity layers, as in Figure 2.1.

2.2 TECHNIQUE OF SURFACE FUNCTIONALIZATION FOR BIOSENSORS

Surface functionalization is a critical aspect of biosensor design, enabling the specific and efficient interaction between the sensor surface and target analytes. The procedures for biosensors differ based on the applications, type, and target analyte. Here are several common techniques for surface functionalization used in biosensors:

2.2.1 Self-Assembled Monolayers

Molecular self-assembly has garnered significant interest in biological domains as a straight-forward and efficient technique for creating biomolecular nanostructures. Self-assembled nanostructures (SANs) can demonstrate significant selectivity and binding affinity by presenting several ligands or surface receptors. The alteration of supramolecular structures upon binding is a compelling method for generating binding signals. In recent decades, researchers have created numerous biosensors based on SANs, utilizing diverse biomolecules such as peptides, DNA, RNA, and lipids, and combining them with materials which are non-biological.

Intermolecular forces are used in the scientific field of supramolecular chemistry to create intricate molecular assemblages with particular structures and functionalities (Kim et al. 2023). Self-assembled monolayers (SAMs), which are created when molecules spontaneously arrange themselves on a substrate surface, are an example of supramolecular chemistry.

Given their distinct features that vary with size and their potential uses in many areas such as medicine regenerative and engineering biomedical, researchers have extensively investigated SAMs. Additionally, they may be constructed from diverse materials, such as polymers, metals, and semiconductors, and can be incorporated into a wide range of shapes and sizes. Altering the forces of intermolecular and composition that affect the process of self-assembly enables the modification of the material's characteristics (Basso et al. 2023). SAMs are molecular assemblies that arrange themselves in an orderly manner on surfaces, usually by the absorption of organic molecules. These molecules frequently possess a head group that exhibits strong interactions with the substrate, such as gold or silver, and a hydrophobic tail that tends to align in a specific orientation. This alignment results in the creation of a monolayer. Self-assembled monolayers (SAM), or surface-to-air missiles, find utility in a wide range of applications, encompassing: Surface modification refers to the process of changing the characteristics of a surface, such as its capacity to repel or attract liquids, its resistance to corrosion, or its compatibility with biological systems. SAMs possess adjustable characteristics that may be modified by altering the length and chemical composition of the molecules employed (Ali et al. 2020; Singh et al. 2020).

SAMs generally consist of two or more distinct molecular types that organize themselves into a specific structure (Ariga et al. 2020). The process of self-assembly is influenced by various forces, such as Van der Waals forces, hydrophobic interactions, and electrostatic interactions. They are attractive for biomedical applications because of their unique qualities. Because SAMs may be designed in a variety of sizes and forms, they can be used in fields like biosensing, tissue engineering, and medication delivery. Furthermore, biocompatible materials can be used to build SAMs, guaranteeing their safety for human use. Different substances, such as proteins and DNA, can functionalize SAMs to allow for certain interactions with tissues and cells.

Ultimately, these have been organized into three-dimensional structures, allowing them to replicate the architecture and functionality of native tissues (Basso et al. 2023; Shojaei et al. 2022). SANs based on biomolecules show promise as materials for biosensor applications. By expressing several receptors on their surface, SANs are able to recognize targets with efficiency. This is due to two main factors: 1) the likelihood of interacting with target molecules is increased when receptors are arranged in different orientations on the surface of the nanostructure; and 2) multivalent interactions typically show much higher binding affinity and selectivity than monovalent interactions (Jeong et al. 2022; Kim et al. 2022; Ko et al. 2023). Biomolecule-based self-assembled nanostructures (BSN) produced through a variety of non-covalent interactions have the ability to alter the supramolecular structure in response to target contact, providing opportunities for signal transduction from sensors and their generation. When self-assembling biomolecules are hybridized with different non-biological materials, the electrical, magnetic, mechanical, and optical properties of BSN-based biosensors are enhanced (Kim et al. 2023).

2.2.1.1 Adhesion Mechanism of SAMs

The mechanism of adhesion of SAMs is heavily reliant on the anchoring group and the surface chemical nature of the substrate. The most widely used SAMs anchoring groups are carboxyl acid, phosphoryl acid, and silane. The compounds commonly involve the following elements:

- Sulfur is frequently employed in thiol molecules, which in turn create SAMs on gold and silver surfaces.
- Silicon is utilized in silane compounds, which have the ability to form SAMs on oxide surfaces.
- Oxygen is found in silane and other chemicals that make up SAMs (Ali et al. 2020).

2.2.1.2 The Molecular Structure of SAMs

SAMs are precisely structured organic molecules that can spontaneously adhere to a substrate's surface through molecular composition in either the vapor or liquid phase. SAMs will exhibit intermolecular interactions at the surface of the substrate, with the strength of these interactions mostly determined by the structural composition of the SAMs and the chemical characteristics of the substrate. Typically, the SAMs consist of anchoring, spacer, and terminal functional groups, each group having distinct responsibilities determined by their chemical composition. The anchoring groups of SAMs adhere to the surface of the substrate and primarily impact the electrical state of the substrate surface by chemical bonding. The often-employed anchoring groups are silanes, carbonyl acids, and phosphate groups. Anchoring groups have a crucial role in promoting the molecular association between SAMs and the surface of the substrate. These groups have a significant impact on important metrics such as the work function and surface contact resistance. The strength of the intermolecular connection between SAMs and the substrate surface mostly relies on the electron affinity of the anchoring groups and the chemical characteristics of the interface. The spacer groups serve as the structural support of SAM molecules by connecting the anchoring and terminal functional groups. They are often made up of alkyl chains. The strength of the interchain Van der Waals contacts, which are essential for the formation of highly organized supramolecular structures, is determined by the length of the alkyl chains in SAM molecules. The length of the branched chain of the spacer group directly influences the shape and surface coverage of the SAMs that are generated. The branched chain structure of the spacer group is advantageous for modifying the charge transport characteristics of SAMs, particularly for the extraction and transport of charges at the interface. Additional terminal functional groups that have been documented include amino, sulfhydryl, and carbonyl groups. The terminal functional groups of SAMs primarily govern the interfacial chemistry between the SAM and the substance it interfaces with (Kim et al. 2020). To enhance the conjugation effect of SAMs, it is necessary to build SAM molecules with a fully conjugated structure by substituting aliphatic chains with conjugated linkage groups (Aktas et al. 2021). Furthermore, utilizing the π-extended structures is a highly effective method for enhancing the stability of SAMs when exposed to light and temperature stressors (Wu et al. 2023).

2.2.1.3 Biomedical Application of Self-Assembling Nanomaterials

Self-assembly refers to the process of spontaneously assembling ingredients to produce a composite nanostructure, without the need for considerable external intervention. Self-assembly can occur in two forms: intermolecular, which involves the arrangement of molecules between different molecules; and intramolecular, which involves the arrangement of molecules within a single molecule. Self-assembly enables the synthesis of structures through two main approaches: top-down fabrication, which involves using specific techniques to create nanoscale structures from bottom-up organization; and larger materials, where structures are arranged into regular patterns through local forces to reach the most stable energy state.

Because it allows for the assembly of small structures or patterns into highly organized arrangements, self-assembly offers substantial advantages and produces a wide range of functionalities for synthesized materials. Combining several weak forces, such as dipole–dipole interactions of ionic chirality, electrostatic connections, hydrogen bonds, Van der Waals forces, π-π stacking, non-covalent interactions, and unique hydrophobic forces, is known as self-assembly. These forces might not be very strong on their own, but when coupled, they have the ability to cause self-assembly. The process of self-assembly enables the structures to be modified, enhancing their compatibility with biological applications and biomedical applications. SANs have desirable characteristics such as enhanced stability, prolonged cellular fluid circulation time, focused dissemination, and improved efficacy in treatments. The process of self-assembly of nanocrystals facilitates the creation of specific nanocrystals with adjustable characteristics. Various self-assembled nanocrystals have been created, including those composed of cadmium selenide, Ag, TiO_2, indium phosphide, platinum, Au, Fe_2O_3, silver sulfide, cobalt, and others. Self-assembled nanocrystal arrays are structured into monolayers, super lattices consisting of nanoclusters, and thin films. The self-assembled arrays are enclosed within small organic covers in order to enhance their stability and longevity (Uniyal et al. 2024; Qi et al. 2018).

2.2.1.4 Peptides as Self-Assembled Monolayers

Proteins are essential components of cells that serve a wide range of purposes, including providing structural support (such as collagen and actin), facilitating biochemical reactions (such as enzymes), defending against foreign substances (such as antibodies), regulating bodily processes (such as hormones), storing molecules (such as ferritin), and transporting substances (such as hemoglobin), among other functions (Redondo-Gómez et al. 2024). The ability to easily modify and adapt chemically is a key feature. Peptide sequences have the ability to include not only proteinogenic amino acids but also synthetic ones that have unique side chains. This makes it possible to add and show redox- or bioactive parts along with non-diffusible ligands, which are very important for studying how biomolecules interact with each other. Peptides can also be added to functionalized SAMs that are already there. These are commonly used to make antifouling surfaces and surgical implants that are functionalized (Wang et al. 2023; Teixeira et al. 2023). In order to screen repulsive electrostatic interactions, such a phenomenon is frequently caused by altering charge states, either by changing the pH or by adding counterions, like calcium as represented in Figure 2.2 (Matthew and Thomas 2021). By adjusting surface charge, establishing host–guest interactions between fibers, or incorporating multivalent binding groups to non-covalently crosslink several fibers at once, self-assembled nanostructure interactions can be adjusted at the molecular level.

2.2.1.5 Thiolates as Self-Assembled Monolayers

Alkanethiols form the SAM on noble metals, which has garnered extensive research and is considered the most studied SAM system. The sulfur head group forms robust

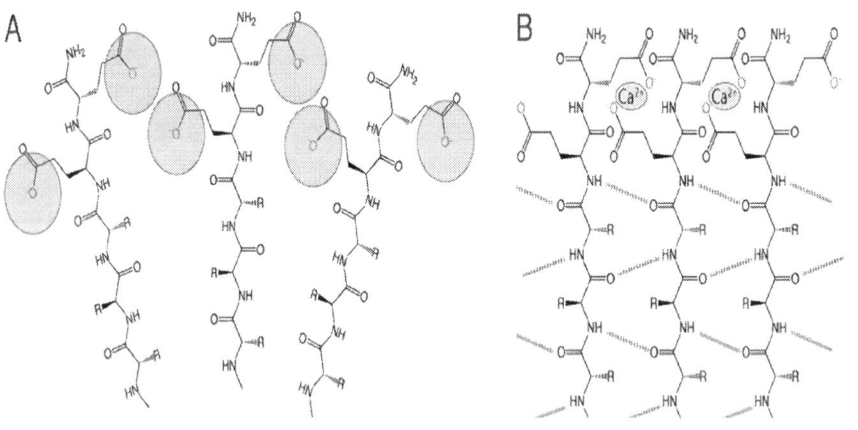

FIGURE 2.2 (A) Self-assembling peptides typically contain hydrophilic domains comprised of charged amino acids, leading to electrostatic repulsion, which may disfavor self-assembly under physiological conditions. (B) These repulsive forces can be overcome to drive self-assembly and/or stabilize assembled nanostructures by changing pH or adding ions, like calcium, to screen and/or bridge charged groups. (Reprinted with permission from Matthew and Thomas, 2021.)

contacts with the surfaces of noble metals, particularly gold. Intra- and intermolecular forces, such as Van der Waals forces, hydrogen bonds, and interaction with hydrophobic and electrostatic molecules, facilitate the self-assembly of thiolates on noble metal surfaces. There is a strong attraction of the sulfur group to metals through semicovalent connections. In addition to the hydrophobic resulting from aromatic chains or alkane, Van der Waals interactions, and the force affinity might facilitate the creation of the self-assembled layer. The driving forces cause the chains of thiol to tilt, which increases the chain interaction and decreases the total surface energy. The carbon-based chain has the ability to regulate the characteristics of the entire structure as well as the organized SAM platforms. Thiolates as typical platform SAM based on comprises three primary components: 1) a sulfur head group that binds the noble metal surface, 2) the aromatic chain or alkane, and 3) the functional terminal group (or tail) that enables the attachment of various functional groups such as carboxylic acids, amides, hydroxyls, halogens, etc. The head groups of sulfur exhibit a high affinity for metals such as nickel, palladium, copper, gold, platinum, and silver, forming strong bonds with them (Ghorbanizamani et al. 2024).

In conclusion, the integration of SAMs with various nanomaterials, the modification of SAMs with functional groups, the application techniques for molecular imprinting, the incorporation of technologies for advanced sensing, and the advancement of novel fabrication methods can address certain limitations of sensors and biosensors. The solutions have facilitated the creation of sensors with high selectivity and sensitivity and biosensors that are applicable across diverse fields, including biomedical, environmental, and food analysis.

2.2.2 Silane Chemistry

Surface functionalization is a critical aspect of biosensor design, enabling the specific and efficient interaction between the sensor surface and target analytes. Various methods have been outlined for the functionalization of hydrogen-terminated silicon. The various strategies each present distinct advantages and disadvantages; however, regardless of the chosen approach, the initial critical step is the removal of the native oxide that covers the surface (Henriksson et al. 2021). Silane chemistry is widely utilized for surface functionalization in biosensors due to its ability to modify surfaces with various functional groups, enhancing biomolecule immobilization and sensor performance. Silane coupling agents such as (3-aminopropyltriethoxysilane) APTES and (3-glycidoxypropyltrimethoxysilane) GPTES are commonly used to covalently link biomolecules to surfaces like glass, silicon, and metal oxides. They possess functional groups (amine, thiol, etc.) that can react with surface hydroxyl groups to form stable bonds. Silane chemistry allows for precise control over surface properties, such as hydrophobicity, charge, and biocompatibility, which are crucial for biosensor functionality. Biosensors benefit from silane-modified surfaces by improving biomolecule immobilization efficiency, minimizing non-specific binding, and enhancing sensor sensitivity and selectivity. Silane chemistry offers versatility, stability, and control over surface properties critical for biosensor performance (Sypabekova et al. 2022). Various strategies can be taken to incorporate the

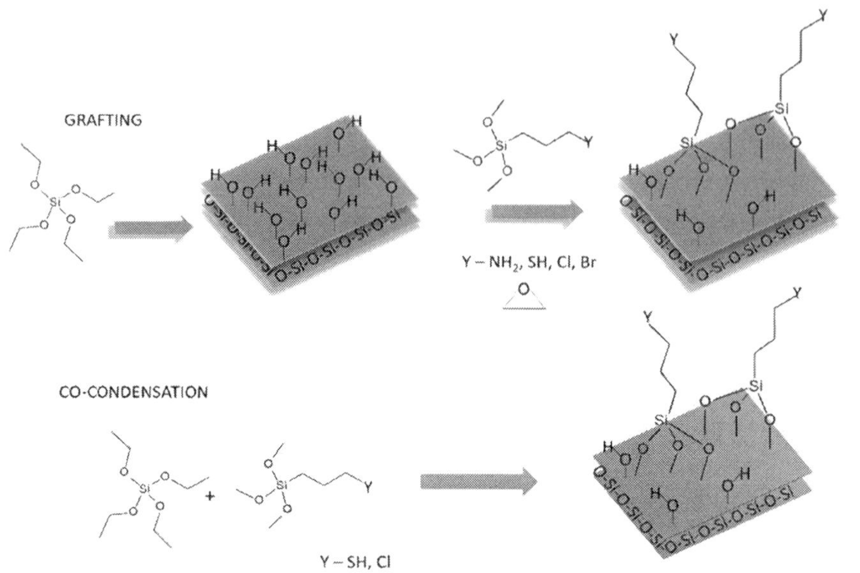

FIGURE 2.3 Scheme of method organosilane fabrication: grafting and co-condensation. (Reprinted with permission from Karolina et al., 2021.)

functionality into the silica structure, as shown in Figure 2.3, and these routes rely on the nature of functionalization, the chemical reactivity of the material, and its physicochemical qualities (Karolina et al. 2021).

Silicon nanowires (SiNWs), as a novel category of one-dimensional semiconductor nanomaterials, demonstrate significant potential for applications in biomedical sensing. SiNWs exhibit superior electronic properties that enhance the sensitivity detection of biosensors. The integration of SiNWs with field effect transistors (FETs) resulted in a specialized biosensor characterized by enhanced sensitivity and precise target selectivity, enabling label-free detection in real time. SiNW-FETs have recently garnered increased attention in biomedical detection fields. Alongside performance factors like sensitivity and specificity, economic considerations are also crucial in the biosensor's development and fabrication. Nanowires are increasingly favored for the fabrication of biomedical sensors owing to their sensitivity to biological targets and their potential for large-scale production (Li et al. 2023). The source electrode, drain electrode, and gate electrode (which can be either a back gate or a solution gate) make up SiNW-FETs, like other electrochemical biosensors. Small biological molecules modify SiNW through diverse methods to ensure alignment with the intended target. The conductance modulation in SiNW-FETs arises from fluctuations in charge density. Additionally, the capture of targets on the sensor surface may result in alterations of the device's internal charge density. Currently, SiNW-FETs represent a significant advancement in biological research according to their exceptional physical properties, containing high mobility carrier, a high switch ratio current, and a subthreshold slope approaching ideal levels (Song et al. 2022).

The materials are also inexpensively available. Researchers have effectively used these methods to detect and recognize specific proteins, DNA, or RNA as markers for cancer or cells (Yang et al. 2018). Several studies have shown big improvements in how SiNW-FETs are made and how their surfaces are changed. Their sensitivity has been validated to achieve the level of attomole. SiNW-FETs demonstrate significant potential for applications in the biosensing and biomedical domains. SiNWs demonstrate efficient electron transfer, leading to a rapid response in detection. Silicon nanowires are utilized for the identification of the virus of dengue. The resistance variation of the silicon nanowires is assessed prior to and following the binding of molecules virus (Li et al. 2023; Tran et al. 2018).

Applications for DNA sensing have used morpholino-functionalized silicon nanowires (SiNWs). Morpholino as a modifier of SiNWs exhibits elevated responsiveness and demonstrates effective hybridization of DNA with morpholino. The conductivity of SiNW decreases as the concentration of complementary DNA increases. This may serve as a label-free biosensor for the analysis of gene expressions. Researchers have utilized silver nanomaterial-coated silicon nanowires to measure amoxicillin in milk. Surface enhanced raman spectroscopy (SERS) has utilized them as a substrate for the detection of proteins and immunoglobulins. SiNW-based FET demonstrates exceptional sensitivity and selectivity in biosensing applications. SiNW can be employed in the development of point-of-care sensors for the detection of diseases such as cardiovascular diseases, infectious diseases, and cancer (Thirugnanasambandan et al. 2024). The remarkable biocompatibility of nanostructures renders them appropriate for biomedical applications. Sol-gel procedures, chemical vapor deposition (CVD), rapid thermal annealing (RTA), ion implantation, carbon-assisted growth, and laser ablation are the primary methods used for nanostructure centering. The usage of array SiNW biosensors for the simultaneous detection of several cancer indicators on a single, flexible substrate was demonstrated in this study. To improve the concurrent of multiple signals detection, an increased number of single nanowires was fabricated in parallel through metal deposition methods and photolithography. The array SiNW comprised three distinct SiNW biosensors, each featuring a unique fluid-based assembly. The biosensors were designed for electrical addressability and were immobilized using three distinct antibodies. The three antibodies were designed to specifically target specific antigen of prostate (APS), antigen of carcinoembryonic (ACE), and mucin ^{1}. All of these biomarkers have undergone clinical validation for the purpose of cancer detection. The SiNW biosensors facilitated the real-time detection of three cancer biomarkers: APS, ACE, and mucin^{-1}. Improvements in the the early-stage detection of cancer may be accomplished through the simultaneous analysis of multiple biomarkers, ensuring sensitivity to a superior level. This advancement would significantly enhance survival prospects for cancer patients (Thirugnanasambandan, Ramanathan, and Gopinath 2024). The electrochemical signal was stronger when the SiO_2 nanoprobes were used with the electroactive labels and DNA probes. The simultaneous detection of sarcosine and prostate-specific antigen (APS), two biomarkers for prostate cancer (PCa), was made possible by this biosensing device (Wang et al. 2023).

2.2.3 Polymer Coatings

Polymer coatings involve the application of polymeric materials to sensor surfaces. These coatings provide functional groups that can be further modified to attach specific biomolecules, enhancing the sensor's functionality and performance. Polymers can be tailored to impart various properties to the sensor surface, making them a versatile choice for surface functionalization. Conductive polymer nanocomposites (CPNs) have increasingly attracted interest in the development of electrochemical biosensors (Zeng et al. 2024). Doping, chemically altering their structure, or mixing them with other additives—such as metals, acids, semiconductors, or surfactants—can all be used to adjust their conductivity. Polythiophene, polypyrrole (PPy), polyaniline (PANI), and poly(3,4-ethylenedioxythiophene) polystyrene sulfonate (PEDOT: PSS) are the most commonly utilized conductive polymers in flexible sensors (Pavel, Lakard, and Lakard 2022).

These materials are used primarily to increase the active surface area of the electrode, enhance electron efficiency and charge transfer, and serve as an immobilization matrix to boost the stability and usability of the biorecognition element (Kalita et al. 2023). The inclusion of components of metal in the composites enhances sensor quantitative capabilities. Additionally, the introduction of functional groups that can be coupled with biomolecules facilitates the attachment to antibody conjugates for antigen detection. Carbon materials enhance the chemical stability and conductivity of composite materials, enabling them to withstand more demanding detection environments and effectively detect metal ions. Carbon materials exhibit superior detection performance in protein detection due to their unique structure compared to other materials (Zeng et al. 2024). Surface modifications have been employed to enhance MXenes. During this process, different kinds of materials are mixed together. These include polymers like PVP, PVA, cellulose, soybean phospholipid (SP), PEG, and mesoporous silica nanoparticles (MSNs) and polyoxometalates (POMs). Polymers which are biocompatible are preferred for their properties such as being non-toxic, having biological fluid compatibility, and the lack of the need for surgical removal post-treatment. A novel biosensor, designated GOx/Au/Ti_3C_2/Nafion/GCE, was developed for the detection of glucose. Adding Nafion makes it easier for enzymes to stick to glassy carbon electrodes (GCE), gets rid of unwanted signals, and makes biosensors more selective (Zahra et al. 2024). Electronic biosensors enable the early detection of cancer cells. A portable biosensor for the rapid detection of the cancer biomarker carcinoembryonic antigen (CEA) was developed at ambient temperature. This biosensor employed a novel poly[(1,4-phenylene)-*alt*-(3,6-(1,2,4,5-tetrazine)/3,6-(1,2,4,5-dihydrotetrazine))] (PhPTz) matrix and reduced graphene oxide (rGO) as a conducting layer for the immobilization of CEA antibodies (Joshi et al. 2022). A straightforward and efficient sensing surface for Curcumin (CUR) detection was developed using a carbon nanotube paste electrode modified with poly(glutamine). With outstanding recovery, the produced sensor device was applied and improved for CUR estimation in a natural food supplement (Girish and Manjunatha 2020).

A photoelectrochemical (PEC) platform biosensor has been constructed utilizing graphene (G) by transforming it into an electroconductive polymer nano sponge (EPNS) and integrating titanium dioxide nanowires (TiO_2 NW), referred to as TiO_2(G)NW@EPNS. TiO_2(G)NW@EPNS works well as an immobilization matrix for the enzyme Cytochrome C (Cyt C), making it a good choice for a PEC biosensor. A very selective and sensitive PEC biosensor for NO_2 was made possible by the fast kinetics of direct electron transfer (DET) to the electrode and the good separation of photogenerated holes and electrons (Muthuchamy, Lee, and Gopalan 2017).

Researchers in various studies made an immunosensor that can find the biomarker IL-6 by connecting black phosphorus (BP) and the multifunctional polyallylamine as a cationic polymer (PAMI) through electrostatic pull. It addresses the BP degrading problem, enhances transfer of charge, and incorporates $-NH_2$ (amine groups) for the attachment of target biomolecules. The cationic characteristics of PAMI improved the immobilized biomolecules' stability by establishing a saline environment. The BP platform coated by PAMI had many sites of binding, which made it very stable, easy to make, and able to be used for point-of-care (POC) cancer analysis (Barman et al. 2021). Since the modified poly thiazole yellow-G (TY-G) layered multi-walled carbon nanotube paste electrode (MWCNTPE) exhibits superior active surface area and electrocatalytic activity, it is a more promising and powerful sensing tool in the analysis of numerous other electro-active molecules and medicinal samples. It also offers excellent formoterol fumarate recovery in a medicinal sample (Hareesha and Manjunatha 2021). To resolve tyrosine (Tyr), uric acid (UA), and ascorbic acid (AA) separately and simultaneously, a modified electrode poly-threonine modified graphite-carbon nanotube paste electrode was created. With very encouraging findings, the new approach might be used to resolve Tyr in pharmaceutical samples. The modified electrode's strong selectivity, low limit of detection, and high current responsiveness for Tyr determination show promise as a useful sensor (Chenthattil et al. 2020).

Owing to their substantial surface-to-volume ratio, one strategy is the non-covalent integration of biocompatible polymers into transition metal dichalcogenides (TMDC) layered materials to produce TMDC/polymer composites, hence enhancing the biosensing potential (Li and Wong 2017). A PEC sensing platform for the detection of glutathione (GSH) was created utilizing PANI and nanoMoS_2 composites as optoelectronic materials (Hun et al. 2018). The protective efficacy of poly(2-methacryloyloxyethyl phosphorylcholine-*co*-glycidyl methacrylate) and polyvinylimidazole-polysulfostyrene was examined when integrated into the biosensor design for the galactose detection of plasma in the human body (Figueiredo et al. 2024).

2.2.4 Electrochemical Deposition

Electrochemical (EC) biosensors are extensively utilized in medical diagnostics and the identification of biowarfare agents because of their cost-effectiveness, low power requirements, and ease of downsizing. The electrode is the crucial element of EC sensors as it interacts with biological components known as "analytes." The use of an electrochemical deposition method is a practical and efficient way to produce metal nanoparticles. This process is advantageous since it is fast, cost-effective, and does not produce a significant number of harmful by-products (Tonelli, Scavetta, and Gualandi 2019). The three characteristics of EC in various applications are 1) selectivity, 2) precision, and 3) compatibility. The amount of material deposited, its composition, size of particles, roughness of surface, and, in some cases, the porosity of electrodeposit can all be precisely controlled using electrodeposition. Single-stranded DNA (ss-DNA), chelators, antibodies, and other receptors can all be combined to create composites (Choi et al. 2023).

Another method of electrodeposition is to apply a thin film or layer of a substance to a surface by using an electric stimulus (current or potential) through an electrolytic solution that contains dissolved precursors of the desired substance. A three-electrode setup is usually used for this method. The working electrode is where, through chemical reaction, metal ions were deposited controlled by current or potential; the counter electrode balances the reaction charge and completes the electrical circuit; and the reference electrode compares the working electrode potential to a standard reference electrode (Samanta, Ghosh, and Kuila 2022). The selection of electrode materials is crucial in all electrochemical systems, including voltammetric biosensors. Electron transport characteristics, surface texture, porosity, and the capacity for chemical functionalization are crucial considerations when building a biosensor. Numerous materials and designs can be used to make these sensors. Carbon, platinum, and gold are often used, and disc electrodes and interdigitated structures are common designs (Lee et al. 2022).

Most of the research on electrodeposited polymers uses in-situ polymerization to create hybrid materials. These can be metal-complex polymers, nanomaterial polymer composites, and biological polymer composites (Napierała et al. 2021) (Kassa, Abebe, and Amare 2021). These materials may make sensing platforms better. Although polymeric sensing layers have significant advantages in selectively detecting target molecules, they are vulnerable to degradation in severe, non-inert conditions, thereby constraining their practical applicability (da Silva, Queiroz, and Brett 2020). In this context, inorganic materials like metals or metal oxides exhibit superior physicochemical stability relative to organic substances. ED and ED-based methods, which provide solution-phase sensing, categorize the direct synthesis of diverse inorganic compounds into three distinct groups: noble metals are designed as microelectrodes of high-surface area, metal alloys, metal oxides, metal-organic frameworks (MOFs), non-noble metals, and MXenes (Choi et al. 2023). For the sensitive and specific detection of cetirizine with paracetamol was created. The suggested electrode is very helpful for determining cetirizine in pharmaceutical, blood serum, and urine samples with acceptable recovery because of its exceptional stability and reproducibility (Pushpanjali et al. 2021). Alloxan (AL) is a poisonous glucose analog that preferentially kills the pancreatic β-cells that produce insulin, making it a powerful diabetogenic inducer. Therefore, for the detection of AL in the presence of anthrone, a carbon paste electrode immobilized with cetyl trimethylammonium bromide (CTAB) was used, which is both sensitive and selective (Amrutha et al. 2020; Amrutha et al. 2021).

2.2.5 Nanoparticle Functionalization

Surface functionalization of nanoparticles for biosensors has seen significant advancements in recent years, involving attaching specific molecules or biomolecules onto the nanoparticle surface to increase surface area, focusing on enhancing sensitivity, specificity, binding capacity, sensing capabilities, and stability for various biomedical applications. Three primary classes are widely acknowledged based on this: the functional groups thiols (–SH), amines (–NH_2), and cyanides (–CN) are most commonly utilized to functionalize noble plasmonic metals as gold and silver; oxides (such as iron oxides) can be improved with hydroxyl and acidic groups using oxygen bonds; binary compounds (elements from Groups 12 to 16) and fluorescent semiconductor NPs exhibit increased affinity toward –OH, –SH, and –NH_2 groups, as shown in Figure 2.4 (Karolina et al. 2021). Sometimes, coating applications in the biosensing and biomedical domains employ gold and silver, sometimes referred to as noble metals (Kumari, Vyas, and Kumar 2022).

The most typical coatings for these metals are cyanides, amines, or thiol groups. With these organic groups, metal nanoparticles have a great inclination to set up covalent connections. An organo-sulfur bond is created when adding a thiol group to the surface of metal NPs. The NP is stable against oxidation because of this organo-sulfur link covering its surface, which prevents it from reacting or clumping together. There are many approaches to achieving this kind of surface modification on metal nanoparticles. The first method entails covering the previously created nanoparticles with a sulfur group. There are many techniques to carry out this, including coating the nanoparticles in organo-functional groups by induction coupled plasma

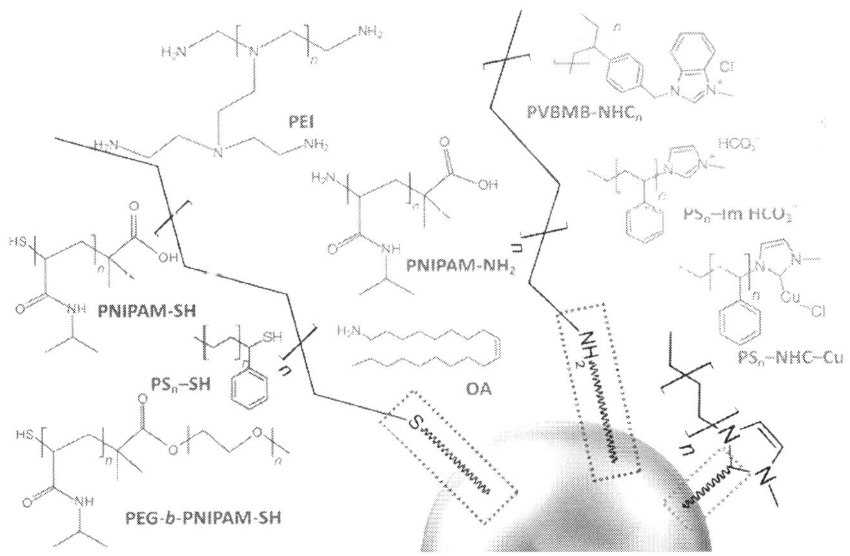

FIGURE 2.4 Surface functionalization of NPs by polymers via metal-S, metal-N, and metal-C bonds. (Reprinted with permission from Karolina et al., 2021.)

technology or a dip coating approach. An alternative method directly synthesizes metal nanoparticles capped with organo-sulfur in a single step. This technique is made easier by the wet chemical method, which enables the protecting group to be simultaneously capped and the metal precursor to be nucleated in a single step. Several studies within the domain of biosensing have explored this particular surface modification (Guerrini, Alvarez-Puebla, and Pazos-Perez 2018).

In the realm of biosensing, metallic oxides as titanium dioxide, silicon dioxide, and iron oxide are frequently utilized. By creating an oxygen link between metal oxides and acidic or hydroxyl groups, coating oxide nanoparticles can be accomplished with simplicity. In general, oxide nanoparticles can be produced in two variations: colloidal solution and powder. For commercial and industrial applications, the powder-form synthesis method, which often involves vapor deposition and precipitation, is more widely demanded. The majority of metallic oxide nanoparticles are hydrophobic by nature; thus, they must be modified to become soluble in water and compatible with cells. To achieve this, their surface layer could be covered in layers of other hydrophilic (water-soluble) polymers, like poly-(styrene sulfonic acid) or poly-(oxyethylene methacrylate), which are both ionic and nonionic. For coating purposes, the following methods are typically used (Uniyal et al. 2024):

1. The metal oxide nanoparticles' surface –OH is activated.
2. They are modified by covering them with the chlorine group (–Cl).
3. Polymers are applied to the surface of nanoparticle using transfer radical of atom polymerization process.

Because there are several carbon allotropic forms, from well-known allotropic phases like amorphous carbon, diamonds, and graphite to more recent developments like graphene oxide (GO), carbon nanotubes (CNTs), and graphene quantum dots (GQDs), carbon-based nanomaterials (CBNs) are gaining interest in science and technology (Kumar, Raghuwanshi, and Kumar 2022; Mazrouaa et al. 2022; Gupta et al. 2023). Functionalization by covalent or non-covalent methods significantly improves CNTs' processibility and solubility, providing a pathway towards the synthesis of emerging nano-conjugates from these altered CNTs. These carbon family members have all been widely employed in many biological applications for a range of reasons, including biosensing, medication transport, tissue engineering, faulty imaging, diagnosis, and related therapy. Graphene-based nanocomposites have appeared as significant substrates for plasma resonance (SPR) among carbon-based nanomaterials because of their appealing qualities, which include efficient biomolecule adsorption, enhanced response signal owing to higher mobility carrier, superior optical properties, and a simpler fabrication process (Gupta et al. 2023). Reduced graphene oxide (rGO)-AuNPs and AuNPs polyaniline (PANI) were used in the biosensor's development as a tracer label for amplification and as a sensing platform, respectively. By speeding up the electron transport, the rGO can improve the electrochemical signal. Moreover, AuNPs can improve rGO's electrical properties and molecular stability due to their high adsorption ability and superior conductivity. PANI has a polymer chain structure and offers excellent electrochemical performance, surface large area, high conductivity, and rapid electron transport kinetics. PANI with AuNPs improves signal amplification and DNA probe immobilization. The recommended method shows a strong ability to assess clinical samples and serves as an effective, versatile, and easy-to-use instrument for precise diagnosis (Eivazzadeh-Keihan et al. 2024).

2.2.6 Photo-Activation and Photo-Lithography

Surface functionalization is a key aspect of biosensor development, allowing for the precise modification of surfaces to enable specific biological interactions. Photoactivation and photolithography are powerful techniques in this domain, offering high resolution and control over the functionalization process.

2.2.6.1 Photo-Activation Techniques

- Photo-induced electron transfer: Photo-induced electron transfer has been employed to change surfaces with functional groups that can enhance the sensitivity of biosensors. Recent research has focused on improving the conditions for photo-activation to achieve better control over surface chemistry and stability. By analyzing the resistance produced under sporadic UV light, the photoactivity-induced performance of the manufactured biosensor was evaluated. By joining the two interdigitated gold electrodes with the ZnO nanorods layer, the transducer component of the created biosensor was evaluated. UV photons produce electrons on the sensing material surface upon analyte exposure to the sensor layer. These electrons may interact directly with the oxygen dissolved in the sample or with the target molecules and their interference. A reaction between electrons and dissolved oxygen may take place, depending on the ZnO energy level and the required voltage for producing oxygen reactive species. Transferring electrons from conduction band of ZnO to oxygen, the electron acceptor improves charge carrier separation. The accessible charge carriers on the surface provide an explanation for the biosensor response and resistance variations. Grain boundary potentials and ZnO–ZnO contacts determine the quantity of accessible charge carriers on the surface (Taleghani and Taghipour 2024).
- Photo-responsive polymers: Photo-responsive polymers have been developed for dynamic surface modification. These materials change their properties in response to light, allowing for reversible and controllable functionalization. These polymers have applications in various fields, including sensors, actuators, drug delivery systems, and smart materials. Here is a brief overview of the types and mechanisms of photo-responsive polymers. Some common types of photo-responsive polymers include:
  - Azobenzene-containing polymers: Azobenzene groups are commonly used in photo-responsive polymers. They undergo a reversible trans-to-cis isomerization upon exposure to ultraviolet or visible light. This change in isomerization leads to alterations in the polymer's physical properties, such as shape or volume. By adding azobenzene functionality to their molecular structures, conjugated polymers show an amazing

ability to respond to a variety of stimuli and show fluorescence quenching properties. Recent decades have witnessed significant advancements in the development of functional azobenzene-based conjugated polymers (azo-CPs), utilizing a variety of molecular design strategies and synthetic methods. Based on their molecular structures, we can broadly classify azo-CPs into three main categories: linear CPs with azobenzene in the side chain, linear CPs with azobenzene in the main chain, and branched CPs with azobenzene moieties. These systems hold great potential for applications in the biomedical field, including biosensors, bioimaging, targeted protein degradation, and cellular (Ma et al. 2024).

- Photochromic polymers are materials composed of polymer chains containing photochromic units that exhibit reversible color changes when exposed to specific wavelengths of irradiation (Zou et al. 2024). According to their composition, photochromic materials can be divided into three categories: organic, inorganic, and organic-inorganic hybrid (Yang et al. 2021). We categorize organic photochromic materials into six groups based on their mechanism of reaction: cis-trans isomerization of azobenzenes (Rad, Balzade, and Mahdavian 2022; Zheng and Yuan 2022), redox reactions of aromatic compounds, thick-ringed, pericyclic reaction systems involving captive arginine anhydride and heterocyclic diarylethene (DAE) groups, proton transfer reciprocal isomerization of salicylaldehyde condensed aniline compounds, heterogeneous cleavage of spiropyran and spirooxazine bonds, and homogeneous cleavage of hexaphenyl-bis-imidazole bonds. Rare earth complexes, metal halides, and transition metal oxides are the three main categories into which inorganic photochromic compounds fall (Wang et al. 2023; Brändler et al. 2023; Li and Zheng 2021; Pan et al. 2021; Zhang et al. 2023).

2.2.6.2 Photo-Lithography Techniques

One proven method for microfabrication procedures is photolithography (Patel 2021). It entails pattern-making on a substrate using substances called photoresists. Photolithography can be used in biosensor manufacture to create microscale features and structures, like fluid flow channels or biosensor platform electrodes for sensing (Kumari et al. 2023). Since resolution is essential for detailed sensor design, this method provides control over feature size and placement, guaranteeing patterning of electrodes, channels, and other components on biosensor substrates (Fruncillo et al. 2021). Additionally, photolithography is very scalable, making it suitable for large-scale production when needed in particular applications. Due to the complexity of photolithography processes, efficient completion of the required procedures requires competent operators. Furthermore, photolithography equipment can be expensive. There are three types of lithography: (Ahuekwe et al. 2024).

- Nanoimprint lithography: Nanoimprint lithography has been refined to achieve high-resolution patterning for biosensor applications. This technique allows for the creation of nanoscale features that can significantly enhance sensor performance.

- Direct laser writing: Direct laser writing has been utilized for creating complex micro- and nanostructures on biosensor surfaces. This technique offers high flexibility and precision, enabling the creation of intricate patterns for improved sensor functionality.
- Photochemical patterning: Photochemical patterning involves the use of light-sensitive chemicals to create specific surface patterns. Recent developments have focused on improving the resolution and repeatability of this technique for biosensor applications.

A unique photolithographic approach was developed to introduce a metal array with a micro-pattern onto flexible substrates without affecting the substrates through solvent, temperature, or UV exposure. A sensor of glucose was effectively created using the technology of micropatterning that was developed. The suggested glucose sensors, which are based on novel photolithographic techniques, have great promise for use as medical sensors in applications like devices for continuous and flexible monitoring of glucose levels (Ma et al. 2020). The fabrication of microarrays using photolithography has garnered a lot of interest lately. An innovative non-enzymatic glucose biosensor was proposed using a hemispherical microarray structure made by thermal melting and photolithography (Tran et al. 2018).

Superior current responsiveness is demonstrated by this biosensor, which also creates new avenues for 3D electrode creation. A number of 3D microarray electrodes with varying diameters and aspect ratios were created using photolithography on silicon wafers, significantly increasing the substrate's specific surface area and surface roughness. A thin film of gold with exceptional conductivity was produced by vacuum sputtering gold nanoparticles. The advantages of the obtained unique non-enzymatic H_2O_2 biosensor include its lack of enzymes, ease of fabrication, affordability, and long-term preservation. Consequently, this electrochemical biosensor may eventually be marketed and applied in medical settings (Li et al. 2021).

2.2.6.3 Hybrid Techniques

Recent studies have explored combining photoactivation and photolithography to achieve dual-mode functionalization of biosensor surfaces. This hybrid approach can offer greater versatility and enhanced sensor performance. This technique requires the addition of a layer of molecules on the surface of the substrate (either with photoactivable moieties or susceptible to degradation). The irradiation of the surface is usually done by exposure to ultraviolet light (UV) through a mask with the desired pattern. It has the advantage that it can be performed with relatively inexpensive devices, under biomolecule-friendly conditions allowing iterative cycles, and it can create numerous features per cycle and with high lateral resolutions (Alvarado-Meza 2018).

2.2.7 Layer-by-Layer Assembly

Thirty years ago, the very adaptable assembly technique layer-by-layer (LbL) was introduced. It involves the sequential assembly of complementary polyelectrolytes or biomolecules to create thin self-assembled films. It is useful due to its versatility to any type of supporting material which act as a carrier, simplicity, and compatibility with biological molecules. The capacity for immobilizing or loading proteins with retained bioactivity has led to the emergence of specific applications in several biological domains (Borges et al. 2024). NaYF4:Yb3+/Er3+ upconversion nanoparticles (UCNPs) modified with cysteine and anti-*E. coli* antibodies were used to create the LbL bio-optical sensor, which detects *E. coli* bacteria. In order to create the UCNPs@Cys system, scientists modified the UCNPs surface by adding cysteine. The second hybrid system, UCNP@ Cys-Ab, was then created by adding anti-*E. coli* antibody (Ab). LbL self-assembly was used to create the films, and polyvinyl sulfate (PVS) was used as the nanoparticles' immobilization matrix. The development of the system of photophysical with improved sensitivity of fluorescence for detecting *E. coli* antibody resulted from the attachment of UCNPs@Cys-Ab using LbL film.

When antibodies are added to the particle system, they make it much easier to recognize *E. coli* bacteria. This makes it easier to find and diagnose bacterial illnesses. These findings have a great deal of potential to advance biosensing applications and diagnostic methods, paving the way for future illness detection that is more precise and effective (Ramin et al. 2024). Fabricated polyelectrolyte multilayers (PEMs) using LbL assembly have been utilized in the biosensor's advancement. A biosensor is a tiny, highly advanced analytical device that uses a biological component as its sensing element. To detect signals, this part is connected to a transducer. Enzymes, receptor proteins, antibodies, and nucleic acids are examples of recognized biological components (Chauhan and Thakur 2023). The fact that biosensors can detect so many different diseases, among them cancer, myocardial infarction, diabetes, cardiovascular disorders, and Alzheimer's disease, is its main technological advantage.

A variety of polysaccharides, such as alginates, carrageenan, and chitosan (CHT), have been used to develop biosensors that use polymer electrolyte membranes (PEMs). The capacity of LbL polysaccharide-based sensors to engage in electrostatic and other interactions (ion-dipole, dipole-dipole, and hydrogen bonds) with biomolecules and proteins leads to their increased selectivity. Biosensors are more affordable, don't require temperature control or sample pretreatment, and are known for collecting data faster than other biological analysis methods like high-performance liquid chromatography with UV detection (Vilsinski et al. 2024). By incorporating glucose oxidase into electrolyte membranes composed of CHT/kappa-carrageenan polymer and gold nanoparticles, a glucose voltammetry biosensor was developed. In the presence of uric acid, creatinine, ascorbic acid, and urea, it showed remarkable stability and repeatability and little interference. Saliva samples spiked with glucose were used to evaluate the biosensor's effectiveness, and the recovery rates were satisfactory (Rassas et al. 2019).

By employing the LbL technique, a voltammetric biosensor was created to detect inosine, a biomarker for diseases such as coronary artery disease. Using purine nucleoside phosphorylase on the screen-printed carbon electrode surface, they electrostatically added CHT to PEMs (Si et al. 2019). To create biosensors, inorganic materials have been used in combination with biopolymers. The LbL approach is used to combine graphene oxide and carbon-based heat transfer (CHT) in order to diagnose beta-amyloid1–42 and identify Alzheimer's disease (Nangare and Patil 2022). Folate receptor alpha (FR-a) is a glycoprotein that functions as a biomarker for diagnosis and is overexpressed on the surfaces of tumor cells, especially in gynecologic cancers.

Currently, fluorescence imaging techniques or positron emission tomography (PET) are used to measure FRa. But it takes a lot of time and money to carry out these procedures. To detect FRa, an electrochemical biosensor based on nanostructured LbL film was developed. Polyallylamine hydrochloride (PAH) charged positively and folic acid (FA) with negative charge, the biorecognition component, alternately assembled to produce multilayer films on indium tin oxide (ITO) electrodes. By measuring FRa on the surface of cancer cells, the biosensor device that uses [PAH/FA] films presents a viable method for simple, rapid, and affordable cancer diagnostics (Correia et al. 2021). In order to detect guanine and adenine simultaneously, the electrochemical biosensor was created layer-by-layer, assembling TiO_2 nanoparticles, poly(L-lysine), and graphene quantum dots (GQDs) on a glassy carbon electrode. For the electrochemical oxidation of guanine and adenine, the TiO_2/poly(L-lysine)/GQDs composite film demonstrated improved conductivity and a significantly larger effective surface area, along with a quasi-reversible electrode process. The electrochemical biosensor's outstanding stability, resistance to interference, and repeatability allowed for its practical use in the effective simultaneous quantification of guanine and adenine in cow milk, urine, and calf thymus DNA samples (Chenthattil et al. 2024).

2.2.8 Bioaffinity Layers

A class of biosensors known as bioaffinity nanoprobes uses the unique binding properties of biological molecules, such as enzymes, antibodies, and nucleic acids, to identify foodborne pathogens. Surface functionalization methods are necessary for enhancing the performance and selectivity of biosensors by facilitating effective interactions between bioaffinity layers and target molecules. The benefits of bioaffinity nanoprobes encompass their capacity to identify minimal concentrations of pathogens, swift analytical processing, and economic efficiency. The possibility of cross-reactivity with other species and the necessity for specialized equipment are, however, limits (Bruce-Tagoe and Danquah 2023). The three main steps in all nanosensing technologies are as follows: element recognition, additionally referred to as probe bioaffinity, recognizes the target by forming a compound with it. The target identification event is converted by the transducer into a measurable signal, such as a voltammetric, colorimetric, impedimetric, or electrochemical signal. A third element transforms the altered signal into a format that the analyst may use for analysis or presentation (Nikhil et al. 2016).

In addition to stability, detection limit, cost, response time, shelf life, linearity, recyclability, and hysteresis, the selectivity, sensitivity, and specificity of the recognition element are the primary determinants of a nanosensor's efficacy (Kaur, Sharma, and Tripathy 2019). The selected bioreceptor for the nano sensor must exhibit a high affinity for the target in order to facilitate detection at low concentrations, given that foodborne pathogens typically exist in trace amounts. Consequently, biomolecules such proteins, aptamers, enzymes, cells, and antibodies are favored as bioaffinity probes in nanosensors or nano-biosensors (Gaudin 2017). To evaluate these bioaffinity probes' effectiveness in real time, particular protocols have been created. Several methods of analysis are used to examine the interactions between bioaffinity probes and their targets, including surface plasmon resonance (SPR), circular dichroism (CD), flow cytometry (FC), and fluorescence resonance energy transfer (FRET) studies. In nanosensors, biomolecules known as bioaffinity nanoprobes function as bioreceptors. After the molecules attach to the target, a process is started which may develop into an understandable signal for examination. These substances consist of non-antibody-binding proteins, aptamers, enzymes, and antibodies (Bruce-Tagoe and Danquah 2023).

2.2.9 Click Chemistry

Click chemistry denotes reaction classes that are thermodynamically favorable, employing "spring-loaded" modules to form new heteroatom–carbon bonds. To accurately define click chemistry, a set of rigorous criteria was devised for a reaction to be classified as a "click reaction." Click chemistry is now equivalent to the copper (I)-catalyzed azide-alkyne cycloaddition (CuAAC) reaction. Recently, the advantages of click reaction classes primarily employed in the design and synthesis of active luminogens (AIEgens) for activity-based sensing (ABS) have been highlighted, which include CuAAC, SPAAC (Strain-promoted Azide-Alkyne Click Chemistry), 1,4-Michael additions, and inverse demand Diels-Alder reactions. Their next use case involves the detection of metal ions, protein tagging and detection, tumor imaging and ablation, assessment of cellular unfolded protein burden, and mitochondrial imaging (Giel and Hong 2023). Electrochemical biosensors garner significant interest owing to the benefits they offer, like affordability, portability, sensitivity, and ease of use (Sekhosana, Majeed, and Feleni 2023) despite experiencing challenges in detection reproducibility and robustness due to variations in engineering and measurement conditions. Clicked electrochemical biosensors, which are advanced electrical devices that utilize biological reactions, are employed for the quantification and detection of an analyte by sending electrical signals (Mahmudiono et al. 2022). These devices demonstrate superiority in complex applications like surgery and bioimaging. Click chemistry is advantageous and far more pragmatic, even under the most severe situations. Consequently, covalent bonding in chemical reactions and at electrode surfaces can be readily accomplished by employing click chemistry methodologies (Fiore et al. 2022; Nontawong et al. 2022).

2.2.9.1 Organometallic Complexes as Clicked Electrochemical Sensors

An electrochemical sensor for ascorbic acid (AA) assay has been developed, exhibiting good selectivity and sensitivity, via a Cu(I) CuAAC. A Cu(I) catalyst facilitates a quantitative reaction between the azide-modified Au electrode surface and terminal propargyl-functionalized ferrocene, resulting in 1,2,3-triazoles. The suggested electrochemical sensor has excellent stability, high selectivity, and a low detection limit for ascorbic acid detection. Furthermore, it illustrates that the suggested sensor is capable of detecting AA in actual urine samples with commendable results (Mishra et al. 2023).

2.2.9.2 Biomolecules as Clicked Electrochemical Sensors

Biomolecules are essential for the advancement of electrochemical biosensors. These molecules engage with analytes that possess a high affinity for biological systems and convey information via electrical signals (Sekhosana, Majeed, and Feleni 2023). In the creation of an electrochemiluminescence biosensor, thiols that selectively engage with metal surfaces were employed. Hairpin DNA (hDNA) was used to immobilize the gold electrode. Following the obstruction of active sites on the gold electrode surface using 6-mercapto-1-hexanol (MCH), we subsequently inserted tobacco mosaic virus (TMV) RNA (tRNA) possessing azide functionality. Click chemistry was used to facilitate the attachment of propargyl-bromoisobutyrate (PBIB) through alkyne–azido interaction. ARGET-ATRP was utilized to finalize the clicked electrochemical system construction intended for numerous binding sites (Guo et al. 2022). A biosensor for gene mutation was created by immobilizing N_3-modified hairpin (P-N3), subsequently submerged in an MCH solution to yield P-N_3/MCH/AuE (Bi et al. 2020).

2.2.9.3 Polymers as Clicked Electrochemical Sensors

A novel, transparent, and economical electrochemical analytical technique utilizing click polymerization for signal enhancement has been established for the extremely sensitive identification of aflatoxin B1 (AFB1). The aptamer functioned as the capture probe for AFB1, self-assembling on the gold electrode surface through gold–sulfur bonding. After the hybridization of azide-modified complementary DNA (c-DNA) with the aptamer, click chemistry integrated a notable click polymer as a signal marker. Ultimately, the acquired AFB1 competed with the c-DNA for binding to the aptamer. This resulted in the dissociation of the c-DNA from surface of the electrode and a reduction in current intensity. The electrochemical aptamer sensor is very dependable, user-friendly, selective, and sensitive for detecting AFB1 in real samples. With the capacity to adapt to trace analysis of diverse analytes, safety detection, environmental monitoring, and disease diagnosis via appropriate capture probes, a platform for AFB1 analysis has been developed that is swift, transparent, and efficient. This platform has demonstrated promising applications in food hygiene and safety (Liu et al. 2022).

2.2.9.4 Clicked Electrochemical Sensors Based on Metal Organic Frameworks

Metal organic frameworks (MOFs) are a class of very promising functional materials formed by the coordination of metal ions with organic molecules containing a minimum of two functionally active moieties. MOFs are significant innovative materials that have exceptional relevance in diverse domains, particularly in electrochemical sensing, owing to their superior electrocatalytic characteristics. Nonetheless, the incorporation of click chemical components remains considerably restricted. Post-modification of MOFs via click chemistry has been evidenced for zirconium-based

MOFs containing azide groups. For instance, click reactions were utilized to synthesize an alkyne-functionalized gold nanoparticle-porphyrin. This electrochemical biosensor can detect metal ions (Sekhosana, Majeed, and Feleni 2023; Qing et al. 2019).

2.3 CONCLUSION

A biosensor is a tool for detecting and quantifying analytes that use biological components. A transducer transforms the biological reaction into a quantifiable signal, whereas a biosensor material interacts with the target analyte. Crucial to the functioning of a biosensor are the following three parts:

1. The utilization of biomolecules and nanoparticles in the conjugation process;
2. The chosen signaling protocol;
3. The process by which the transducer and the target analyte interact.

The production and functionalization of nanomaterials for detection and sensing creates specific problems. The amalgamation of nanotechnology with several disciplines, the inclusion of chemistry, biology, physics, and medicine, positions it as the leading platform for biomedical applications, including biosensing, bioengineering, gene therapy, and drug delivery. The possible hazards linked to these structures in the human body will be a significant concern for their therapeutic applications. Future research will concentrate on creating modified nanostructures that exhibit enhanced sensitivity and selectivity relative to current technologies, while ensuring human safety is not compromised.

2.4 TRENDS AND CHALLENGES OF BIOSENSOR TECHNOLOGY

Various significant advancements have occurred in the field of biosensors. The obstacles to developing enhanced and dependable gadgets are still there. Current and future trends in this domain are shaped by the integration of emerging innovations such as genetic engineering and nanotechnology, advancements in manufacturing methods such as economical mass production and miniaturization, and the amalgamation of wireless communication systems with information and communication technology (ICT) via concurrent multi-sensing devices. There is a concerted effort underway to use nanotechnology for the creation of biosensors and sensors with novel uses. There is a need for the large-scale manufacturing of low-cost sensors. The multi-sensing device is specifically engineered to detect many substances and to track different immune reactions simultaneously. The miniaturization of analytical systems offers several advantages in biosensing, such as reduced sample and reagent consumption. Improvements in state-of-the-art technology, such as the integration of wireless communication systems with sensing tasks and the application of genetic engineering to develop novel biosensor materials, will influence the future.

REFERENCES

Ahuekwe, E. F., A. F. Akinyele, A. E. Benson, M. I. Oniha, and O. Oziegbe. 2024. A Review of Fabrication Techniques and Optimization Strategies for Microbial Biosensors. *IOP Conference Series: Earth and Environmental Science* 1342: 12015.

Aktas, E., N. Phung, H. Köbler, et al. 2021. Understanding the Perovskite/Self-Assembled Selective Contact Interface for Ultra-Stable and Highly Efficient p–i–n Perovskite Solar Cells. *Energy & Environmental Science* 14 (7): 3976–85.

Ali, F., C. Roldán-Carmona, M. Sohail, and M. K. Nazeeruddin. 2020. Applications of Self-assembled Monolayers for Perovskite Solar Cells Interface Engineering to Address Efficiency and Stability. *Advanced Energy Materials* 10 (48): 2002989.

Alvarado-Meza, R. 2018. *Functionalization Strategies for the Development of Innovative Biochips*. Université Grenoble Alpes.

Amrutha, B. M., J. G. Manjunatha, and S. B. Aarti. 2020. Design of a Sensitive and Selective Voltammetric Sensor Based on a Cationic Surfactant-Modified Carbon Paste Electrode for the Determination of Alloxan. *ACS Omega* 5: 23481–90.

Amrutha, B. M., J. G. Manjunatha, and H. Nagarajappa. 2021. An Overview of Recent Developments of Carbon-based Sensors for the Analysis of Drug Molecules. *Journal of Electrochemical Science and Engineering* 11 (3): 161–77.

Ariga, K., X. Jia, J. Song, et al. 2020. Nanoarchitectonics beyond Self-assembly: Challenges to Create Bio-like Hierarchic Organization. *Angewandte Chemie International Edition* 59 (36): 15424–46.

Barman, S. C., M. Sharifuzzaman, M. A. Zahed, et al. 2021. A Highly Selective and Stable Cationic Polyelectrolyte Encapsulated Black Phosphorene Based Impedimetric Immunosensor for Interleukin-6 Biomarker Detection. *Biosensors and Bioelectronics* 186: 113287.

Basso, C. R., B. P. Crulhas, G. R. Castro, and V. A. Pedrosa. 2023. Recent Advances in Functional Nanomaterials for Diagnostic and Sensing Using Self-Assembled Monolayers. *International Journal of Molecular Sciences* 24 (13): 10819.

Bi, Q., X. Gan, R. Yuan, and Y. Xiang. 2020. Copper-Free Click Chemistry-Mediated Cyclic Ligation Amplification for Highly Sensitive and Non-Label Electrochemical Detection of Gene Mutation. *Journal of The Electrochemical Society* 167 (2): 27535.

Borges, J., J. Zeng, X. Q. Liu, et al. 2024. Recent Developments in Layer-by-Layer Assembly for Drug Delivery and Tissue Engineering Applications. *Advanced Healthcare Materials* 13 (8): 2302713.

Brändler, L., L. Niklaus, M. Schott, and P. Löbmann. 2023. Photochemical Precharging of Tungsten Trioxide for Enhanced Transmittance Modulation in Flexible Electrochromic Devices. *Advanced Materials Technologies* 8 (11): 2201903.

Bruce-Tagoe, T. A., and M. K. Danquah. 2023. Bioaffinity Nanoprobes for Foodborne Pathogen Sensing. *Micromachines* 14 (6): 1122.

Chauhan, S., and A. Thakur. 2023. Chitosan-Based Biosensors-A Comprehensive Review. *Materials Today: Proceedings* 50.

Chen, Y. T., Y.-C. Lee, Y.-H. Lai, Jin-Chun Lin, Nien-Tsu Huang, Chih-Ting Lin, and Jian-Jang Huang. 2020. Review of Integrated Optical Biosensors for Point-of-Care Applications. *Biosensors* 10 (12): 209.

Chenthattil, R., S. Luo, H. Ao, W. Xiao, and H. Ju. 2024. Layer-by-Layer Assembled Film of TiO2/Poly (l-Lysine)/Graphene Quantum Dots for Simultaneous Electrochemical Determination of Guanine and Adenine. *Journal of Electroanalytical Chemistry* 967: 118482.

Chenthattil, R., J. G. Manjunathaa, K. R. Doddarasinakere, F. Santosh, S. Gurumallappa, and N. Lingappa. 2020. Validated Electrochemical Method for Simultaneous Resolution of Tyrosine, Uric Acid, and Ascorbic Acid at Polymer Modified Nano-Composite Paste Electrode. *Surface Engineering and Applied Electrochemistry* 56 (4): 415–26.

Choi, E J., N. P. Drago, N. J. Humphrey, et al. 2023. Electrodeposition-Enabled, Electrically-Transduced Sensors and Biosensors. *Materials Today* 62: 129–50.

Correia, A. R., I. Sampaio, E. J. Comparetti, N. C. S. Vieira, and V. Zucolotto. 2021. Optimized PAH/Folic Acid Layer-by-Layer Films as an Electrochemical Biosensor for the Detection of Folate Receptors. *Bioelectrochemistry* 137: 107685.

Eivazzadeh-Keihan, R., Z. Saadatidizaji, M. Mahdavi, A. Maleki, M. Irani, and I. Zare. 2024. Recent Advances in Gold Nanoparticles-Based Biosensors for Tuberculosis Determination. *Talanta* 275: 126099.

Figueiredo, C., C. Psotta, K. Jayakumar, et al. 2024. Effect of Protection Polymer Coatings on the Performance of an Amperometric Galactose Biosensor in Human Plasma. *Biosensors* 14 (4): 167.

Fiore, L., R. Vitiello, A. Perna, G. Maccauro, and F. Arduini. 2022. Fast and Reliable Infection Diagnosis during Orthopaedic Surgery Using Bluetooth-Assisted Miniaturized-Electrochemical Sensor. *Microchemical Journal* 183: 108061.

Fruncillo, S., X. Su, H. Liu, and L. S. Wong. 2021. Lithographic Processes for the Scalable Fabrication of Micro-and Nanostructures for Biochips and Biosensors. *ACS Sensors* 6 (6): 2002–24.

Gaudin, V. 2017. Advances in Biosensor Development for the Screening of Antibiotic Residues in Food Products of Animal Origin—A Comprehensive Review. *Biosensors and Bioelectronics* 90: 363–77.

Ghorbanizamani, F., E. G. Celik, H. Moulahoum, and S. Timur. 2024. Self-Assembled Monolayer–Based Nanoscaled Surfaces. In A. Denizli (ed.), *Biophysics at the Nanoscale*. Elsevier, 1–25.

Giel, M. C., and Y. Hong. 2023. Click Chemistry in the Design of AIEgens for Biosensing and Bioimaging. *Aggregate* 4 (4): e336.

Girish, T., and J. G. Manjunatha. 2020. Poly(glutamine) Film-coated Carbon Nanotube Paste Electrode for the Determination of Curcumin with Vanillin: Electroanalytical Approach. *Monatshefte für Chemie - Chemical Monthly* 151: 1681–8.

Guerrini, L., R. A. Alvarez-Puebla, and N. Pazos-Perez. 2018. Surface Modifications of Nanoparticles for Stability in Biological Fluids. *Materials* 11 (7): 1154.

Guo, X., L. Chen, P. Li, X. Li, J. Wang, L. Guo, and H. Yang. 2022. Construction of Electrochemiluminescence Biosensor via Click Chemistry and ARGET-ATRP for Detecting Tobacco Mosaic Virus RNA. *Analytical Biochemistry* 655: 114834.

Gupta, V. K., K. Choudhary, S. K. Raghuwanshi, and S. Kumar. 2023. Lossy Mode Resonance Based Single Mode Fiber Structure for Detection of Ascorbic Acid. In B. L. Miller (ed.), *Frontiers in Biological Detection: From Nanosensors to Systems XV, 12397.* SPIE, 30–7.

Hareesha, N., and J. G. Manjunatha. 2021. Electro-oxidation of Formoterol Fumarate on the Surface of Novel Poly (thiazole yellow-G) Layered Multi-walled Carbon Nanotube Paste Electrode. *Scientific Reports* 11: 12797.

Henriksson, A., P. Neubauer, and M. Birkholz. 2021. Functionalization of Oxide-Free Silicon Surfaces for Biosensing Applications. *Advanced Materials Interfaces* 8 (23): 2100927.

Hun, X., Y. Li, S. Wang, Y. Li, J. Zhao, H. Zhang, and X. Luo. 2018. Photoelectrochemical Platform for Cancer Cell Glutathione Detection Based on Polyaniline and NanoMoS_2 Composites Modified Gold Electrode. *Biosensors and Bioelectronics* 112: 93–9.

Jeong, W., J. Bu, R. Jafari, et al. 2022. Hierarchically Multivalent Peptide–Nanoparticle Architectures: A Systematic Approach to Engineer Surface Adhesion. *Advanced Science* 9 (4): 2103098.

Joshi, S., K. A. Raj, M. R. Rao, and R. Ghosh. 2022. An Electronic Biosensor Based on Semiconducting Tetrazine Polymer Immobilizing Matrix Coated on RGO for Carcinoembryonic Antigen. *Scientific Reports* 12 (1): 3006.

Kalita, N., S. Gogoi, S. D. Minteer, and P. Goswami. 2023. Advances in Bioelectrode Design for Developing Electrochemical Biosensors. *ACS Measurement Science Au* 3 (6): 404–33.

Karolina, W., S. Katarzyna, J. W. Marta, L. Jagoda, M. M. Barbara, and J. Stefan. 2021. Surface Functionalization—The Way for Advanced Applications of Smart Materials. *Coordination Chemistry Reviews* 436: 213846.

Kassa, A., A. Abebe, and M. Amare. 2021. Synthesis, Characterization, and Electropolymerization of a Novel Cu (II) Complex Based on 1, 10-Phenanthroline for Electrochemical Determination of Amoxicillin in Pharmaceutical Tablet Formulations. *Electrochimica Acta* 384: 138402.

Kaur, R., S. K. Sharma, and S. K. Tripathy. 2019. Advantages and Limitations of Environmental Nanosensors. In *Advances in Nanosensors for Biological and Environmental Analysis*. Elsevier, 119–32.

Kim, J., J. H. Kang, W. H. Kwon, et al. 2023. Self-Assembling Biomolecules for Biosensor Applications. *Biomaterials Research* 27 (1): 127.

Kim, J. Y., J. Y. Lee, H. Y. Park, et al. 2022. Combination of Peptides with Biological, Organic, and Inorganic Materials for Synergistically Enhanced Diagnostics and Therapeutics. *Peptide Science* 114 (3): e24233.

Kim, S. Y., S. J. Cho, S. E. Byeon, X. He, and H. J. Yoon. 2020. Self-assembled Monolayers as Interface Engineering Nanomaterials in Perovskite Solar Cells. *Advanced Energy Materials* 10 (44): 2002606.

Ko, S., J. Y. Kim, J. Y. Park, et al. 2023. Modulating the Folding and Binding of Peptides Using a Stimuli-Responsive Molecular Tweezer. *Chemical Science* 14 (35): 9600–7.

Kumar, V., S. K. Raghuwanshi, and S. Kumar. 2022. Recent Advances in Carbon Nanomaterials Based SPR Sensor for Biomolecules and Gas Detection—A Review. *IEEE Sensors Journal* 22 (16): 15661–72.

Kumari, A., V. Vyas, and S. Kumar. 2022. Synthesis, Characterization, and Applications of Gold Nanoparticles in Development of Plasmonic Optical Fiber-Based Sensors. *Nanotechnology* 34 (4): 42001.

Kumari, R., A. Singh, U. P. Azad, and P. Chandra. 2023. Insights into the Fabrication and Electrochemical Aspects of Paper Microfluidics-Based Biosensor Module. *Biosensors* 13 (9): 891.

Lee, J., H. N. Suh, H. Park, Y. M. Park, H. J. Kim, and S. H. Kim. 2022. Regenerative Strategy of Gold Electrodes for Long-Term Reuse of Electrochemical Biosensors. *ACS Omega* 8 (1): 1389–400.

Li, H., D. Li, H. Chen, et al. 2023. Application of Silicon Nanowire Field Effect Transistor (SiNW-FET) Biosensor with High Sensitivity. *Sensors* 23 (15): 6808.

Li, X., M. Xu, Q. Wu, W. Wei, and X. Liu. 2021. Photolithographic 3D Microarray Electrode-Based High-Performance Non-Enzymatic H_2O_2 Sensor. *Colloids and Surfaces A: Physicochemical and Engineering Aspects* 628: 127249.

Li, X. X., and S. T. Zheng. 2021. Three-Dimensional Metal-Halide Open Frameworks. *Coordination Chemistry Reviews* 430: 213663.

Li, Z., and S. L. Wong. 2017. Functionalization of 2D Transition Metal Dichalcogenides for Biomedical Applications. *Materials Science and Engineering: C* 70: 1095–106.

Liu, Z., J. Xue, L. Chen, et al. 2022. A Signal-off Aptamer Sensor Based on Competition with Complementary DNA and Click Polymerization for Electrochemical Detection of AFB1. *Microchemical Journal* 181: 107775.

Ma, D., S. Chon, S. Cho, et al. 2020. A Novel Photolithographic Method for Fabrication of Flexible Micro-Patterned Glucose Sensors. *Journal of Electroanalytical Chemistry* 876: 114720.

Ma, Z., J. Wu, Y. Tan, and C. Tan. 2024. Azobenzene-Based Conjugated Polymers: Synthesis, Properties, and Biological Applications. *Macromolecular Rapid Communications*: 2400048.

Mahmudiono, T., D. O. Bokov, S. A. Jasim, W. K. Abdelbasset, and D. M. Khashirbaeva. 2022. State-of-the-Art of Convenient and Low-Cost Electrochemical Sensor for Food Contamination Detection: Technical and Analytical Overview. *Microchemical Journal* 179: 107460.

Matthew J. W., and E. P. Thomas. 2021. (Macro)molecular Self-assembly for Hydrogel Drug Delivery. *Advanced Drug Delivery Reviews* 172: 275–95.

Mazrouaa, A. M., M. G. Mohamed, N. A. Mansour, et al. 2022. Properties and Application of Surface Modified Carbon Nanotubes. In *Surface Modified Carbon Nanotubes Volume 1: Fundamentals, Synthesis and Recent Trends*. ACS Publications, 91–109.

Mishra, D. P., B. Acharya, S. Tripathy, B. Barik, and P. K. Sahu. 2023. An Overview of the Biosensing Potential of Organometallic Compounds. *Chemical Physics Impact*: 100326.

Muthuchamy, N., K. P. Lee, and A. I. Gopalan. 2017. Enhanced Photoelectrochemical Biosensing Performances for Graphene (2D)–Titanium Dioxide Nanowire (1D) Heterojunction Polymer Conductive Nanosponges. *Biosensors and Bioelectronics* 89: 390–99.

Nangare, S., and P. Patil. 2022. Chitosan Mediated Layer-by-Layer Assembly Based Graphene Oxide Decorated Surface Plasmon Resonance Biosensor for Highly Sensitive Detection of β-Amyloid. *International Journal of Biological Macromolecules* 214: 568–82.

Napierała, S., M. Kubicki, V. Patroniak, and M. W. Chorab. 2021. Electropolymerization of [2× 2] Grid-Type Cobalt (II) Complex with Thiophene Substituted Dihydrazone Ligand. *Electrochimica Acta* 369: 137656.

Nikhil, B., J. Pawan, F. Nello, and E. Pedro. 2016. Introduction to Biosensors. *Essays Biochem* 60 (1): 1–8.

Nontawong, N., P. Ngaosri, S. Chunta, et al. 2022. Smart Sensor for Assessment of Oxidative/Nitrative Stress Biomarkers Using a Dual-Imprinted Electrochemical Paper-Based Analytical Device. *Analytica Chimica Acta* 1191: 339363.

Pan, Q. Y., M. E. Sun, C. Zhang, et al. 2021. A Multi-Responsive Indium-Viologen Hybrid with Ultrafast-Response Photochromism and Electrochromism. *Chemical Communications* 57 (86): 11394–97.

Patel, B. A. 2021. Microelectrodes and Nanoelectrodes. In A. Editor and B. Editor (eds.), *Electrochemistry for Bioanalysis*. Elsevier, 51–72.

Pavel, I. A., S. Lakard, and B. Lakard. 2022. Flexible Sensors Based on Conductive Polymers. *Chemosensors* 10 (3): 97.

Pushpanjali, P. A., J. G. Manjunatha, N. Hareesha, Edwin S. D' Souza, M. M. Charithra, and N. S. Prinith. 2021. Voltammetric Analysis of Antihistamine Drug Cetirizine and Paracetamol at poly(L-Leucine) Layered Carbon Nanotube Paste Electrode. *Surfaces and Interfaces* 24: 101154.

Qi, GB., Y. J. Gao, L. Wang, and H. Wang. 2018. Self-assembled Peptide-based Nanomaterials for Biomedical Imaging and Therapy. *Advanced Materials* 30 (22): 1703444.

Qing, M., S. Chen, S. Xie, Y. Tang, J. Zhang, and R. Yuan. 2019. Encapsulation and Release of Recognition Probes Based on a Rigid Three-Dimensional DNA 'Nanosafe-Box' for Construction of a Electrochemical Biosensor. *Analytical Chemistry* 92 (2): 1811–17.

Rad, J. K., Z. Balzade, and A. R. Mahdavian. 2022. Spiropyran-Based Advanced Photoswitchable Materials: A Fascinating Pathway to the Future Stimuli-Responsive Devices. *Journal of Photochemistry and Photobiology C: Photochemistry Reviews* 51: 100487.

Ramin, B. B. S., W. G. Santos, Y. Messaddeq, E. Deffune, M. L. Moraes, and S. J. L. Ribeiro. 2024. Layer-by-Layer Assembly of NaYF4: Yb3+/Er3+ UCNPs@ Cystein and UCNPs@ Cys-Ab as Hybrid Bio-Optical Sensors for E. Coli Bacteria. *Journal of Luminescence* 271: 120590.

Rassas, I., M. Braiek, A. Bonhomme, et al. 2019. Voltammetric Glucose Biosensor Based on Glucose Oxidase Encapsulation in a Chitosan-Kappa-Carrageenan Polyelectrolyte Complex. *Materials Science and Engineering: C* 95: 152–59.

Redondo-Gómez, C., P. Parreira, M. C. L. Martins, and H. S. Azevedo. 2024. Peptide-Based Self-Assembled Monolayers (SAMs): What Peptides Can Do for SAMs and Vice Versa. *Chemical Society Reviews* 53: 3714–73.

Rohaizad, N., C. C. Mayorga-Martinez, M. Fojtů, N. M. Latiff, and M. Pumera. 2021. Two-Dimensional Materials in Biomedical, Biosensing and Sensing Applications. *Chemical Society Reviews* 50 (1): 619–57.

Samanta, P., S. Ghosh, and T. Kuila. 2022. *Manganese (IV) Oxide-Based Supercapacitors*. Elsevier, 475–85

Sandhyarani, N. 2019. Surface Modification Methods for Electrochemical Biosensors. In A. P. F. Turner and M. Mascini (eds.), Electrochemical Biosensors. Elsevier, 45–75.

Sekhosana, K. E., S. A. Majeed, and U. Feleni. 2023. Click Chemistry in the Electrochemical Systems: Toward the Architecture of Electrochemical (Bio) Sensors. *Coordination Chemistry Reviews* 491: 215232.

Shojaei, T. R., S. Soltani, and M. Derakhshani. 2022. Synthesis, Properties, and Biomedical Applications of Inorganic Bionanomaterials. In *Fundamentals of Bionanomaterials*, 139–174.

Si, Y., J. W. Park, S. Jung, et al. 2019. Voltammetric Layer-by-Layer Biosensor Featuring Purine Nucleoside Phosphorylase and Chitosan for Inosine in Human Serum Solutions. *Sensors and Actuators B: Chemical* 298: 126840.

da Silva, W., AC.Queiroz, and C. M. A. Brett. 2020. Nanostructured Poly (Phenazine)/ Fe2O3 Nanoparticle Film Modified Electrodes Formed by Electro polymerization in Ethaline-Deep Eutectic Solvent. Microscopic and Electrochemical Characterization. *Electrochimica Acta* 347: 136284.

Singh, M., N. Kaur, and E. Comini. 2020. The Role of Self-Assembled Monolayers in Electronic Devices. *Journal of Materials Chemistry C* 8 (12): 3938–55.

Song, X., T. Zhang, L. Wu, et al. 2022. Highly Stretchable High-Performance Silicon Nanowire Field Effect Transistors Integrated on Elastomer Substrates. *Advanced Science* 9 (9): 2105623.

Sypabekova, M., A. Hagemann, D. Rho, and S. Kim. 2022. 3-Aminopropyltriethoxysilane (APTES) Deposition Methods on Oxide Surfaces in Solution and Vapor Phases for Biosensing Applications. *Biosensors* 13 (1): 36.

Taleghani, N., and F. Taghipour. 2024. Photo-Induced Metal-Oxide Biosensor for Analysis of Biofluids. *Talanta* 280: 126668.

Teixeira, G. T. L., J. P. L. do Nascimento, R. V. Gelamo, J. A. Moreto, and N. B. L. Slade. 2023. Strategies for Functionalization of Metallic Surfaces with Bioactive Peptides: A Mini Review. *International Journal of Peptide Research and Therapeutics* 29 (2): 24.

Thirugnanasambandan, T., S. Ramanathan, and S. C. B. Gopinath. 2024. Revolutionizing Biosensing through Cutting-Edge Nanomaterials: An In-Depth Exploration of Recent Technological Advances. *Nano-Structures & Nano-Objects* 38: 101128.

Tonelli, D., E. Scavetta, and I. Gualandi. 2019. Electrochemical Deposition of Nanomaterials for Electrochemical Sensing. *Sensors* 19 (5): 1186.

Tran, D. P., T. T. T. Pham, B. Wolfrum, A. Offenhäusser, and B. Thierry. 2018. CMOS-Compatible Silicon Nanowire Field-Effect Transistor Biosensor: Technology Development toward Commercialization. *Materials* 11 (5): 785.

Uniyal, S., K.Choudhary, S. Sachdev, and S. Kumar. 2024. Nano-Bio Fusion: Advancing Biomedical Applications and Biosensing with Functional Nanomaterials. *Optics & Laser Technology* 168: 109938.

Verma, A., and B. C. Yadav. 2024. Comprehensive Review on Two Dimensional Nanomaterials for Optical Biosensors: Present Progress and Outlook. *Sustainable Materials and Technologies* 40: e00900.

Vilsinski, B H., A. C. de Oliveira, P. R. Souza, and A. F. Martins. 2024. Polysaccharide-Based Polyelectrolyte Multilayers Fabricated via Layer-by-Layer Approach: From Preparation to Applications. *Progress in Organic Coatings* 196: 108720.

Wang, F., Y. Xie, W. Zhu, and T. Wei. 2023. Recent Advances in Functionalization Strategies for Biosensor Interfaces, Especially the Emerging Electro-Click: A Review. *Chemosensors* 11 (9): 481.

Wang, X., X. Yang, Q. Wang, and D. Meng. 2023. Unnatural Amino Acids: Promising Implications for the Development of New Antimicrobial Peptides. *Critical Reviews in Microbiology* 49 (2): 231–55.

Wang, Y., L. Huang, X. Wang, et al. 2023. Photoresponsive Triple Shape Memory Polymers with a Self-Healing Function Based on Poly (Lactic Acid)/Polycaprolactone Blends. *Polymer Testing* 120: 107966.

Wu, T., X. Xu, L. K. Ono, et al. 2023. Graphene-Like Conjugated Molecule as Hole-Selective Contact for Operationally Stable Inverted Perovskite Solar Cells and Modules. *Advanced Materials* 35 (21): 2300169.

Yang, R., Y. Jiao, B. Wang, B. Xu, and W. Tian. 2021. Solid-State Reversible Dual Fluorescent Switches for Multimodality Optical Memory. *The Journal of Physical Chemistry Letters* 12 (4): 1290–94.

Yang, X., A. Gao, Y. Wang, and T. Li. 2018. Wafer-Level and Highly Controllable Fabricated Silicon Nanowire Transistor Arrays on (111) Silicon-on-Insulator (SOI) Wafers for Highly Sensitive Detection in Liquid and Gaseous Environments. *Nano Research* 11: 1520–29.

Zahra, T., U. Javeria, H. Jamal, M M. Baig, F. Akhtar, and U. Kamran. 2024. A Review of Biocompatible Polymer-Functionalized Two-Dimensional Materials: Emerging Contenders for Biosensors and Bioelectronics Applications. *Analytica Chimica Acta* 1316: 342880.

Zeng, H., Y. Xie, T. Liu, Z. Chu, E. Dempsey, and W. Jin. 2024. Conductive Polymer Nanocomposites: Recent Advances in the Construction of Electrochemical Biosensors. *Sensors & Diagnostics* 3: 165–80

Zhang, T., H. Kang, B. Li, et al. 2023. Dynamic CPL Switching Realized in Chiral Mn-Based Metal Halides with Reversible Thermochromism. *Journal of Materials Chemistry C* 11 (16): 5461–68.

Zheng, M., and J. Yuan. 2022. Polymeric Nanostructures Based on Azobenzene and Their Biomedical Applications: Synthesis, Self-Assembly and Stimuli-Responsiveness. *Organic & Biomolecular Chemistry* 20 (4): 749–67.

Zou, J., J. Liao, Y. He, et al. 2024. Recent Development of Photochromic Polymer Systems: Mechanism, Materials and Applications. *Research* 7: 0392

3 Microfluidics and Lab-on-a-Chip Technologies

Ramon Rial and Luciano Benedini

3.1 INTRODUCTION

3.1.1 Background and Importance of Microfluidics

3.1.1.1 Definition and Brief History

3.1.1.1.1 What Is Microfluidics?

Microfluidics is the science and technology of controlling and manipulating small volumes of fluids, typically ranging from nanoliters to microliters, within microscale networks of channels (Whitesides 2006). These channels, often in the order of tens to hundreds of micrometers in size, significantly affect fluid behavior due to surface forces, capillary effects, and minimized inertia, which dominate over volumetric forces like gravity (Chen et al. 2021). Microfluidic systems enable precise fluid control, making them crucial in a wide range of applications such as biochemical analysis, drug delivery, diagnostics, and environmental monitoring (Zhang et al. 2023; Aubry, Lee, and Lu 2023; Wang, Yu, and Xie 2020). The ability to miniaturize complex laboratory processes into small, chip-sized platforms offers advantages such as reduced sample volume, cost-effectiveness, faster reaction times, and enhanced sensitivity (Temiz et al. 2015; Cooper 2022; Keçili, Büyüktiryaki, and Hussain 2020).

Microfluidics is inherently interdisciplinary, bridging the fields of physics, chemistry, biology, engineering, and materials science. It allows for the integration of multiple laboratory functions—such as mixing, separation, chemical reaction, and detection—into a single device, reducing the need for expensive and bulky laboratory equipment. As a result, microfluidics has become an essential technology in the life sciences, healthcare, and biotechnology industries (Deliorman, Ali, and Qasaimeh 2023a; Kumari et al. 2023).

The development of microfluidics is deeply rooted in the convergence of multiple scientific disciplines and technological advancements. As highlighted by George M. Whitesides, one of the foremost pioneers in the field, four primary domains—molecular analysis, biodefense, molecular biology, and microelectronics—significantly contributed to its evolution (Whitesides 2006). The trajectory of microfluidics mirrors the broader history of technological miniaturization, beginning with fundamental breakthroughs in the mid-20th century.

DOI: 10.1201/9781003536239-3

3.1.1.1.2 Origins and Pioneering Contributions

The birth of microfluidics can be traced to the invention of the transistor in 1947 by John Bardeen, Walter Brattain, and William Shockley at Bell Labs (Warner 2001). This milestone revolutionized electronics, making miniaturization and integration possible. It demonstrated the feasibility of creating compact, efficient devices, inspiring analogous developments in fluid handling systems. The photolithography techniques developed for semiconductor manufacturing subsequently became essential for fabricating the intricate microchannels used in microfluidic devices.

3.1.1.1.3 Early Analytical Techniques

Concurrently, advancements in molecular analysis provide a critical impetus. The emergence of gas-phase chromatography (GPC) and high-pressure liquid chromatography (HPLC) during the 1950s and 1960s allowed for the separation of complex mixtures with exceptional sensitivity and minimal sample volumes (Terry, Herman, and Angell 1979). By the 1980s, capillary electrophoresis (CE) further transformed molecular analysis, enabling high-resolution separation of biomolecules (Karger, Chu, and Foret 1995; Nielsen and Foley 1998). These developments highlighted the advantages of small-scale systems, laying the groundwork for integrating these techniques into microfluidic platforms.

3.1.1.1.4 Biodefense and Government Investment

The geopolitical climate of the late twentieth century, marked by the Cold War and the subsequent focus on biodefense, significantly shaped microfluidic research. Recognizing the need for portable systems capable of detecting chemical and biological threats, the U.S. Defense Advanced Research Projects Agency (DARPA) funded numerous projects in the 1990s (Whitesides 2006). This directed funding spurred innovation, resulting in the development of compact, robust microfluidic devices for field applications.

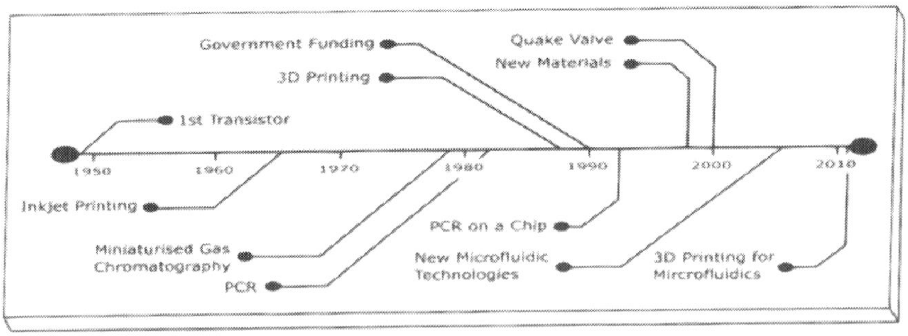

FIGURE 3.1 Timeline highlighting the main advances in the field of microfluidics. (Reprinted with Open Access permission from reference (Convery and Gadegaard 2019.)

3.1.1.1.5 Molecular Biology and Genomics

The rapid advancements in genomics, particularly the Human Genome Project (HGP) initiated in 1990, also propelled microfluidic technologies forward (Collins, Morgan, and Patrinos 2003). High-throughput techniques for DNA sequencing and amplification demanded systems that could manage minute volumes with high precision. Microfluidics emerged as an ideal solution, integrating sample preparation, reaction, and detection within a single chip.

3.1.1.1.6 Material Innovations and Soft Lithography

Initially, microfluidic devices were fabricated using silicon and glass, leveraging the well-established techniques from the semiconductor industry. However, these materials proved expensive and cumbersome, particularly for applications involving biological samples. The transition to using novel polymers, especially poly(dimethylsiloxane) (PDMS), marked a paradigm shift (McDonald et al. 2000; McDonald and Whitesides 2002). PDMS is inexpensive, optically transparent, and easy to mold, making it a preferred material for prototyping and experimentation. The advent of soft lithography, pioneered by Whitesides and others, enabled rapid fabrication of complex microfluidic structures, democratizing access to microfluidic technology (Xia and Whitesides 1998).

3.1.1.1.7 Integration and Expansion in the Twenty-First Century

The rise of micro total analysis systems (μTAS) (Manz, Graber, and Widmer 1990) and lab-on-a-chip (LoC) technologies (Figeys and Pinto 2000) in the 1990s also marked a significant milestone. μTAS devices aimed to miniaturize and integrate all the functions of a traditional laboratory into a single chip, streamlining processes and reducing costs. This concept gained traction in various fields, including clinical diagnostics and environmental monitoring.

In the twenty-first century, microfluidics diversified further with innovations like droplet microfluidics (Tawfik and Griffiths 1998), paper-based devices (Martinez et al. 2007), and open microfluidics (Juncker, Schmid, and Delamarche 2005). These advances underscored the versatility and transformative potential of microfluidics, broadening its scope and further expanded its applicability, from high-throughput screening in drug discovery to low-cost diagnostics in resource-limited settings.

3.1.1.1.8 Organ-on-a-Chip: A Transformative Frontier

One of the most transformative applications of microfluidics is the development of organ-on-a-chip systems (Leung et al. 2022; Zhang et al. 2018). These devices simulate the physiological functions of human organs, providing a more accurate model than traditional 2D cell cultures or animal testing. By incorporating features like perfused microchannels and mechanical cues, organ-on-a-chip platforms enable studies of drug effects, disease mechanisms, and personalized medicine. While challenges remain—such as material limitations and complex fabrication protocols—the potential to replace animal models and reduce drug development costs underscores the transformative impact of this technology (Wu et al. 2020).

3.1.1.1.9 Conclusion: Microfluidics as a Cornerstone of Modern Science

The journey of microfluidics, from its inception in the mid-twentieth century to its current role as a cornerstone of modern analytical and diagnostic techniques, illustrates the power of interdisciplinary innovation. By leveraging the principles of miniaturization and integration, microfluidics continues to redefine possibilities in the life sciences, healthcare, and beyond. The field remains poised for further breakthroughs, particularly as new materials and fabrication methods address existing challenges, solidifying microfluidics' role as a transformative technology of the twenty-first century (Hajam and Khan 2024; Battat, Weitz, and Whitesides 2022).

3.1.1.2 Relevance to Biomaterials and Biophysics

As previously exposed, microfluidics has become a key platform for advancing research in biomaterials and biophysics by replicating biological microenvironments with high precision. In biomaterials, microfluidic devices are used to study how materials interact with cells and tissues under simulated physiological conditions, such as blood flow. By controlling material properties, such as stiffness and surface chemistry, researchers can investigate their effects on cell behavior and develop organ-on-a-chip models for tissue engineering and regenerative medicine. In biophysics, microfluidic systems enable the study of cellular and molecular interactions by generating controlled gradients of chemical, mechanical, or thermal stimuli. This is essential for understanding processes like cell motility, shear stress, and tissue morphogenesis. A significant advantage of microfluidics is its ability to create dynamic, in vivo-like environments, which traditional static cell culture systems cannot replicate. These dynamic conditions improve the accuracy of in vitro models and are valuable for both basic research and drug development. In summary, microfluidics enables precise simulation of physiological conditions, making it an essential tool in both biomaterials and biophysics. It facilitates the study of material–cell interactions and biophysical forces, driving further advancements in biological research and biomedical technologies.

3.2 FUNDAMENTALS OF MICROFLUIDICS

3.2.1 Basic Principles of Microfluidics

As exposed, microfluidics is a field focused on the precise control of fluids at the microscale; therefore, it involves unique fluid behaviors governed by the interactions between fluid dynamics, material properties, and external forces. Due to the small scales involved, the behavior of fluids in microfluidic devices deviates significantly from macroscale fluid behavior, influenced heavily by viscosity, surface tension, and other surface forces (Fernandes 2024). Understanding these principles is essential for effectively designing and optimizing microfluidic systems for various applications. Below are the basis and principles that form the foundation of microfluidic systems.

3.2.1.1 Fluid Dynamics at the Microscale

Fluid dynamics at the microscale are governed by unique physical interactions, with surface forces playing a dominant role and inertial forces becoming negligible. This shift leads to a predictable and steady flow, particularly in narrow channels. In microfluidics, Reynolds numbers are typically low, favoring laminar flow, which occurs in layers with minimal mixing between them. This is especially useful in applications requiring precise control over fluid interactions, such as chemical assays or mixing small reagent volumes.

At the microscale, viscosity, the force that resists fluid motion, is a primary factor affecting fluid flow. Viscous forces dominate inertial forces because the small scale of the channels limits the development of turbulence. A high surface-to-volume ratio also means that surface-related phenomena like capillary action and interfacial tension become central to fluid manipulation in microfluidic systems. These effects are advantageous in diagnostics and biochemical assays, allowing small samples to be controlled with precision.

Another critical characteristic of microscale fluid flow is the dominance of diffusion over convection (Squires and Quake 2005). In larger systems, convection is generally the primary mechanism for mixing and transporting solutes. However, in microfluidic channels, where flow is primarily laminar, molecular diffusion is often the main transport mechanism. The Péclet number (Pe) is used to quantify the relative importance of convection versus diffusion and is defined by the equation:

$$Pe = \frac{uL}{D} \tag{3.1}$$

where u is the velocity of the fluid, L is the characteristic length scale, and D is the diffusion coefficient of the solute. In microfluidics, low Péclet numbers indicate that diffusion dominates over convection, affecting the design of devices that rely on rapid mixing, such as droplet-based microfluidics.

3.2.1.2 Laminar Flow, Reynolds Number, and Scaling Laws

The behavior of fluids in microfluidic systems differs considerably from macroscopic systems due to the dominance of laminar flow and the impact of scaling laws. At the microscale, fluid flow typically exhibits ordered, streamlined movement known as laminar flow, rather than turbulent mixing. This characteristic of microfluidic systems is crucial for applications that require precise control of fluid streams, such as mixing, separation, and chemical reactions.

3.2.1.2.1 Laminar Flow

In laminar flow, fluid layers move parallel to one another with minimal intermixing. This flow regime is highly predictable and can be mathematically modeled with relative ease, making it advantageous for analytical and diagnostic applications. Laminar flow arises when viscous forces dominate over inertial forces, which is generally the case in microfluidic systems due to the small dimensions involved.

In microfluidic channels, laminar flow is characterized by a parabolic velocity profile (Kuang et al. 2009), with the highest velocity at the center of the channel and zero velocity at the walls (no-slip boundary condition). This velocity distribution ($u(y)$) across a channel of height h is described by:

$$u(y) = \frac{4U_{\text{avg}}}{h^2}\left(\frac{h}{2} - y\right)y \tag{3.2}$$

where U_{avg} is the average velocity of the fluid, and y is the distance from the channel wall.

The parabolic profile has significant implications for applications like mixing and separation in microfluidic channels, where fluid streams can flow side by side without turbulence-induced mixing.

3.2.1.2.2 *Reynolds Number*

The Reynolds number (Re) is a dimensionless parameter used to predict flow regimes in fluid systems (Stone and Kim 2001). It is defined as the ratio of inertial forces to viscous forces and is given by:

$$Re = \frac{\rho UL}{\mu} \tag{3.3}$$

where ρ is the fluid density, U is the characteristic velocity, L is the characteristic length (such as the hydraulic diameter of the microchannel), and μ is the dynamic viscosity of the fluid.

In most microfluidic systems, Re remains low (typically $Re < 2300$), indicating laminar flow. For Re values below approximately 2300, the flow is generally considered laminar, while values above 4000 indicate turbulence. Due to the small characteristic lengths in microfluidic systems, even high flow rates result in low Reynolds numbers, making turbulent flow rare in these devices.

The low Reynolds number regime of microfluidic systems has both advantages and limitations. For instance, the predictable and stable nature of laminar flow allows for controlled reactions and transport processes. However, it also means that mixing is slow and limited to diffusion, posing a challenge for applications that require efficient mixing over short distances.

3.2.1.2.3 *Scaling Laws in Microfluidics*

Scaling laws have an essential impact in microfluidics by explaining how physical forces scale with system size. In microfluidics, as the characteristic dimensions decrease, certain forces become more prominent while others diminish (Bruus 2014). Three key scaling effects are the following:

- *Viscous dominance*: Viscous forces scale with the surface area (proportional to L^2), while inertial forces scale with volume (proportional to L^3). As a result, the ratio of viscous to inertial forces increases as the characteristic dimension (L) decreases, favoring laminar flow and reducing the likelihood of turbulence in microscale systems.

- *Surface area-to-volume ratio*: Surface area scales as L^2, while volume scales as L^3. Therefore, as channel dimensions shrink, the surface area-to-volume ratio increases. This scaling has important implications, as surface effects such as capillary action and adsorption become more significant. The high surface area-to-volume ratio is particularly advantageous for chemical and biological assays, where reactions at surfaces or interfaces are common.
- *Diffusion time*: Diffusion time (t_{diff}) for a particle across a length (L) is given by:

$$t_{\text{diff}} = \frac{L^2}{D} \tag{3.4}$$

where D is the diffusion coefficient. This equation indicates that diffusion time scales with the square of the distance, so as L decreases, the time required for molecules to diffuse across the channel decreases significantly. This rapid diffusion over small distances is beneficial in microfluidic systems where efficient mixing and reaction rates are critical.

3.2.1.2.4 Implications for Microfluidic Design and Operation

The predominance of laminar flow and the effects of scaling laws impose specific constraints and opportunities in microfluidic design. The lack of turbulence in laminar flow makes it challenging to achieve rapid mixing of reactants, which is critical for chemical and biological processes. As a result, alternative methods such as staggered herringbone mixers, chaotic advection, and diffusion-based strategies are often employed to enhance mixing in microfluidic devices. Moreover, the increased surface area-to-volume ratio allows for efficient heat and mass transfer in microfluidic systems, which can be advantageous in applications like cell culture, biochemical assays, and chemical synthesis. The high control over fluid flow also enables the formation of highly reproducible concentration gradients, which are essential for studies involving cell signaling and drug delivery.

3.2.1.3 Surface Interactions and Capillary Effects

Surface interactions and capillary effects are fundamental to the behavior of fluids in microfluidic systems. The physics behind these interactions governs fluid dynamics at small scales, where forces such as surface tension and molecular adhesion dominate over bulk forces like gravity or inertia. Understanding these effects is crucial for designing and optimizing microfluidic devices for applications ranging from chemical analysis to biological sensing.

3.2.1.3.1 Surface Interactions: The Physics of Wetting and Adhesion

At the microscale, surface interactions become the dominant factor in determining the flow behavior of fluids (Darhuber and Troian 2005). These interactions can be classified as adhesive forces (between the liquid and the surface) and cohesive forces (within the liquid). The balance between these forces dictates whether the fluid wets the surface or forms droplets, as well as the shape and movement of these droplets.

The key parameter governing surface interactions is the *contact angle* (θ), which quantifies the wetting behavior of a liquid on a solid surface. It is defined as the angle between the tangent to the liquid interface and the solid surface at the contact point. The contact angle is determined by the *Young-Laplace equation*:

$$\gamma_{SL} - \gamma_{SV} = \gamma_{LV} \cos(\theta) \tag{3.5}$$

where γ_{SL} is the solid–liquid interfacial tension, γ_{SV} is the solid–vapor interfacial tension, γ_{LV} is the liquid–vapor interfacial tension (surface tension of the liquid), and θ is the contact angle.

A small contact angle ($\theta < 90^\circ$) implies that the liquid wets the surface (hydrophilic), while a large contact angle ($\theta > 90^\circ$) indicates poor wetting (hydrophobic). The nature of the surface (whether rough, smooth, hydrophilic, or hydrophobic) significantly affects the flow of fluids, droplet formation, and even the efficiency of mixing or separation processes in microfluidic devices.

In microfluidic channels, surfaces are often engineered to control the contact angle and alter fluid behavior (Fernandes 2024). For example, plasma treatment or chemical modification of surfaces can be used to make them more hydrophilic or hydrophobic, which directly influences fluid retention, droplet formation, and even the speed of capillary rise.

3.2.1.3.2 The Influence of Channel Geometry on Capillary Flow

In microfluidics, the channel geometry is a key factor in modulating the capillary effects. Capillary flow is heavily influenced by the channel dimensions, surface roughness, and wettability. The *Capillary number (Ca)* is an important dimensionless number used to describe capillary flow behavior. It is defined as the ratio of viscous forces to surface tension forces:

$$Ca = \frac{\mu v}{\gamma} \tag{3.6}$$

where μ is the dynamic viscosity of the fluid, v is the characteristic velocity of the fluid, and γ is the surface tension.

For low capillary numbers (where $Ca \ll 1$), the flow is dominated by surface tension, and capillary effects govern the fluid motion. In contrast, at higher Ca values (where $Ca \gg 1$), viscous forces dominate, and the flow becomes pressure-driven. In microfluidic systems, capillary forces are generally more significant at low flow rates, where inertial forces are negligible. Thus, the geometry of the microchannel, such as its length and width, can control whether capillary-driven flow or pressure-driven flow dominates. For instance, in a *Y-junction* microchannel, capillary forces can dictate the direction of fluid flow and determine how droplets merge or split at junctions.

3.2.1.3.3 Capillary Effects and Fluid Movement

Capillary effects refer to the spontaneous movement of liquids within narrow spaces due to surface tension. In microfluidic systems, these effects become much more pronounced due to the small dimensions of the channels (Chakraborty 2005). As fluid interacts with the walls of the channel, surface tension generates forces that drive the liquid through the channel without the need for external pumps. This is often referred to as *capillary-driven flow* (Olanrewaju et al. 2018).

The height to which a liquid rises in a capillary tube or microchannel due to capillary action can be predicted by the *capillary rise equation*:

$$h = \frac{2\gamma \cos(\theta)}{\rho g r} \tag{3.7}$$

where h is the height of the liquid rise, γ is the surface tension of the liquid, θ is the contact angle, ρ is the density of the liquid, g is the acceleration due to gravity, and r is the radius of the capillary channel.

At the microscale, the effect of capillary forces is more significant because the capillary rise is inversely proportional to the radius of the channel. This means that in microchannels (where r is on the order of micrometers), capillary forces are strong enough to move liquids over significant distances even without external pressure. Capillary effects are not limited to fluid rise; they also influence the behavior of fluids in various operations, such as droplet generation, mixing, and transport. In applications like droplet microfluidics, capillary forces drive the formation and movement of droplets, where surface tension dictates the size and uniformity of droplets and the dynamics of droplet-based reactions.

3.2.1.3.4 Droplet Microfluidics and Capillary Effects

Capillary effects are also fundamental to droplet-based microfluidics, where tiny droplets of liquid are generated, transported, and manipulated within microchannels. The generation of droplets relies on the interplay between surface tension, channel geometry, and flow conditions (Tice et al. 2003). At the junctions of microchannels, capillary forces lead to the pinching off of a droplet from a continuous fluid stream, and the size of the droplet is determined by the balance between these forces and the fluid flow rate. The formation and manipulation of droplets in microfluidics are governed by the *Laplace pressure* (ΔP) across the interface of a droplet (Simon et al. 2012), which is related to the radius R of the droplet and the surface tension of the liquid:where ΔP is the pressure difference across the droplet interface, γ is the surface tension, and R is the radius of the droplet.

This pressure difference can influence the stability and size of droplets, as well as their interactions with other droplets. Capillary forces also affect droplet transport, merging, splitting, and mixing within microfluidic systems.

3.2.1.3.5 The Role of Surface Roughness and Functionalization

The roughness of the channel surface further modifies capillary effects by creating local variations in the contact angle, which in turn influences the flow characteristics (Darhuber and Troian 2005). *Surface roughness* can either enhance or hinder wetting depending on the roughness scale relative to the liquid molecular scale (Kim et al. 2005). A *superhydrophobic* surface, which exhibits a high contact angle, can prevent wetting and enable the formation of high-speed flows, while a *superhydrophilic* surface promotes wetting and results in slower flows. Surface functionalization techniques, such as chemical coatings or plasma treatment, can further control the interaction between the liquid and the channel, allowing for fine-tuned control over droplet dynamics, flow behavior, and even the transport of cells or nanoparticles in biological applications (Gandhiraman et al. 2010; Pedersen et al. 2010; Kitsara and Ducrée 2013).

3.2.1.4 Driving Forces in Microfluidics

Microfluidic devices are categorized based on the primary driving forces used to control fluid flow within their channels. These forces govern the movement of fluids in microchannels and are crucial for a variety of microfluidic applications such as lab-on-a-chip devices, diagnostic platforms, and chemical synthesis. Below is an overview of the primary driving forces used in microfluidic systems, with a comparison of their advantages and limitations.

3.2.1.4.1 Pressure Force–Driven Flow

Pressure-driven flow is one of the most widely used mechanisms in microfluidic devices (Tesař 2007). It relies on the generation of pressure gradients across microchannels, leading to fluid motion. The flow within these devices typically follows a laminar profile, where fluids remain in distinct layers without substantial mixing. Pressure can be applied using external sources like syringes, pumps, or integrated micropumps, or through internal mechanisms like gas expansion or pneumatic displacement of membranes. The flow rate Q in a microchannel driven by pressure difference ΔP is governed by the Hagen–Poiseuille equation:

$$Q = \frac{\pi r^4 \Delta P}{8 \mu L} \tag{3.9}$$

where r is the channel radius, μ is the dynamic viscosity of the fluid, and L is the channel length. This method is suitable for high precision flow control but can cause challenges, such as pressure fluctuations, which might damage delicate biological samples.

TABLE 3.1
Comparison of the Most Common Primary Driving Forces in Microfluidic Systems

Driving Force	Advantages	Limitations
Pressure	Precise control of flow rates	Requires high pressure gradients
	Suitable for various fluid types	Can damage biological samples
	Relatively simple to implement	Sensitive to pressure variations
	Compatible with standard equipment	Risk of bubbles or blockages
Capillary	No external power required	Not suitable for high flow rates
	Gentle on cells and biological samples	Difficult-to-control flow rate
	Can be used in small channels	Limited force for long distances
	Simple and low-cost	Poor compatibility with viscous fluids
Centrifugal	High flow rates achievable	Requires bulky rotating devices
	Simple to implement	Not suitable for large height differences
	Can be automated for high throughput	Can damage biological samples
	Efficient mixing and separation	Requires precise rotation control
Acoustic	High flow rates achievable	Difficult-to-control sound waves
	Gentle on biological samples	Can cause cavitation damage
	Non-contact method	Limited penetration depth
	Minimal contamination risk	Requires specialized equipment
Magnetic	Can manipulate fluids in small channels	Difficult-to-control magnetic fields
	Non-contact and biocompatible	Requires magnetic particles in fluid
	Can separate specific particles or cells	Limited range of forces
	Precise fluid control	Not suitable for all fluids
Electric	Can manipulate fluids in small channels	Requires difficult-to-control electric fields
	Suitable for particle separation	Limited to conductive fluids
	Non-contact, precise manipulation	Can damage biological samples
	Biocompatible	Risk of electrolysis and contamination

Adapted from: https://pubs.rsc.org/en/content/articlehtml/2004/pb/d3bm01463k.

3.2.1.4.2 Capillary Force–Driven Flow

Capillary-driven microfluidics utilizes the natural tendency of liquids to rise in small-diameter tubes due to surface tension. The motion is driven by the creation of a pressure difference across the liquid–air interface, which can be enhanced by the design of microchannel dimensions. This method does not require external power sources, making it ideal for portable or compact systems, such as diagnostic tools and lab-on-a-chip (LOC) devices. The capillary rise in a microchannel can be described by the capillary pressure equation:

$$\Delta P = \frac{2\gamma\cos\theta}{r} \tag{3.10}$$

where γ is the surface tension of the liquid, θ is the contact angle, and r is the radius of the capillary tube. However, this technique is limited in its ability to handle large flow rates and relatively high viscous fluids.

3.2.1.4.3 Centrifugal Force–Driven Flow

Centrifugal microfluidics, often referred to as lab-on-a-disc (LOD) (Glass et al. 2012), uses rotational motion to manipulate fluids. As the disc rotates, centrifugal forces push fluids radially outward, enabling various functions like mixing, separation, and fluid transportation. The rotational speed and angular velocity control the flow rate and enable operations like cell sorting and sample separation. The centrifugal force is given by:

$$F_c = m \cdot \omega^2 \cdot r \tag{3.11}$$

where m is the mass of the fluid element, ω is the angular velocity, and r is the radial distance from the center. This method provides high throughput and rapid manipulation of samples but requires complex fabrication and precise rotational control.

3.2.1.4.4 Acoustic Force–Driven Flow

Acoustofluidics uses sound waves, typically in the ultrasonic range, to control fluid flow and manipulate particles in microchannels. The application of acoustic waves can generate localized pressure variations, leading to fluid movement without direct contact, which is ideal for biocompatible applications. Acoustic forces can be harnessed to mix fluids, separate particles, and achieve rapid manipulation (Zhang et al. 2020). The relationship between pressure variations and fluid velocity can be expressed as:

$$v = \frac{p}{\eta k} \tag{3.12}$$

where v is the fluid velocity, p is the acoustic pressure, η is the fluid viscosity, and k is the wave number. Acoustic force is advantageous for non-invasive operations but can be limited by the frequency of the sound waves and the potential for cavitation.

3.2.1.4.5 Magnetic Force–Driven Flow

Magnetic fields can be used to manipulate fluids containing magnetic particles, which are common in applications like cell sorting and lab-on-a-chip diagnostics. Magnetic force can also aid in the mixing of fluids and the separation of specific particles, providing a non-contact method of fluid control (Yaman et al. 2018). The force on a magnetic material is given by:

$$F_m = \nabla(\mathbf{m} \cdot \mathbf{B}) \tag{3.13}$$

where **m** is the magnetic moment and **B** is the magnetic field. Magnetic microfluidics is particularly useful for manipulating cells or particles that are either magnetic or paramagnetic, although its range is limited compared to other forces.

3.2.1.4.6 Electric Field–Driven Flow

Electric fields are used to manipulate fluids and particles based on electrokinetic effects such as electrophoresis, dielectrophoresis, and electroosmosis (Hossan et al. 2018). These techniques enable precise control over charged particles in microchannels, making them ideal for applications like DNA sequencing, cell sorting, and particle separation. The movement of fluid under an electric field can be described by:

$$v = \mu E \tag{3.14}$$

where v is the fluid velocity, μ is the electrophoretic mobility of the fluid, and E is the applied electric field. This method is non-contact and biocompatible, but it requires careful control of the electric field to avoid electrolysis or damage to samples (Pan, Karns, and Herr 2014).

3.2.2 Materials and Fabrication Techniques

3.2.2.1 Common Materials Used in Microfluidic Devices

Microfluidic devices require specific material properties to achieve the desired functionality, such as optical transparency, chemical compatibility, mechanical strength, and ease of fabrication. Several materials are commonly used in the fabrication of microfluidic devices, each with distinct advantages and limitations:

- *Polydimethylsiloxane (PDMS)*: PDMS is one of the most commonly used materials due to its transparency, flexibility, ease of molding, and biocompatibility (McDonald et al. 2000; McDonald and Whitesides 2002). It is widely employed in prototyping and for devices used in biological applications. PDMS is particularly well-suited for soft lithography techniques.
- *Glass*: Glass offers excellent optical clarity and chemical resistance, making it ideal for devices that require precise optical detection or use harsh chemicals (Whitesides 2018; Qi et al. 2018). However, it is more challenging to fabricate and bond compared to PDMS, which limits its use in rapid prototyping.
- *Polymers (e.g., PMMA, Polycarbonate)*: These thermoplastics are used when optical clarity and cost-effectiveness are important (Xia et al. 1997). Polycarbonate (PC) and polymethyl methacrylate (PMMA) are particularly useful for mass-produced devices. PMMA offers good optical transparency, but it lacks the flexibility of PDMS. Polycarbonate, on the other hand, is robust and resistant to impact.

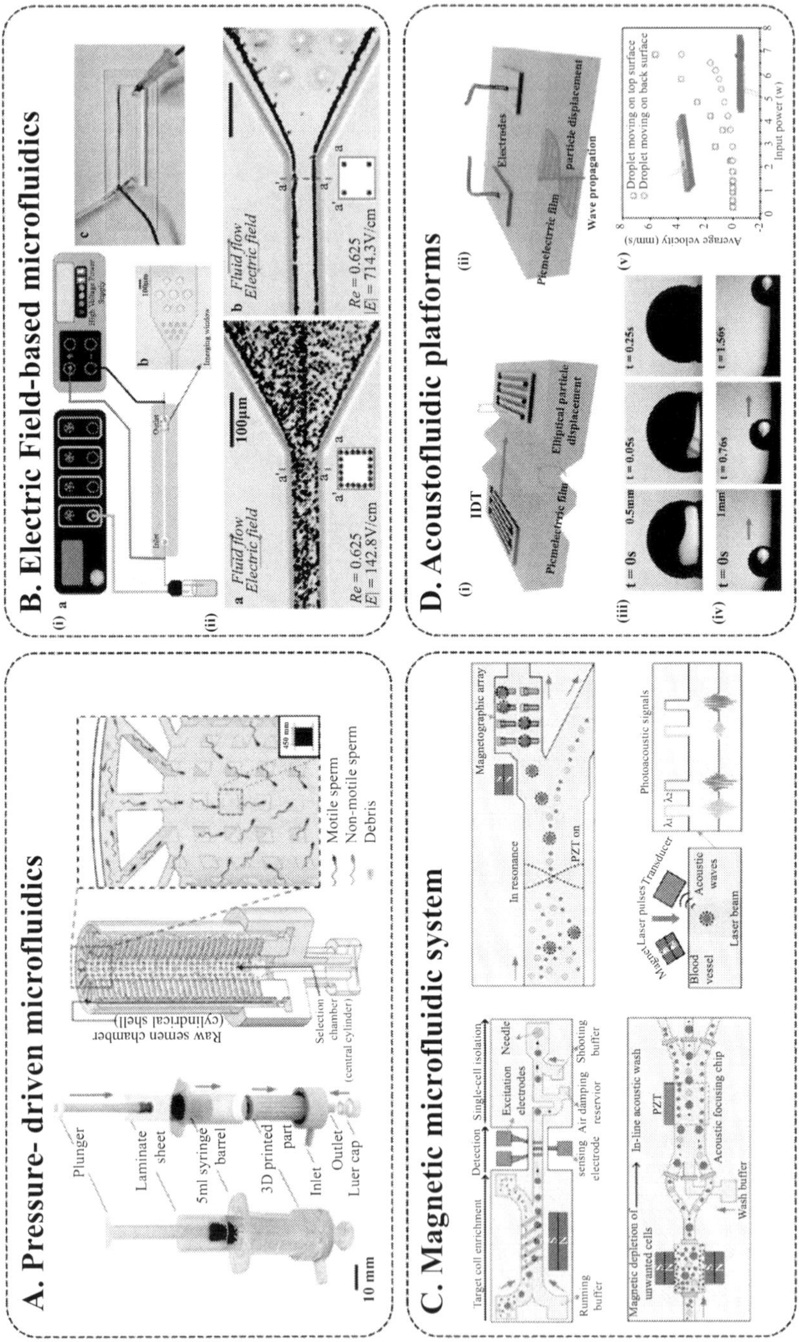
A. Pressure- driven microfluidics
Plunger
Laminate sheet
5ml syringe barrel
3D printed part
Inlet
Outlet
Luer cap
10 mm
Raw semen chamber (cylindrical shell)
Selection chamber (central cylinder)
Motile sperm
Non-motile sperm
Debris
B. Electric Field-based microfluidics
Fluid flow
Electric field
100μm
D. Acoustofluidic platforms
IDT
Piezoelectric film
Elliptical particle displacement
Electrodes
Particle displacement
Wave propagation
Droplet moving on top surface
Droplet moving on back surface
Input power (w)
Average velocity (mm/s)
C. Magnetic microfluidic system
Target cell enrichment
Detection
Single-cell isolation
Running buffer
Excitation electrodes
Needle
Air damping reservior
Shooting buffer
In resonance
PZT on
Magnetographic array
Magnetic depletion of unwanted cells
In-line acoustic wash
Acoustic focusing chip
Wash buffer
PZT
Laser pulses
Transducer
Acoustic waves
Laser beam
Blood vessel
Photoacoustic signals

- *Silicon*: Silicon is used in applications requiring robust mechanical properties and integration with electronics. While its use in microfluidics is less common due to its cost and difficulty in bonding with other materials, it is valuable in specialized applications, such as sensor integration (Kalkandjiev et al. 2011).
- *Cyclic Olefin Copolymers (COC)*: COC materials offer good optical clarity and chemical resistance and are used for disposable devices (Agha et al. 2022). They are gaining popularity due to their ability to be thermally bonded, enabling high-throughput manufacturing.

3.2.2.2 Fabrication Methods

The field of microfluidics has evolved significantly due to advancements in fabrication technologies, enabling the production of diverse devices used in various applications such as drug development, diagnostics, and environmental monitoring. Important breakthroughs in fabrication methods have greatly contributed to the miniaturization and functionalization of microfluidic systems. The selection of a fabrication technique depends on several factors, including material properties, design complexity, cost-effectiveness, and scalability, making it essential to choose a method that aligns with the specific requirements of the microfluidic platform.

Fabrication techniques are broadly classified into two categories: methods that remove material (removing techniques) and methods that deposit material (deposition techniques). These methods can be further categorized based on the nature of the processes involved, including chemical, mechanical, laser-based, and additive processes. A selection of these techniques is discussed below.

FIGURE 3.2 Different driving forces in microfluidics, highlighting their applications and mechanisms. A. 3D-printed microfluidic "sperm syringe" designed for high-quality sperm selection. The device features 560 parallel microchannels integrated into a 3D-printed component, a modified 5 mL syringe barrel, a laminate sheet to prevent sample mixing, a plunger for sample handling, and a luer cap for secure closure. (Reprinted with Open Access permission from Yazdan Parast, O'Bryan, and Nosrati, 2022.) B. Experimental setup and sidewall focusing under varied electric field conditions. i) Schematic of the experimental setup, including the imaging window and a photograph of the microfluidic chip with its fluidic and electric connections. ii) Sidewall focusing of particles under different electric field strengths: (a) Low electric field strength with fluid flow and electric field aligned, and (b) high electric field strength with fluid flow and electric field opposing each other. Insets depict particle cross-sectional equilibrium positions, with scale bars representing 100 μm. (Reprinted with Open Access permission from Abdorahimzadeh et al., 2023.) C. Magnetic microfluidic systems isolate circulating tumor cells (CTCs) using magnetophoretic principles, supported by functionalized magnetic nanoparticles and advanced magnetic field sources for enhanced cancer diagnostics. (Reprinted with Open Access permission from Luo and He 2020.) D. Acoustofluidic platforms utilize hybrid wave modes for multifunctionality, enabling liquid mixing, transport, and biosensing, demonstrated here for chemotherapeutic monitoring. (Reprinted with Open Access permission from Tao et al., 2020.)

3.2.2.2.1 Chemical Processes

Chemical processes have long been a cornerstone in the fabrication of glass and silicon-based microfluidic devices. Notable chemical methods include the following: i) *Wet etching*: this widely-used technique involves the removal of material using chemical solutions, typically hydrofluoric acid for glass. Wet etching offers rapid material removal but tends to result in isotropic (non-directional) etching, which may limit the precision of microfluidic structures. Additionally, it involves the use of highly corrosive chemicals, which can pose environmental and safety risks. ii) *Dry etching*: known as reactive ion etching, this method uses ionized gases to etch precise, anisotropic (directional) features onto substrates. Dry etching allows for more controlled feature definition, particularly on transparent substrates, although it is slower and less selective compared to wet etching (Garra et al. 2002; Balakrisnan, Patil, and Smela 2009). iii) *Electrochemical discharge machining*: this non-conventional method involves generating an electrochemical spark between two electrodes immersed in an electrolyte to remove material through thermal or chemical processes. It is suitable for non-conductive materials like ceramics and glass (Zou et al. 2022; Benilov et al. 2008).

3.2.2.2.2 Mechanical Processes

As already explained in the introduction section, mechanical processes were among the first used for fabricating microfluidic devices, particularly in semiconductor industries. These techniques are advantageous for their flexibility and low cost, but they are less precise than lithographic methods. Key mechanical processes include the following: i) *Micromachining*: this technique, adapted from semiconductor manufacturing, uses mechanical tools to remove material and create microscale features. While it allows for the production of crack-free surfaces and complex 3D structures, it can struggle with precision and productivity at the microscale (Fiorini and Chiu 2005). ii) *Micro-milling*: this high-precision technique involves using a rotary cutting tool controlled by a computer numerical system. It is highly effective for creating complex 3D shapes and structures (Guckenberger et al. 2015). iii) *Injection molding*: this high-throughput technique involves injecting a thermoplastic material into a mold, where it solidifies into the desired shape (Chien 2006). While suitable for mass production of microfluidic devices, it is limited to thermoplastics and has challenges with mold fabrication costs and resolution. iv) *Hot embossing*: similar to injection molding, hot embossing shapes thermoplastic materials under heat and pressure to form microfluidic devices (Chien 2006). It is advantageous for creating delicate designs with reduced shrinkage but also restricted to thermoplastics. v) *Soft lithography*: a widely used method for biomedical applications, soft lithography involves creating a master mold from a hard material, followed by pouring a polymer like PDMS into the mold. The resulting flexible polymer replica can then be used to create microfluidic structures. This method offers high-resolution replication and is relatively low-cost, but it is limited by the need for cleanroom conditions to create the hard master mold (Kim et al. 2008; Urbanski et al. 2006; Kane et al. 2002).

3.2.2.2.3 Laser-Based Processes

Laser-based fabrication methods are gaining popularity due to their flexibility and reduced need for cleanroom facilities. Lasers are capable of directly etching microfluidic features on various materials, and common laser-based processes include the following: i) *Laser ablation*: in this process, a focused laser beam removes material through thermal degradation. Laser ablation is rapid and flexible but can suffer from poor reproducibility, surface damage, and limitations in throughput (Gao et al. 2019). ii) *Stereolithography*: a type of 3D printing that uses laser light to cure photopolymer resins layer by layer (Kuo et al. 2019). It is particularly useful for creating intricate microfluidic devices with fine features. iii) *Two-photon polymerization*: this advanced laser technique uses ultrashort pulses to cure resin at very fine resolutions (Perrucci et al. 2018). Although it offers exceptional precision, it is time-consuming and limited by the need to cure each voxel individually.

3.2.2.2.4 3D Printing

3D printing is a rapidly developing technique in microfluidic device fabrication, enabling the creation of complex geometries that are difficult to achieve using traditional methods. Key 3D printing technologies include the following: i) *Fused deposition modeling (FDM)*: this common 3D printing technique uses a thermoplastic filament that is melted and extruded to form the desired structure. FDM is cost-effective and suitable for creating multi-material devices but has limitations such as susceptibility to compressive stress fractures and poor resolution in small features (Kotz et al. 2020). ii) *Multi-jet printing*: in this process, droplets of photopolymer resin are ejected and cured using a light source. It offers high precision and the ability to print multi-material devices, making it suitable for complex microfluidic applications (Childs et al. 2020). iii) *Inkjet printing*: originally used for depositing inks, inkjet printing has been adapted for microfluidic fabrication by depositing a variety of materials such as metals and polymers. Inkjet printing is low-cost, rapid, and highly precise, but it faces challenges in producing hollow features (Bassous, Taub, and Kuhn 1977).

3.2.2.2.5 Hybrid Technologies

Hybrid fabrication techniques combine multiple methods to overcome the limitations of individual processes. For example, combining 3D printing with micromachined laser lamination has allowed for the production of intricate transparent microfluidic devices without the need for expensive cleanrooms, while enhancing precision in the lamination process.

3.2.2.3 Emerging Materials

The field of microfluidics is expanding beyond traditional materials, with emerging materials offering new possibilities for advanced applications such as tissue engineering, diagnostics, and drug delivery. Two particularly promising categories of materials are hydrogels and paper-based systems: i) *Hydrogels* are materials that absorb large amounts of water, making them ideal for biomedical applications, including organ-on-a-chip devices, cell culture, and drug delivery systems. These materials can mimic the mechanical properties of living tissues, which is important for creating more realistic environments for cells and tissues. Hydrogels are also used in 3D cell culture platforms, where they provide a scaffold for cells to grow and interact with one another in ways that are more similar to how they behave in the body. In microfluidic devices, hydrogels can be integrated as soft materials for creating tunable microenvironments (Nie, Fu, and He 2020). They can also be used for controlled release systems, where drugs or growth factors are embedded within the hydrogel matrix and gradually released in response to changes in environmental conditions (e.g., pH, temperature, or electrical fields). ii) *Paper-based microfluidics* is an innovative, low-cost, and highly portable alternative for microfluidic devices. These systems use paper as a substrate and rely on capillary action to drive the flow of liquids through microchannels etched or printed onto the paper (Martinez et al. 2007). Paper-based devices are particularly promising for point-of-care diagnostics, where cost and accessibility are critical factors. One of the most well-known applications of paper-based microfluidics is the development of paper analytical devices (PADs) (Krishna et al. 2023a), which are used for rapid, on-site detection of biomarkers or pathogens. These devices can be produced cheaply, are lightweight, and can be stored for long periods without refrigeration, making them ideal for use in resource-limited settings.

3.3 MICROFLUIDIC DEVICES AND THEIR COMPONENTS

Microfluidic systems rely on several key components to achieve precise control over the manipulation of fluids at the microscale. This section covers the core components of microfluidic devices, including microchannels, pumps, valves, and detection mechanisms, highlighting the design considerations and functions of each.

3.3.1 Core Components of Microfluidic Systems

Microfluidic systems rely on several core components that work in tandem to achieve precise fluid control, mixing, separation, and analysis within small-scale channels. The effectiveness of these components directly impacts the system's performance in applications ranging from biochemical analysis to drug testing. Below is a detailed overview of these essential components, starting with the key elements that manage fluid flow.

3.3.1.1 Microchannels and Network Designs

Microchannels form the backbone of microfluidic devices, enabling fluid transport and interaction within confined spaces (Hajam and Khan 2024). The design of these channels, including their shape, layout, and flow characteristics, is crucial for optimizing the system's performance.

3.3.1.1.1 Design Considerations for Optimizing Performance

When designing microfluidic devices, it is essential to consider factors such as overall device geometry, channel layout, and fluidic behavior to ensure efficient fluid transport and mixing (Prakash and Kumar 2014). These design parameters include:

- *Geometry and surface area*: Minimizing dead volumes, maximizing surface area-to-volume ratios, and ensuring compatibility with the target fluid and sample volumes are critical for efficient mixing and transport. Shapes like serpentine, circular, or rectangular channels may be chosen based on the application's requirements.

- *Channel layout*: The channel width, height, and length must be designed to achieve optimal fluidic behavior, such as laminar flow and efficient mixing. Features like pillars or grooves can further enhance mixing, and the layout must provide precise control overflow rate and direction.
- *Fluidic behavior*: External fields (e.g., electric or magnetic fields) and surface coatings can be used to manipulate fluid flow or particle movement. Additionally, reducing fouling and enhancing mixing can be achieved through careful selection of material coatings and surface modifications.

3.3.1.1.2 Types of Microchannels

Microfluidic channels are categorized based on their geometry and the application they serve (Prakash and Kumar 2014). Various channel types offer distinct advantages and limitations:

- *Straight channels*: Straight channels are among the simplest and most versatile microfluidic designs, often used for fundamental applications such as material compatibility testing, reagent optimization, and basic fluid transport. These channels are particularly effective for tasks where straightforward fluid handling is required, such as separating components based on natural capillary forces or basic filtration processes.
- *Y-junction channels*: Y-junction channels are frequently employed to split or combine fluid flows, making them ideal for applications like particle sorting or fluid mixing. They are widely used in systems that require the separation of different phases or components, allowing precise control over flow direction and enhancing flow characteristics in more complex systems.
- *T-junction channels*: T-junctions are essential for generating droplets, a core feature of droplet-based microfluidics. This design enables the precise formation of distinct fluid phases, such as the creation of emulsions for chemical reactions. These channels are commonly used in applications requiring accurate control over droplet size and formation, which is critical in fields like chemical synthesis and biological assays.
- *Spiral channels*: Spiral channels utilize the principles of Dean flow to separate particles or cells based on size, shape, or inertia. These channels are particularly useful in applications where high separation efficiency is necessary, such as isolating cells from blood samples or concentrating particles. Spiral designs enable fine-tuned control over particle movement and alignment, offering high precision in sorting and focusing at low flow rates.
- *Cross junction channels*: Cross junctions are utilized to enable complex flow dynamics, such as flow focusing or creating multi-compartmentalized fluid systems. These channels are often employed in applications such as single-cell analysis, droplet-based barcoding, or the creation of materials with intricate, multi-functional properties.

The choice of channel type depends on the application, with trade-offs between mixing efficiency, fabrication complexity, and operational demands.

3.3.1.1.3 Mixing and Separation Techniques

Microfluidic systems employ a range of techniques for efficient mixing and separation (Salafi, Zeming, and Zhang 2016; Ward and Fan 2015). Methods such as passive mixing (via serpentine channels or pillars) and active mixing (through the application of electric or magnetic fields) are commonly used. Additionally, separation techniques such as centrifugation, size exclusion, and magnetic or electrokinetic forces can be used to isolate different components based on their physical properties (size, charge, etc.).

3.3.1.2 Pumps and Valves for Microfluidic Systems

Efficient fluid manipulation in microfluidic systems requires specialized pumps and valves that can accurately control flow rates, directions, and the timing of fluid delivery.

3.3.1.2.1 Types of Pumps

- *Syringe pumps*: Commonly used for precise and continuous flow, syringe pumps are ideal for maintaining steady fluidic pressure. While they are highly accurate, they can be too bulky for miniaturized systems.
- *Peristaltic pumps*: These pumps provide controlled flow through rotating rollers that compress flexible tubing. Peristaltic pumps are widely used for delivering fluids in a continuous flow but can be less efficient in handling high-pressure applications.
- *Electrokinetic pumps*: Using electric fields to manipulate charged fluids, electrokinetic pumps are well-suited for applications where precise control of fluid movement is required at a small scale, such as particle focusing or separation.
- *Capillary pumps*: Driven by surface tension, capillary pumps use the natural flow of liquids through narrow channels to achieve flow, eliminating the need for external pumping devices.

TABLE 3.2
Comparison of Different Microfluidic Channel Types: Advantages and Limitations

Channel Type	Pros	Cons
Straight	Simple design, easy fabrication, low pressure drops	Limited functionality, no active mixing or separation
Y-junction	Versatile (merges or splits flow), enhanced control	Uneven flow distribution, more complex fabrication
T-junction	Efficient mixing, good for droplet formation	High pressure drops, potential droplet merging
Spiral	Enhanced mixing, efficient heat/mass transfer	High pressure drops, complex fabrication
Cross junction	High mixing efficiency, useful for chemical reactions	Complicated fabrication, high pressure drops

- *Centrifugal pumps*: Integrated into disk-based systems, they use rotational force to drive fluid flow and are commonly found in diagnostic assays like enzyme-linked immunosorbent assays.
- *Piezoelectric pumps*: These generate flow by displacing fluids with piezoelectric actuators. They are often used in droplet microfluidics and applications requiring rapid, precise flow adjustments.

3.3.1.2.2 Valve Mechanisms

Valves are critical for regulating fluid flow within microfluidic devices. Various valve mechanisms are employed, including:

- *Mechanical valves*: These valves rely on mechanical movements to open and close flow channels. While they offer high precision, they can be prone to wear over time.
- *Electrovalves*: These valves are controlled by electric fields, allowing for precise, on-demand switching in microfluidic systems.
- *Pneumatic valves*: By applying pressure through external air or fluids, pneumatic valves control the flow of liquids, providing a high degree of control in more complex systems.

The choice of valve type is determined by the specific requirements of the application, such as speed, reliability, and the need for multi-layer integration.

3.3.1.3 Detection and Sensing Mechanisms

Detection and sensing mechanisms are essential for monitoring and analyzing the fluid's properties and behavior within the microfluidic system.

3.3.1.3.1 Types of Sensors

Several types of sensors are utilized in microfluidic systems to provide real-time data on fluid properties and system performance:

- *Optical sensors*: These sensors are based on light transmission or scattering and are commonly used for detecting cell count, concentration, and particle size.
- *Electrochemical sensors*: These sensors measure the electrochemical properties of the fluid, providing data on parameters like pH, ion concentration, and chemical composition.
- *Impedance sensors*: Impedance sensors monitor changes in electrical resistance, which can be used for detecting cell behavior or fluid composition.
- *Flow sensors*: These sensors are used to measure the rate and volume of fluid flow within the channels, ensuring optimal performance in systems requiring precise flow control.

Each sensor type is chosen based on the specific requirements of the application, such as sensitivity, accuracy, and the ability to integrate with other components within the system. In Figure 3.3, a scheme of components of microfluidic devices is shown.

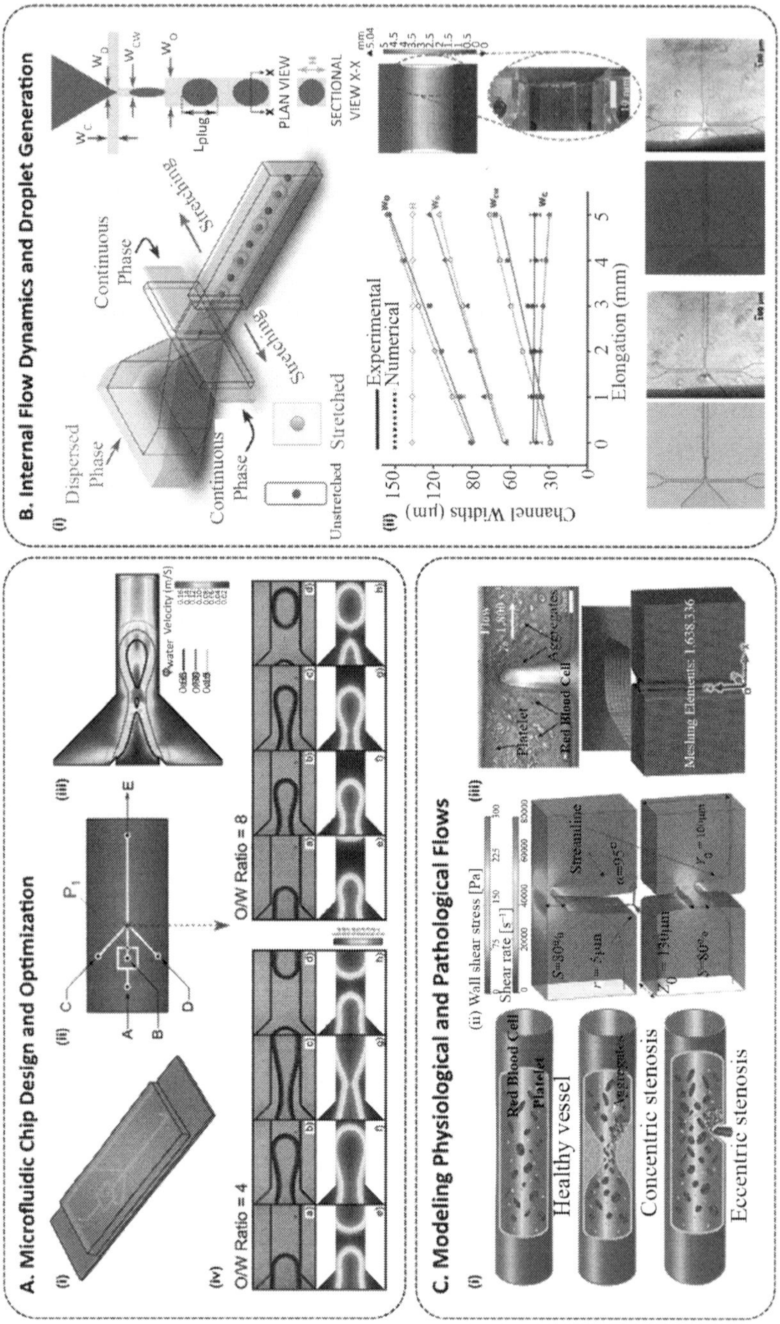
A. Microfluidic Chip Design and Optimization
O/W Ratio = 4
O/W Ratio = 8
B. Internal Flow Dynamics and Droplet Generation
Dispersed Phase
Continuous Phase
Stretching
Unstretched
Stretched
PLAN VIEW
SECTIONAL VIEW X-X
Experimental
Numerical
Channel Widths (μm)
Elongation (mm)
C. Modeling Physiological and Pathological Flows
(ii) Wall shear stress [Pa]
Shear rate [s⁻¹]
Healthy vessel
Concentric stenosis
Eccentric stenosis
Red Blood Cell
Platelet
Aggregates
Streamline
Meshing Elements: 1,638,336

3.4 MICROFLUIDIC COMPUTATIONAL DESIGN AND OPTIMIZATION

The design and optimization of microfluidic systems benefit significantly from computational methodologies, which provide powerful tools to enhance precision and efficiency in predicting and understanding fluid behavior at the microscale. Computational approaches, such as fluid dynamics simulations and data-driven frameworks, are increasingly pivotal in addressing the complex challenges associated with microfluidic device design.

3.4.1 Role of Computational Fluid Dynamics (CFD) in Microfluidics

Computational fluid dynamics (CFD) is a transformative technology in microfluidics, enabling the detailed simulation of fluid flow, heat transfer, and other transport phenomena within microchannels. By solving the Navier-Stokes equations and related governing equations, CFD provides a deeper understanding of fluid behavior in complex geometries, where experimental methods can be challenging or resource-intensive. CFD is particularly valuable in microfluidics due to the predominance of laminar flow and the sensitivity of microscale systems to geometric and operational parameters (Razavi Bazaz et al. 2020; Somoza et al. 2023). For instance, CFD simulations can optimize microchannel designs to improve mixing efficiency, minimize pressure drops, or enhance heat dissipation. It also facilitates the study of phenomena such as droplet formation, particle sorting, and multiphase flows, offering insights that guide the development of high-performance devices (Amini et al. 2023). By significantly reducing the reliance on physical prototyping, CFD enables rapid design iterations and cost-effective innovation. Its application extends across various fields, including biomedical diagnostics, chemical processing, and environmental monitoring, making it an indispensable tool in microfluidics research and development.

FIGURE 3.3 This figure demonstrates diverse applications of computational fluid dynamics (CFD) in microfluidic systems for design, analysis, and optimization. A. Visualization and simulation results of a microfluidic device, showcasing design geometry, flow pathways, and computational insights. (Reprinted with Open Access permission from Rial et al. 2021.) B. i) Schematic representation of channel dimension changes and droplet generation characteristics resulting from stretching the microfluidic device, along with key parameters considered for theoretical analysis. ii) Dimensional changes in a flow-focusing microfluidic device under lateral stretching, showing experimentally measured and simulated microchannel deformation, including geometry and flow-focusing arrangement before and after a 5mm stretch. The simulation and experimental images illustrate the displacement field and deformed channel structure. (Reprinted with Open Access permission from Roshan et al., 2025.) C. Simulations of blood flow in microfluidic models mimicking healthy and stenosed vessels, providing insights into wall shear stress, shear rates, and platelet aggregation. (Reprinted with Open Access permission from Zhao et al., 2021.)

3.4.2 Simulation Tools for Microfluidic Design

Simulation tools are essential for optimizing microfluidic device design, enabling researchers to predict system behavior under different conditions without the need for physical prototypes. These tools integrate multiphysics capabilities, allowing for the simulation of fluid dynamics, heat transfer, and chemical reactions. They are particularly valuable for optimizing factors like flow stability and mixing efficiency.

By facilitating multi-objective optimization, simulation tools help balance performance metrics such as throughput and sensitivity. They accelerate development cycles, reduce costs, and improve the robustness of devices. Ultimately, these tools provide critical insights and streamline the design process, ensuring microfluidic systems perform efficiently in real-world applications(Ferreira et al. 2024).

3.4.3 Data-Driven Approaches and Machine Learning in Design Optimization

As microfluidic systems grow more sophisticated, data-driven approaches and machine learning (ML) techniques are emerging as transformative tools for design and optimization. These methodologies enable to navigate the complex parameter spaces of microfluidic devices and accelerate innovation. ML techniques leverage large datasets from experiments and simulations to uncover patterns, predict outcomes, and optimize designs (Zheng et al. 2021; McIntyre et al. 2022). They can automate the identification of optimal configurations and provide insights that would be difficult or impossible to derive through traditional methods. Applications of ML in microfluidic design include:

- *Predictive modeling*: ML algorithms, trained on experimental or simulated data, predict system behaviors such as droplet size, flow distribution, or mixing efficiency based on input parameters like channel dimensions and flow rates. These models significantly reduce computational costs compared to full-scale simulations.
- *Design optimization*: Machine learning enables multi-objective optimization, where multiple performance metrics—such as throughput, efficiency, and sensitivity—are simultaneously maximized. Generative design algorithms propose innovative layouts tailored to specific applications, exploring possibilities beyond conventional designs.
- *Surrogate modeling*: By creating fast, approximate models of computationally expensive simulations, surrogate models allow for rapid exploration of design spaces. These models are particularly useful in iterative optimization processes.
- *Real-time feedback and control*: Machine learning integrates with experimental setups to adjust flow rates, pressures, or temperatures dynamically, ensuring optimal device performance during operation.

3.4.3.1 Integration of Machine Learning with Microfluidics

Combining ML with traditional simulation tools creates a synergistic framework for microfluidic innovation. For example, hybrid workflows can use ML to pre-screen design parameters and feed promising candidates into detailed simulations for validation. This reduces computational overhead while maintaining accuracy. In addition to machine learning, broader data-driven approaches involve using statistical methods, design-of-experiments (DOE), and optimization algorithms to refine device designs (Rebollo et al. 2022). By systematically varying design parameters and analyzing outcomes, researchers can identify trends and fine-tune microfluidic systems to meet specific goals.

3.5 MICROFLUIDICS-BASED NANOBIOSENSORS

Fast diagnostic tools help users to make decisions on the fly and can greatly assist in managing medical treatments. This fact improves the efficacy of the treatments and can be the difference between life and death in some cases, such as glucose monitoring devices. One of the crucial roles of nanobiosensors based on microfluidics is the point-of-care (POC) diagnosis of various diseases and their function in healthcare monitoring. POC bioanalysis is a concept that evolved via a microfluidic lab-on-a-chip (LOC). These platforms have rapidly advanced over the last two decades and demonstrate versatile applications in different fields. In this sense, microfluidic devices are designed to work with small volumes of samples suitable for clinical diagnosis.

The rapid detection of analytes is guaranteed by microfluidic systems working together with nano- and biotechnologies. This combination of strategies gives rise to biosensors of nanometric size, nanobiosensors, which are designed for detecting biomolecules of clinical diagnosis through matching with specific receptors. This interaction generates an electrical signal modulated by a physicochemical cascade, resulting in a measure of the parameter in an electric display or electronic device. The possibility of detecting selected molecules can play a significant role in the early identification and treatment of diseases or can be used to control them. Microfluidics-based nanobiosensors can detect and quantify biological molecules of clinical interest such as proteins, DNA, and sugars by combining them with specific molecules such as antibodies or nanomaterials, enabling sensitive detection of targeted molecules. These nanobiosensors detect insignificant concentrations of biomarkers in real time, thereby reducing the possibility of false positives and negatives in the testing process. This section addresses the detection method of biomolecules of clinical interest by nanobiosensors based on microfluidic platforms and their application (Kumari et al. 2024).

3.5.1 NANOBIOSENSORS

Biosensors are essentially, devices to measure the presence of biological interest molecules, DNA, RNA, small single molecules, pathogen cells, or metabolites, by a specific interaction of some of these structures with others associated with a signal transduction. Beyond healthcare applications, biosensors have also been utilized for evaluating food quality and monitoring the environment (Adam et al. 2023). Due to the advances in nanotechnology, different components of biosensors for enhancing sensory phenomena have emerged, including nanomaterials, nanoparticles, nanotubes, and nanorods, improving the detection process and reproducibility of measurements (Malik et al. 2021). The new materials obtained by nanotechnology applications have differentiated from the previous ones because they have achieved high electrical conductivity and a sensitive response, in addition to mechanical properties such as increased shock resistance (Ramesh et al. 2022). Nanobiosensors can be developed in different manners: top-down, bottom-up, and self-assembly. The first approach implies the disintegration of materials to produce smaller nanostructures of the initial material. The second entails the combination of smaller sections (molecules or atoms) to create bigger nanostructures, and the self-assembly process involves the association of molecules or clusters generating a defined order or arrangement resulting in a nanostructure (Malik et al. 2021).

Biological detection after unambiguous interaction generates a physicochemical effect that is turned into an electric signal through the signal transductor, which can be visualized in a display. Nanobiosensors consist of the following components: bioreceptors, transducers, and displays. The selected analyte is detected by the bioreceptor, which is the primary recognition structure, and can be formed by enzymes, cells, aptamers, DNA, RNA, and antibodies. The analyte–bioreceptor interaction generates a physicochemical signal such as light, heat, pH, charge, or mass change. Then, the transducer converts one form of energy (physicochemical) into another (electric), which is easy to measure (Kumari et al. 2024; Bhalla, Chiang, and Shen 2018).

The advantages of nanobiosensors are related to specificity and sensitivity detections of biomolecules in real time for disease management. However, several challenges must be faced in the fabrication of these devices. The nanobiosensor has to show high accuracy in detecting biomolecules (among a pull of many molecules) with similar structures or those in a very low concentration. Additionally, the materials used for in vivo applications should be biocompatible and non-toxic, and must not cause immune response. Another important point is related to detection capability. To ensure consistent results, the device must show reproducibility and the affectation of environmental conditions, and interfering substances in the outcomes should be minimal to avoid differences in the interpretation. The search for this standard leads to high manufacturing costs. Added to this cost, extensive tests and documentation processes to demonstrate safety and efficacy must be carried out to be approved by the regulatory agencies.

There are different detection methods used in healthcare monitoring nanobiosensors based on microfluidics. Absorbance, fluorescence, and surface plasmon resonance (SPR) spectroscopy are widely used optical detection methods. They are applied to detect distinct biomolecules from humans, bacteria, or viruses, including proteins and genetic material. Another detection method frequently used is based on the measurement of electrochemical parameters, such as amperometry and potentiometry.

Electrochemical biosensors transform chemical stimulation into an electric signal. Electrochemical detection methods can achieve high sensitivity and specificity for detecting biomolecules such as glucose, hemoglobin, blood ketones, cholesterol, and diverse types of proteins.

3.5.2 Detection Methods

Microfluidic biosensors use conventional optical detection methods to establish the presence of analytes in a sample. These methods include absorbance, fluorescence, chemiluminescence, interferometry, and surface plasmon resonance (Gai, Li, and Yeung 2011). The miniaturization of the detection methods implies a profound challenge. This strategy avoids the requirement of expensive hardware to detect the analyte. However, lower-length scales show many inconveniences. The sensitivity of the detector depends on the optical path lengths. Lower paths decrease sensitivity. Additionally, a higher surface-to-volume ratio can increase the noise of the determination due to the non-specific absorption to the cell walls of the device. The improvement of sensitivity and the reduction of costs can be achieved by the incorporation of optoelectronic elements, filters, and other devices into the microfluidic chambers. Among the optic components that have been integrated are laser diodes, photodiodes, and more. Optical systems have been used not only for detection but also for achieving a tunable and reconfigurable microfluidic device (Myers and Lee 2008).

3.5.2.1 Optical detection

3.5.2.1.1 Fluorescence

The most relevant detection methods for determining molecular presence in microfluidic devices are the optical ones. Fluorescence-based methods are commonly used because of their ease of integration into microfluidic devices, high sensitivity, low cost, immediate analysis, and the management of low sample volumes. Numerous examples of fluorescence-based methods are considered for microfluidic devices such as surface-enhanced Raman spectroscopy, high-throughput single-molecule imaging, fluorescence lifetime imaging, multicolor analyses, and surface plasmon resonance (Baker et al. 2009; Li et al. 2023). Fluorescence-based detection assays are applied to the determination of nucleic acids (RNAs) of pathogens viruses such as SARS-COV (Cheong et al. 2020; Kang et al. 2020). The use of labeled fluorophores produces a detectable signal that increases or quenches the fluorescence. In the case of DNA detection and quantification, fluorescent dyes embedded in DNA are used. These dyes are added to the PRC to bind with the amplified DNA, and thus the complex shows fluorescence. While the quantities of DNA increase by amplification cycle by cycle, the intensity increases, allowing the simultaneous quantification of DNA. A modified DNA oligonucleotide probe that emits fluorescence in the presence of DNA confers an attractive alternative as a detection method because it is a real-time method that shows higher sensitivity than the DNA amplified method (Li et al. 2023).

The application of fluorescence techniques and sensors in microfluidics has improved the detection sensitivity of DNA, and DNA curtain is an example. This fluorescence-based method uses a dye (YOYO-1) as a visualizing method. The multichannel microfluidic DNA curtain technique is a high-throughput single-molecule imaging platform based on poly(dimethylsiloxane) to investigate complex biological reactions. This curtain is formed by a lipid bilayer which avoids the nonspecific adsorption of proteins to the surface of the slide and allows the migration of DNA under hydrodynamic forces caused by the membrane fluidity (Kang et al. 2020). Therefore, when DNA associated with lipid molecules reaches the diffusion barrier, it accumulates at the curtain. In case of low initial concentration of DNA portions, the flow carries them, and hundreds of DNA moieties reach the barrier, achieving high-throughput observation of DNA molecules (Visnapuu et al. 2008). Therefore, when DNA associated with lipid molecules reaches the diffusion barrier, it accumulates at the curtain. In case of low initial concentration of DNA portions, the flow carries them, and hundreds of DNA moieties reach the barrier, achieving high-throughput detection of DNA. In the initial versions of the DNA curtains, the lipid molecules associated with DNA cannot pass through the barrier due to the high bending energy. This fact interferes with an accurate determination of protein–DNA interaction sites. Diffusion barriers carried out by nano-sized chromium have improved the arrangement of DNA (Kang et al. 2020).

3.5.2.1.2 UV/Vis Spectroscopy

Due to its simplicity, sensitivity, and adequate detection limit, UV/Vis absorption spectroscopy is one of the most applied detection methods associated with microfluidic processes. The rapid identification and accurate confirmation of molecules such as marijuana have aroused great interest in academic and governmental domains. In this sense, the portable device development for detecting these types of molecules has been proposed considering costs, handleability, and versatility. Thus, a microfluidic portable device based on paper chromatography fulfills all the required criteria. Krishna et al. (2023b) have described a simple procedure used to estimate the content of cannabinoids in a sample. The sample reacts with the reagent Fast Blue B (FBB) immobilized in an area delimited in the paper. Then the area containing the sample and FBB is removed and the content is extracted with methanol and read at 450 nm with a UV/Vis spectrometer for the complex FBB-Cannabinoids. In this work, the presence of the molecules was carried out after the reaction in a traditional UV/Vis spectrometer. There are other examples where the determination is produced in an integrated UV/Vis spectrometer in microfluidic devices such as reported by Yue et al. (2013). In this work, the authors show the coupling of microreactors with UV/Vis spectroscopy for online process study under segmented flow conditions.

3.5.2.1.3 Surface Plasmon Resonance (SPR)

Surface plasmon resonance (SPR) is a highly sensitive, real-time sensing technique that does not require labeling with external markers or dyes (label-free). It is widely applied in biomedical analysis and used as a biosensor. SPR is based on the principle that resonance conditions of surface plasmons show high sensitivity to the dielectric property of an interface when interacting with biomolecules. The detection method is based on the collective oscillations of electrons on a metal surface (usually gold) when illuminated by light under certain conditions (Wang and Fan 2016). SPR devices can be utilized to measure interactions between different molecules (antibody–antigen, ligand–receptor, and enzyme–substrate reaction) and show an excellent application in biological, biomedicine, and diagnostics-related fields. Additionally, it has been utilized in biosensors to monitor binding interactions between biomolecules, such as proteins and DNA. It is important to notice that the detection limit of an SPR biosensor is on the order of 10 pg/ml (Nguyen et al. 2015).

3.5.2.2 Electrochemical Detection Methods

Microfluidics is based on the downsizing principle and integration. Reducing the size of devices implies reducing the size of analytical sensors coupled. Electrochemical detection sensors integrated into microfluidic devices are attractive due to their simplicity for coupling, allowing miniaturization and the possibility to integrate different sensors in a single platform. Several methods for fabricating microbiochemical sensors have created the next biochemical microsystems applied in microfluidics. These microsensors have been based on electrochemical principles measuring changes in current, voltage, or impedance due to the register of redox reactions occurring in the sample and by detecting ions (Adam et al. 2023). In these cases, electrodes form the main components of the devices, which must integrate with other parts of the device. The approach for applying electrochemical principles to biosensors designed for biochemical determinations based on affinity can integrate into microfluidic chips for inline microelectrode functioning, biomarker detection, and sensor regeneration on organ-on-a-chip systems. This design allows for continuous, in situ, and non-invasive quantification of soluble biomarkers on organ platforms (Aleman et al. 2021; Fuchs et al. 2021).

One of the main detection techniques applied to microfluidics is potentiometry, and this is the most common electrochemical detection method in flow injection analysis (FIA). These sensors measure a voltage or potential difference between two electrodes (an ion-selective and a reference electrode). This parameter is associated with the concentration of specific ions in the solution. The variation of potential with the ion concentration follows the Nernst equation. Significant advances have reached the integration of electrochemical sensors, based on potentiometry measurements, into a digital microfluidic platform. This portable device allows continuous monitoring of multiple ions for home applications. Simple automatization, longer sensor life, minimal sample consumption, and user-definable or reconfigurable sensor configuration are some advantages of this system (Farzbod and Moon 2018). The creation of a selective and low-cost analysis device is due to the possibility of combining the fabrication of ion-selective electrodes (ISEs) in small sizes associated with the simplicity of potentiometry measurements. In this sense, Cui et al. (2014) have developed a new potentiometric sensor using a paper-based microfluidic sampling device. The authors show a procedure where two electrodes, on one hand, a solid-contact ion-selective electrode, and on the other, a solid-contact reference electrode, are pressed one against the other through a common filter paper in which the analyte (ions) solution is absorbed. This is a low-cost and ecological device because the paper shows both functions: as a sampling unit and a sample container during potentiometric sensing.

In addition to potentiometry, voltammetry is another technique widely used as an electrochemical determining method. Voltammetry measures the current produced between two electrodes while applying an electrical potential. An on-chip electrochemical detector associated with a capillary microelectrophoresis device performed by a gold electrode directly sputtering on the outlet of the capillary has been reported by Wang et al. (Wang, Tian, and Sahlin 1999; Wang et al. 2000). An on-chip electrochemical detector associated with a capillary microelectrophoresis device performed by a gold electrode directly sputtering on the outlet of the capillary has been reported by Wang et al. The importance of this detector lies in the lack of necessity of microfabrication or decoupling mechanisms for the preparation. This device has demonstrated a linear response for detecting dopamine from 20 to 200 μM and a sensitivity of 52 pA/μM. Due to the low concentrations of neurotransmitters in living systems, their recognition using microfluidics could evidence a good approximation of the behavior of these molecules by a lab-on-a-chip–based method. Skrzypek et al. (2005) have reported the fabrication of a device based on the voltametric determination of famotidine in urine due to this drug being excreted 30% unchanged in this way. There are many reports where different types of voltammetry methods were applied for determining active pharmaceutical ingredients such as formoterol (Hareesha and Manjunatha 2021), dopamine (Manjunatha et al. 2011), cetirizine (Pushpanjali et al. 2021), and ascorbic acid (Chenthattil Raril et al. 2020), among others (Manjunatha, Pushpanjali, and Hareesha 2021; Tigari and Manjunatha 2020; Monnappa, Manjunatha, and Bhatt 2020). The main techniques applied are cyclic voltammetry, linear sweep voltammetry, differential pulse voltammetry, or square wave voltammetry. In this context, Skrzypek et al. (2005) have described an alternative method for detecting active compounds in urine to HPLC. The method was proposed as square wave adsorption stripping voltammetry which is sensitive, simple, and rapid (Erk 2003; Skrzypek et al. 2005).

There are several differences in the sensitivity and applications of these two methods, and it is important to remark on them. Potentiometry is applied to ion assessments in solutions using a glass electrode and an ion-selective one. On the other hand, voltammetry is used for substance trace detection. It is used to seek traces in food or establish biochemical processes where redox reactions are commonly required (Kim et al. 2022).

Amperometry is a sensitive and specific quantification method for establishing the presence of analytes or interest molecules in a sample. This technique implies measuring a current when a potential is applied to an electrode. The intensity of this current is proportional to the concentration of the detected analyte (Agustini et al. 2020). Amperometry is widely used in biomedical research and diagnostics because it is applied in biomolecule detection such as neurotransmitters. It is used in particular for detecting electrochemical reactions during the process because it obeys Faraday's law. Once the current is produced and a dynamic response is established, the application of a voltage generates the electron transfer of the analytes, and the magnitude of the current variation is quantified as a function of the analyte concentration. Due to the similarities between platforms for the preparation of samples and detection, microfluidic systems can be used to study high-throughput sample processes. In many cases, electrodes are deposited in the bottom of thick microfluidic channels, immobilizing a bioreceptor, which could be a chemical species or an enzyme. In this context, Agustini et al. (2020) have described a cheap microfluidic device based on a miniaturized Arduino connected by USB or Bluetooth to a computer or smartphone. The proposed device permits the electrochemical detection (by amperometric measurements) of different substances by interaction with three electrodes. The graphite electrode is used for the determination of uric acid; the carbon-printed electrode to quantify hydrochlorothiazide (diuretic drug, for hypertension); and gold electrodes sputtered into the microfluidic channel for, first, the chromatographic separation of two molecules of ascorbic acid and epinephrine, and then, their electrochemical determination. The authors have validated this device by comparing the electrochemical signals obtained here with those obtained from a commercial potentiostat, reaching good analytical performance at much lower costs.

3.5.2.3 Thermal Detection

Calorimetric devices operate on the principle that chemical or biochemical reactions involve changes in enthalpy, which can be detected as heat produced or absorbed by a reaction or phase change. This method is used to detect chemical reactions or to monitor the reaction kinetics in different systems. One of the advantages of this method relies on its ability to assess the thermodynamics of a certain process without labeling or analyte immobilization. The integration of the calorimetric method with microfluidics allows work with reduced volumes of sample due to devices such as "chip calorimeters." Lee et al. (2009) have reported the fabrication of closed-chamber calorimeters associated with microfluidics that increase the sensitivity in the enthalpic process determinations. Thus, the calorimeter is embedded within a thin-film parylene microfluidic system being thermally isolated by an on-chip vacuum encapsulation. With an accurate reaction control system formed by pumps, valves, and flow channels, it is possible to control samples of 1–2 nl in volume. The fabrication of a vacuum-insulated chamber for sample detection in a microfluidic device allows the determination of calorimetric processes such as isothermal titration calorimetry, differential scanning calorimetry, and flow calorimetry.

In another work, Zhang and Tadigadapa (2004) have developed a microfluidic device for measuring, in real time, the enthalpy changes of different biochemical reactions and thermal properties of biological fluids. The device is formed by a glass reaction chamber (Pyrex 7740) associated with a microthermopile p-type polysilicon/gold of 2mcm thick disposed onto a membrane with a sensitivity of 0.94 V/W. This system was conceived initially to measure the heat of enzymatic reactions involving the catalytic action of glucose oxidase, catalase, and urease on glucose, hydrogen peroxide, and urea, respectively. However, they can be adapted to other non-enzymatic reactions. By continuous flow, in an enclosed fluidic reaction chamber, and open air, the experiments reached a sensitivity of 53.5 μV/M for glucose, 26.5 μV/M for hydrogen peroxide, and 17 μV/M for urea. This device shows a capacity to determine enthalpic heat in a continuous mode for enzymatic and non-enzymatic reactions.

3.5.2.4 Raman Spectroscopy

Raman spectroscopy (RS) provides molecular fingerprinting by detecting the inelastic scattering of the monochromatic light from molecules. It shows high specificity and low detection limits in sample detection for identifying chemical compounds and biomolecules in different systems (Yue, Fang, and Xu 2022). In RS, almost all scattered photons have the same energy as those of the incident beam. However, a small fraction of those is scattered with a change in energy, and the difference can be attributed to vibrational energy. As the system loses or gains energy, the frequency shifts are called Stokes and anti-Stokes shifts, respectively. To be detected by RS, molecules have to show non-zero polarizability (Chrimes et al. 2013).

RS detects and quantifies biomolecules, especially proteins, identifying structures and compositions. This technique allows us to evidence the conformational changes in the secondary structure of proteins associated with specific atom bonds such as C-O or C-H (Kayani et al. 2012). Owing to different molecules having different vibrational modes, sample identification occurs because of the spectrum of their inelastically scattered light. In this sense, RS associated with microfluidics can analyze liquid, solid suspensions, dissolved gases, and interfaces (Chrimes et al. 2013).

To improve sensitivity and spectral resolution when RS is used as a detection method in microfluidics, surface-enhanced Raman spectroscopy, surface-enhanced resonance Raman spectroscopy, and other techniques have been designed as variations of RS.

3.5.2.5 Mass Spectrometry

Mass spectroscopy (MS) has demonstrated high sensitivity and the advantage of being a "label-free" analytical technique for determining sample composition and concentration. The acquisition data speed depends on many features inherent to the equipment and operator. The automatization of this method has improved data harvest and has been advanced and coupled to higher throughput analysis. Microfluidics can be coupled with MS for high-throughput analysis (HTS) of proteins, peptides, metabolites, and other small molecules. The small volumes handled in microfluidics are ideal for coupling with MS for sensitive and accurate detection. Owing to the highest acquisition speeds without compromising resolution, time of flight (TOF) MSs have been suitable for HTS applications. In this context, Kempa et al. (2020) have reported the coupling of HTS microfluidics to MS for liquid analytes determination. In this work, the challenge was focused on the need to extract or divert the aqueous analyte-containing phase from the non-aqueous separative phase before MS infusion. This aim was reached by using fast scanning acquisition IM-Q-TOF instrumentation. The designed device is readily interfaced with commercially available nanoelectro spray sources with slight modifications, allowing for an infusion of microdroplets up to a rate of 9 Hz. This permits the acquisition of chromatograms from which mass spectra for each droplet.

3.6 MOLECULAR AND CELLULAR BIOPHYSICS

Owing to their capability to replicate the biological environment, microfluidic devices provide controlled environments for studying tissue formation and cellular processes, Therefore, they have been applied as excellent tools in the advances of biophysics and cellular culture.

3.6.1 Techniques for Single-Cell Trapping and Analysis

There is a change of trend in the comprehension of cell behavior in which the study of cells in cell culture (associations of cells, groups of cells) has been replaced by the knowledge of cell behavior as a single entity also involving extracellular structures and the attached matrix. In this way, the researchers pretend to reveal intercellular interactions, cellular and external environmental effects, and intracellular signaling pathways. Due to mammalian eukaryotic cells having a volume at fl order and a diameter of 1–10 mcm, the demand for real-time assays and dynamic methods has led to the emergence of new analysis techniques and methods downsizing the scale to a range of tens to several hundreds of micrometers. This breakthrough is produced by lab-on-a-chip (LOC) technology because it allows a good manipulation of small-volume liquids, such as cell isolation, localization, and capture (Dittrich and Manz 2006).

Once the cells are captured, these procedures aim to study fusion, cell pairing, lysis, intracellular composition, and other functional and structural features. Microfluidic single-cell trapping can be divided into three categories to perform these analyses. The first is surface contact capture. Here, the capture is produced by chemical and hydrodynamic methods. The second is non-surface contact capture. Physical properties are applied in this group for reaching the separation such as electrophoretic mobility, magnetic and acoustic forces, sound waves, and others. The last one is capture between surface contact and non-surface contact. This separation method is based on the application of colloidal systems such as hydrogels (Carr et al. 2011; Takayama et al. 2003). Some special features must be considered in single-cell trapping methods based on fluid dynamics. The construction of channels to fix the cells depends on the size of the cells, and the formation of channels or cavities plays different roles, making cell behavior quite different. The flow must be accurately controlled. This reduces the cell culture time and costs and improves the statistically significant data because it provides a suitable number of individual cells to analyze, revealing cell properties (individual and/or differences among cells). Additionally, the concentration of nutrients and metabolites around the cells, affects the growth of cells (very conditioned by cell adhesion) and the apoptosis time (Gravina et al. 2024). For these reasons, the microenvironment developed in the chip is determining. In this sense, Deng et al. (2019) have designed and developed three different microfluidic devices for single-cell trapping and demonstrate their ability to capture considering the controllability and feasibility of cell growth and single-cell observation. The authors have developed a microcavity cell-trapping structure, a U-shaped cell-trapping structure, and a flow shortcut structure. The microcavity structure shows less waste on the sample and a large flux. However, a low number of cells can be cultured due to cell overlapping, which also complicates single-cell observation. This structure has the disadvantage of collapsing during chip fabrication. On the other hand, the U-shaped cell-trapping structure increases the number of cell survival days, but shows an inhomogeneous distribution of cells that makes the observation of cell-single properties difficult. The last, the flow shortcut structure, makes it easy to observe and analyze individual cell properties but results in low cell density and high sample wastage. For these reasons, the authors conclude that the U-type structure was the most efficient.

3.6.2 Studying Protein–Protein and Protein–DNA Interactions

The development of omics analysis (genome, transcriptome, proteome, metabolome) has enabled researchers to uncover molecular mechanisms related to diseases, discover biomarkers, identify drug targets, and enhance diagnostics and precision medicine. As we discussed in the previous section, for more accurate biological interpretations, it is important to recognize the impact of cell heterogeneity on biological results, which leads to the omics characterization of single cells. In this context, microfluidic devices are highlighted as ideal platforms for omics analysis due to their ability to handle small sample volumes efficiently. This technology has recently arisen in proteomic analysis (Gebreyesus et al. 2022). In this context,

Lazar et al. (2020) have successfully integrated microfluidic technologies with high-throughput sequencing and mass spectrometry for proteomic analysis. This integration improves sensitivity and allows for rapid analysis of protein samples, making it possible to identify thousands of peptides in a short time frame. The application of mass spectrometry (MS) in proteomics is crucial for identifying protein sequences, isoforms, mutations, and post-translational modifications (PTMs), and its integration with microfluidic platforms enhances sample processing through multiplexing and high-throughput capabilities. The data acquisition speed has increased with new devices. They can process a high number of peptides within a short time frame (e.g., 10,000–15,000 peptides in 15 min). The design of microfluidic devices is crucial for efficient sample ionization and delivery to MS. Two main ionization techniques—electrospray ionization (ESI) and matrix-assisted laser desorption ionization (MALDI)—are discussed by the authors, with ESI being more commonly used for interfacing with separation platforms.

The advancements in proteomic analysis are focused on the miniaturization of enzymatic reactors and improved sample preparation techniques for speeding up protein digestion and analysis. In this context, Kecskemeti et al. (2017) have developed an immobilized enzymatic reactor (IMER) by adsorbing trypsin onto a porous serpentine channel within a PDMS/glass chip. This reactor allows for rapid digestion of human tear proteins in about 1 min and demonstrates comparable protein identification to traditional methods but with enhanced sequence coverage. The IMER can be multiplexed with up to 8 channels, and the trypsin layer can be regenerated, indicating potential for clinical applications. IMERs have improved their digestion capability by using polymer monolithic materials. They achieved a reproducible digestion of dried blood spots within 5 min, yielding around 150 protein IDs per measurement. The electrochemical cleavage strategy is an innovative method that introduces employing a microfluidic cell to cleave proteins at specific sites, which may offer better control and faster analysis compared to traditional enzymatic methods. However, further validation in complex samples is needed (van den Brink et al. 2016).

3.6.3 Microfluidic-Based on Extracellular Vesicles

The main objective of precision medicine is to deliver personalized treatments adapted to the unique molecular profiles of individual patients. Traditionally, cancer profiling has relied on tissue biopsies obtained through surgical procedures. However, this approach has several disadvantages such as restricted sampling methods, challenges accessing deep tumor tissues, intra-tumor heterogeneity, and difficulties characterizing multiple tumor sites. In contrast, liquid biopsy is a non-invasive technique that assesses disease biomarkers in various biofluids such as blood, urine, saliva, ascites, and breast milk (Rolfo et al. 2014; Shbeer and Robadi 2024). This innovative method allows for the monitoring of tumor progression and genomic evolution over time, as well as the assessment of treatment responses for timely adjustments. Liquid biopsy is recognized as one of the most groundbreaking techniques in cancer diagnostics, opening up new possibilities for patient care. The key components of liquid biopsy samples include circulating tumor cells, circulating tumor DNAs, and circulating exosomes released from primary tumors and metastases.

Extracellular vesicles (EVs) are nano-sized structures released from cells. They are a polydisperse group mainly divided into four categories: exosomes (30–150 nm), ectosomes (100–1000 nm), microvesicles (100–1000 nm), and apoptotic bodies (1000–5000 nm). They are implicated in many physiological processes, such as cellular homeostasis, and engaged in pathological ones such as cardiovascular diseases, infections, and cancer (D'Elía et al. 2024). Tumor-derived exosomes carry oncogenic materials from cancer cells to both other cancerous and normal cells, influencing gene expression and modifying the biological functions of the recipient cells. As a result, exosomes play a significant role in tumor progression, survival, and metastasis. Exosomes stand out as promising biomarkers due to several advantages over other liquid biopsy targets in tumor detection and screening because they derive from nearly all cell types and are present in most biofluids, including human plasma, saliva, and urine. The concentration of exosomes is significantly higher than other components of tumors, reaching over 109 vesicles/ml. Additionally, exosomes carry a variety of bioactive molecules that facilitate intercellular communication, providing extensive information about the genetic and epigenetic changes in their cells of origin. Finally, the liposome-like membrane structures of exosomes offer inherent stability that protects them from enzymatic degradation and preserves the integrity of their molecular contents. These unique characteristics have generated considerable research interest in exosome-based liquid biopsy approaches (Lin et al. 2021).

Independently of the isolation techniques to obtain a suitable concentration of exosomes, the detection methods used for their quantification face many challenges, such as the low level of exosome markers, their non-homogeneous subpopulation, and others that complicate their use as biomarkers. Microfluidic methods are applied to improve these issues and increase the detection levels that increase sensitivity and selectivity (Xie et al. 2023). Additionally, it has advantages over conventional detection methods because of the low consumption of reagents, higher throughput, and shorter analysis times, as well as easy integration and automatization. This technique has demonstrated a suitable performance for exosome analysis and detection. The main strategies for exosome detection by microfluidic devices are based on fluorescence, colorimetric, electrochemical, and surface plasmon resonance (SPR). The application of one or another method depends on the different signal transduction pathways (Lin et al. 2021).

3.7 APPLICATIONS OF MICROFLUIDICS IN BIOMATERIALS AND BIOPHYSICS

3.7.1 Cell Culture and Tissue Engineering

Today, the importance of developing cell-based assays has become a fundamental tool for understanding the mechanisms of cell assembly in tissues and organs and for improving the comprehension of the behavior of healthy conditions or those disrupted by an agent or a disease (Syahruddin, Anggraeni, and Ana 2023). In this context, one of the most promising applications of microfluidics in biomedical sciences is based on the versatility of developing 2D and 3D cell cultures by reproducing a specific body organ. This strategy, named organ-on-a-chip, has emerged in recent years as a suitable alternative to in vivo and in vitro models because it can closely reproduce the human cell microenvironment, reduces the need for tests on animals, and permits tailored drug assays, reinforcing the concept of personalized medicine.

The search for materials applied to bone reparation is still a challenge. Bone is a dynamic and adaptable tissue that continuously engages in self-repair to maintain its functionality throughout life. This process includes modeling during childhood and adolescence, balancing the formation and resorption of bone, and ongoing remodeling. Bone has four primary cell types: osteoblasts, osteoclasts, bone lining cells, and osteocytes, which perform several essential functions, including providing structural support for the body and soft tissues, acting as the central site for hematopoiesis in adults, and participating in mineral homeostasis. Despite its remarkable regenerative capabilities, bone tissue can suffer damage due to various factors, including congenital issues, age-related changes, or acquired conditions such as trauma, inflammation, infections, or surgical interventions that result in bone loss (Benedini and Messina 2024). Some of the most prevalent bone-related issues include osteoporosis, osteosarcoma, and bone metastases. Osteoporosis is a systemic disease that poses a significant public health concern due to its high prevalence. Osteosarcoma remains the most common type of bone malignancy. Additionally, bones are the third most frequent site for metastases after the lungs and liver. When talking about bone, we cannot forget dental diseases. In dentistry, periodontitis ranks as the second most common oral health issue worldwide, following dental caries. This multifactorial chronic inflammatory disease is closely linked to biofilm dysbiosis (Benedini 2023). By understanding these aspects of bone health and disease, we can better address the challenges of maintaining skeletal integrity throughout life.

3.7.2 Microfluidics for Pathogen Detection

Microfluidics has demonstrated diverse applications in the detection of pathogens in food science. These organisms belong to a wide classification very different from each other (domain) but with the same results, causing human diseases. This group is formed by microorganisms such as protozoa, fungi, prions, viruses, and bacteria. The primary contributors to food safety issues are foodborne diseases caused by food consumption contaminated with harmful microorganisms. Notable pathogenic bacteria responsible for foodborne infections include *Salmonella* spp., *Escherichia coli* (*E. coli*), *Staphylococcus aureus* (*S. aureus*), Shiga toxin-producing *E. coli*, *Listeria monocytogenes*, *Campylobacter* spp., *Enterobacter sakazakii*, and *Clostridium botulinum*. The main indicators of foodborne pathogens in humans include food poisoning, dysentery, diarrhea, and even fatalities. The importance of detection lies in the fact that projected bacterial infections are growing and have tragic results. Given this context, it is crucial to identify and detect foodborne pathogens effectively. Traditional methods for identifying these bacteria include culturing techniques, nucleic acid-based methods, and immunological analyses such as PCR-ELISA. However, these approaches have significant drawbacks for point-of-care applications: they are often time-consuming, expensive, and require specialized equipment. Common detection methods involve labor-intensive processes such as non-selective and selective enrichment cultures, plate separation, purification stages, biochemical reactions, and serological identification. Unfortunately, conventional techniques do not meet the urgent need for rapid diagnosis and effective food safety monitoring. Recent advancements in food technology have led to the development of automated identification systems and point-of-care (POC) technologies for detecting specific bacteria. However, these methods still face limitations, including the need for culture purification and enrichment of foodborne bacteria. Additionally, food products may harbor multiple microorganisms and pathogens simultaneously; thus, a platform capable of detecting several target microorganisms is desirable. In this sense, microfluidics-based electrochemical biosensors offer a promising solution for the rapid detection of foodborne diseases, enabling the simultaneous identification of multiple pathogens more efficiently (Kulkarni, Ayachit, and Aminabhavi 2023).

Microfluidics-based electrochemical biosensors are increasingly utilized to detect foodborne pathogenic bacteria. These biosensors offer a powerful solution for identifying pathogens, including *Escherichia coli* (E. coli), *Listeria monocytogenes*, *Salmonella*, *Staphylococcus aureus* (*S. aureus*), and others. Notably, some microfluidics-based electrochemical sensors can detect pathogenic bacteria at remarkably low concentrations, with a 1 CFU/ml limit. To enhance their performance, the sensing electrodes of these biosensors are fabricated and modified using specialized nanomaterials that leverage their advantageous properties (Shaibani et al. 2018; Dutta et al. 2021). In this context, many devices using electrochemical biosensors have been reported. Lu et al. (2016) have applied amperometric detection techniques for *Listeria* determinations. The authors created a highly sensitive and specific immunosensor by immobilizing horseradish peroxidase-labeled antibodies targeting *Listeria monocytogenes* onto the surface of innovative multi-walled carbon nanotube fibers. Scanning electron microscopy was used to characterize the sensor's morphology, while its electrochemical behavior was analyzed through cyclic voltammetry. A linear relationship was established between *L. monocytogenes* concentration, ranging from 10^2 to 10^5 CFU/ml, with a detection limit of 1.07×10^2 CFU/ml. It is important to remark that the immunosensor successfully detected *L. monocytogenes* in a mixed bacterial milk sample (1.51×10^3 CFU/ml), demonstrating excellent specificity. Additionally, the immunosensor exhibited good storage stability and reproducibility, highlighting its potential for rapid detection of *L. monocytogenes*.

3.7.3 Microfluidics for Neuronal Cell and Circuit Engineering

The neural circuits in the human central nervous system (CNS) have various excitatory and inhibitory neuronal cell types, each with unique biophysical and functional characteristics. While astrocytes and oligodendrocytes also play essential support roles in the brain, neurons are the primary units responsible for information processing and forming neural circuits. The mapping of the anatomical and functional features of the brain is a complex challenge for scientists. Neuronal loss and dysfunction are linked to numerous neurological disorders. For these reasons, it becomes crucial to understand the underlying cellular and circuit-level pathophysiology to develop effective therapies (Latifi et al. 2016). Currently, the primary methods for studying brain function include using native neural circuits in their natural in vivo environment, analyzing brain slices ex vivo, and creating in vitro circuits that mimic in vivo conditions. The latter approach allows researchers to simplify the complexities of living systems and examine circuit functionality under controlled experimental settings. However, traditional in vitro neuronal cultures grown on flat surfaces often do not accurately replicate the structure and organization of in vivo circuits and typically fail to capture essential microenvironmental cues. In this regard, microfluidic devices offer a valuable tool for engineering advanced neuronal circuits that more closely resemble in vivo counterparts (Cromwell 2019).

One of the most important goals of microfluidics applied to neuronal cells is based on the study of sorting properties. Briefly, the sorting methods can be categorized into two main types: label-free and fluorescent-based techniques. In the first one, several innovative platforms have been developed for label-free sorting neuronal cell types based on their intrinsic biophysical properties. For example, viscoelastic tuning and flow rate adjustments in microchannels have successfully separated neuronal and glial cells from rat spinal cord samples. An inertial microfluidic platform has also been utilized to differentiate primary neuronal and glial cells, achieving over 80% purity by directing larger neuronal cells to the center of the channel while smaller glial cells are pushed to the sides. Additionally, spiral-shaped channels have been employed to isolate neuronal cells from large clusters. Beyond hydrodynamic methods, techniques such as dielectrophoresis (DEP) and acoustophoresis have been explored. DEP uses nonuniform electric fields to manipulate cells based on their membrane capacitance, allowing for label-free sorting regardless of cell size (Han et al. 2024). This method has successfully separated murine neurogenic and astrogenic progenitor cells by modulating alternating current frequencies. Acoustophoresis, on the other hand, utilizes ultrasound radiation forces to separate cells based on their size and compressibility. The second, fluorescence-activated cell sorting (FACS), is a widely used method that sorts cells based on specific marker expression. Cells can be labeled using genetically encoded fluorescent proteins or fluorophore-conjugated antibodies. In FACS, labeled cells are passed through a flow cell where lasers excite the fluorophores, allowing for precise identification and sorting of targeted cell types. Modern FACS systems can analyze multiple markers simultaneously but tend to be expensive and require skilled operators. Microfluidic FACS (μFACS) offers a cost-effective alternative by integrating sample pumping and sorting into lab-on-a-chip devices; however, its application in sorting neuronal cells must be fully tested.

Recent advancements include devices that sort cells based on their membrane capacitance and cytoplasmic conductivity, which can monitor changes during neuronal stem cell differentiation. Microfluidic platforms can also assess cellular responses to stimuli, enabling the collection of specific cell populations for further analysis. For instance, one study designed a microfluidic device to track olfactory neurons' responses to L-lysine, facilitating subsequent transcriptional profiling of responsive sensory neurons. These microfluidics-based techniques enhance the ability to sort and analyze various cell types efficiently, providing valuable insights into cellular behavior and differentiation processes (Habibey et al. 2022).

3.8 APPLICATIONS OF MICROFLUIDICS IN DRUG DELIVERY AND SCREENING

The high costs and lengthy, complex processes involved in bringing a new active pharmaceutical ingredient from concept to market reinforces the need for more efficient preclinical studies to evaluate the efficacy and toxicity of drug candidates. Recently, microfluidic solutions on-chip have emerged to enhance the drug screening process and bring preclinical and clinical studies closer. The use of organoids—formed by three-dimensional cell aggregates derived from primary or stem cells that replicate organ functionality—is specifically applied to drug screening. These organoids are part of microphysiological systems (MPS) that create analytical models mimicking the environmental and dynamic conditions of in vivo systems and are combined with high-throughput screening (HTS) methods that allow for the parallel processing of numerous chemical or biological tests. These advanced models facilitate the evaluation of interactions between drug candidates and receptors or targets (proteins, membranes, cells, and organs), making them a powerful and efficient tool for drug testing. Despite existing challenges and limitations, they have the potential advantage of replacing and/or reducing reliance on animal testing (Feng et al. 2023). In this context, there are three main strategies of MPS technology on-chip for the microenvironment analysis: first, in-vitro 3D tissue models using stem cells or patient-specific stem cells; second, organ-on-a-chip models; and third, the development of complex systems on a chip such as human/body or small animals-on-a-chip (Akagi et al. 2013; Musah et al. 2017), reproducing the inter-organ interactions.

3.8.1 High-Throughput Drug Screening

Recent advancements in technology have opened up new possibilities for assays utilizing cells and simple organisms. In these systems, the selected entity can be confined independently in traps or encapsulated in droplets, allowing for further processing through robotics, liquid handling devices, and sensor array processing software, collectively referred to as high-throughput screening (HTS) Yoon et al. 2024). HTS is a crucial strategy in the drug discovery process, enabling the simultaneous processing of thousands of reactions and facilitating the rapid analysis and assessment of numerous potential active pharmaceutical ingredients (De Stefano, Bianchi, and Dubini 2022). Traditional technologies often require complex and expensive equipment for liquid and plate handling, laborious data acquisition, and application of automatized processes by robotics (Feng et al. 2023). In contrast, microfluidics offers a promising alternative, as it supports HTS with benefits such as low sample consumption, straightforward handling of nanoliter (or femtoliter), and reduction of analysis costs. Microfluidics can significantly enhance the understanding of drug interactions with proteins, membranes, and cells through three primary methods: continuous flow, droplet microfluidics (Payne et al. 2020), and microfluidic microarrays in an HTS format. The first method was applied to evaluate the cytotoxicity of different xenobiotics by screening against mammalian cell lines (Wang et al. 2007), and it was used for studying the synergic effect of a systemic antifungal with more than 50,000 molecules to improve its efficacy against *Candida* sp. (Qiang et al. 2019). The second has the advantage of a high processing number of droplets associated with detection methods such as fluorescence or spectroscopic techniques and was applied to the study of virus diseases (Payne et al. 2020). The third has the advantage of producing many samples from different origins. In this context, it has shown high efficiency in managing 1,080 individual cell cultures of 20 nl each of alginate gels in one chip for screening drug candidates' toxicity, reducing the reagents consumption by more than 2,000-fold. This procedure has shown similar results to those obtained with conventional microplate assays. As a disadvantage, microarray chips require precise and in some cases expensive devices for small volumes of sample manipulation (Lee et al. 2008).

3.8.2 Drug Delivery

In the search for new drugs, whether based on the discovery of new molecules or through the repositioning of existing drugs as new lead compounds, numerous challenges arise, such as high costs and lengthy research and clinical and preclinical trial times. Once a validated hit or a new lead has been selected through high-throughput screening, it is necessary to enhance its therapeutic properties and reduce its toxicity. This is achieved through chemical modifications of the molecules. Therefore, once the molecule with appropriate pharmacological properties is identified, preclinical trials must be conducted. These trials are performed on animals (Elvira 2021). Most lead compounds do not reach the market when subjected to these tests due to pharmacokinetic issues. For this reason, the incorporation of microfluidics into drug development allows for the simulation of pharmacokinetic trials, minimizing the use of laboratory animals. The development of a platform resembles the function of human organs, and their connection has become a suitable alternative for studying drug delivery replacing animal models. In this context, different ADME (absorption, distribution, metabolism, and excretion) processes can be simulated by organ-on-a-chip methodologies (Sunildutt et al. 2023). As well as these processes, the recreation of molecular mechanisms or hormonal feedback loops by multi-organ-on-a-chip techniques has been positioned as crucial for the study of drug delivery systems. For example, Maoz et al. (2018) have designed a multi-organ on-chip system for studying the interaction of neurons and blood–brain barrier (BBB) microvascular cells. The possibility to evaluate the behavior of drugs against BBB helps to predict if the drug can reach the central nervous system after administration. Finally, different organs have been simulated, such as lung, liver, kidney, heart, gut, breast, and blood vessels, mimicking the physiological conditions for studying the behavior of drugs in an isolated organ or a multi-organ procedure (Bhise et al. 2014).

3.9 MICROFLUIDICS AND INTERNET OF THINGS (IOT)

As the name suggests, the Internet of Things (IoT), defines the interaction (communication) among devices (things) via the internet. The devices comprise equipment that generates different signal types, such as sensors (nanobiosensors), detectors, and signal transducers. IoT's keys are connectivity, automation, and real-time data processing. For that reason, the integration of IoT with microfluidics seems to be suitable for achieving more precise and automatization of biological in vitro experiments to POC procedures (Voitiuk et al. 2024). The interaction of both disciplines reduces human errors by automatization reducing time and costs, and, in the case of drug design, for example, enabling improved high-throughput screening and analysis. Additionally, for cell response experiments, the automatic regulation of flux, gradient of reagents, or their concentration by sensing the requirements and generation of an automatic correction of those parameters can improve the results of the protocols (Torres-Acosta, Lye, and Dikicioglu 2022; Sun et al. 2023). Therefore, microfluidics integrated with IoT can generate dynamic communication applicable to different fields and could be a suitable approach to overcome rapid and precise occasional drawbacks (Deliorman, Ali, and Qasaimeh 2023b).

3.10 PERSPECTIVES AND CONCLUSION

Microfluidics has opened a new door to understanding key concepts in different science fields. It is well understood that many distinct concepts, from basic science to applications such as point-of-care (POC) diagnosis devices, have been developed. Microfluidics can play a variety of roles and fulfill different activities; for that reason, a strict design must be followed. Additionally, each device, detection system, and sample processing approach have specific considerations that should not be overlooked. In this chapter, we have addressed numerous aspects related to microfluidics, from its historical perspective, conception, and theoretical considerations to the main material used for developing a lab on a chip and its wide diversity of applications. Insight into the capacity and potential application of microfluidics we thought that the next step of microfluidics is jumping outside the laboratory or research groups and getting into the lives of citizens. POC for diagnosis is one of them, but the next step is the Internet of Things (IoT), which could contribute positively to the next generation of microfluidic systems.

REFERENCES

Abdorahimzadeh, Seyedamirhosein, Feby W. Pratiwi, Seppo J. Vainio, Henrikki Liimatainen, and Caglar Elbuken. 2023. "Interplay of Electric Field and Pressure-Driven Flow Inducing Microfluidic Particle Migration." *Chemical Engineering Science* 276 (July): 118754. https://doi.org/10.1016/j.ces.2023.118754.

Adam, Hussaini, Subash C. B. Gopinath, M. K. Md Arshad, Tijjani Adam, Uda Hashim, Zaliman Sauli, Makram A. Fakhri, et al. 2023. "Integration of Microfluidic Channel on Electrochemical-Based Nanobiosensors for Monoplex and Multiplex Analyses: An Overview." *Journal of the Taiwan Institute of Chemical Engineers* 146 (May): 104814. https://doi.org/10.1016/j.jtice.2023.104814.

Agha, Abdulrahman, Waqas Waheed, Nahla Alamoodi, Bobby Mathew, Fadi Alnaimat, Eiyad Abu-Nada, Aissa Abderrahmane, and Anas Alazzam. 2022. "A Review of Cyclic Olefin Copolymer Applications in Microfluidics and Microdevices." *Macromolecular Materials and Engineering* 307 (8): 2200053.

Agustini, Deonir, Lucas Fedalto, Douglas Agustini, Luis Gustavo de Matos dos Santos, Craig E. Banks, Márcio F. Bergamini, and Luiz Humberto Marcolino-Junior. 2020. "A Low Cost, Versatile and Chromatographic Device for Microfluidic Amperometric Analyses." *Sensors and Actuators B: Chemical* 304 (February): 127117. https://doi.org/10.1016/j.snb.2019.127117.

Akagi, Jin, Khashayar Khoshmanesh, Chris J. Hall, Jonathan M. Cooper, Kathryn E. Crosier, Philip S. Crosier, and Donald Wlodkowic. 2013. "Fish on Chips: Microfluidic Living Embryo Array for Accelerated in Vivo Angiogenesis Assays." *Sensors and Actuators B: Chemical* 189 (December): 11–20. https://doi.org/10.1016/j.snb.2012.11.036.

Aleman, Julio, Tugba Kilic, Luis S. Mille, Su Ryon Shin, and Yu Shrike Zhang. 2021. "Microfluidic Integration of Regeneratable Electrochemical Affinity-Based Biosensors for Continual Monitoring of Organ-on-a-Chip Devices." *Nature Protocols* 16 (5): 2564–93. https://doi.org/10.1038/s41596-021-00511-7.

Amini, Younes, Valiyollah Ghazanfari, Mehran Heydari, Mohammad Mahdi Shadman, A. Gh Khamseh, Mohammad Hassan Khani, and Amin Hassanvand. 2023. "Computational Fluid Dynamics Simulation of Two-Phase Flow Patterns in a Serpentine Microfluidic Device." *Scientific Reports* 13 (1): 1–11. https://doi.org/10.1038/s41598-023-36672-6.

Aubry, Guillaume, Hyun Jee Lee, and Hang Lu. 2023. "Advances in Microfluidics: Technical Innovations and Applications in Diagnostics and Therapeutics." *Analytical Chemistry* 95 (1): 444–67. https://doi.org/10.1021/ACS.ANALCHEM.2C04562/ASSET/IMAGES/LARGE/AC2C04562_0008.JPEG.

Baker, Christopher A., Cindy T. Duong, Alix Grimley, and Michael G. Roper. 2009. "Recent Advances in Microfluidic Detection Systems." *Bioanalysis* 1 (5): 967–75. https://doi.org/10.4155/bio.09.86.

Balakrisnan, Bavani, Sheetal Patil, and Elisabeth Smela. 2009. "Patterning PDMS Using a Combination of Wet and Dry Etching." *Journal of Micromechanics and Microengineering* 19 (4): 047002.

Bassous, E., H. H. Taub, and L. Kuhn. 1977. "Ink Jet Printing Nozzle Arrays Etched in Silicon." *Applied Physics Letters* 31 (2): 135–37. https://doi.org/10.1063/1.89587.

Battat, Sarah, David A. Weitz, and George M. Whitesides. 2022. "An Outlook on Microfluidics: The Promise and the Challenge." *Lab on a Chip* 22 (3): 530–36. https://doi.org/10.1039/D1LC00731A.

Benedini, Luciano. 2023. "Periodontal Drug Delivery: Advances and Challenges." In *Advanced and Modern Approaches for Drug Delivery*, 751–82. Elsevier. https://doi.org/10.1016/B978-0-323-91668-4.00019-8.

Benedini, Luciano, and Paula Messina. 2024. "Advances in Polymer/Ceramic Composites for Bone Tissue Engineering Applications." In *Bioceramics: Status in Tissue Engineering and Regenerative Medicine (Part 1)*, edited by Saeid Kargozar and Francesco Baino, 231–51. Bentham Science Publishers. https://doi.org/10.2174/9789815238396124010012.

Benilov, A., V. Skryshevsky, Y. Robach, and M. Cabrera. 2008. "Micro and Nano Electrical Discharge Machining in Microfluidics and Micro Nanotechnology." *International Journal of Material Forming* 1: 1315–18.

Bhalla, Nikhil, Hung-Ju Chiang, and Amy Q. Shen. 2018. "Cell Biology at the Interface of Nanobiosensors and Microfluidics." In *Methods in Cell Biology*, 203–27. https://doi.org/10.1016/bs.mcb.2018.09.009.

Bhise, Nupura S., João Ribas, Vijayan Manoharan, Yu Shrike Zhang, Alessandro Polini, Solange Massa, Mehmet R. Dokmeci, and Ali Khademhosseini. 2014. "Organ-on-a-Chip Platforms for Studying Drug Delivery Systems." *Journal of Controlled Release* 190 (September): 82–93. https://doi.org/10.1016/j.jconrel.2014.05.004.

Brink, Floris T. G. van den, Tao Zhang, Liwei Ma, Johan Bomer, Mathieu Odijk, Wouter Olthuis, Hjalmar P. Permentier, Rainer Bischoff, and Albert van den Berg. 2016. "Electrochemical Protein Cleavage in a Microfluidic Cell with Integrated Boron Doped Diamond Electrodes." *Analytical Chemistry* 88 (18): 9190–98. https://doi.org/10.1021/acs.analchem.6b02413.

Bruus, Henrik. 2014. "Governing Equations in Microfluidics." In *Microscale Acoustofluidics*, edited by Thomas Laurell and Andreas Lenshof. The Royal Society of Chemistry. https://doi.org/10.1039/9781849737067-00001.

Carr, John A., Archana Parashar, Richard Gibson, Alan P. Robertson, Richard J. Martin, and Santosh Pandey. 2011. "A Microfluidic Platform for High-Sensitivity, Real-Time Drug Screening on C. Elegans and Parasitic Nematodes." *Lab on a Chip* 11 (14): 2385. https://doi.org/10.1039/c1lc20170k.

Chakraborty, Suman. 2005. "Dynamics of Capillary Flow of Blood into a Microfluidic Channel." *Lab on a Chip* 5 (4): 421–30. https://doi.org/10.1039/B414566F.

Chen, L., C. Yang, Y. Xiao, X. Yan, L. Hu, M. Eggersdorfer, D. Chen, D. A. Weitz, and F. Ye. 2021. "Millifluidics, Microfluidics, and Nanofluidics: Manipulating Fluids at Varying Length Scales." *Materials Today Nano* 16 (December): 100136. https://doi.org/10.1016/J.MTNANO.2021.100136.

Cheong, Jiyong, Hojeong Yu, Chang Yeol Lee, Jung-uk Lee, Hyun-Jung Choi, Jae-Hyun Lee, Hakho Lee, and Jinwoo Cheon. 2020. "Fast Detection of SARS-CoV-2 RNA via the Integration of Plasmonic Thermocycling and Fluorescence Detection in a Portable Device." *Nature Biomedical Engineering* 4 (12): 1159–67. https://doi.org/10.1038/s41551-020-00654-0.

Chenthattil Raril, Jamballi G. Manjunatha, Doddarasinakere K. Ravishankar, Santosh Fattepur, Gurumallappa Siddaraju, and Lingappa Nanjundaswamy. 2020. "Validated Electrochemical Method for Simultaneous Resolution of Tyrosine, Uric Acid, and Ascorbic Acid at Polymer Modified Nano-Composite Paste Electrode." *Surface Engineering and Applied Electrochemistry* 56 (4): 415–26. https://doi.org/10.3103/S1068375520040134.

Chien, Rean-Der. 2006. "Hot Embossing of Microfluidic Platform." *International Communications in Heat and Mass Transfer* 33 (5): 645–53.

Childs, Elizabeth H., Andrew V. Latchman, Andrew C. Lamont, Joshua D. Hubbard, and Ryan D. Sochol. 2020. "Additive Assembly for PolyJet-Based Multi-Material 3D Printed Microfluidics." *Journal of Microelectromechanical Systems* 29 (5): 1094–96.

Chrimes, Adam F., Khashayar Khoshmanesh, Paul R. Stoddart, Arnan Mitchell, and Kourosh Kalantar-zadeh. 2013. "Microfluidics and Raman Microscopy: Current Applications and Future Challenges." *Chemical Society Reviews* 42 (13): 5880. https://doi.org/10.1039/c3cs35515b.

Collins, Francis S., Michael Morgan, and Aristides Patrinos. 2003. "The Human Genome Project: Lessons from Large-Scale Biology." *Science* 300 (5617): 286–90. https://doi.org/10.1126/SCIENCE.1084564/ASSET/5D5E540C-EB40-46FB-86B2-035C61E51E8A/ASSETS/GRAPHIC/SE1431421001.JPEG.

Convery, Neil, and Nikolaj Gadegaard. 2019. "30 Years of Microfluidics." *Micro and Nano Engineering* 2 (March): 76–91. https://doi.org/10.1016/j.mne.2019.01.003.

Cooper, Jonathan M. 2022. "Challenges in Lab-on-a-Chip Technology." *Frontiers in Lab on a Chip Technologies* 1. https://www.frontiersin.org/journals/lab-on-a-chip-technologies/articles/10.3389/frlct.2022.979398.

Cromwell, Howard Casey. 2019. "Translating Striatal Activity from Brain Slice to Whole Animal Neurophysiology: A Guide for Neuroscience Research Integrating Diverse Levels of Analysis." *Journal of Neuroscience Research* 97 (12): 1528–45. https://doi.org/10.1002/jnr.24480.

Cui, Jingwen, Grzegorz Lisak, Sylwia Strzalkowska, and Johan Bobacka. 2014. "Potentiometric Sensing Utilizing Paper-Based Microfluidic Sampling." *The Analyst* 139 (9): 2133–36. https://doi.org/10.1039/C3AN02157B.

D'Elía, Noelia L., A. Noel Gravina, Luciano A. Benedini, and Paula V. Messina. 2024. "A Commentary: Harnessing Vesicles Power with New Scenes of Membrane-Based Devices for Drug Delivery." *Biocell* 48 (10): 1401–3. https://doi.org/10.32604/biocell.2024.055512.

Darhuber, Anton A., and Sandra M. Troian. 2005. "Principles of Microfluidic Actuation by Modulation of Surface Stresses." *Annual Review of Fluid Mechanics* 37 (2005): 425–55. https://doi.org/https://doi.org/10.1146/annurev.fluid.36.050802.122052.

Deliorman, Muhammedin, Dima Samer Ali, and Mohammad A. Qasaimeh. 2023a. "Next-Generation Microfluidics for Biomedical Research and Healthcare Applications." *Biomedical Engineering and Computational Biology* 14 (November): 11795972231214388–11795972231214388. https://doi.org/10.1177/11795972231214387.

Deliorman, Muhammedin, Dima Samer Ali, and Mohammad A. Qasaimeh. 2023b. "Next-Generation Microfluidics for Biomedical Research and Healthcare Applications." *Biomedical Engineering and Computational Biology* 14 (January). https://doi.org/10.1177/11795972231214387.

Deng, Bing, Heyi Wang, Zhaoyi Tan, and Yi Quan. 2019. "Microfluidic Cell Trapping for Single-Cell Analysis." *Micromachines* 10 (6): 409. https://doi.org/10.3390/mi10060409.

Dittrich, Petra S., and Andreas Manz. 2006. "Lab-on-a-Chip: Microfluidics in Drug Discovery." *Nature Reviews Drug Discovery* 5 (3): 210–18. https://doi.org/10.1038/nrd1985.

Dutta, Prakhar, Yi-Jung Lu, Han-Yu Hsieh, Tyng-Yuh Lee, Yi-Tzu Lee, Chao-Min Cheng, and Yu-Jui Fan. 2021. "Detection of *Candida albicans* Using a Manufactured Electrochemical Sensor." *Micromachines* 12 (2): 166. https://doi.org/10.3390/mi12020166.

Elvira, Katherine S. 2021. "Microfluidic Technologies for Drug Discovery and Development: Friend or Foe?" *Trends in Pharmacological Sciences* 42 (7): 518–26. https://doi.org/10.1016/j.tips.2021.04.009.

Erk, Nevin. 2003. "Differential Pulse Anodic Voltammetric Determination of Pantoprazole in Pharmaceutical Dosage Forms and Human Plasma Using Glassy Carbon Electrode." *Analytical Biochemistry* 323 (1): 48–53. https://doi.org/10.1016/j.ab.2003.08.023.

Farzbod, Ali, and Hyejin Moon. 2018. "Integration of Reconfigurable Potentiometric Electrochemical Sensors into a Digital Microfluidic Platform." *Biosensors and Bioelectronics* 106 (May): 37–42. https://doi.org/10.1016/j.bios.2018.01.048.

Feng, Jianguo, Jiri Neuzil, Andreas Manz, Ciprian Iliescu, and Pavel Neuzil. 2023. "Microfluidic Trends in Drug Screening and Drug Delivery." *TrAC Trends in Analytical Chemistry* 158 (January): 116821. https://doi.org/10.1016/j.trac.2022.116821.

Fernandes, Pedro. 2024. "Basic Principles of Microfluidics." In *Microfluidics in Pharmaceutical Sciences: Formulation, Drug Delivery, Screening, and Diagnostics*, edited by Dimitrios A Lamprou and Edward Weaver, 1–26. Cham: Springer Nature Switzerland. https://doi.org/10.1007/978-3-031-60717-2_1.

Ferreira, Mariana, Violeta Carvalho, João Ribeiro, Rui A. Lima, Senhorinha Teixeira, and Diana Pinho. 2024. "Advances in Microfluidic Systems and Numerical Modeling in Biomedical Applications: A Review." *Micromachines* 15 (7): 873. https://doi.org/10.3390/MI15070873/S1.

Figeys, Daniel, and Devanand Pinto. 2000. "Lab-on-a-Chip: A Revolution in Biological and Medical Sciences." *Analytical Chemistry* 72 (9). https://doi.org/10.1021/AC002800Y.

Fiorini, Gina S, and Daniel T Chiu. 2005. "Disposable Microfluidic Devices: Fabrication, Function, and Application." *BioTechniques* 38 (3): 429–46.

Fuchs, Stefanie, Sofia Johansson, Anders Ø. Tjell, Gabriel Werr, Torsten Mayr, and Maria Tenje. 2021. "In-Line Analysis of Organ-on-Chip Systems with Sensors: Integration, Fabrication, Challenges, and Potential." *ACS Biomaterials Science & Engineering* 7 (7): 2926–48. https://doi.org/10.1021/acsbiomaterials.0c01110.

Gai, Hongwei, Yongjun Li, and Edward S. Yeung. 2011. "Optical Detection Systems on Microfluidic Chips." In *Topics in Current Chemistry*, 171–201. https://doi.org/10.1007/128_2011_144.

Gandhiraman, Ram P., Cedric Volcke, Vladimir Gubala, Colin Doyle, Lourdes Basabe-Desmonts, Christian Dotzler, Michael F. Toney, Marcello Iacono, Robert I. Nooney, and Stephen Daniels. 2010. "High Efficiency Amine Functionalization of Cycloolefin Polymer Surfaces for Biodiagnostics." *Journal of Materials Chemistry* 20 (20): 4116–27.

Gao, Kexin, Jingji Liu, Yiqiang Fan, and Yajun Zhang. 2019. "Ultra-Low-Cost Fabrication of Polymer-Based Microfluidic Devices with Diode Laser Ablation." *Biomedical Microdevices* 21: 1–7.

Garra, J., T. Long, J. Currie, T. Schneider, R. White, and M. Paranjape. 2002. "Dry Etching of Polydimethylsiloxane for Microfluidic Systems." *Journal of Vacuum Science & Technology A: Vacuum, Surfaces, and Films* 20 (3): 975–82.

Gebreyesus, Sofani Tafesse, Asad Ali Siyal, Reta Birhanu Kitata, Eric Sheng-Wen Chen, Bayarmaa Enkhbayar, Takashi Angata, Kuo-I Lin, Yu-Ju Chen, and Hsiung-Lin Tu. 2022. "Streamlined Single-Cell Proteomics by an Integrated Microfluidic Chip and Data-Independent Acquisition Mass Spectrometry." *Nature Communications* 13 (1): 37. https://doi.org/10.1038/s41467-021-27778-4.

Glass, Nick R., Richie J. Shilton, Peggy P. Y. Chan, James R. Friend, and Leslie Y. Yeo. 2012. "Miniaturized Lab-on-a-Disc (MiniLOAD)." *Small* 8 (12): 1881–88.

Gravina, A. Noel, Noelia D'Elía, Luciano A. Benedini, and Paula Messina. 2024. "A Commentary on the Interplay of Biomaterials and Cell Adhesion: New Insights in Bone Tissue Regeneration." *Biocell* 48 (11): 1517–20. https://doi.org/10.32604/biocell.2024.055513.

Guckenberger, David J., Theodorus E. De Groot, Alwin M. D. Wan, David J. Beebe, and Edmond W. K. Young. 2015. "Micromilling: A Method for Ultra-Rapid Prototyping of Plastic Microfluidic Devices." *Lab on a Chip* 15 (11): 2364–78.

Habibey, Rouhollah, Jesús Eduardo Rojo Arias, Johannes Striebel, and Volker Busskamp. 2022. "Microfluidics for Neuronal Cell and Circuit Engineering." *Chemical Reviews* 122 (18): 14842–80. https://doi.org/10.1021/acs.chemrev.2c00212.

Hajam, Mohammad Irfan, and Mohammad Mohsin Khan. 2024. "Microfluidics: A Concise Review of the History, Principles, Design, Applications, and Future Outlook." *Biomaterials Science* 12 (2): 218–51. https://doi.org/10.1039/D3BM01463K.

Han, Wenbo, Hongyuan Zou, Yiwen Zheng, Yu Liu, Xin Wang, Wei Li, Yuqing Sun, and Hongpeng Zhang. 2024. "Model-Based Investigation of a Dielectrophoretic Microfluidic Device for the Separation of Polystyrene Particles." *Microgravity Science and Technology* 36 (6): 68. https://doi.org/10.1007/s12217-024-10150-8.

Hareesha, N., and J. G. Manjunatha. 2021. "Electro-Oxidation of Formoterol Fumarate on the Surface of Novel Poly(Thiazole Yellow-G) Layered Multi-Walled Carbon Nanotube Paste Electrode." *Scientific Reports* 11 (1): 12797. https://doi.org/10.1038/s41598-021-92099-x.

Hossan, Mohammad R., Diganta Dutta, Nazmul Islam, and Prashanta Dutta. 2018. "Electric Field Driven Pumping in Microfluidic Device." *Electrophoresis* 39 (5–6): 702–31.

Juncker, David, Heinz Schmid, and Emmanuel Delamarche. 2005. "Multipurpose Microfluidic Probe." *Nature Materials* 4 (8): 622–28. https://doi.org/10.1038/nmat1435.

Kalkandjiev, K., L. Riegger, D. Kosse, M. Welsche, L. Gutzweiler, R. Zengerle, and P. Koltay. 2011. "Microfluidics in Silicon/Polymer Technology as a Cost-Efficient Alternative to Silicon/Glass." *Journal of Micromechanics and Microengineering* 21 (2): 025008.

Kane, Ravi S., Abraham D. Stroock, Noo Li Jeon, Donald E. Ingber, and George M. Whitesides. 2002. Chapter 18 "Soft Lithography and Microfluidics." *Optical Biosensors*: 571–95. https://doi.org/10.1016/B978-044450974-1/50018-5

Kang, Yujin, Na Young Cheon, Jongjin Cha, Ayoung Kim, Hyung-il Kim, Luda Lee, Kang O. Kim, Kyubong Jo, and Ja Yil Lee. 2020. "High-throughput Single-molecule Imaging System Using Nanofabricated Trenches and Fluorescent DNA-binding Proteins." *Biotechnology and Bioengineering* 117 (6): 1640–48. https://doi.org/10.1002/bit.27331.

Karger, B. L., Y. H. Chu, and F. Foret. 1995. "Capillary Electrophoresis of Proteins and Nucleic Acids." *Annual Review of Biophysics and Biomolecular Structure* 24 (1995): 579–610. https://doi.org/10.1146/ANNUREV.BB.24.060195.003051/CITE/REFWORKS.

Kayani, Aminuddin A., Khashayar Khoshmanesh, Stephanie A. Ward, Arnan Mitchell, and Kourosh Kalantar-zadeh. 2012. "Optofluidics Incorporating Actively Controlled Micro- and Nano-Particles." *Biomicrofluidics* 6 (3). https://doi.org/10.1063/1.4736796.

Keçili, Rüstem, Sibel Büyüktiryaki, and Chaudhery Mustansar Hussain. 2020. "Micro Total Analysis Systems with Nanomaterials." In *Handbook of Nanomaterials in Analytical Chemistry: Modern Trends in Analysis*, January, 185–98. https://doi.org/10.1016/B978-0-12-816699-4.00008-6.

Kecskemeti, Adam, Cynthia Nagy, Eva Csosz, Gergo Kallo, and Attila Gaspar. 2017. "The Application of a Microfluidic Reactor Including Spontaneously Adsorbed Trypsin for Rapid Protein Digestion of Human Tear Samples." *PROTEOMICS—Clinical Applications* 11 (11–12). https://doi.org/10.1002/prca.201700055.

Kempa, Emily E., Clive A. Smith, Xin Li, Bruno Bellina, Keith Richardson, Steven Pringle, James L. Galman, Nicholas J. Turner, and Perdita E. Barran. 2020. "Coupling Droplet Microfluidics with Mass Spectrometry for Ultrahigh-Throughput Analysis of Complex Mixtures up to and above 30 Hz." *Analytical Chemistry* 92 (18): 12605–12. https://doi.org/10.1021/acs.analchem.0c02632.

Kim, Bongjoon, Wenliang Yao, Jong Won Rhie, and Honggu Chun. 2022. "Microfluidic Potentiometric Cytometry for Size-Selective Micro Dispersion Analysis." *BioChip Journal* 16 (4): 471–79. https://doi.org/10.1007/s13206-022-00083-y.

Kim, Pilnam, Keon Woo Kwon, Min Cheol Park, Sung Hoon Lee, Sun Min Kim, and Kahp Yang Suh. 2008. "Soft Lithography for Microfluidics: A Review."

Kim, Young-Min, Woo-Seung Kim, Sang-Hoon Lee, and Ju-Yeoul Baek. 2005. "Effects of Surface Roughness on the Flow Characteristics in PDMS Microchannels." In *2005 3rd IEEE/EMBS Special Topic Conference on Microtechnology in Medicine and Biology*, 292–95. https://doi.org/10.1109/MMB.2005.1548452.

Kitsara, Maria, and Jens Ducrée. 2013. "Integration of Functional Materials and Surface Modification for Polymeric Microfluidic Systems." *Journal of Micromechanics and Microengineering* 23 (3): 033001.

Kotz, Frederik, Markus Mader, Nils Dellen, Patrick Risch, Andrea Kick, Dorothea Helmer, and Bastian E. Rapp. 2020. "Fused Deposition Modeling of Microfluidic Chips in Polymethylmethacrylate." *Micromachines* 11 (9): 873.

Krishna, Rohith, Ketan Patil, Anirudha Dixit, Jilja Joseph, and Astha Pandey. 2023a. "A Novel Paper-based Microfluidic Device and UV-visible Spectroscopy Coupled Method for the Field Detection and Analysis of Seized Marijuana Samples." *Applied Research* 2 (4). https://doi.org/10.1002/appl.202200068.

Krishna, Rohith, Ketan Patil, Anirudha Dixit, Jilja Joseph, and Astha Pandey. 2023b. "A Novel Paper-based Microfluidic Device and UV-visible Spectroscopy Coupled Method for the Field Detection and Analysis of Seized Marijuana Samples." *Applied Research* 2 (4). https://doi.org/10.1002/appl.202200068.

Kuang, Cuifang, Wei Zhao, Fang Yang, and Guiren Wang. 2009. "Measuring Flow Velocity Distribution in Microchannels Using Molecular Tracers." *Microfluidics and Nanofluidics* 7 (4): 509. https://doi.org/10.1007/s10404-009-0411-z.

Kulkarni, Madhusudan B., Narasimha H. Ayachit, and Tejraj M. Aminabhavi. 2023. "Recent Advances in Microfluidics-Based Electrochemical Sensors for Foodborne Pathogen Detection." *Biosensors* 13 (2): 246. https://doi.org/10.3390/bios13020246.

Kumari, Monika, Verruchi Gupta, Natish Kumar, and Ravi Kumar Arun. 2023. "Microfluidics-Based Nanobiosensors for Healthcare Monitoring." *Molecular Biotechnology* 66 (3): 378–401. https://doi.org/10.1007/S12033-023-00760-9.

Kumari, Monika, Verruchi Gupta, Natish Kumar, and Ravi Kumar Arun. 2024. "Microfluidics-Based Nanobiosensors for Healthcare Monitoring." *Molecular Biotechnology* 66 (3): 378–401. https://doi.org/10.1007/s12033-023-00760-9.

Kuo, Alexandra P., Nirveek Bhattacharjee, Yuan-Sheng Lee, Kurt Castro, Yong Tae Kim, and Albert Folch. 2019. "High-precision Stereolithography of Biomicrofluidic Devices." *Advanced Materials Technologies* 4 (6): 1800395.

Latifi, Shahrzad, Ali Tamayol, Rouhollah Habibey, Reza Sabzevari, Cyril Kahn, David Geny, Eftekhar Eftekharpour, et al. 2016. "Natural Lecithin Promotes Neural Network Complexity and Activity." *Scientific Reports* 6 (1): 25777. https://doi.org/10.1038/srep25777.

Lazar, Iulia M., Nicholas S. Gulakowski, and Alexandru C. Lazar. 2020. "Protein and Proteome Measurements with Microfluidic Chips." *Analytical Chemistry* 92 (1): 169–82. https://doi.org/10.1021/acs.analchem.9b04711.

Lee, Moo-Yeal, R. Anand Kumar, Sumitra M. Sukumaran, Michael G. Hogg, Douglas S. Clark, and Jonathan S. Dordick. 2008. "Three-Dimensional Cellular Microarray for High-Throughput Toxicology Assays." *Proceedings of the National Academy of Sciences* 105 (1): 59–63. https://doi.org/10.1073/pnas.0708756105.

Lee, Wonhee, Warren Fon, Blake W. Axelrod, and Michael L. Roukes. 2009. "High-Sensitivity Microfluidic Calorimeters for Biological and Chemical Applications." *Proceedings of the National Academy of Sciences* 106 (36): 15225–30. https://doi.org/10.1073/pnas.0901447106.

Leung, Chak Ming, Pim de Haan, Kacey Ronaldson-Bouchard, Ge Ah Kim, Jihoon Ko, Hoon Suk Rho, Zhu Chen, et al. 2022. "A Guide to the Organ-on-a-Chip." *Nature Reviews Methods Primers* 2 (1): 1–29. https://doi.org/10.1038/s43586-022-00118-6.

Li Zheng, Xiaojian Xu, Dou Wang, and Xingyu Jiang. 2023. "Recent Advancements in Nucleic Acid Detection with Microfluidic Chip for Molecular Diagnostics." *TrAC Trends in Analytical Chemistry* 158 (January): 116871. https://doi.org/10.1016/j.trac.2022.116871.

Lin, Bingqian, Yanmei Lei, Junxia Wang, Lin Zhu, Yuqi Wu, Huimin Zhang, Lingling Wu, Peng Zhang, and Chaoyong Yang. 2021. "Microfluidic-Based Exosome Analysis for Liquid Biopsy." *Small Methods* 5 (3). https://doi.org/10.1002/smtd.202001131.

Lu, Ying, Yongling Liu, Yong Zhao, Wenjiao Li, Longbing Qiu, and Li Li. 2016. "A Novel and Disposable Enzyme-Labeled Amperometric Immunosensor Based on MWCNT Fibers for Listeria Monocytogenes Detection." *Journal of Nanomaterials* 2016: 1–8. https://doi.org/10.1155/2016/3895920.

Luo, Laan, and Yongqing He. 2020. "Magnetically Driven Microfluidics for Isolation of Circulating Tumor Cells." *Cancer Medicine* 9 (12): 4207–31. https://doi.org/10.1002/cam4.3077.

Malik, Parth, Rachna Gupta, Vibhuti Malik, and Rakesh Kumar Ameta. 2021. "Emerging Nanomaterials for Improved Biosensing." *Measurement: Sensors* 16 (August): 100050. https://doi.org/10.1016/j.measen.2021.100050.

Manjunatha, J. G., B. E. Kumara, G. P. Swamy, Mamatha, O. Gilbert, and B. S. Sherigara. 2011. "Poly(Maleic Acid) Modified Carbon Paste Electrode for Simultaneous Detection of Dopamine in the Presence of Uric Acid: A Cyclic Voltammetric Study." *Analytical and Bioanalytical Electrochemistry* 3 (2): 146–59.

Manjunatha, J. G., Pemmatte A. Pushpanjali, and Nagarajappa Hareesha. 2021. "An Overview of Recent Developments of Carbon-Based Sensors for the Analysis of Drug Molecules." *Journal of Electrochemical Science and Engineering* 11 (3): 161–77. https://doi.org/10.5599/jese.999.

Manz, A., N. Graber, and H. M. Widmer. 1990. "Miniaturized Total Chemical Analysis Systems: A Novel Concept for Chemical Sensing." *Sensors and Actuators B: Chemical* 1 (1–6): 244–48. https://doi.org/10.1016/0925-4005(90)80209-I.

Maoz, Ben M., Anna Herland, Edward A. FitzGerald, Thomas Grevesse, Charles Vidoudez, Alan R. Pacheco, Sean P. Sheehy, et al. 2018. "A Linked Organ-on-Chip Model of the Human Neurovascular Unit Reveals the Metabolic Coupling of Endothelial and Neuronal Cells." *Nature Biotechnology* 36 (9): 865–74. https://doi.org/10.1038/nbt.4226.

Martinez, Andres W., Scott T. Phillips, Manish J. Butte, and George M. Whitesides. 2007. "Patterned Paper as a Platform for Inexpensive, Low-Volume, Portable Bioassays." *Angewandte Chemie International Edition* 46 (8): 1318–20. https://doi.org/10.1002/ANIE.200603817.

McDonald, J. Cooper, David C. Duffy, Janelle R. Anderson, Daniel T. Chiu, Hongkai Wu, Olivier J. A. Schueller, and George M. Whitesides. 2000. "Fabrication of Microfluidic Systems in Poly(Dimethylsiloxane)." *Electrophoresis* 21 (1): 27–40. https://doi.org/10.1002/(SICI)1522-2683(20000101)21:1<27::AID-ELPS27>3.0.CO;2-C.

McDonald, J. Cooper, and George M. Whitesides. 2002. "Poly(Dimethylsiloxane) as a Material for Fabricating Microfluidic Devices." *Accounts of Chemical Research* 35 (7): 491–99. https://doi.org/10.1021/AR010110Q/ASSET/IMAGES/LARGE/AR010110QF00008.JPEG.

McIntyre, David, Ali Lashkaripour, Polly Fordyce, and Douglas Densmore. 2022. "Machine Learning for Microfluidic Design and Control." *Lab on a Chip* 22 (16): 2925–37. https://doi.org/10.1039/D2LC00254J.

Monnappa, Amrutha B., Jamballi G. Manjunatha, and Aarti S. Bhatt. 2020. "Design of a Sensitive and Selective Voltammetric Sensor Based on a Cationic Surfactant-Modified Carbon Paste Electrode for the Determination of Alloxan." *ACS Omega* 5 (36): 23481–90. https://doi.org/10.1021/acsomega.0c03517.

Musah, Samira, Akiko Mammoto, Thomas C. Ferrante, Sauveur S. F. Jeanty, Mariko Hirano-Kobayashi, Tadanori Mammoto, Kristen Roberts, et al. 2017. "Mature Induced-Pluripotent-Stem-Cell-Derived Human Podocytes Reconstitute Kidney Glomerular-Capillary-Wall Function on a Chip." *Nature Biomedical Engineering* 1 (5): 0069. https://doi.org/10.1038/s41551-017-0069.

Myers, Frank B., and Luke P. Lee. 2008. "Innovations in Optical Microfluidic Technologies for Point-of-Care Diagnostics." *Lab on a Chip* 8 (12): 2015. https://doi.org/10.1039/b812343h.

Nguyen, Hoang, Jeho Park, Sebyung Kang, and Moonil Kim. 2015. "Surface Plasmon Resonance: A Versatile Technique for Biosensor Applications." *Sensors* 15 (5): 10481–510. https://doi.org/10.3390/s150510481.

Nie, Jing, Jianzhong Fu, and Yong He. 2020. "Hydrogels: The Next Generation Body Materials for Microfluidic Chips?" *Small* 16 (46): 2003797. https://doi.org/https://doi.org/10.1002/smll.202003797.

Nielsen, Kurt R., and Joe P. Foley. 1998. "Capillary Electrophoresis THEORY and PRACTICE." (January 1998): 136–81. https://books.google.com/books/about/Capillary_Electrophoresis.html?hl=es&id=K-PFJ6tDpqYC.

Olanrewaju, Ayokunle, Maïwenn Beaugrand, Mohamed Yafia, and David Juncker. 2018. "Capillary Microfluidics in Microchannels: From Microfluidic Networks to Capillaric Circuits." *Lab on a Chip* 18 (16): 2323–47.

Pan, Yuchen, Kelly Karns, and Amy E. Herr. 2014. "Microfluidic Electrophoretic Mobility Shift Assays for Quantitative Biochemical Analysis." *Electrophoresis* 35 (15): 2078–90.

Payne, Emory M., Daniel A. Holland-Moritz, Shuwen Sun, and Robert T. Kennedy. 2020. "High-Throughput Screening by Droplet Microfluidics: Perspective into Key Challenges and Future Prospects." *Lab on a Chip* 20 (13): 2247–62. https://doi.org/10.1039/D0LC00347F.

Pedersen, R. H., M. Hamzah, S. Thoms, P. Roach, M. R. Alexander, and N. Gadegaard. 2010. "Electron Beam Lithography Using Plasma Polymerized Hexane as Resist." *Microelectronic Engineering* 87 (5–8): 1112–14.

Perrucci, F., V. Bertana, S. L. Marasso, Giorgio Scordo, S. Ferrero, C. F. Pirri, M. Cocuzza, A. El-Tamer, U. Hinze, and B. N. Chichkov. 2018. "Optimization of a Suspended Two Photon Polymerized Microfluidic Filtration System." *Microelectronic Engineering* 195: 95–100.

Prakash, Shashi, and Subrata Kumar. 2014. "Fabrication of Microchannels: A Review." 229 (8): 1273–88. https://doi.org/10.1177/0954405414535581.

Pushpanjali, P. A., J. G. Manjunatha, N. Hareesha, Edwin S. D'Souza, M. M. Charithra, and N. S. Prinith. 2021. "Voltammetric Analysis of Antihistamine Drug Cetirizine and Paracetamol at Poly(L-Leucine) Layered Carbon Nanotube Paste Electrode." *Surfaces and Interfaces* 24 (June): 101154. https://doi.org/10.1016/j.surfin.2021.101154.

Qi, ZhenBang, Lining Xu, Yi Xu, Junjie Zhong, Ali Abedini, Xiang Cheng, and David Sinton. 2018. "Disposable Silicon-Glass Microfluidic Devices: Precise, Robust and Cheap." *Lab on a Chip* 18 (24): 3872–80. https://doi.org/10.1039/C8LC01109E.

Qiang, Le, Jing Guo, Yingkuan Han, Jianfeng Jiang, Xiaowen Su, Hong Liu, Qingguo Qi, and Lin Han. 2019. "A Novel Anti *Candida albicans* Drug Screening System Based on High-Throughput Microfluidic Chips." *Scientific Reports* 9 (1): 8087. https://doi.org/10.1038/s41598-019-44298-w.

Ramesh, Manickam, Ravichandran Janani, Chinnaiyan Deepa, and Lakshminarasimhan Rajeshkumar. 2022. "Nanotechnology-Enabled Biosensors: A Review of Fundamentals, Design Principles, Materials, and Applications." *Biosensors* 13 (1): 40. https://doi.org/10.3390/bios13010040.

Razavi Bazaz Sajad, Ali Mashhadian, Abbas Ehsani, Suvash Chandra Saha, Timm Krüger, and Majid Ebrahimi Warkiani. 2020. "Computational Inertial Microfluidics: A Review." *Lab on a Chip* 20 (6): 1023–48. https://doi.org/10.1039/C9LC01022J.

Rebollo, René, Feras Oyoun, Yohann Corvis, Mazen M. El-Hammadi, Bruno Saubamea, Karine Andrieux, Nathalie Mignet, and Khair Alhareth. 2022. "Microfluidic Manufacturing of Liposomes: Development and Optimization by Design of Experiment and Machine Learning." *ACS Applied Materials and Interfaces* 14 (35): 39736–45. https://doi.org/10.1021/ACSAMI.2C06627/ASSET/IMAGES/LARGE/AM2C06627_0009.JPEG.

Rial, Ramón, Natalia Hassan, Zhen Liu, and Juan M. Ruso. 2021. "The Design and Green Nanofabrication of Noble Hydrogel Systems with Encapsulation of Doped Bioactive Hydroxyapatite toward Sustained Drug Delivery." *Journal of Molecular Liquids* 343 (December): 117598. https://doi.org/10.1016/j.molliq.2021.117598.

Rolfo, Christian, Marta Castiglia, David Hong, Riccardo Alessandro, Inge Mertens, Geert Baggerman, Karen Zwaenepoel, et al. 2014. "Liquid Biopsies in Lung Cancer: The New Ambrosia of Researchers." *Biochimica et Biophysica Acta (BBA) - Reviews on Cancer* 1846 (2): 539–46. https://doi.org/10.1016/j.bbcan.2014.10.001.

Roshan, Uditha, Yuchen Dai, Ajeet Singh Yadav, Samith Hettiarachchi, Amith Mudugamuwa, Jun Zhang, and Nam-Trung Nguyen. 2025. "Flexible Droplet Microfluidic Devices for Tuneable Droplet Generation." *Sensors and Actuators B: Chemical* 422 (January): 136617. https://doi.org/10.1016/j.snb.2024.136617

Salafi, Thoriq, Kerwin Kwek Zeming, and Yong Zhang. 2016. "Advancements in Microfluidics for Nanoparticle Separation." *Lab on a Chip* 17 (1): 11–33. https://doi.org/10.1039/C6LC01045H.

Shaibani, Parmiss Mojir, Hashem Etayash, Keren Jiang, Amirreza Sohrabi, Mahtab Hassanpourfard, Selvaraj Naicker, Mohtada Sadrzadeh, and Thomas Thundat. 2018. "Portable Nanofiber-Light Addressable Potentiometric Sensor for Rapid Escherichia Coli Detection in Orange Juice." *ACS Sensors* 3 (4): 815–22. https://doi.org/10.1021/acssensors.8b00063.

Shbeer, Abdullah M., and Ibrahim Ahmed Robadi. 2024. "Liquid Biopsy Holds a Promising Approach for the Early Detection of Cancer: Current Information and Future Perspectives." *Pathology - Research and Practice* 254 (February): 155082. https://doi.org/10.1016/j.prp.2023.155082.

Simon, Melinda G., Robert Lin, Jeffrey S. Fisher, and Abraham P. Lee. 2012. "A Laplace Pressure Based Microfluidic Trap for Passive Droplet Trapping and Controlled Release." *Biomicrofluidics* 6 (1): 014110. https://doi.org/10.1063/1.3687400.

Skrzypek, S., W. Ciesielski, A. Sokolowski, S. Yilmaz, and D. Kazmierczak. 2005. "Square Wave Adsorptive Stripping Voltammetric Determination of Famotidine in Urine." *Talanta* 66 (5): 1146–51. https://doi.org/10.1016/j.talanta.2005.01.017.

Somoza, Manuel, Ramón Rial, Zhen Liu, Iago F. Llovo, Rui L. Reis, Jesús Mosqueira, and Juan M. Ruso. 2023. "Microfluidic Fabrication of Gadolinium-Doped Hydroxyapatite for Theragnostic Applications." *Nanomaterials* 13 (3): 501. https://doi.org/10.3390/NANO13030501/S1.

Squires, Todd M., and Stephen R. Quake. 2005. "Microfluidics: Fluid Physics at the Nanoliter Scale." *Reviews of Modern Physics* 77 (3): 977–1026. https://doi.org/10.1103/RevModPhys.77.977.

Stefano, Paola De, Elena Bianchi, and Gabriele Dubini. 2022. "The Impact of Microfluidics in High-Throughput Drug-Screening Applications." *Biomicrofluidics* 16 (3). https://doi.org/10.1063/5.0087294.

Stone, H. A., and S. Kim. 2001. "Microfluidics: Basic Issues, Applications, and Challenges." *AIChE Journal* 47 (6): 1250–54. https://doi.org/https://doi.org/10.1002/aic.690470602.

Sun, Hao, Wantao Xie, Jin Mo, Yi Huang, and Hui Dong. 2023. "Deep Learning with Microfluidics for On-Chip Droplet Generation, Control, and Analysis." *Frontiers in Bioengineering and Biotechnology* 11 (June). https://doi.org/10.3389/fbioe.2023.1208648.

Sunildutt, Naina, Pratibha Parihar, Abdul Rahim Chethikkattuveli Salih, Sang Ho Lee, and Kyung Hyun Choi. 2023. "Revolutionizing Drug Development: Harnessing the Potential of Organ-on-Chip Technology for Disease Modeling and Drug Discovery." *Frontiers in Pharmacology* 14 (April). https://doi.org/10.3389/fphar.2023.1139229.

Syahruddin, Muhammad Hidayat, Rahmi Anggraeni, and Ika Dewi Ana. 2023. "A Microfluidic Organ-On-A-Chip: Into the Next Decade of Bone Tissue Engineering Applied in Dentistry." *Future Science OA* 9 (10). https://doi.org/10.2144/fsoa-2023-0061.

Takayama, Shuichi, Emanuele Ostuni, Philip LeDuc, Keiji Naruse, Donald E. Ingber, and George M. Whitesides. 2003. "Selective Chemical Treatment of Cellular Microdomains Using Multiple Laminar Streams." *Chemistry & Biology* 10 (2): 123–30. https://doi.org/10.1016/S1074-5521(03)00019-X.

Tao, Ran, Julien Reboud, Hamdi Torun, Glen McHale, Linzi E. Dodd, Qiang Wu, Kai Tao, et al. 2020. "Integrating Microfluidics and Biosensing on a Single Flexible Acoustic Device Using Hybrid Modes." *Lab on a Chip* 20 (5): 1002–11. https://doi.org/10.1039/C9LC01189G.

Tawfik, Dan S., and Andrew D. Griffiths. 1998. "Man-Made Cell-like Compartments for Molecular Evolution." *Nature Biotechnology* 16 (7): 652–56. https://doi.org/10.1038/nbt0798-652.

Temiz, Yuksel, Robert D. Lovchik, Govind V. Kaigala, and Emmanuel Delamarche. 2015. "Lab-on-a-Chip Devices: How to Close and Plug the Lab?" *Microelectronic Engineering* 132 (January): 156–75. https://doi.org/10.1016/J.MEE.2014.10.013.

Terry, Stephen C., John H. Herman, and James B. Angell. 1979. "A Gas Chromatographic Air Analyzer Fabricated on a Silicon Wafer." *IEEE Transactions on Electron Devices* 26 (12): 1880–86. https://doi.org/10.1109/T-ED.1979.19791.

Tesař, Václav. 2007. *Pressure-Driven Microfluidics*. Artech House Norwood.

Tice, Joshua D., Helen Song, Adam D. Lyon, and Rustem F. Ismagilov. 2003. "Formation of Droplets and Mixing in Multiphase Microfluidics at Low Values of the Reynolds and the Capillary Numbers." *Langmuir* 19 (22): 9127–33. https://doi.org/10.1021/la030090w.

Tigari, Girish, and J. G. Manjunatha. 2020. "Poly(Glutamine) Film-Coated Carbon Nanotube Paste Electrode for the Determination of Curcumin with Vanillin: An Electroanalytical Approach." *Monatshefte Für Chemie - Chemical Monthly* 151 (11): 1681–88. https://doi.org/10.1007/s00706-020-02700-8.

Torres-Acosta, Mario A., Gary J. Lye, and Duygu Dikicioglu. 2022. "Automated Liquid-Handling Operations for Robust, Resilient, and Efficient Bio-Based Laboratory Practices." *Biochemical Engineering Journal* 188 (December): 108713. https://doi.org/10.1016/j.bej.2022.108713.

Urbanski, John Paul, William Thies, Christopher Rhodes, Saman Amarasinghe, and Todd Thorsen. 2006. "Digital Microfluidics Using Soft Lithography." *Lab on a Chip* 6 (1): 96–104.

Visnapuu, Mari-Liis, Teresa Fazio, Shalom Wind, and Eric C. Greene. 2008. "Parallel Arrays of Geometric Nanowells for Assembling Curtains of DNA with Controlled Lateral Dispersion." *Langmuir* 24 (19): 11293–99. https://doi.org/10.1021/la8017634.

Voitiuk, Kateryna, Spencer T. Seiler, Mirella Pessoa de Melo, Jinghui Geng, Sebastian Hernandez, Hunter E. Schweiger, Jess L. Sevetson, et al. 2024. "A Feedback-Driven IoT Microfluidic, Electrophysiology, and Imaging Platform for Brain Organoid Studies." https://doi.org/10.1101/2024.03.15.585237.

Wang, Da-Shin, and Shih-Kang Fan. 2016. "Microfluidic Surface Plasmon Resonance Sensors: From Principles to Point-of-Care Applications." *Sensors* 16 (8): 1175. https://doi.org/10.3390/s16081175.

Wang, Joseph, Ronen Polsky, Baomin Tian, and Madhu Prakash Chatrathi. 2000. "Voltammetry on Microfluidic Chip Platforms." *Analytical Chemistry* 72 (21): 5285–89. https://doi.org/10.1021/ac000484h.

Wang, Joseph, Baomin Tian, and Eskil Sahlin. 1999. "Integrated Electrophoresis Chips/Amperometric Detection with Sputtered Gold Working Electrodes." *Analytical Chemistry* 71 (17): 3901–4. https://doi.org/10.1021/ac9904720.

Wang, Ting, Cecilia Yu, and Xing Xie. 2020. "Microfluidics for Environmental Applications." *Advances in Biochemical Engineering/Biotechnology* 179: 267–90. https://doi.org/10.1007/10_2020_128.

Wang, Zhanhui, Min-Cheol Kim, Manuel Marquez, and Todd Thorsen. 2007. "High-Density Microfluidic Arrays for Cell Cytotoxicity Analysis." *Lab on a Chip* 7 (6): 740. https://doi.org/10.1039/b618734j.

Ward, Kevin, and Z. Hugh Fan. 2015. "Mixing in Microfluidic Devices and Enhancement Methods." *Journal of Micromechanics and Microengineering* 25 (9): 094001. https://doi.org/10.1088/0960-1317/25/9/094001.

Warner, Jr. 2001. "Microelectronics: Its Unusual Origin and Personality." *IEEE Transactions on Electron Devices* 48 (11): 2457–67. https://doi.org/10.1109/16.960368.

Whitesides, George. 2018. "Microfluidics in Late Adolescence." *ArXiv Preprint ArXiv:1802.05595.*

Whitesides, George M. 2006. "The Origins and the Future of Microfluidics." *Nature* 442 (7101): 368–73. https://doi.org/10.1038/nature05058.

Wu, Qirui, Jinfeng Liu, Xiaohong Wang, Lingyan Feng, Jinbo Wu, Xiaoli Zhu, Weijia Wen, and Xiuqing Gong. 2020. "Organ-on-a-Chip: Recent Breakthroughs and Future Prospects." *BioMedical Engineering OnLine* 19 (1): 1–19. https://doi.org/10.1186/S12938-020-0752-0.

Xia, Younan, Jabez J. McClelland, Rajeev Gupta, Dong Qin, Xiao-Mei Zhao, Lydia L. Sohn, Robert J. Celotta, and George M. Whitesides. 1997. "Replica Molding Using Polymeric Materials: A Practical Step toward Nanomanufacturing." *Advanced Materials* 9 (2): 147–49. https://doi.org/https://doi.org/10.1002/adma.19970090211.

Xia, Younan, and George M Whitesides. 1998. "Soft Lithography." *Angewandte Chemie International Edition* 37 (5): 550–75. https://doi.org/https://doi.org/10.1002/(SICI)1521-3773(19980316)37:5<550::AID-ANIE550>3.0.CO;2-G.

Xie, Yujiao, Xiawei Xu, Jing Wang, Jie Lin, Yong Ren, and Aiguo Wu. 2023. "Latest Advances and Perspectives of Liquid Biopsy for Cancer Diagnostics Driven by Microfluidic On-Chip Assays." *Lab on a Chip* 23 (13): 2922–41. https://doi.org/10.1039/D2LC00837H.

Yaman, Sena, Muge Anil-Inevi, Engin Ozcivici, and H Cumhur Tekin. 2018. "Magnetic Force-Based Microfluidic Techniques for Cellular and Tissue Bioengineering." *Frontiers in Bioengineering and Biotechnology* 6:192.

Yazdan Parast, Farin, Moira K. O'Bryan, and Reza Nosrati. 2022. "Sperm Syringe: 3D Sorting Platform for Assisted Reproduction." *Advanced Materials Technologies* 7 (9). https://doi.org/10.1002/admt.202101291.

Yoon, Sunhee, Dilara Kilicarslan You, Uiechan Jeong, Mina Lee, Eunhye Kim, Tae-Joon Jeon, and Sun Min Kim. 2024. "Microfluidics in High-Throughput Drug Screening: Organ-on-a-Chip and C. Elegans-Based Innovations." *Biosensors* 14 (1): 55. https://doi.org/10.3390/bios14010055.

Yue, Jun, Floris H. Falke, Jaap C. Schouten, and T. Alexander Nijhuis. 2013. "Microreactors with Integrated UV/Vis Spectroscopic Detection for Online Process Analysis under Segmented Flow." *Lab on a Chip* 13 (24): 4855. https://doi.org/10.1039/c3lc50876e.

Yue, Shuai, Jin Fang, and Zhangrun Xu. 2022. "Advances in Droplet Microfluidics for SERS and Raman Analysis." *Biosensors and Bioelectronics* 198 (February): 113822. https://doi.org/10.1016/j.bios.2021.113822.

Zhang, Boyang, Anastasia Korolj, Benjamin Fook Lun Lai, and Milica Radisic. 2018. "Advances in Organ-on-a-Chip Engineering." *Nature Reviews Materials* 3 (8): 257–78. https://doi.org/10.1038/s41578-018-0034-7.

Zhang, Huan, Jie Yang, Rongze Sun, Songren Han, Zhaogang Yang, and Lesheng Teng. 2023. "Microfluidics for Nano-Drug Delivery Systems: From Fundamentals to Industrialization." *Acta Pharmaceutica Sinica B* 13 (8): 3277–99. https://doi.org/10.1016/J.APSB.2023.01.018.

Zhang, Peiran, Hunter Bachman, Adem Ozcelik, and Tony Jun Huang. 2020. "Acoustic Microfluidics." *Annual Review of Analytical Chemistry* 13 (1): 17–43.

Zhang, Yuyan, and Srinivas Tadigadapa. 2004. "Calorimetric Biosensors with Integrated Microfluidic Channels." *Biosensors and Bioelectronics* 19 (12): 1733–43. https://doi.org/10.1016/j.bios.2004.01.009.

Zhao, Yunduo Charles, Parham Vatankhah, Tiffany Goh, Rhys Michelis, Kiarash Kyanian, Yingqi Zhang, Zhiyong Li, and Lining Arnold Ju. 2021. "Hemodynamic Analysis for Stenosis Microfluidic Model of Thrombosis with Refined Computational Fluid Dynamics Simulation." *Scientific Reports* 11 (1): 6875. https://doi.org/10.1038/s41598-021-86310-2.

Zheng, Jiahao, Tim Cole, Yuxin Zhang, Jeeson Kim, and Shi Yang Tang. 2021. "Exploiting Machine Learning for Bestowing Intelligence to Microfluidics." *Biosensors and Bioelectronics* 194 (December): 113666. https://doi.org/10.1016/J.BIOS.2021.113666.

Zou, Zhixiang, Zhongning Guo, Kai Zhang, Yingjie Xiao, Taiman Yue, and Jiangwen Liu. 2022. "Electrochemical Discharge Machining of Microchannels in Glass Using a Non-Newtonian Fluid Electrolyte." *Journal of Materials Processing Technology* 305: 117594.

4 Analysis of Textile Dye Pollutants Using Nanosensors

Swati Shrivastava, Nimisha Jadon and Puja Tomar

4.1 INTRODUCTION

With the onset of globalization, the textile industries positively affect the economic development worldwide as the quantity of clothing produced and discarded has skyrocketed within just a few decades. China is the leading exporter of textiles, followed by the European Union, India, and the United States (Rahman, Shahriar, and Kea 2019). However, with such a huge demand, one major issue associated with textile factories is the discharge of harmful effluents, especially dyes, which are challenging to degrade. Every year, over 70 million tons of synthetic dyes are manufactured globally, with textile industries consuming more than 10,000 tons of these dyes, and about 2.8 k tones of textile dyes are disposed of as trash in textile effluents around the world (Dutta et al. 2024). Along with the dyes, textile wastes also include organic particles, aromatic compounds, chemical nutrients, pathogenic microbes, and heavy metals (Bhatia 2017).

The discharge of textile dye effluents into the environment is a serious concern as they alter water quality parameters such as biological oxygen demand, chemical oxygen demand, total suspended solids, and optical characteristics due to the presence of poisonous, carcinogenic, and mutagenic agents that significantly impact the growth of aquatic flora and fauna (Valli Nachiyar et al. 2023). These effluents degrade ecosystems, contribute significantly to environmental pollution, and adversely affect aquatic organisms, plants, and even humans, as illustrated in Figure 4.1. They often persist in entire food chains, leading to biomagnification, the process whereby higher trophic levels accumulate more contaminants as compared to their prey (Sandhya 2010; Newman 2014). These effluents are primary water contaminants and can enter bodies of water like streams, lakes, and oceans and disrupt the aquatic ecosystem as their associated color components reduce light penetration, affecting photosynthesis in aquatic plants and depleting dissolved oxygen levels, which harms aquatic flora and fauna (Lellis et al. 2019). An elevated biological oxygen demand and the presence of toxic dye compounds significantly impair the photosynthetic activity of aquatic flora, thereby disrupting the food supply for aquatic organisms and negatively impacting the entire ecosystem (Sharma, Sharma, and Soni 2021).

DOI: 10.1201/9781003536239-4

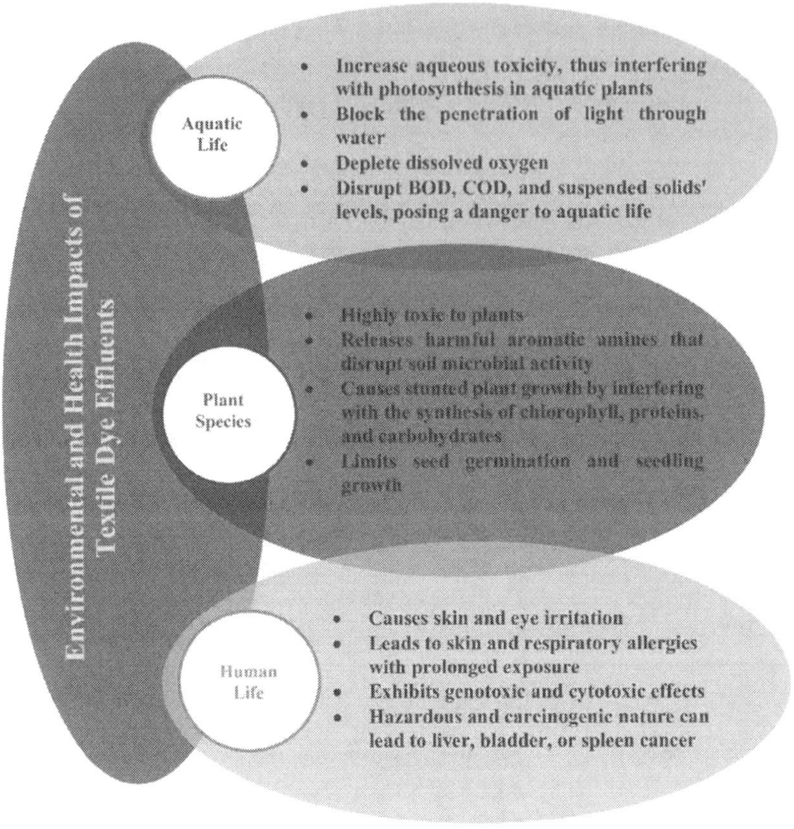

FIGURE 4.1 Harmful effects of textile dye effluents on humans, plants, and aquatic ecosystems.

Textile dye effluents negatively impact plants also; in particular when contaminated water is used for irrigation, they interfere with the synthesis of chlorophyll, protein, and carbohydrate, leading to stunted growth due to their mutagenic, teratogenic, and carcinogenic nature (Sornaly et al. 2024). Metal dyes consisting of chromium, copper, cobalt, nickel, and so on are detrimental for plant growth and reduce seed germination and seedling growth while also harming soil microbial activity (Wang et al. 2019). Azo dyes, commonly used in textiles, are especially harmful as they release aromatic amines that alter soil microbial communities and inhibit plant growth (Imran et al. 2016). The human health impact of textile dyes includes respiratory issues, skin disorders, and severe diseases such as bladder cancer and liver carcinoma (Khan and Malik 2018). Workers exposed to these dyes often suffer from allergic reactions and respiratory complications (Hunger 2003). Contaminated water and food introduce these toxins into human systems, causing nervous system dysfunction and genotoxic and cytotoxic effects (Al-Tohamy et al. 2022). Certain azo dyes have been identified as confirmed carcinogens, linked to cancers of the bladder, spleen, and liver (Alsantali et al. 2022).

The discussed toxic nature of textile dye effluents highlights the critical need for effective wastewater treatment and robust regulatory measures. Achieving the United Nations' Sustainable Development Goal 12 (SDG 12) requires collective action, and many countries have already implemented stringent environmental regulations within the textile industry to address this pressing issue (Zor 2023). Environmental monitoring plays a pivotal role, demanding highly sensitive techniques to detect and quantify dyes in water bodies and soil, as even trace levels pose significant threats to human health and aquatic ecosystems. In industrial processes, identifying specific dyes and measuring their concentrations are essential for pre-treatment, quality control, and process optimization. To tackle these analytical demands, traditional methods such as spectroscopic and chromatographic techniques coupled with advanced detectors have been widely utilized, but their application is often limited to laboratory settings due to high costs, lengthy procedures, and the requirement for skilled professionals (Gabrielle Sutanto et al. 2024; Coldibeli and Sartori 2024; Shen et al. 2024).

In contrast, nanosensor-based detection methods are rapidly emerging as innovative, cost-effective, and efficient solutions for textile dye analysis (Nalini et al. 2024). These nanosensors, typically fabricated from nanomaterials, can be integrated into sensing platforms to monitor parameters such as changes in current or refractive index. The incorporation of nanomaterials enhances the sensors' capability to provide real-time information due to their high sensitivity, selectivity, and accuracy. The primary aim of this chapter is to provide a concise overview of natural and synthetic dyes and recent research on nanosensors for detecting textile dye pollutants. In this context, a brief introduction of nanomaterials and their classification is presented first; then, the chapter explores various sensing techniques used in textile dye analysis. Subsequent sections discuss the latest advancements in nanomaterial-based sensors specifically designed to detect textile dyes. Finally, the chapter addresses the future perspective of using green nanomaterial as a sensor platform and the integration of smartphones and nanosensors to assess pollutant levels, supporting more effective remediation efforts.

4.2 TEXTILE DYES: IMPACT AS POLLUTANT

The textile dyeing industry has a rich history that spans over 4,000 years, with earlier dyes being derived from natural resources like plants, insects, and minerals (Ferreira et al. 2004). In 1856, William Henry Perkin accidentally discovered *mauveine*, the first synthetic dye, while trying to synthesize quinine, marking the birth of the modern chemical dye industry. Dyes are intricate organic molecules. Structurally, they consist of three key components: chromophores, auxochromes, and a conjugated double-bond system connecting the chromophores and auxochromes. Chromophores, which impart color to the fabric by absorbing light within the visible region of the electromagnetic spectrum, include groups such as azo, anthraquinone, nitro, nitroso, indigoid, diphenylmethane, triphenylmethane, phthalocyanine, and xanthene (Tkaczyk, Mitrowska, and Posyniak 2020). Furthermore, the fixation of these colors to materials is facilitated by auxochromes, as they enhance the affinity of the dye towards fibers and can be either acidic (COOH, SO_3, OH) or basic (NH_2, NHR, NR_2) in nature (Sahu and Poler 2024). Dyes can be classified based on their origin, as natural and synthetic dyes; toxicity associated with these dye molecules is discussed here on the basis of their origin and structure.

4.2.1 Natural Dyes

Natural dyes are abundantly found in nature and can be derived from plant extracts, microorganisms, animals, or minerals. Among these sources, plants are considered a more sustainable option due to their abundant biomass and the higher quantities of dye molecules they contain. While it is commonly believed that natural dyes are less toxic than synthetic dyes, research indicates that natural dyes can sometimes be equally toxic—or even more so—when chronic effects are considered. Given the growing concern over dye pollution and its impact on the environment, several studies are now simultaneously evaluating the toxicity of both natural and synthetic dyes to better understand and mitigate their environmental risks (Farias et al. 2023).

4.2.2 Synthetic Dyes

Synthetic dyes are man-made chemical compounds widely used to impart color to materials such as textiles, plastics, paper, and food. They are often favored over natural dyes due to their broader color spectrum, greater consistency, shade reproducibility, ease of production, and lower cost, which have led to a significant decline in the use of natural dyes. However, the chemical structure of synthetic dyes raises serious concerns about their environmental impact, as many of these compounds are non-biodegradable, toxic, and persistent in ecosystems. Further discussion explores the structurally different synthetic dyes and their impact in the environment.

4.2.2.1 Azo Dyes

Azo dyes are among the most popular and widely used dyes in the textile industry, accounting for 60%–70% of all dyestuffs in this category. Reflecting their extensive demand, the textile dye database lists 896 azo dyes (Sahu and Poler 2024; Brüschweiler and Merlot 2017). Among these dyes, methyl orange, congo red, methyl red, acid red, reactive red, and direct blue are the most commonly used azo compounds.

However, the biotransformation of azo dyes can lead to the release of aromatic amines (AAs), which are often employed as intermediates during their synthesis. The environmental release of aromatic amines is a major concern, as many of these compounds are classified as mutagens or carcinogens. In the European Union, certain azo dyes are banned due to their potential to release 22 regulated carcinogenic aromatic amines, highlighting the significant health and environmental risks associated with their use (Brüschweiler and Merlot 2017).

4.2.2.2 Indigo Dyes

Indigo, a cornerstone of denim dyeing, plays a crucial role in the textile industry but presents significant environmental and health challenges. Despite its annual demand growth of 5%, the dyeing process remains highly inefficient, with 10%–30% of the dye remaining unfixed and discharged into wastewater, along with by-products from reducing agents like sodium dithionite (Sarafpour, Alihosseini, and Bayat 2022; Chowdhury et al. 2020).

This leads to the generation of dark, highly alkaline effluents with elevated chemical oxygen demand (COD), which are particularly difficult to biodegrade due to indigo's stable aromatic structure. Such discharges pose serious threats to aquatic ecosystems and can cause health hazards in humans, including severe skin and eye irritation as well as potential carcinogenic effects (Zhang et al. 2023). To mitigate these impacts and ensure sustainability in indigo production, the implementation of effective wastewater treatment strategies is essential.

4.2.2.3 Anthraquinone Dyes

Anthraquinone dyes belong to the broader category of quinone dyes, which include benzoquinones, naphthoquinones, and anthraquinones. Among these, anthraquinone dyes are of particular interest as they are emerging as potential alternatives to synthetic dyes. These dyes are chemically stable molecules derived from anthracene, containing a quinone structure as the chromophore to which hydroxyl or amino groups can be attached (Berradi et al. 2019).

However, anthraquinone dyes pose significant environmental challenges, as their reinforced structure makes them highly resistant to natural degradation. In the textile industry, they are the second most widely used dye category, surpassed only by azo dyes. Despite their advantageous applications, certain anthraquinone dyes, such as emodin, alizarin red, and reactive blue, are known for their toxic and/or mutagenic effects, raising concerns about their environmental and health impacts (Mathew Tharayil and Chinnaiyan 2023).

4.2.2.4 Nitro and Nitroso Dyes

Nitrated and nitrosated dyes represent a relatively small and historically significant class of dyes. The oldest nitro dye, picric acid, was first synthesized in 1771 by Woulfe through the treatment of indigo with nitric acid. In 1908, Lithol Fast Yellow became the first nitro dye to be used as a pigment (Raue and Corbett 2000). Despite their age, these dyes remain in use primarily due to their low cost, which is attributed to the simplicity of their molecular structure. -Nitro dyes are defined by the presence of a nitro group ($-NO_2$) conjugated with an electron-donating group through an aromatic system. In most industrially significant nitro dyes, the nitro group and the electron-donating group, such as a hydroxyl or amino group are positioned ortho to each other. Additionally, a further donor group may also be present. In contrast, nitroso dyes consistently feature a hydroxyl group as the electron-donating group. However, exposure to these dyes beyond certain levels poses serious health risks, including liver disease and cancer in both animals and humans (Garg and Chopra 2021).

4.2.2.5 Triarylmethane Dyes

Triarylmethane dyes, renowned for their brilliant colors and high tinctorial value, encompass a wide range of dyes widely used in the textile industry, including crystal violet, methyl violet, malachite green, brilliant green, fuchsine, and Victoria blue. As their name suggests, these dyes feature a central carbon atom bonded to three aromatic rings, which act as chromophores. The sulfonation of triphenylmethane produces acid dyes, while the incorporation of OH-auxochromes or neighboring carboxyl groups leads to the formation of metallizable dyes. The aromatic structure of triarylmethane dyes imparts exceptional stability, rendering them non-biodegradable and recalcitrant in nature. Therefore, when they are released into the aquatic environment they pose significant health risks to organisms, and their uptake can cause severe health disorders, including hypertension, intermittent fever, kidney damage, and cramps. Furthermore, dyes such as crystal violet and brilliant blue are recognized for their genotoxic and carcinogenic properties, underscoring their potential to harm both ecosystems and human health (Mitra et al. 2022).

4.2.2.6 Xanthene Dyes

These dyes are widely favored as markers in various fields such as maritime accidents, molecular imaging, and bioimaging due to their tendency to fluoresce. Among the most commonly used xanthene dyes in the textile industry are fluorescein and rhodamine. Structurally, these dyes are nearly identical; however, in rhodamine, the oxygen atom in fluorescein is replaced with nitrogen. This substitution enhances the brightness and photostability of rhodamine dyes, making them particularly effective in various applications. Despite their advantages, rhodamines exhibit highly carcinogenic and neurotoxic properties and have been associated with irritation and other health disorders, as reported in previous research (Tahir and Saad 2021). Another significant xanthene dye, Eosin Y, is also used in textiles but is known for its high toxicity, poor degradability, and potential to contaminate water sources, posing serious environmental and health risks (Nandhini, Rajeshkumar, and Mythili 2019).

4.3 NANOSENSORS: SENSING OF TEXTILE DYES

A nanosensor is a highly specialized device designed to detect minute changes in physical or chemical parameters and transduce these changes into measurable signals by analyzing the collected data. The unique physical, chemical, and mechanical properties of nanomaterials grant them remarkable potential across various fields, making them essential for the development of advanced sensors. The fabrication of a nanosensor for a specific analyte involves three critical stages:

I. Synthesis and characterization of nanomaterials: The selection and synthesis of nanomaterials for specific analyte detection is a crucial first step, followed by the establishment of their structural, optical, and electrochemical properties, which are essential for sensor performance.
II. Sensor construction: This involves the integration of the tailored nanomaterials into a functional device for precise analyte detection.
III. Validation of the sensor: Rigorous testing is carried out to confirm the device's efficiency and reliability in practical applications.

Evaluating the performance of a newly developed nanosensor for a specific analyte typically entails the study of the following parameters: 1) concentration range, 2) accuracy, 3) sensitivity, 4) selectivity, 5) linearity, 6) stability, 7) repeatability, 8) reproducibility, and 9) response time. This structured approach not only ensures precise analyte detection but also underlines the transformative potential of nanosensors in cutting-edge technological applications.

Nanosensors have emerged as pivotal tools, capturing significant attention across diverse fields such as industrial, environmental, and agricultural monitoring along with healthcare diagnostics (Dhahi et al. 2024). These cutting-edge devices leverage nanomaterials with dimensions typically under 100 nanometers, harnessing their remarkable properties to achieve unparalleled sensitivity and precision. Nanomaterials encompass a broad spectrum of substances classified based on dimensions, morphology, chemical composition, and their applicable states, as illustrated in Figure 4.2. The selection of an appropriate nanomaterial for a specific application depends on several critical factors, including its size, shape, and unique physicochemical properties.

For instance, the optical characteristics of nanomaterials can vary significantly with their morphology. Gold nanourchins, with their spiky, irregular surfaces, exhibit distinct optical properties compared to spherical gold nanoparticles of the same core diameter. The uneven, spiked surface of nanourchins causes a red shift in the surface plasmon resonance peak and generates a more substantial enhancement of electromagnetic fields at the tip of the spikes. These unique attributes make them especially valuable in applications such as enhanced sensing and imaging. Similarly, nanourchins and nanoflower are almost similar nanostructures, but nanourchins have high density spikes which gives them optical properties and a larger surface area (Perwez et al. 2023). By tailoring nanomaterials to specific requirements, nanosensors continue to revolutionize technology and redefine possibilities across multidisciplinary domains. The use of sensors based on nanomaterials is an advanced analytical approach for quantifying and detecting textile dyes. In the last decade, various nanosensor-based methods have been developed to detect textile dyes from environmental samples, including electrochemical, enzyme-based, surface-enhanced Raman spectroscopy (SERS); colorimetric, fluorescence-based, surface plasmon-resonance (SPR); and quartz crystal microbalances sensors (QCM), a schematic display shown in Figure 4.3. The following section discusses the integration of nanomaterials and these sensing techniques utilized for analyzing textile dyes at trace levels.

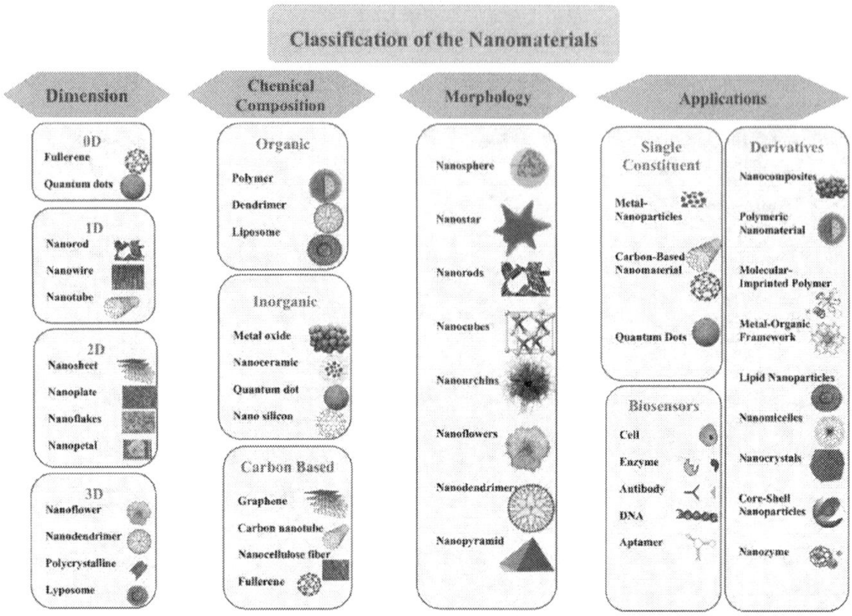

FIGURE 4.2 Classification of nanomaterials on the basis of their dimensions, morphology, chemical composition, and applications.

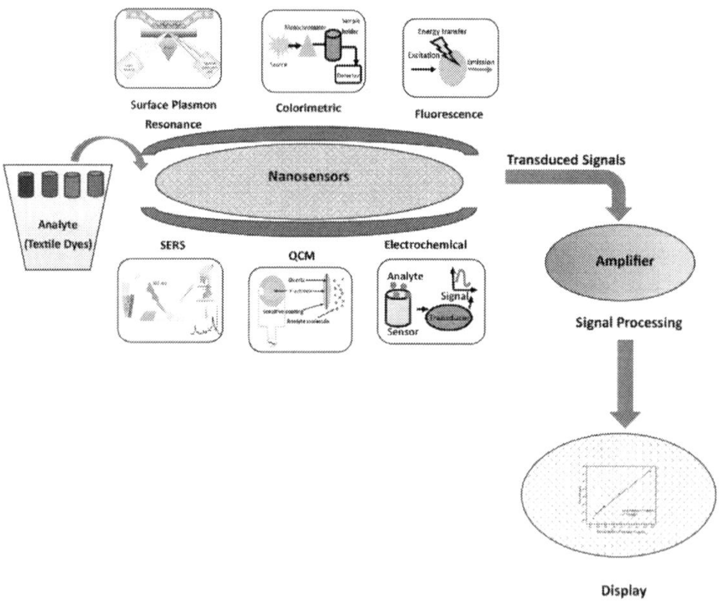

FIGURE 4.3 Schematic display of sensing and detecting a linear concentration range of textile dyes based on reported analytical techniques.

4.3.1 Electrochemical Sensors

Electrochemical nanosensors have emerged as a highly versatile and efficient method for dye detection, owing to their exceptional selectivity and sensitivity. Electrochemical analysis provides a straightforward, cost-effective approach for both the qualitative and quantitative detection of electroactive species in solutions. The rapid progress in electrochemical sensing methods is propelled by their suitability for miniaturization, ease of field portability, quick response times, and extremely low detection limits (Shrivastava, Jadon, and Jain 2016; Jadon et al. 2016; Jadon et al. 2024). In textile dye analysis, voltammetric techniques such as cyclic voltammetry (CV), differential pulse voltammetry (DPV), and square wave voltammetry (SWV) are widely employed. These methods monitor the reaction rate of electroactive compounds by measuring the current change in response to an applied potential.

Typically, electrochemical sensors are configured with either two-electrode systems, comprising a working and reference electrode, or three-electrode systems, which include an additional auxiliary electrode. The choice of working electrode depends on the target analyte, and in dye analysis, it ranges from carbon paste electrodes to enzymatic sensors, and electrodes modified with molecularly imprinted polymers (MIPs), carbon nanomaterials, metal-organic frameworks (MOFs), and metal nanoparticles (MNPs) (Liu et al. 2014; Hassan et al. 2015; Shetti et al. 2019; Amrutha et al. 2019; Khosropour, Rezaei, and Ensafi 2019; Li et al. 2021; Hayat et al. 2022; Aslam et al. 2023; Rafaqat et al. 2024; Ganash et al. 2024). The integration of nanomaterials significantly enhances the performance of these electrodes by increasing their surface roughness, surface area, and electrocatalytic activity toward electroactive compounds. As a result, electrochemical nanosensors continue to redefine precision and efficiency in dye detection, offering a sustainable and innovative alternative for real-world applications.

4.3.2 Surface-Enhanced Raman Spectroscopy (SERS) Sensors

Surface-enhanced Raman spectroscopy (SERS) is a highly valuable vibrational spectroscopy technique that has gained significant attention in environmental monitoring, food safety, chemical analysis, and biomedical research due to its ability to rapidly collect data, high sensitivity, and non-destructive detection capability. Light scattering by molecules occurs in two forms: elastic scattering (Rayleigh scattering) and inelastic scattering (Raman scattering). Raman signals, which arise from this process, are inherently weak, particularly when using visible light excitation, resulting in a low number of scattered photons available for detection. To address this limitation, surface-enhanced Raman scattering (SERS) is employed to amplify weak Raman signals using nanoscale roughened metal surfaces, typically composed of gold or silver. Molecules that are absorbed on or lie close to this enhanced field experience a significant amplification in their Raman signals. The observed Raman enhancement in SERS arises from two primary mechanisms: the electromagnetic effect involves the enhancement of the local electromagnetic field near the roughened metal surface; and the chemical effect is is based on charge transfer interactions between the absorbed molecules and the metal surface (Dinh et al. 2016). Together, these mechanisms enable SERS to provide a powerful approach for detecting and analyzing molecules with enhanced sensitivity.

However, the effectiveness of SERS is highly dependent on the affinity of analyte molecules to the metallic surface. Poor absorption can lead to suboptimal SERS signals, limiting its practical application. To address this challenge, recent advancements have explored the development of hybrid nanostructures, combining metal surfaces with other nanomaterials to enhance signal intensity and improve detection efficiency (Liu et al. 2022). This continuous evolution of SERS underscores its transformative potential as an analytical technique in textile dye analysis.

4.3.3 Fluorescence Sensors

Fluorescent nanosensors have become powerful tools for qualitative and quantitative analysis across biological, pharmaceutical, and industrial domains, offering superior sensitivity, real-time monitoring, and reduced matrix interference compared to conventional UV-visible spectroscopy. Traditionally, fluorescence sensors have relied on down-conversion luminescent materials, such as fluorescent dyes and quantum dots (QDs), which utilize UV-visible light as their excitation source. However, the advent of upconversion luminescent nanoparticles (UCNPs), a next-generation material, is revolutionizing fluorescence-based detection (Ge et al. 2020). Unlike conventional materials, UCNPs are excited by low-energy near-infrared (NIR) light and emit higher-energy visible or UV light through an anti-Stokes mechanism involving multi-photon absorption or energy transfer processes.

In the context of textile dye analysis, lanthanide-doped UCNPs have emerged as exceptional fluorescent probes. Their unique advantages—biocompatibility, suppression of background noise, and multi-spectral response—make them highly effective for precise and sensitive dye detection (Li et al. 2021a; Qi, Xu, and Lei 2024a). This cutting-edge approach highlights the transformative potential of UCNPs in advancing fluorescence sensing technologies.

4.3.4 Colorimetric Sensors

Colorimetric nanosensors are innovative tools that translate detectable interactions between nanomaterials and analytes into analytically relevant information based on changes in light intensity. They operate by measuring the intensity required to induce a color change in the nanomaterial-based solution, triggered by a chemical reaction upon the addition of the textile dye analyte (Hussain et al. 2024). This reaction renders the analyte easily identifiable under visible light, as distinct absorption bands emerge in the UV-Vis spectrum.

4.3.5 Surface Plasmon Resonance (SPR) Sensors

Surface plasmon resonance is an optical technique to study molecular interactions and the recognition of toxic and biochemical molecules in real time with high sensitivity. It measures the refractive index changes in the vicinity of thin gold layers in response to biomolecular interactions; when an analyte binds to the gold surface, the resonance angle shifts, providing a measurable signal for analyte recognition. To enhance the sensitivity and selectivity of the SPR sensors in real world applications, the gold layer can be modified by an organic, inorganic, or carbon nanomaterial (D'Agata, Bellassai, and Spoto 2024). These modifications enable enhanced performance by tailoring the sensor's surface properties for specific analytes. The working principle of SPR sensors relies on the variation in the refractive index and the progressive increase in layer thickness during interactions with analytes. In the analysis of textile dyes, the real part of the refractive index typically decreases as the thickness of the adsorption layer increases.

4.3.6 Quartz Crystal Microbalance (QCM) Sensors

The quartz crystal microbalance (QCM) sensor is a highly sensitive instrument that enables real-time measurement of mass changes at the nanoscale, driven by molecular interactions. The QCM is built around a quartz disc, which is a piezoelectric material that is made to vibrate at a specific frequency by applying a suitable voltage via the working electrode. Mass adsorption on the surface, ranging from nanograms to micrograms, causes measurable frequency shifts, with a decrease in frequency correlating to an increase in mass.

QCM sensors leverage this principle to detect subtle molecular interactions by monitoring frequency shifts induced by molecules adhering to an electrode surface coated with sensing materials. The integration of nanomaterials as a key component enhances sensor performance, offering an expansive surface area, numerous recognition sites, and a composite structure that facilitates high molecular interaction sensitivity (Chen et al. 2025).

4.4 REAL-WORLD APPLICATIONS OF NANOSENSORS FOR DETECTION OF THE TEXTILE POLLUTANT DYES

The revolutionary integration of nanomaterials with advanced sensing techniques has propelled the development of a wide array of nanosensors capable of detecting textile dyes at trace levels. Building on this progress, this section highlights the most recently developed nanosensors for the analysis of textile dyes in water effluents, as summarized in Table 4.1.

Among the various techniques, electrochemical methods have been extensively studied for their ability to detect diverse dyes across different sample matrices. These methods stand out for their cost-effectiveness and exceptional sensitivity, making them a preferred choice for many applications. Following this, surface-enhanced Raman scattering (SERS) has also demonstrated remarkable potential, offering selective and highly sensitive detection of dyes with noteworthy results. Additionally, other nanosensor technologies, including fluorometric, surface-plasmon resonance, quartz crystal microbalance systems, and colorimetric approaches,

TABLE 4.1
Nanosensors Used for the Detection of Textile Pollutant Dyes

Nanomaterials	Target Dye	Detection Technique	Limit of Detection	Concentration Range	References
Fe_2O_3 /MWCNTs-COOH/OP/CPE	Orange G orange II	Voltammetric	0.05 μM 0.1 μM	0.1–20.0 μM 0.2–50.0 μM	Liu et al. (2014)
Ibuprofen coated AuNPs	MB	Voltammetric	3.9 nM	0.01–1.1 μ M	Hassan et al. (2015)
GO	Congo red	Voltammetric	0.24 μM	0.01–0.2 μM	Shetti et al. (2019)
TX-100 modified CNT paste	Alizarin red tartrazine	Voltammetric	1 μM	2–10 μM 15–35 μM	Amrutha et al. (2019)
RuO_2/reduced-GO/carboxylated MWCNT	Acid blue 40 Acid blue 29	Voltammetric	0.01 μM 0.1 μM	0.05–50.0μM 0.5–75.0 μM	Khosropour, Rezaei, and Ensafi (2019)
MR-MIP@Co-FeC	Methyl red	Voltammetric	3.3 pM	0.01–500 nM	Li et al. (2021b)
COOH-MWCNTs /NH_2-MWCNTs	Rhodamine B	Voltammetric	0.057 nM	0.1 nM–0.05 mM	Hayat et al. (2022)
Fe-doped ZnO NPs/MWCNT	Brilliant green	Voltammetric	0.40 nM	—	Aslam et al. (2023)
MnP/CS/Carbon felt	RR 195	Voltammetric biosensor	10 μg/L	—	Rafaqat et al. (2024)
CuBDC MOF@GCE	Direct violet 1	Voltammetric	0.19 mg/L	0.99 mg/L–249 mg/L	Ganash et al. (2024)
AgNPs	MG	SERS	1–2 μg/L	1–100 μg/L	Lee et al. (2007)
MWCNT-AgNPs	MB	SERS	1 mg/L	—	Dinh et al. (2016)
Fe_3O_4/Cu_2O/Au NC	MB MG	SERS	0.01 nM	—	Huang et al. (2022)
Fe_3O_4/mTiO_2/ AgNPs	MB	SERS	1 nM	1 μM–0.1 nM	Wen and Li (2023)
Ag@Fe_3O_4	MB	SERS	0.3 nM	100–0.1 nM	Hoang et al. (2023)
Au-NPs-Co_3O_4 NWs/carbon fiber	MB	SERS	0.142 nM	10 μM–0.1 nM	Barveen et al. (2024)

(Continued)

TABLE 4.1 (CONTINUED)
Nanosensors Used for the Detection of Textile Pollutant Dyes

Nanomaterials	Target Dye	Detection Technique	Limit of Detection	Concentration Range	References
$Ni(OH)_2$/AgNPs	MB	SERS	10 nM	—	Wen et al. (2024)
Polypyrrole-CS-calcium ferrite	MB MO	SPR	0.01 mg/L	—	Sadrolhosseini et al. (2020)
NiCo-layered double hydroxide	MB MO	SPR	0.005 mg/L	—	Sadrolhosseini et al. (2023)
Nanomicelles	Methyl orange	Ratiometric	190 nM	0.1~0.8 μM	Wang et al. (2025)
LDUNs	Methyl orange	Fluorescence	1.5 μg/mL	2–75 μg/mL	Li et al. (2021a)
LDUNc	Methyl orange	Fluorescence	0.25 μM	0–8 μM	Qi, Xu, and Lei (2024b)
CS-hydroxyapatite	Congo red	Fluorescence	101.52 nM	5–50 μ M	Senthilkumar et al. (2024)
Carbon quantum dots	Methyl red Methyl orange BG MB	Fluorescence	26.4 μg/L, 214.5 μg/L, 46.2 μg/L 29.7 μg/L	0–100 μM	Vyas et al. (2024)
CS/PVA/GO/Ce-ZnO NC	Methyl orange	QCM sensor	25 ng	—	Katowah et al. (2023)
Iron nanoparticles	Methyl orange	Colorimetric	13 μM	10–350 μM	Hussain et al. (2024)

Abbreviation: Fe_2O_3/MWCNTs-COOH/OP/CPE: Fe_2O_3 nanomaterials/oxygen functionalized multi-walled carbon nanotubes/triton X-100 modified carbon paste electrode; MB: methylene blue; GO: graphene oxide nanoparticles; CNT: carbon nanotubes; MWCNT: multi-walled carbon nanotube; MIP@Co-FeC: molecular imprinted polymer and cobalt-doped iron carbide nanoparticles; COOH-MWCNTs/NH_2-MWCNTs: carboxylic and amino group functionalized multi-walled carbon nanotubes; MnP: manganese peroxidase; CS: chitosan; RR: reactive red; CuBDC MOF@GCE: copper 1,4-benzene dicarboxylate metal-organic framework-glassy carbon electrode; NC: nanocomposite; SERS: surface-enhanced Raman spectroscopy; MS: malachite green; SPR: surface plasmon resonance; MO: methylene orange; LDUNs: lanthanide-doped upconversion nanosheets; LDUNc: lanthanide-doped upconversion nanocrystals; BG: bromocresol green; CS/PVA/GO/Ce-ZnO NC: chitosan-poly(vinyl alcohol)-GO nanosheets-cerium-doped zinc oxide nanoparticles; QCM: quartz crystal micro-balance.

have shown enhanced capabilities for dye quantification. These methods leverage the unique properties of nanomaterials to achieve greater precision and reliability in dye detection.

The development of electrochemical sensors for quantification of redox active textile dyes at trace level have attracted an increasing interest in the last two decades. In 1999, Alf et al. studied the electrochemical properties of a number of azo-salicylic acid derivatives at a bare glassy carbon electrode (GCE) using cyclic and hydrodynamic voltammetry (Eriksson and Nyholm 1999). Afterwards, numerous modification of electrodes took place for sensing the textile dyes; in that context, Li et al. (Li et al. 2021b) fabricated an iron paste modified molecularly imprinted electrochemical sensor using a mixture of cobalt-doped iron carbide nanoparticles and liquid paraffin for the detection of methyl red, while Hassan et al. (Hassan et al. 2015) fabricated a nanosensor with ibuprofen coated gold nanoparticles for the analysis of methylene blue, where during the synthesis of this nanomaterial, ibuprofen was used as a reducing and capping agent.

Carbon-based nanomaterials have emerged as a versatile and widely used platform for the electrochemical sensing of textile dyes. These modifications offer several advantages, including a fast electron transfer, great electrical conductivity, simple surface renewal process and high surface area (Gabrielle Sutanto et al. 2024, Pemmatte A. Pushpanjali, Manjunatha, and Hareesha 2021). Common carbon-based nanomaterials include graphene, carbon nanotubes, and fullerenes. Carbon nanotubes are particularly popular for use in textile dye nanosensors, due to their excellent electrical conductivity. CNTs are substances with a 1D cylindrical carbon structure that can be divided into single-walled CNTs (SWCNTs) and multi-walled CNTs (MWCNTs) depending on the number of layers (Hareesha and Manjunatha 2021; Pushpanjali et al. 2021; Tigari and Manjunatha 2020). Hayat et al. developed a carboxylic and amino group functionalized MWNCNT modified glassy carbon sensor for the analysis of Rhodamine B dye (Hayat et al. 2022). This dual functionalization significantly improved the sensitivity of the sensor to detect Rhodamine B dye up to picomolar concentration. Another group of researchers combined the carboxylated MWCNT with reduced graphene oxide and ruthenium dioxide to develop a nanosensor for the simultaneous determination of acid blue 40 and acid blue 29 (Khosropour, Rezaei, and Ensafi 2019). A good recovery was achieved for the individual and simultaneous detection of these dyes in a real wastewater sample with the developed sensor. Aslam et al. developed an electrochemical sensor for the detection of brilliant green dye based on a glassy carbon electrode modified with Fe-doped ZnO nanoparticles and MWCNTs (Aslam et al. 2023).

Further, the unique properties of CNT nanomaterial were combined with the use of surfactants to achieve a higher quality sensor. The use of surfactants to enhance detection efficiency is very demanding for certain electrochemical sensing applications. A surfactant is an amphiphilic linear molecule which has a hydrophilic group at one end and a hydrophobic group at the other end, and these molecules will be spontaneously absorbed at the interface of the electrodes (Coldibeli and Sartori 2024, Monnappa, Manjunatha, and Bhatt 2020). Some studies have utilized the beneficial effects of using triton X-100 (TX-100) as a surfactant in textile dye analysis. For instance, Amrutha et al. developed a TX-100 modified carbon nanotube paste electrode for the analysis of alizarin red, an anthraquinone dye (Amrutha et al. 2019). Liu et al. developed another carbon paste electrode using Fe_2O_3 nanomaterials with oxygen functionalized MWCNT and triton X-100 for simultaneous detection of orange G and orange II (Liu et al. 2014).

Shetti et al. (2019) reported an electrochemical method for the sensitive and rapid analysis of Congo red using graphene oxide nanoparticles as the sensing film. Recently, metal-organic frameworks (MOFs) have garnered significant interest among researchers for their electrode modification. MOFs comprise organic ligands and metal ions/clusters and play a crucial role in electroanalysis due to their crystalline ordered arrangement, ultra-high porosity, remarkable thermal stability, and large surface area. Ganash and his co-workers recently designed a nanosensor based on a copper 1,4-benzene dicarboxylate metal-organic framework at a glassy carbon electrode for the analysis of direct violet 1 dye (Ganash et al. 2024). The sensor was developed by electrochemical deposition through the chronoamperometry method. The developed sensor showed a very low detection limit, high sensitivity, and a broad linear range with a good recovery for the detection of direct violet 1 in real samples. The proposed electrooxidation mechanism of direct violet 1 at the surface of the MOF modified sensor is shown in Figure 4.4 (A).

Rafaqat et al. (2024) created a biosensor for the detection and toxicological monitoring of two azo dyes, that is, reactive red 195 (RR 195) and reactive blue 221 (RB 221). Purified manganese peroxidase (MnP) enzyme was crosslinked to a carbon felt microelectrode. For the toxicology studies, the impact of these dyes was investigated on fungal growth and enzyme activity was monitored, where RR195 showed a highly toxic inhibitory effect on MnP activity and RB221 showed a benign, non-inhibitory influence. In a nutshell, the developed enzyme-based biosensor showed a low detection limit for the detection of RR 195, acceptable reproducibility, and minimum susceptibility towards interfering agents, and it maintained a very good stability even after 100 days.

Malachite green and methylene blue are two cationic dyes, widely used in industries such as the textile, paper, leather, pulp, wool, and fish farming industries for coloring and disinfecting purposes. Methylene blue is known to release potentially carcinogenic aromatic amines, such as benzidine and methylene, highlighting its toxic nature (Al-Tohamy et al. 2022). To address these concerns, a diverse array of SERS-based sensors has been developed for the analysis of these dyes in water effluents using silver/gold nanomaterials, either individually (Lee et al. 2007) or combined with other nanomaterials (Dinh et al. 2016; Huang et al. 2022; Wen and Li 2023; Hoang et al. 2023; Barveen et al. 2024; Wen et al. 2024). For instance, Dinh et al. designed a SERS sensor based on the MWCNTs-Ag nanocomposite for the detection of methylene blue and found a LOD of ~1 ppm (Dinh et al. 2016). Barveen et al. decorated Au nanoparticles onto the Co_3O_4 nanowires grafted onto the carbon fiber cloth for the detection of methylene blue with detection limit of 0.142 nM. The practical applicability of the developed sensor was also assessed by detecting the same dye in fish surfaces through the swab-sampling technique, as shown in Figure 4.4(B) (Barveen et al. 2024). Similarly, Wen et al. (2024) have developed a

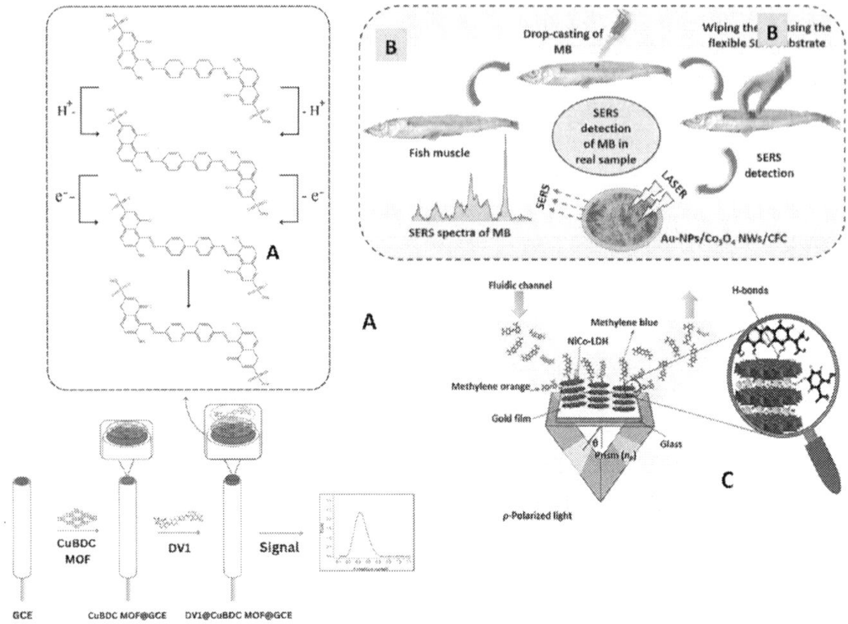

FIGURE 4.4 (A) Voltammetric detection of direct violet 1 @ CuBDC MOF@GCE with their proposed mechanism of action; (B) SERS sensor for the detection of methylene blue in fish sample; (C) surface plasmon resonance method-adsorption of methylene orange and methylene blue on the surface of NiCO-LDH sensing layer (Ganash et al. 2024; Barveen et al. 2024; Sadrolhosseini et al. 2023).

flower-like $Ni(OH)_2$@AgNPS on silicon wafer as an SERS platform for methylene blue analysis.

Magnetic nanomaterials have also gained much attention in the design of SERS sensors, due to their unique magnetic field response, enabling simple sample processing and convenient recycling. A magnetic shell acts as a physical barrier to protect the metallic core from irreversible aggregation (this occurs due to the non-uniform shape of metallic NPs and the magnetic properties of the core). They also help in regulating the chemical and optical properties of the nanocomposites that overall enhance the SERS performance. Hoang et al. (2023) utilized the beneficial properties of magnetic NPs in developing a SERS sensor based on Ag@Fe_3O_4 core-shell nanoparticles for the analysis of methylene blue. Further, researchers modified the magnetic-metal core structure with the addition of a photocatalyst such as Cu_2O, mesoporous TiO_2 to achieve high sensitivity of the SERS platform by inducing the synergistic effect that led to the enhancement of the localized electromagnetic field on the surfaces of noble metals induced by localized surface plasmon resonance and the charge transfer (Huang et al. 2022; Wen and Li 2023).

To enhance the sensitivity of surface plasmon resonance sensors for the determination of methylene blue and methylene orange dyes, researchers utilized the beneficial properties of nanocomposites and modified the gold surface with the polypyrrole-chitosan-magnetic nanoparticle (Sadrolhosseini et al. 2020) and NiCo-layered double hydroxides (Sadrolhosseini et al. 2023). The modification of the gold layer was achieved by an electrodeposition method. Adsorption of dye was reported as a function of the thickness of the modified layer and the concentration of dye in the solution. Both of the developed methods showed high sensitivity for methylene blue as compared to methylene orange. Figure 4.4 C shows the proposed mechanism of formation of H-bonds among the dye molecules and -OH groups of layered double hydroxides.

The remarkable features of nanomaterials have revolutionized fluorescence-based sensing applications in multiple forms such as nanocrystals, nanosheets, and nanomicelles (Li et al. 2021a; Qi, Xu, and Lei 2024a; Wang et al. 2025). To detect methyl orange (MO), a very popular azo dye, researchers developed the fluorescence sensors based on the upconversion luminescence systems that showed a very sensitive detection of MO with negligible background noise. These studies involve the anti-Stokes effect that converts low energy near infrared (NIR) photons to visible light, which can suppress auto fluorescence in the biological medium, potentially allowing background-free detection. For instance, Li et al. demonstrated the use of lanthanide-doped upconversion nanosheets, that is, $NaBiF_4$:Yb/Tm, to detect methyl orange dye with a very low detection limit of 1.5 μg/mL (Li et al. 2021). Moreover, another lanthanide based fluorescent probe based on $BaYF_5$:Yb/Tm core nanocrystals was studied, where $BaYF_5$ was a host, included Yb^{3+} as a sensitizer and Tm^{3+} as an activator with doping concentrations of about 20 mol% and 0.5 mol%. The upconversion intensity of the developed $BaYF_5$:20Yb/0.5 Tm core NCs was greatly enhanced 425 times by constructing a heterogeneous $BaYF_5$:20Yb/0.5Tm @$SrYF_5$ core-shell structure. The developed nanocrystals demonstrated excellent capability in selective MO sensing, with detection limits as low as 0.25 μM (Qi, Xu, and Lei 2024b).

Recently, Wang et al. (2025) developed a ratiometric nanosensor based on triplet–triplet annihilation upconversion luminescence (TTA-UCL). The study utilized platinum octaethylporphyrin (PtOEP) as a sensitizer and 9,10-diphenylanthracene (DPA) as an annihilator; both of these probes were encapsulated in nanomicelles. Under excitation of a 532 nm laser, triplet–triplet annihilation occurred, where lower energy photons absorbed by a sensitizer were converted through a series of electronic transfers into a high energy photon. In the presence of MO, UC blue light was variably quenched, while the red phosphorescence remained unchanged, establishing a connection between the relative intensity of blue/red light and the MO concentration. The developed sensor demonstrated a good linear relationship between the ratiometric signal and the MO concentration (R^2 = 0.996) with a LOD of 155 nM. Additionally, a smartphone-assisted colorimetric detection method based on the ratiometric sensor was also developed, enabling rapid and convenient detection of MO without the need for a spectrometer as depicted in Figure 4.5. Vyas et al. (2024) used the quantum dots as the down-conversion luminescent materials for the analysis of textile dyes.

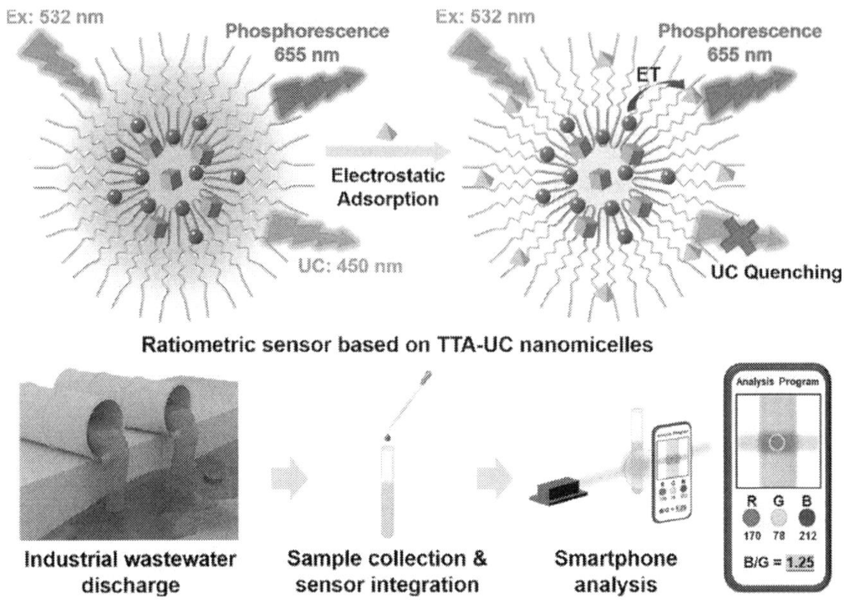

FIGURE 4.5 Ratiometric detection of methyl orange based on TTA-UC nanomicelles and smartphone integrated detection method in real water samples (Wang et al. 2025).

The demand for green nanomaterials extracted from plants that enhance sensitivity while minimizing environmental impact has been steadily increasing, marking a pivotal shift toward greener, high-performance technologies (Javaid and Mudavath 2024). Among emerging biological modifiers, polysaccharides such as chitosan, cellulose, and pectin have gained notable attention for their applications in sustainable sensor development (Jadon et al. 2024). For the analysis of congo red dye, a fluorescent sensor based on chitosan and hydroxyapatite was recently reported. Both chitosan and hydroxyapatite simultaneously were isolated from the same waste, that is, edible garden snail shells. The developed sensor showed a selective detection of congo red dye in presence of other azo dyes in water medium and obtained a linear range of detection from 10 to 100 μM with a LOD of 101.52 nM (Senthilkumar et al. 2024). Similarly, Hussain et al. (2024) synthesized nanoscaled iron nanoparticles using green and sustainable approach from *Prunus armeniaca* Gum (PAG). The developed PAG-FeNPs were used as colorimetric probe for the sensing of MO and demonstrated excellent linearity in the concentration range of 10–350 μM with a LOD of 13 μM.

Katowah et al. (2023) developed a QCM sensor where two nanocomposites, chitosan-poly(vinyl alcohol)-graphene oxide and chitosan-poly(vinyl alcohol)-graphene oxide, were prepared with cerium-doped ZnO crystals. Then they compared their sensitivity for the analysis of methyl orange dye by the QCM technique. The authors found that adding cerium-doped ZnO crystals increases the number of active free sites for the adsorption of the dye and therefore enhanced the sensing performance of the developed QCM sensor.

4.5 FUTURE PROSPECTS AND CHALLENGES

The future of textile dye detection is poised to be revolutionized by nanosensor technology, which offers unmatched sensitivity, real-time monitoring, and cost-effective solutions for environmental management. Inroads into analyzing textile dyes through nanosensor fabrication involve various sensing techniques aimed at accelerating nanosensor development. This entails improving the surface area of the sensing device using various combinations of nanomaterial, resulting in a lower limit of detection. However, challenges like sensor stability, selectivity, and reproducibility remain significant hurdles. The futuristic scope in dealing the nanosensor development should include the following:

- The materials used for nanosensor development play a prominent role based on the sensing mechanism; generally, metal nanoparticles and carbon-based nanomaterials are used for the formation of these sensors. In the future, more sophisticated recognition elements, such as aptamers, metal-organic frameworks, or molecularly imprinted polymers, tailored specifically for individual dyes or classes of dyes, need to be developed for onsite-based detection.
- To ensure sustainability, future research should explore green nanomaterials for sensor fabrication. Utilizing biological resources and waste to produce eco-friendly, cost-effective nanomaterials can mitigate environmental impact while advancing nanosensor technology.
- The translation of nanosensors from laboratory research to commercial applications is hindered by issues of stability, reproducibility, and cost. Developing standardized testing protocols and reference materials, coupled with regulatory alignment, is essential for widespread adoption. Collaboration among nanotechnology experts, academic researchers, and industry stakeholders is vital to overcome these barriers and create efficient, reliable, and portable nanosensors.
- Integrating nanosensors with smartphone technology offers the potential for rapid and smart detection methods. This approach can facilitate real-time monitoring and provide user-friendly interfaces for broader accessibility.

4.6 CONCLUSION

In conclusion, this chapter highlights the current state of textile dye pollution and showcases the remarkable advancements made by the scientific community in recent years for the sustainable detection of textile dyes using nanosensors in an economical manner. According to the literature, electrochemical methods have been extensively investigated, ranging from simple carbon-based electrodes to advanced nanocomposites that integrate graphene, carbon nanotubes, metal nanoparticles, metal-organic frameworks, and molecularly imprinted polymers. SERS and optical sensors also excel at detecting textile dyes with significant sensitivity, effectively identifying these dyes even at concentrations as low as the nanomolar or subnanomolar level. Reported nanosensors have demonstrated broad linear dynamic ranges and impressive applicability to real-world environmental samples. However, despite these notable advancements, challenges persist, particularly regarding sensor stability, reproducibility, and the ability to conduct multiplexed detection in complex matrices. The integration of nanomaterials has significantly improved sensitivity and selectivity, while molecularly imprinted polymers and metal-organic frameworks have effectively addressed the challenge of differentiating between structurally similar dyes.

The integration of advanced degradation technologies and cutting-edge nanosensors capable of detecting trace levels of toxic dyes in effluents is crucial for mitigating the environmental and health risks posed by these pollutants. Together, these advancements pave the way for a more sustainable and safer future. To achieve the United Nations' Sustainable Development Goal 12 (SDG 12), collaboration among researchers, environmentalists, industrialists, and policymakers is essential. Such partnerships will help translate these scientific findings into practical solutions in making effluent treatment plans, including the development of portable, real-time nanosensors that can be readily adopted to ensure the safe management of textile dye effluents.

REFERENCES

Alsantali, Reem I., Qandeel Alam Raja, Abdullah Y. A. Alzahrani, Amina Sadiq, Nafeesa Naeem, Ehsan Ullah Mughal, Munirah M. Al-Rooqi, Nizar El Guesmi, Ziad Moussa, and Saleh A. Ahmed. 2022. "Miscellaneous Azo Dyes: A Comprehensive Review on Recent Advancements in Biological and Industrial Applications." *Dyes and Pigments*. Elsevier Ltd. doi:10.1016/j.dyepig.2021.110050.

Al-Tohamy, Rania, Sameh S. Ali, Fanghua Li, Kamal M. Okasha, Yehia A. G. Mahmoud, Tamer Elsamahy, Haixin Jiao, Yinyi Fu, and Jianzhong Sun. 2022. "A Critical Review on the Treatment of Dye-Containing Wastewater: Ecotoxicological and Health Concerns of Textile Dyes and Possible Remediation Approaches for Environmental Safety." *Ecotoxicology and Environmental Safety*. Academic Press. doi:10.1016/j.ecoenv.2021.113160.

Amrutha, B. M., J. G. Manjunatha, A. S. Bhatt, C. Raril, and P. A. Pushpanjali. 2019. "Electrochemical Sensor for the Determination of Alizarin Red-S at Non-Ionic Surfactant Modified Carbon Nanotube Paste Electrode." *Physical Chemistry Research* 7 (3). Iranian Chemical Society: 523–533. doi:10.22036/pcr.2019.185875.1636.

Aslam, Faiza, Afzal Shah, Naimat Ullah, and Shamsa Munir. 2023. "Multiwalled Carbon Nanotube/Fe-Doped ZnO-Based Sensors for Droplet Electrochemical Detection and Degradation Monitoring of Brilliant Green." *ACS Applied Nano Materials* 6 (7). doi:10.1021/acsanm.3c00488.

Barveen, Nazar Riswana, Sathishkumar Chinnapaiyan, Bo Yi Lee, and Yu Wei Cheng. 2024. "Flexible SERS Sensor Based on the Photodecoration of Au-NPs on Co_3O_4 NWs/Carbon Fiber Cloth for the Ultrasensitive Detection of Methylene Blue in the Curved Fish Surfaces." *Analytica Chimica Acta* 1307 (June). Elsevier B.V. doi:10.1016/j.aca.2024.342629.

Berradi, Mohamed, Rachid Hsissou, Mohammed Khudhair, Mohammed Assouag, Omar Cherkaoui, Abderrahim El Bachiri, and Ahmed El Harfi. 2019. "Textile Finishing Dyes and Their Impact on Aquatic Environs." *Heliyon*. doi:10.1016/j.heliyon.2019.e02711.

Bhatia, S. C. 2017. "Pollution Control in Textile Industry." doi:10.1201/9781315148588.

Brüschweiler, Beat J., and Cédric Merlot. 2017. "Azo Dyes in Clothing Textiles Can Be Cleaved into a Series of Mutagenic Aromatic Amines Which Are Not Regulated Yet." *Regulatory Toxicology and Pharmacology* 88. doi:10.1016/j.yrtph.2017.06.012.

Chen, Qiao, Yao Yao, Jin Ao, Xingling Yu, Decheng Wu, Mengjie Shou, Rui Li, and Pingan Yang. 2025. "Advances in Quartz Crystal Microbalance Relative Humidity Sensors: A Review." *Measurement* 243: 116415. doi:https://doi.org/10.1016/j.measurement.2024.116415.

Chowdhury, Mir Ferdous, Shahjalal Khandaker, Forkan Sarker, Aminul Islam, Mir Tamzid Rahman, and Md Rabiul Awual. 2020. "Current Treatment Technologies and Mechanisms for Removal of Indigo Carmine Dyes from Wastewater: A Review." *Journal of Molecular Liquids*. doi:10.1016/j.molliq.2020.114061.

Coldibeli, Bruna, and Elen Romão Sartori. 2024. "Voltammetric Sensing of Hormones Employing Surfactant Approach." In *Surfactant Based Electrochemical Sensors and Biosensors*. doi:10.1016/B978-0-443-15493-5.00009-9.

D'Agata, Roberta, Noemi Bellassai, and Giuseppe Spoto. 2024. "Exploiting the Design of Surface Plasmon Resonance Interfaces for Better Diagnostics: A Perspective Review." *Talanta*. doi:10.1016/j.talanta.2023.125033.

Dhahi, Th S., Alaa Kamal Yousif Dafhalla, Omer Elsier Tayfour, Azath Mubarakali, Abdulrahman Saad Alqahtani, Amira Elsir Tayfour Ahmed, Mohamed Elshaikh Elobaid, Tijjani Adam, and Subash C.B. Gopinath. 2024. "Advances in Nano Sensors for Monitoring and Optimal Performance Enhancement in Photovoltaic Cells." *IScience*. doi:10.1016/j.isci.2024.109347.

Dinh, Ngo Xuan, Tran Quang Huy, Le Van Vu, Le Thi Tam, and Anh Tuan Le. 2016. "Multiwalled Carbon Nanotubes/Silver Nanocomposite as Effective SERS Platform for Detection of Methylene Blue Dye in Water." *Journal of Science: Advanced Materials and Devices* 1 (1). Elsevier B.V.: 84–89. doi:10.1016/j.jsamd.2016.04.007.

Dutta, Sohini, Satadal Adhikary, Suchandra Bhattacharya, Dipsikha Roy, Sovona Chatterjee, Aritra Chakraborty, Diyasha Banerjee, Abhratanu Ganguly, Sayantani Nanda, and Prem Rajak. 2024. "Contamination of Textile Dyes in Aquatic Environment: Adverse Impacts on Aquatic Ecosystem and Human Health, and Its Management Using Bioremediation." *Journal of Environmental Management*. doi:10.1016/j.jenvman.2024.120103.

Eriksson, Alf, and Leif Nyholm. 1999. "Comparison of the Electrochemical Properties of Some Azosalicylic Acids at Glassy Carbon Electrodes by Cyclic and Hydrodynamic Voltammetry." *Electrochimica Acta* 44 (23). doi:10.1016/S0013-4686(99)00170-X.

Farias, Natália Oliveira de, Anjaina Fernandes de Albuquerque, Amanda dos Santos, Gabriela Cristina Fonseca Almeida, Harold Stanley Freeman, Riikka Räisänen, and Gisela de Aragão Umbuzeiro. 2023. "Is Natural Better? An Ecotoxicity Study of Anthraquinone Dyes." *Chemosphere* 343. doi:10.1016/j.chemosphere.2023.140174.

Ferreira, Ester S.B., Alison N. Hulme, Hamish McNab, and Anita Quye. 2004. "The Natural Constituents of Historical Textile Dyes." *Chemical Society Reviews* 33 (6). doi:10.1039/b305697j.

Gabrielle Sutanto, Laurencia, Syarifa Sabilla, Brasstira Yuva Wardhana, Anggi Ramadani, Anis Puspita Sari, Qonita Kurnia Anjani, Wan Jeffrey Basirun, Tahta Amrillah, Ilma Amalina, and Prastika Krisma Jiwanti. 2024. "Carbon Nanomaterials as Electrochemical Sensors for Theophylline: A Review." *RSC Adv.* 14 (39). The Royal Society of Chemistry: 28927–28942. doi:10.1039/D4RA03585B.

Ganash, Aisha, Wejdan Shafei, Aisha Al-Moubaraki, Heba Alnajjar, and Nuha Alsmani. 2024. "Design an Electrochemical Sensor by Electrodeposition of CuBDC Metal-Organic Framework on a Glassy Carbon Electrode: A Promising Theoretical and Practical Approach for Enhanced Electrochemical Sensing of Direct Violet 1 Dye." *Journal of Industrial and Engineering Chemistry*. Korean Society of Industrial Engineering Chemistry. doi:10.1016/j.jiec.2024.09.037.

Garg, Arushi, and Lalita Chopra. 2021. "Dye Waste: A Significant Environmental Hazard." In *Materials Today: Proceedings*. Vol. 48. doi:10.1016/j.matpr.2021.09.003.

Ge, Jianxian, Qianyi Zhang, Jianfeng Zeng, Zi Gu, and Mingyuan Gao. 2020. "Radiolabeling Nanomaterials for Multimodality Imaging: New Insights into Nuclear Medicine and Cancer Diagnosis." *Biomaterials*. doi:10.1016/j.biomaterials.2019.119553.

Hareesha, N., and J. G. Manjunatha. 2021. "Electro-Oxidation of Formoterol Fumarate on the Surface of Novel Poly(Thiazole Yellow-G) Layered Multi-Walled Carbon Nanotube Paste Electrode." *Scientific Reports* 11 (1). doi:10.1038/s41598-021-92099-x.

Hassan, Syeda S., Ayman Nafady, Sirajuddin, Amber R. Solangi, Muhammad S. Kalhoro, Muhammad I. Abro, and Syed Tufail H. Sherazi. 2015. "Ultra-Trace Level Electrochemical Sensor for Methylene Blue Dye Based on Nafion Stabilized Ibuprofen Derived Gold Nanoparticles." *Sensors and Actuators, B: Chemical* 208 (March). Elsevier: 320–326. doi:10.1016/j.snb.2014.11.021.

Hayat, Mazhar, Afzal Shah, Muhammad Kamran Hakeem, Muhammad Irfan, Abdul Haleem, Sher Bahadar Khan, and Iltaf Shah. 2022. "A Designed Miniature Sensor for the Trace Level Detection and Degradation Studies of the Toxic Dye Rhodamine B." *RSC Advances* 12 (25). Royal Society of Chemistry: 15658–15669. doi:10.1039/d2ra01722a.

Hoang, Van Tuan, Lemma Teshome Tufa, Jaebeom Lee, Mai Quan Doan, Nguyen Ha Anh, Van Tan Tran, and Anh Tuan Le. 2023. "Tunable SERS Activity of Ag@Fe3O4 Core-Shell Nanoparticles: Effect of Shell Thickness on the Sensing Performance." *Journal of Alloys and Compounds* 933 (February). Elsevier Ltd. doi:10.1016/j.jallcom.2022.167649.

Huang, Jie, Tianxiang Zhou, Hui Zheng, Jian Wang, Yuhong Jiang, Yong Zhang, and Yang Liu. 2022. "Construction of Ternary Multifunctional Fe3O4/Cu2O/Au Nanocomposites: SERS Detection and Visible Light Driven Photocatalysis for Organic Dyes." *Ceramics International* 48 (17). Elsevier Ltd: 25413–25423. doi:10.1016/j.ceramint.2022.05.216.

Hunger, Klaus Dr. 2003. "Industrial Dyes : Chemistry, Properties, Applications." https://api.semanticscholar.org/CorpusID:92072875.

Hussain, Amjad, Manzoor Hussain, Mehboob Alam, Muhammad Yasir, Shela Mohammad Khan, Saif Ud Din, Nadeem Hussain, and Ejaz Hussain. 2024. "Biogenic PAG-FeNPs from Prunus Armeniaca (Gum) as Robust Colorimetric Sensors for Methyl Orange Detection." *Results in Chemistry*, October. Elsevier BV, 101789. doi:10.1016/j.rechem.2024.101789.

Imran, Muhammad, Muhammad Arshad, Fayek Negm, Azeem Khalid, Baby Shaharoona, Sabir Hussain, Sajid Mahmood Nadeem, and David E. Crowley. 2016. "Yeast Extract Promotes Decolorization of Azo Dyes by Stimulating Azoreductase Activity in Shewanella Sp. Strain IFN4." *Ecotoxicology and Environmental Safety* 124. doi:10.1016/j.ecoenv.2015.09.041.

Jadon, Nimisha, Rajeev Jain, Swati Sharma, and Kshiti Singh. 2016. "Recent Trends in Electrochemical Sensors for Multianalyte Detection—A Review." *Talanta*. doi:10.1016/j.talanta.2016.08.084.

Jadon, Nimisha, Puja Tomar, Swati Shrivastava, Batoul Hosseinzadeh, S. Irem Kaya, and Sibel A. Ozkan. 2024. "Monitoring of Specific Phytoestrogens by Dedicated Electrochemical Sensors: A Review." *Food Chemistry*. Elsevier Ltd. doi:10.1016/j.foodchem.2024.140404.

Javaid, Aaqib, and Shyam Lal Mudavath. 2024. "Chapter Two - Nanoparticles Derived from Plants and Their Various Applications." In *Edible Nanomaterials*, edited by Sandeep Kumar Verma and Ashok Kumar Das, vol. 107, 27–48. Comprehensive Analytical Chemistry. Elsevier. doi:https://doi.org/10.1016/bs.coac.2024.06.003.

Katowah, Dina F., Sameh H. Ismail, Ahmed H. Sadek, and Mohammed M. Rahman. 2023. "Ultrasensitive QCM Sensor Development for Monitoring of Methyl Orange Dye in Aqueous Phase Based on Novel Cross-Linked Chitosan/PVA/GO/Ce-ZnO Nanocomposite Film." *Materials Science and Engineering: B* 297 (November). Elsevier Ltd. doi:10.1016/j.mseb.2023.116804.

Khan, Sana, and Abdul Malik. 2018. "Toxicity Evaluation of Textile Effluents and Role of Native Soil Bacterium in Biodegradation of a Textile Dye." *Environmental Science and Pollution Research* 25 (5). doi:10.1007/s11356-017-0783-7.

Khosropour, Hossein, Behzad Rezaei, and Ali A. Ensafi. 2019. "A Selective and Sensitive Detection of Residual Hazardous Textile Dyes in Wastewaters Using Voltammetric Sensor." *Microchemical Journal* 146 (May). Elsevier Inc.: 548–556. doi:10.1016/j.microc.2019.01.055.

Lee, Sangyeop, Junghyun Choi, Lingxin Chen, Byungchoon Park, Jin Burm Kyong, Gi Hun Seong, Jaebum Choo, et al. 2007. "Fast and Sensitive Trace Analysis of Malachite Green Using a Surface-Enhanced Raman Microfluidic Sensor." *Analytica Chimica Acta* 590 (2): 139–144. doi:10.1016/j.aca.2007.03.049.

Lellis, Bruno, Cíntia Zani Fávaro-Polonio, João Alencar Pamphile, and Julio Cesar Polonio. 2019. "Effects of Textile Dyes on Health and the Environment and Bioremediation Potential of Living Organisms." *Biotechnology Research and Innovation* 3 (2). doi:10.1016/j.biori.2019.09.001.

Li, Xiang, Yun Wang, Panfeng Zhang, and Wanyin Ge. 2021b. "Highly Sensitivity, Selectivity Chemosensor for Methyl Orange Using Upconversion $NaBiF_4$:Yb/Tm Nanosheets." *Journal of Solid State Chemistry* 301 (September). Academic Press Inc. doi:10.1016/j.jssc.2021.122307.

Li, Xiang, Pai Yu, Yifan Feng, Qiaoran Yang, Yangguang Li, and Bang Ce Ye. 2021b. "Specific Adsorption and Highly Sensitive Detection of Methyl Red in Wastewater Using an Iron Paste Electrode Modified with a Molecularly Imprinted Polymer." *Electrochemistry Communications* 132 (November). Elsevier Inc. doi:10.1016/j.elecom.2021.107144.

Liu, Chao, Di Xu, Xuan Dong, and Qing Huang. 2022. "A Review: Research Progress of SERS-Based Sensors for Agricultural Applications." *Trends in Food Science and Technology*. doi:10.1016/j.tifs.2022.07.012.

Liu, Zhenping, Haiyun Zhai, Zuanguang Chen, Qing Zhou, Zhixian Liang, and Zihao Su. 2014. "Simultaneous Determination of Orange g and Orange Ii in Industrial Wastewater by a Novel Fe2O3/MWCNTs-COOH/OP Modified Carbon Paste Electrode." *Electrochimica Acta* 136 (August). Elsevier Ltd: 370–376. doi:10.1016/j.electacta.2014.05.090.

Mathew Tharayil, Jeevan, and Prakash Chinnaiyan. 2023. "Sustainable Waste Valorisation: Novel Areca Catechu L. Husk Biochar for Anthraquinone Dye Adsorption—Characterization, Modelling, Kinetics, and Isotherm Studies." *Results in Engineering* 20. doi:10.1016/j.rineng.2023.101624.

Mitra, Ruchira, Jing Han, Hua Xiang, Ajay Kumar, and Surojit Bera. 2022. "Microbial Removal of Triarylmethane Dyes: A Sustainable Approach for the Aquatic Ecosystem Services." In *Relationship Between Microbes and the Environment for Sustainable Ecosystem Services, Volume 2: Microbial Mitigation of Waste for Sustainable Ecosystem Services*, vol. 5. doi:10.1016/B978-0-323-89937-6.00003-6.

Monnappa, Amrutha B., Jamballi G. Manjunatha, and Aarti S. Bhatt. 2020. "Design of a Sensitive and Selective Voltammetric Sensor Based on a Cationic Surfactant-Modified Carbon Paste Electrode for the Determination of Alloxan." *ACS Omega* 5 (36). doi:10.1021/acsomega.0c03517.

Nalini, N., K. S. Thangamani, V. Geethalakshmi, and S. Nithyashree. 2024. "14 - Innovative Nanosensors for Detection of Dyes." In *Nanotechnology-Based Sensors for Detection of Environmental Pollution*, edited by Fernanda Maria Policarpo Tonelli, Arpita Roy, Munir Ozturk, and H C Ananda Murthy, 265–275. Elsevier. https://doi.org/10.1016/B978-0-443-14118-8.00014-0.

Nandhini, N. T., S. Rajeshkumar, and S. Mythili. 2019. "The Possible Mechanism of Eco-Friendly Synthesized Nanoparticles on Hazardous Dyes Degradation." *Biocatalysis and Agricultural Biotechnology*. doi:10.1016/j.bcab.2019.101138.

Newman, Michael C. 2014. *Fundamentals of Ecotoxicology: The Science of Pollution*, 4th ed. doi:10.1201/b17658.

Perwez, Mohammad, Sie Yon Lau, Danish Hussain, Shamini Anboo, Mohammad Arshad, and Pankaj Thakur. 2023. "Nanozymes and Nanoflower: Physiochemical Properties, Mechanism and Biomedical Applications." *Colloids and Surfaces B: Biointerfaces*. doi:10.1016/j.colsurfb.2023.113241.

Pushpanjali, Pemmatte A., Jamballi G. Manjunatha, and Nagarajappa Hareesha. 2021. "An Overview of Recent Developments of Carbon-Based Sensors for the Analysis of Drug Molecules." *Journal of Electrochemical Science and Engineering*. doi:10.5599/JESE.999.

Pushpanjali, P. A., J. G. Manjunatha, N. Hareesha, Edwin S. D. Souza, M. M. Charithra, and N. S. Prinith. 2021. "Voltammetric Analysis of Antihistamine Drug Cetirizine and Paracetamol at Poly(L-Leucine) Layered Carbon Nanotube Paste Electrode." *Surfaces and Interfaces* 24. doi:10.1016/j.surfin.2021.101154.

Qi, Taoxin, Shiqing Xu, and Lei Lei. 2024a. "Enhancing Tm3+ Upconversion via Engineering Interfacial Strain for Selective and Sensitive MO Sensing." *Optical Materials* 147. doi:10.1016/j.optmat.2023.114682.

Rafaqat, Shumaila, Bushra Perveen, Raqba, Warda Imran, Arshad Hussain, and Naeem Ali. 2024. "Development of Manganese Peroxidase Based Voltammetric Biosensor for Detection of Textile Azo Dyes RR 195 & RB 221." *Materials Chemistry and Physics* 312 (January). Elsevier Ltd. doi:10.1016/j.matchemphys.2023.128647.

Rahman, Redwanur, Saleh Shahriar, and Sokvibol Kea. 2019. "Determinants of Exports: A Gravity Model Analysis of the Bangladeshi Textile and Clothing Industries." *FIIB Business Review* 8 (3). doi:10.1177/2319714519872643.

Raue, Roderich, and John F. Corbett. 2000. "Nitro and Nitroso Dyes." In *Ullmann's Encyclopedia of Industrial Chemistry*. doi:10.1002/14356007.a17_383.

Sadrolhosseini, Amir Reza, Ebrahim Ghasemi, Azam Pirkarimi, Seyedeh Mehri Hamidi, and Reza Taheri Ghahrizjani. 2023. "Highly Sensitive Surface Plasmon Resonance Sensor for Detection of Methylene Blue and Methylene Orange Dyes Using NiCo-Layered Double Hydroxide." *Optics Communications* 529 (February). Elsevier B.V. doi:10.1016/j.optcom.2022.129057.

Sadrolhosseini, Amir Reza, Pooria Moozarm Nia, Suhaidi Shafie, Kamyar Shameli, and Mohd Adzir Mahdi. 2020. "Polypyrrole-Chitosan-CaFe2O4 Layer Sensor for Detection of Anionic and Cationic Dye Using Surface Plasmon Resonance." *International Journal of Polymer Science* 2020. Hindawi Limited. doi:10.1155/2020/3489509.

Sahu, Abhispa, and Jordan C. Poler. 2024. "Removal and Degradation of Dyes from Textile Industry Wastewater: Benchmarking Recent Advancements, Toxicity Assessment and Cost Analysis of Treatment Processes." *Journal of Environmental Chemical Engineering*. https://api.semanticscholar.org/CorpusID:271750069.

Sandhya, S. 2010. "Biodegradation of Azo Dyes Under Anaerobic Condition: Role of Azoreductase." doi:10.1007/698_2009_43.

Sarafpour, Mojtaba, Farzaneh Alihosseini, and Maryam Bayat. 2022. "New Laccase-Mediated System Utilized for Bio-Discoloration of Indigo-Dyed Denim Fabrics." *Applied Biochemistry and Biotechnology* 194 (12). doi:10.1007/s12010-022-04066-5.

Senthilkumar, Muthupandi, Sathees Kumar Ramachandran, Karthika Lakshmi Servarayan, Ananthappan Periyasamy, Vasantha Vairathevar Sivasamy, and Ellairaja Sundaram. 2024. "Isolation of Chitosan and Hydroxyapatite from Waste Edible White Garden Snail Shells and Their Sensing Applications towards Industrial Congo Red Dye Detection: Greener Approach." *International Journal of Biological Macromolecules* 275 (August). Elsevier B.V. doi:10.1016/j.ijbiomac.2024.133483.

Sharma, Jyotshana, Shubhangani Sharma, and Vineet Soni. 2021. "Classification and Impact of Synthetic Textile Dyes on Aquatic Flora: A Review." *Regional Studies in Marine Science*. doi:10.1016/j.rsma.2021.101802.

Shen, Ying, Shichao Zhao, Yanfei Lv, Fei Chen, and Li Fu. 2024. "Acid Red Dyes and the Role of Electrochemical Sensors in Their Determination." *Microchemical Journal* 207: 111705. doi:https://doi.org/10.1016/j.microc.2024.111705.

Shetti, Nagaraj P., Shweta J. Malode, Ramesh S. Malladi, Shachindra L. Nargund, Shyam S. Shukla, and Tejraj M. Aminabhavi. 2019. "Electrochemical Detection and Degradation of Textile Dye Congo Red at Graphene Oxide Modified Electrode." *Microchemical Journal* 146 (May). Elsevier Inc.: 387–392. doi:10.1016/j.microc.2019.01.033.

Shrivastava, Swati, Nimisha Jadon, and Rajeev Jain. 2016. "Next-Generation Polymer Nanocomposite-Based Electrochemical Sensors and Biosensors: A Review." *TrAC - Trends in Analytical Chemistry*. Elsevier B.V. doi:10.1016/j.trac.2016.04.005.

Sornaly, Hafsa Hena, Sabbir Ahmed, Kanis Fatema Titin, Md Nahidul Islam, Anzana Parvin, Md Azizul Islam, Hossain Md Faruquee, et al. 2024. "The Utility of Bioremediation Approach over Physicochemical Methods to Detoxify Dyes Discharges from Textile Effluents: A Comprehensive Review Study." *Sustainable Chemistry and Pharmacy*. doi:10.1016/j.scp.2024.101538.

Tahir, Hajira, and Muhammad Saad. 2021. "Using Dyes to Evaluate the Photocatalytic Activity." In *Interface Science and Technology*, vol. 32. doi:10.1016/B978-0-12-818806-4.00005-X.

Tigari, Girish, and J. G. Manjunatha. 2020. "Poly(Glutamine) Film-Coated Carbon Nanotube Paste Electrode for the Determination of Curcumin with Vanillin: An Electroanalytical Approach." *Monatshefte Fur Chemie* 151 (11). doi:10.1007/s00706-020-02700-8.

Tkaczyk, Angelika, Kamila Mitrowska, and Andrzej Posyniak. 2020. "Synthetic Organic Dyes as Contaminants of the Aquatic Environment and Their Implications for Ecosystems: A Review." *Science of the Total Environment*. doi:10.1016/j.scitotenv.2020.137222.

Valli Nachiyar, C., A. D. Rakshi, S. Sandhya, N. Britlin Deva Jebasta, and Jayshree Nellore. 2023. "Developments in Treatment Technologies of Dye-Containing Effluent: A Review." *Case Studies in Chemical and Environmental Engineering* 7. doi:10.1016/j.cscee.2023.100339.

Vyas, Tanmay, Hritik Kumar, Sandeep Choudhary, and Abhijeet Joshi. 2024. "Carbon Quantum Dot (CQD)-Dithizone-Based Thin-Film Chemical Sensors for the Specific Detection of Lead Ions in Water Resources." *Environmental Science: Water Research & Technology* 10 (11). The Royal Society of Chemistry: 2858–2868. doi:10.1039/D4EW00452C.

Wang, Hui, Yunchao Zhong, Huimin Yu, Paolo Aprea, and Shiyou Hao. 2019. "High-Efficiency Adsorption for Acid Dyes over CeO2·xH2O Synthesized by a Facile Method." *Journal of Alloys and Compounds* 776. doi:10.1016/j.jallcom.2018.10.228.

Wang, Yuyang, Shuoran Chen, Zongyi Xu, Xiwei Bao, Yuxiang Dong, Lin Li, Yizhong Shi, Shigang Wan, and Changqing Ye. 2025. "Smartphone-Assisted Detection of Trace Methyl Orange in Water by Ratiometric Nanosensors Based on down/up-Conversion Luminescence." *Spectrochimica Acta - Part A: Molecular and Biomolecular Spectroscopy* 324 (January). Elsevier B.V. doi:10.1016/j.saa.2024.125039.

Wen, Xiaojun, Huan Cheng, Weilong Zhang, Lijun You, and Jumei Li. 2024. "Multifunctional Ni(OH)2/Ag Composites for Ultrasensitive SERS Detection and Efficient Photocatalytic Degradation of Ciprofloxacin and Methylene Blue." *Talanta* 266 (January). Elsevier B.V. doi:10.1016/j.talanta.2023.125140.

Wen, Xiaojun, and Jumei Li. 2023. "Synthesis of Recyclable SERS Platform Based on Magnetic Fe3O4/MTiO2/Ag Composites for Photodegradation and Identification of Organic Pollutants in Different Water Samples." *Vibrational Spectroscopy* 129 (November). Elsevier B.V. doi:10.1016/j.vibspec.2023.103600.

Zhang, Wei, Mengdi Zhang, Jiming Yao, and Jiajie Long. 2023. "Industrial Indigo Dyeing Wastewater Purification: Effective COD Removal with Peroxi-AC Electrocoagulation System." *Arabian Journal of Chemistry* 16 (4). doi:10.1016/j.arabjc.2023.104607.

Zor, Shutter. 2023. "Conservation or Revolution? The Sustainable Transition of Textile and Apparel Firms under the Environmental Regulation: Evidence from China." *Journal of Cleaner Production*. https://api.semanticscholar.org/CorpusID:254317483.

5 Optical Biosensors
Fundamentals and Advances

Maya Devi, Swetapadma Praharaj and Dibyaranjan Rout

5.1 INTRODUCTION

Biosensors are self-contained integrated devices that use a biological recognition element in direct-spatial contact with a transductional element to deliver specific qualitative or semi-quantitative analytical information. In other words, these biosensors are tools that can sense changes in living things and turn that information into electrical signals. The first biosensor was developed in 1950 by the American biochemist L.L. Clark, and it was used to measure oxygen levels in the blood utilizing a sensor named the Clark electrode or oxygen electrode. Subsequently, a gel with a glucose-oxidized enzyme was applied to the oxygen electrode to measure blood sugar levels. Similarly, enzyme urease was employed with a specially designed electrode for detecting NH4+ ions to calculate urea in urine, blood, and other bodily fluids. There are three generations of biosensors currently available on the market. The first type operates by allowing the product's reaction to disperse to the sensor, triggering an electrical response. The second type incorporates specific mediators that enhance the interaction between the sensor and the response, resulting in improved performance. In the third type, the response initiates the reaction, with no mediators directly involved. These biosensors are advanced in stability, economic, sensitivity, and reproducibility compared with any other presently existing diagnostic device. Hence, in recent years, biosensors have found a wide range of applications, from biomedical diagnosis and environmental monitoring to drug discovery and forensic research [1].

A biosensor consists of three major components: a bioreceptor, a transducer, and an electric signal or display, as shown in the schematic diagram (Figure 5.1). The first component of the biosensor is the sensor or detector, primarily composed of a biological element. This biochemical receptor interacts specifically with the analyte, signaling changes in its composition through an electrical output. The second component is the transducer, a physical element that amplifies the biochemical signal generated by the sensor. It converts this signal into an electrical format, making the data accessible and easy to interpret. The third component is an electrical circuit, which comprises several key elements: a signal conditioning unit, a processor or microcontroller, and a display unit. Primarily, the tested analyte interacts with a

DOI: 10.1201/9781003536239-5

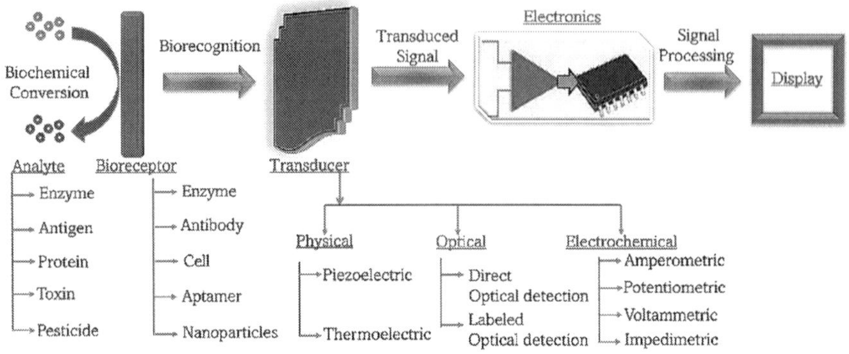

FIGURE 5.1 Major components of a biosensor and its types based on transducer operation.

specific enzyme (bio-receptor), changing the enzyme's biochemical properties. This change subsequently generates an electrical response through an electro-enzymatic approach. The electrical signal is then transformed into a visual display, enabling accurate analysis and effective representation. Generally, biosensors can be grouped into three primary types: electrochemical, physical, and optical biosensors based on the transducer element used (Figure 5.1). The electrochemical biosensor operates through enzymatic catalysis, which involves the consumption or generation of electrons. Enzymes that facilitate these reactions are referred to as redox enzymes. A schematic diagram (Figure 5.1) shows the basic principle of an electrochemical biosensor.

Typically, this biosensor incorporates three types of electrodes: counter, reference, and working electrodes. The working electrodes are designed to improve selectivity and limit the biosensor detection. Recently, several publications have been reported based on the carbon and modified nano-composite-based electrodes used for detecting or sensing different analytes [2–7]. The electrochemical biosensors are classified into amperometric, potentiometric, voltammetric, and impedimetric types, based on their operating principle. Similarly, a physical biosensor responds to physical stimuli like mass, temperature, or acoustic waves. Moreover, optical biosensors work based on the interaction between light and target analyte. They consist of an optical transducer integrated with a biosensing element. The interaction between the analyte and the biorecognition element induces a change in the optical properties of the system. The change is converted to the electrical signal by the transducer used to detect the analyte. The signal produced by the biosensor is proportional to the concentration of the analyte [8]. Out of the above three biosensors, optical biosensors are the most widely used owing to their high specificity, sensitivity, real-time and continuous monitoring of analytes, small size, and cost-effectiveness [8]. These devices have diverse applications, including food and environmental monitoring and healthcare diagnostics.

5.2 CLASSIFICATION OF OPTICAL BIOSENSORS

Optical biosensors are classified into different types based on their modes of operation, design aspects, and detection mechanism. Depending on the modes of operation, optical biosensors can be classified into two major categories. One is a label-free mode in which the analyte directly interacts with the transducer to produce a signal. Label-free optical biosensors can be categorized into several types: evanescent wave fluorescence sensors, photonic crystal-based sensors, surface plasmon resonance (SPR) sensors, and optical waveguide-based sensors [9]. The other is label-based mode, where a label is used, and the signal is produced by the fluorescent or calorimetric method. Based on their design aspects, optical biosensors are further classified as SPR, fluorescence/luminescence, optical waveguide type, ring resonator, interferometer, and optical fiber biosensors. Similarly, depending on the detection mechanism, optical biosensors are of different types. Optics-based biosensing technologies encompass a variety of advanced methods, including surface-enhanced Raman scattering (SERS), surface plasmon resonance (SPR), localized surface plasmon resonance (LSPR), fiber optic surface plasmon resonance (FOSPR), and surface-enhanced fluorescence (SEF), among others. A schematic diagram in Figure 5.2 summarizes different optics-based biosensing technologies [10].

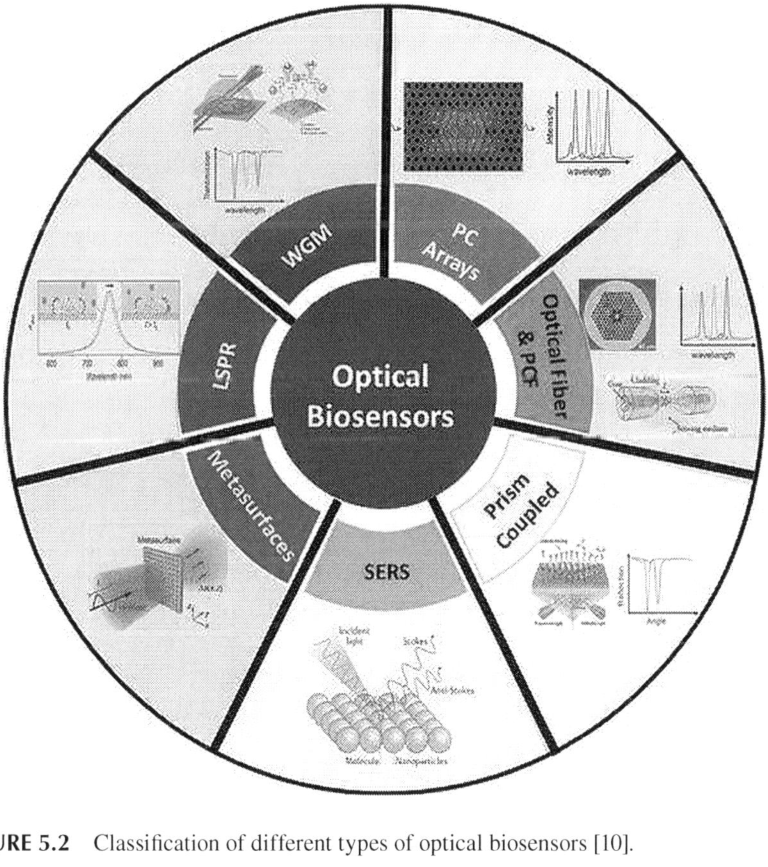

FIGURE 5.2 Classification of different types of optical biosensors [10].

5.3 OPERATION PRINCIPLES OF OPTICAL BIOSENSORS

The principles governing the operation of different optical biosensors involve the interaction of light with biological materials. These sensors utilize various optical techniques to detect and quantify biological substances, relying on changes in light properties such as intensity, wavelength, or phase. By leveraging the unique optical characteristics of biomolecules, these biosensors can provide sensitive and specific measurements, making them invaluable tools for medical diagnostics, environmental monitoring, and food safety.

5.3.1 Spectrophotometry

The light beam from a source is split into a spectrum of different wavelengths after passing through a prism or grating. After passing through a wavelength selector, it is incident on the analyte. Beer Lambert's law analyzes the reflected or transmitted light intensity on the detector. The absorbance of light in the ultraviolet, visible, or infrared regions is directly related to the concentration of the analyte present in the sample [11].

5.3.2 Plasmon-Based Resonance

Optical biosensors that utilize plasmon resonance identify analytes by assessing variations in the refractive index. When polarized light hits a metal surface, the photon energy is transferred to the electrons present on the metal surface, creating plasmons, and these plasmons move parallel to the metal surface. These plasmons enhance the absorption of light, resulting in a reduction of the intensity of reflected light measurable at specific angles. An SPR biosensor consists of immobilized probe molecules on the sensor surface. The affinity interaction between the probe targets due to their contact results in changes in the refractive index at the SPR sensor surface. The change in refractive index is related to the volume concentration of the analyte. The refractive index–related parameters can detect and quantify the target molecules. Under resonance conditions, the surface plasmons in the metal film generate an evanescent field that extends into the volume near the surface of the metal film. Under resonance conditions, the intensity of reflected radiation diminishes. Any alteration in the refractive index of the sample near the surface of the metal film affects the resonance condition, thereby shifting the position of the dip in the reflection spectrum. Thus, observing the shift in dip position, the quantity of analyte can be measured as shown in Figure 5.3 [12]. Recently, a fiber-based SPR was developed. When the target analyte attaches to the bioreceptor, it excites the surface plasmons, resulting in a noticeable dip in the output light intensity at resonance. So, measuring the angular wavelength shift at resonance, the concentration of the analyte can be measured as shown in Figure 5.3. Another type of SPR used is localized surface plasmon resonance (LSPR). Here metallic nanoparticles measuring few nanometers are used in place of metallic films. A dip is observed in the output spectra at the resonance angle. The change in the resonance angle or shift in the wavelength indicates

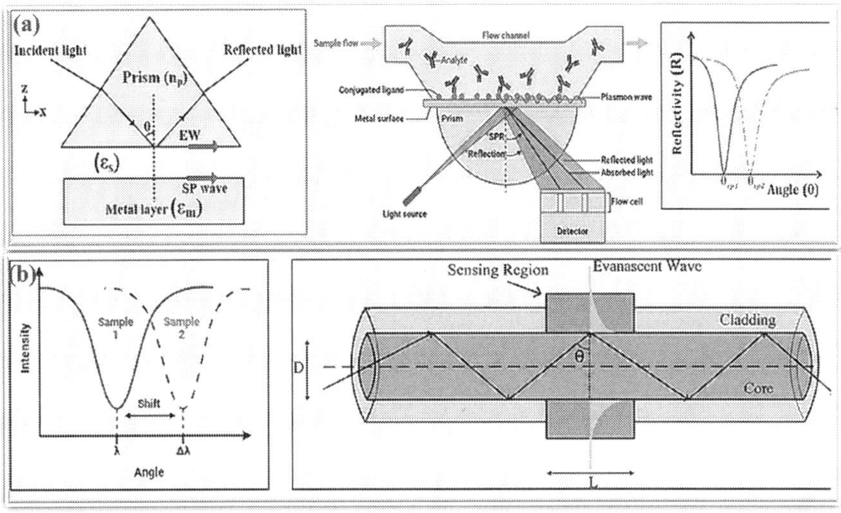

FIGURE 5.3 (a) Prism-based SPR sensor and (b) fiber-based SPR sensors [12].

the binding of the target to the biorecognition element. This shift can be detected by a detector [13]. These LSPRs are more sensitive as they have more localization capability of surface plasmons and can also detect small volumes of analyte. In SPR, the excitation of surface plasmons occurs at the thin metal film surface and the electric field is confined to 200–1000 nm, whereas in LSPR, the excitation of the plasmons occurs at the metal nanoparticle, and the field is confined to 20–40 nm. The SPR spectra change with the thickness of the metal film, angle of incidence, wavelength of incident light, and polarization of light, but in LSPR, the spectra depend on the shape and size of the metal nanoparticles. SPR is used to detect the target molecule on the surface, near the surface, and in the bulk sample, but LSPR can detect the target molecule only up to a few nanometers. The LSPR has better sensitivity and selectivity in comparison to SPR [12].

5.3.3 Colorimetry-Based Biosensors

In this type of biosensor, the analyte is observed by viewing the color change due to interaction [14]. The transformation is done by the use of smart materials like gold nanoparticles (AuNPs), metal oxide NPs, carbon nanotubes (CNTs), graphene oxide (GO), and conjugated polymers. The size-dependent optical properties of Au NPs help to detect the analyte in a colloidal solution. The aggregation or dispersion of nanoparticles in a colloidal solution produces a color change which helps in the detection of the analyte. Metal oxide NPs, CNT, and GO can catalyze the reaction with the analyte to give a color change. The conjugated polymers undergo conformational transformation or aggregation, giving rise to a color change. The distinctive characteristics of these materials enable the conversion of the reaction into a colorimetric signal, which can be easily observed by the naked eye, eliminating the requirement for sophisticated instruments. Additionally, the presence of an analyte can be verified. [15].

5.3.4 Luminescence-Based Biosensors

These biosensors operate via the light emitted from molecules of various materials through fluorescence, chemiluminescence, and bioluminescence. In the fluorescence method, the light is emitted by the molecule at a certain wavelength. In the chemiluminescence method, a chemical reaction elevates the molecule to a higher excited state, resulting in the emission of light as it returns to a lower energy state. In the bioluminescence process, light is emitted by the biochemical reaction occurring inside the living organism. These biosensors don't require any excitation device. These systems are exceptionally sensitive and selective, exhibiting minimal background noise [16].

5.3.5 Interferometry-Based

In this method, the small change in path length or wavelength of the light or refractive index of the media in which the ray is traveling due to the presence of an analyte can be measured using an optical interferometer. The interferometer compares the path length of the two nearly equivalent light rays. One is the reference light and the other passes through the medium in which the refractive index changes due to the bioconjugate reaction. The transition between the two phases is assessed using an interferometer. This device comprises an optical cavity equipped with a configuration of mirrors and optical components that create a resonant cavity for light waves. Interference occurs inside the resonant cavity, giving rise to patterns that can be modified with the parameters of the resonant cavity [17].

5.3.6 Surface-Enhanced Raman Scattering (SERS)

In this technique, low traces of analyte can be detected with high accuracy by analyzing the enhanced Raman scattered signals from the molecules absorbed on the nanostructured metallic surfaces [18]. Lasers are used to excite vibrational transitions in molecules adsorbed on metallic nanoparticles such as gold/silver [19]. The free electrons on the surface which oscillate collectively are called plasmons. When the light frequency matches with the oscillating frequency, resonance occurs. At plasmon frequency, the metal becomes highly polarizable so the local electric field near the metal surface increases significantly, increasing the Raman scattering of the nearby molecules. Even small increases in the electric field can lead to significant increases in the intensity of the emitted Raman signal, as the intensity is directly proportional to the square of the applied electric field at the molecular level. The Raman emission radiation increases the excitation of the surface plasmons [20]. A schematic diagram of the experimental setup of the SERS sensor is shown in Figure 5.4. The irregularly shaped plasmonic nanoparticles create a non-uniform distribution of the

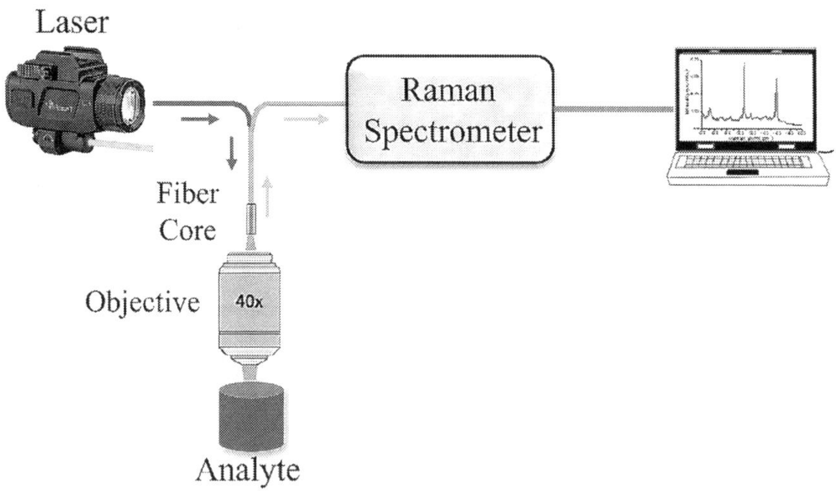

FIGURE 5.4 Schematic diagram representing experimental setup of SERS sensor.

electric field. The electromagnetic intensity at the sharp tips of these rough nanoparticles is amplified due to the lightning rod effect, as illustrated in Figure 5.5 [21]. The enhancement of the Raman signal resulting from this phenomenon is approximately 10^6–10^8 times greater than the traditional Raman signal amplification. Another approach to surface-enhanced Raman spectroscopy (SERS) enhancement involves chemical methods, where electron transfer occurs between the modified molecule and the nanoparticles.

This is possible when the distance between the molecule and the nanoparticle is less than 10nm. However, the enhancement due to chemical transfer is much smaller in comparison to electromagnetic transfer [22, 23].

The development of a SERS active substrate for direct measurement in liquid samples is critical for in situ measurement. To tackle this, fiberoptic remote sensors are developed. In this type of SERS, the laser excitation is transmitted to the probe by a single optical fiber, and the scattered radiation from the sample is collected by another fiber. After passing through the bandpass filter, the laser beam is focused using a microscope objective lens at one end of the excitation fiber. This end of the excitation fiber is positioned close to the SERS substrate. The collection and excitation fibers are placed on both sides of the SERS substrate. In input radiation from the collector, the fiber is fed to a spectrometer [22, 24].

5.4 ADVANCEMENTS IN OPTICAL BIOSENSORS

Advances in new materials and variations in fabrication technologies have resulted in the evolution of advanced sensors. Out of the two classes of optical biosensors, the label-free methods are of much interest as time-consuming and costly biomolecular labeling can be avoided. Most optical biosensors utilize the evanescent field to detect

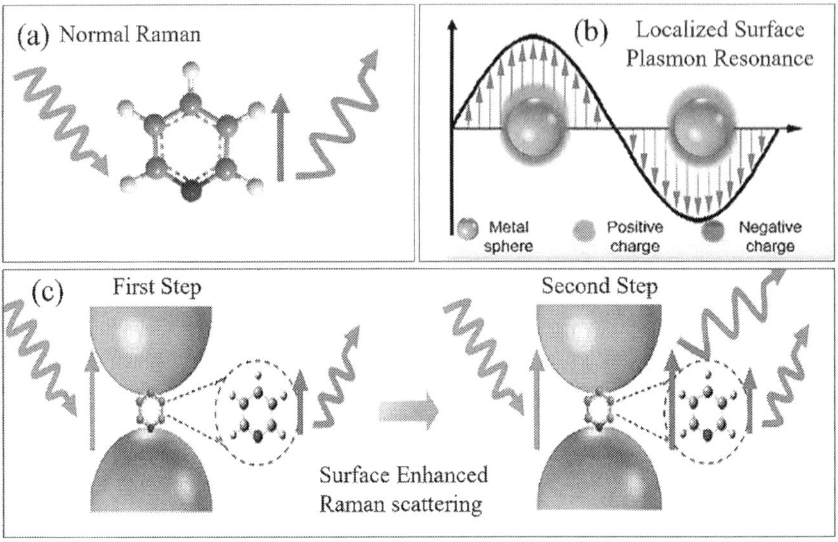

FIGURE 5.5 Schematic diagrams of (a) normal Raman, (b) LSPR (localized surface plasmon resonance), and (c) two-step enhancement mechanism in SERS [21].

target analytes, with the receptor being immobilized on the waveguide's surface. The change in the effective mode index occurs when the analyte interacts with the receptor. The widely used optical biosensor is the SPR type. Refractive index changes as a result of molecular interactions at the metal surface, which are influenced by surface plasmon (SP) waves, have been observed [25]. The surface plasmon waves move at the boundary of the metal-dielectric and attenuate to the dielectric. Specifically, the wave vector of the SP wave is greater than that of the incoming light wave, preventing direct excitation. The increase in the wave number of the incident light can be achieved using a coupler. The widely used couplers are prism couplers, grating couplers, or fiber couplers [26, 27]. From the different configurations of prism couplers, the Kretshmann configuration is preferred due to its structure, which allows light to easily pass through it. It consists of a glass prism and a gold film with no space between them. When light traverses the prism and undergoes total internal reflection at its base, the resulting evanescent field penetrates the gold film. By carefully adjusting the angle of incidence, the evanescent wave can be effectively coupled to the surface plasmon (SP) wave at the gold interface, facilitating the excitation of these surface plasmon waves. Similarly, grating couplers can be used because of the easy fabrication method. A waveguide coupler works on the same principle as a prism coupler. Light passes into the waveguide, which is covered with metal film. The SP waves move along the interface of the metal film. The coupling condition is satisfied when the SP and guided modes' propagation constants are equivalent. The detection limit of such types of devices is of the order of 1pg/mm^3 [28]. To improve the resolution and sensitivity of conventional surface plasmon resonance (SPR), various technologies such as localized surface plasmon resonance (LSPR) and long-range surface plasmon (LRSP) are employed. In LSPR, the SPs are confined within nanoparticles whose dimensions are of the order of the incident wavelength, allowing for stronger light–matter interactions and improved detection capabilities. This confinement results in significant field enhancement, making LSPR particularly effective for sensing applications at the nanoscale. The LSPR is influenced by the shape, size, and composition of the nanoparticles [29]. The interaction between two nearby particles through their near field and the coupling of plasmon oscillation with them improves the signal, which helps in the detection of the analyte. The substantial scattering cross-section of nanoparticles (NPs) results in an intensified scattered spectrum that significantly enhances the spatial resolution for detecting target molecules. This heightened intensity allows for more precise localization and characterization of the molecules under investigation, enabling improved sensitivity in various sensing applications [30]. Functionalizing the surfaces of nanoparticles (NPs) with Raman-active molecules and leveraging localized surface plasmon resonance (LSPR) significantly amplifies the Raman scattering signal from these molecules. This enhancement, referred to as surface-enhanced Raman scattering (SERS), can remarkably improve detection limits, enabling sensitivity levels as low as 10 fM. This powerful combination enhances signal intensity and facilitates the detection and analysis of trace molecules, making SERS an invaluable tool in various fields, including biochemical sensing and clinical diagnostics [31]. For the detection of ultrasensitive bioassays, fluorescence enhancement, along with the effects of surface-enhanced Raman scattering (SERS), has been utilized. This fluorescence enhancement takes place close to metals, likely due to interactions with the exciting laser light and the emitting fluorophores. [32]. The interactions enhance the radiative quantum efficiency near the metal surface, with the emission intensity depending on the distance between the metal and the fluorophore. Usually, metal nanorods, nanospheres, and metals with irregular shapes functionalized with fluorescent materials are considered for the detection of the analyte [33–35]. The detection limit in such types of methods extends to the pM range [35]. Recently the focus has shifted to the development of nanoplatforms for promoting multimodal biodetection using LSPR, SERs, and SEF. Andreea Campu et al. reported on such a paper-based platform [36]. The modulated LSPR responses were deposited on cellulose fiber using gold nano bipyramids with a commercial pen by calligraphy method as shown in Figure 5.6. The synthesized p-amino-thiophenol@Biotin system was immobilized with the nanoplatforms capable of detecting streptavidin binding. The target protein could be detected by three methods. By employing localized surface plasmon resonance (LSPR), we observed changes in the refractive indices at each stage of the interaction. This enabled us to detect the presence of the protein through a measurable redshift in the wavelength. This redshift serves as a reliable indicator of alterations in the local environment surrounding the nanoparticles, thereby enhancing both the sensitivity and specificity of our detection method. Using the SER technique, the protein fingerprint band in the spectrum confirmed the presence of biotin–streptavidin. In the SEF method, the incorporation of the biotin-streptavidin complex as a spacer between the Au-BP surface and the Alexa 680 fluorophore significantly enhanced fluorescence emission. This strategic design led to a twofold increase in

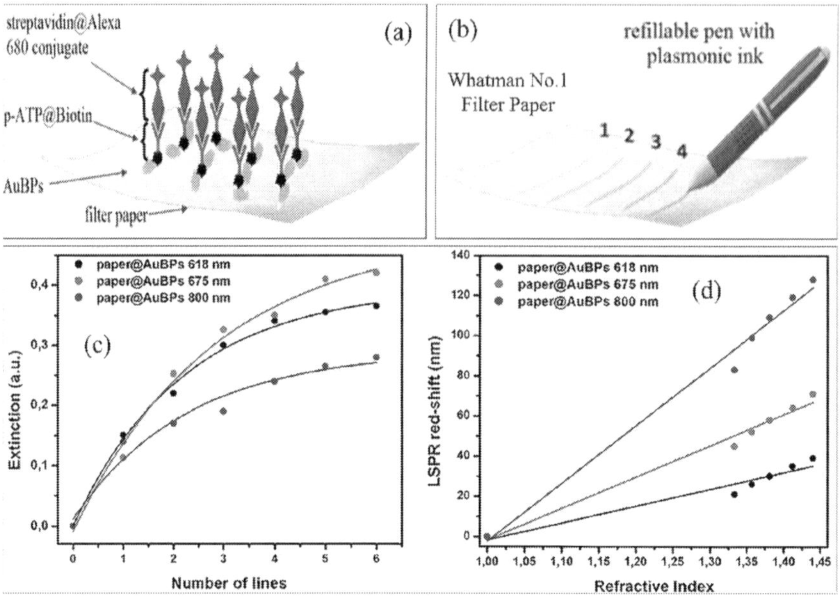

FIGURE 5.6 (a) Paper-based nanoplatforms with gold nano bipyramid deposited on cellulose fiber showing modulated LSPR response; (b) calligraphy method to draw the plasmonic lines; (c) the extinction intensity of the longitudinal LSPR band vs. number of plasmonic lines for the tuneable plasmonic paper based nanoplatforms; (d) the variation of LSPR as a function of bulk refractive index for the tuneable plasmonic paper based nanoplatforms [36].

fluorescence intensity compared to the unmodified paper surface, underscoring the effectiveness of the biotin–streptavidin interaction in amplifying signal detection. So, this paved the way for developing a single flexible and plasmonic nanoplatform, making point-of-care diagnostics possible by using multiple methods.

Optical waveguide-based biosensors work based on the principle of an evanescent wave detection mechanism. As the optical wave propagates through the waveguide film, its evanescent tail extends into the cladding layer, which is surrounded by the bulk solution containing the target molecule. This interaction between the evanescent field and the surrounding medium induces a change in the refractive index, enabling enhanced sensitivity for detecting the target molecule. The material and dimension of the waveguide affect the evanescent field strength [25]. According to the construction, waveguides are classified into different types such as stripe, ridge/rib, buried, and diffused waveguides [37]. Of these, the rib waveguides are usually used as biosensors. The rib waveguide consists of a slot waveguide and interferometer with a sensing arm. The slot waveguide is made from a combination of two slab waveguides placed very close to each other. They have high refractive index values. The space between them is filled with a low-refractive-index medium, which effectively confines the light and facilitates its propagation within the waveguide [38]. The optical wave that traverses the slot differs from the wave that propagates within the conventional waveguide. The gap region can confine a strong field and allow propagation with a low loss. So, this waveguide is integrated into a resonator and interferometer that makes a lab-on-a-chip platform available. Barrios et al. were the pioneers in reporting a resonator that utilized a slot waveguide for bulk sensing applications [39]. Similarly, the multiple-slot and cross-slot waveguides are developed for biosensing purposes [40–42]. Slot waveguides are also used in optical interferometry. The Mach-Zehnder interferometer, which is used as a sensing arm for bulk sensing, uses double-slot hybrid plasmonic. Here two nano slots are formed between a silicon layer and two silver layers as shown in Figure 5.7 [43]. Interferometric waveguides work based on the combination of waveguide and interferometric techniques. When a guided mode propagates through a waveguide, variations in the refractive index result in a phase shift of the mode. The phase-shifted mode interacts with a reference mode, generating an interference signal that is measured at the output. This signal is then processed to determine the concentration of the analyte. There are three different types of interferometers, the Mach-Zehnder (MZ), Hartman, and Young interferometers, which find applications in biosensing. In the MZ configuration, light from a monochromatic source is divided into two separate beams that traverse independent paths. One beam interacts with the test sample, while the other passes through a reference arm filled with a buffer solution. This arrangement allows for the comparison of phase shifts induced by the sample against a stable reference, facilitating sensitive measurements of analyte concentrations. The interaction between the two beams during their recombination results in a variation in field intensity. This change can

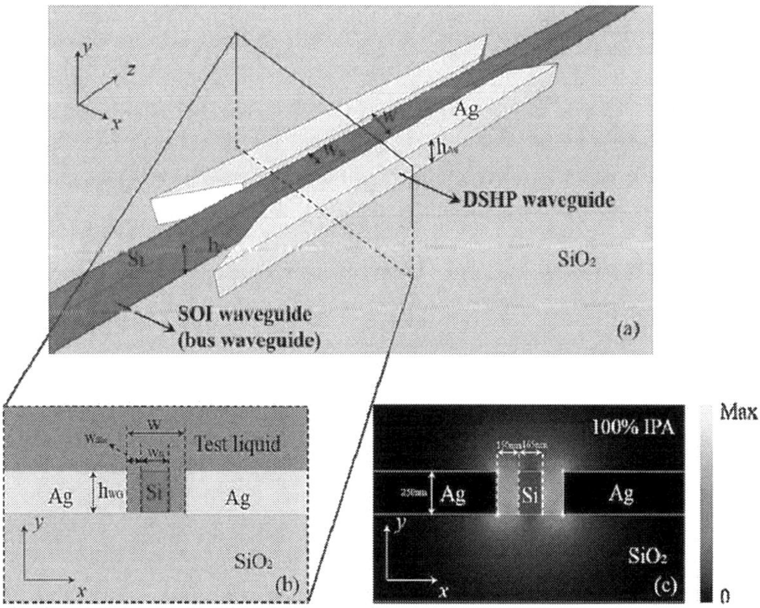

FIGURE 5.7 Double-slot hybrid plasmonic waveguide and cross-sectional view [43].

be utilized to determine the concentration of the analyte [44]. In the Hartman configuration, light is directed into the multi-sensing region of the waveguide through a broad input grating. An optical element is then employed to gather the light that propagates through the waveguide, resulting in the production of interference [45]. In the Young interferometer, the reference layer is strategically positioned vertically beneath the sensing waveguide. This allows the beams coming from both layers to interfere, and the pattern is obtained in the far field, which can be analyzed to determine the concentration of the analyte. The interference pattern can be captured using a CCD camera and analyzed through a Fast Fourier Transform (FFT) algorithm. The detection limit is approximately in the picomolar range [46].These sensors are not cost-effective because of the complexity of the sensor surface and the requirement of extra cooling units [25]. So, the focus shifts to optical resonator-based biosensors. In resonator-based biosensors, light is confined to move in a cavity and interferes with itself, giving rise to resonance.

The microcavity functions as an optical transducer due to alterations in its geometry resulting from deformation or heating. This causes a change in the resonance parameters and is transformed as a change in light intensity. There are three primary types of optical cavities: photonic crystal cavities (PCC), whispering gallery mode (WGM) cavities, and Fabry-Perot (FP) cavities. The optical cavity is evaluated by its Q factor. The WGM resonators have wide applications. Cavity geometry can be designed in different shapes using various materials such as glass, polymer, or semiconductors [47]. Whispering gallery modes exist within the cavity due to the phenomenon of total internal reflection of light. In this context, the refractive index of the dielectric cavity is greater than that of the surrounding medium. When the angle of incidence surpasses the critical angle, light becomes trapped within the cavity through total internal reflection. The interaction of the WG modes with incoming light results in enhanced resonance. Various parameters like transmitted power, free spectrum range, full width at half maxima, and Q factors describe the efficiency of the cavity in the detection of the analyte. The transmitted power is related to the coupling coefficient, which gives the idea of the amount of energy coupled to the cavity. The free spectrum range gives the spectral distance between two adjacent resonance peaks, which is dependent on the dimension of the cavity. Full width at half maxima is related to the optical losses in the cavity. All these factors determine the efficiency of the detection of the analyte [48]. WGM resonators with different configurations can be fabricated by different materials from silica and silicon nitride to quartz [49–51]. The evanescent tails of the resonant modes in dielectric whispering gallery modes (WGM) extend into the analyte, producing a moderate level of sensitivity [48]. To enhance the sensitivity for low-concentration detection, the LSPR nanostructures can be kept close to dielectric WGM. These are called hybrid WGM resonators [52, 53]. They can have two types of configurations such as a micro ring in a metal-dielectric slot [54] or a micro ring with nanostrips along the circumference [55]. In hybrid WGM biosensors, although the sensitivity increases, the Q-factor reduces significantly due to the broadening of the resonance peaks. The main challenge in this type of resonator is the damping in the metal surface. Materials that show little loss in the visible and infrared spectrum are thought to be good substitutes for metals [56]. M. D. Baaske et al. demonstrated a single-nucleic-acid biosensing with a plasmon-enhanced WGM microcavity sensor of unparalleled sensitivity as shown in Figure 5.8(a) [57]. With developments in data acquisition speeds possibly reaching the sub-microsecond range, plasmon-enhanced WGM biosensors offer remarkable time resolution for probing single molecules and their interactions. Photonic crystal-based biosensors are based on photonic crystals (PC), which consist of materials of different refractive indices arranged periodically in space and manipulate light at an optical wavelength range [58]. A complete photonic bandgap in the PC exists, representing a frequency range in which light propagation is forbidden. The introduction of defects disrupts the periodic structure, leading to the formation of localized photonic states within the gap [59]. So, a PC can provide strong field confinement with low extinction loss. Various PC structures are proposed for the detection of multiple analytes. Park et al. introduced a PC structure that utilizes hydrogel as its foundational material for the detection of immunoglobulin G (IgG) [60]. Upon the introduction of the analyte IgG to the sensor surface, a noticeable change in reflective color (green to dark orange) occurs, accompanied by a shift in the photonic bandgap (PBG), as illustrated in Figure 5.8(b). In recent years, PC structures have been applied in microfluidics, telemedicines, and flexible sensors [61]. This facilitates the detection with low sample volume and in multiple channels. PC integrated with microfluidics facilitates the design of point-of-care devices for sample detection. Nowadays, smartphones are intensively connected with sensors. The presence of high-resolution cameras in smartphones helps in the detection of clear images that can be translated into the transmission spectrum. Gallegos et al.

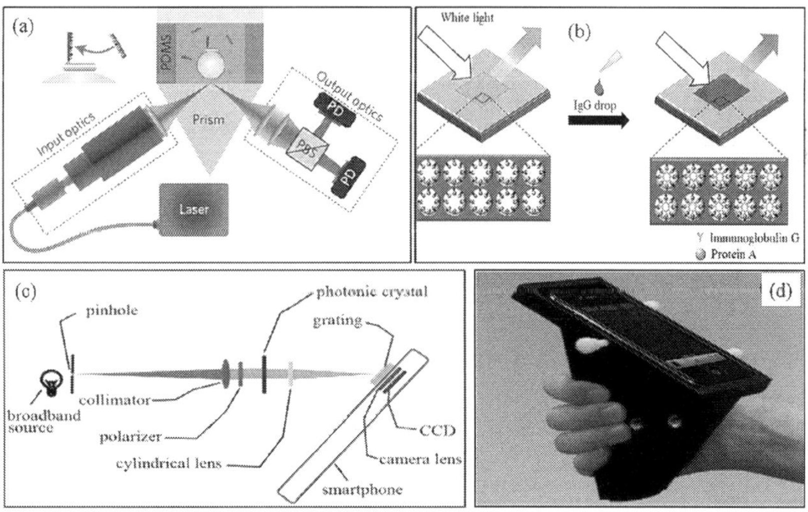

FIGURE 5.8 (a) A hybrid WGM biosensor (glass microsphere decorated with gold nanorods platform) for detection of nucleic acid [57]; (b) Nanoporous hydrogel PC biosensor for IgG detection [60]; (c) and (d) the components in a smartphone cradle used for detection of spectral shift by a PC biosensor [62].

used the built-in camera of smartphones to detect the spectral shift created by a PC biosensor. The intensified polarized light from an external source was collected by the camera of a smartphone held in a cardle at a fixed orientation as shown in Figure 5.8(c, d). Using specialized software, the camera captured images that were converted into transmission spectra [62]. This handheld biosensor paved the path for the design of point-of-care devices in the future. The invention of smart materials like graphene, carbon nanotubes, and hydrogels has resulted in their effective use in sensors. The incorporation of these materials into the PC structure helps in fast detection and enhanced sensitivity. The emergence of flexible materials opens the possibility of the development of wearable sensors in the shape of wrist bands, skin, and fabric patches. Xu et al. developed a microcavity sensor using a flexible PC sensor showing deformation-independent sensitivity [63].

5.5 OPTICAL FIBER-BASED BIOSENSORS

These biosensors work on the principle of the total internal reflection (TIR) of light. For TIR to take place, the refractive index of the core must exceed that of the cladding, and the incident angle at the core–cladding boundary must be greater than the critical angle. When light propagates within the optical fiber, most of it remains confined to the core, while only the evanescent waves extend into the cladding region. To enhance surface plasmon resonance, the evanescent wave must interact with the metal-dielectric layer. Modification of certain parts of the cladding layer with plasmonic material can help in the excitation of surface plasmons. When resonance occurs, the wave vector of the incident light must align with that of the surface plasmon polariton. Variations in the sensing medium lead to changes in resonance, facilitating the detection of the analyte [10]. Recently, various plasmonic fiberoptics-based biosensors have developed; for example, the bio-sensor shown in Figure 5.9(a) was developed by Kaushik et al. to detect *Escherichia coli* (*E.Coli*). The optical fiber sensor is coated with Au/immobilized MoS_2 materials. The immobilized antibodies functionalized MoS_2 nanomaterials and enhanced the detection limit of *E.Coli* [64]. In photonic crystal fiber (PCF) biosensors, the air holes in the fibers are integrated into the fiber structure, which helps to broaden the spectrum of waveguides. Generally, PCFs are classified into two types, index guiding and photonic band gap, as per the guiding mechanism. The index guiding fiber works based on modified TIR [65], and photonic band gap PCF confine light within the hollow core of the porous structure [66]. The performance of PCF sensors is influenced by the choice of materials, the arrangement of structural components, and the functional materials incorporated within them. Figure 5.9(b) illustrates a PCF core biosensor designed with a gold-coated, side-polished hexagonal structure [67].

In numerous plasmonic sensing techniques, the intrinsic losses linked to metals limit detection capabilities. Nevertheless, these constraints can be alleviated by utilizing metasurfaces alongside an insulating material. The resonance in metasurfaces is characterized by a dependence on both the geometric structure and the effective refractive index, as well as the resonance wavelength. Here mainly Mie-type

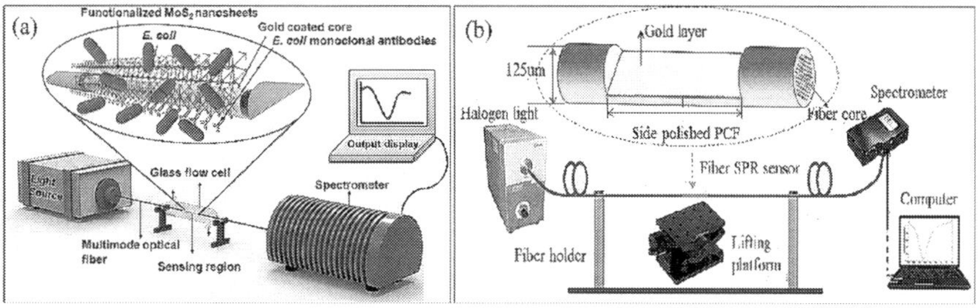

FIGURE 5.9 (a) Fiber optics plasmonic biosensor for detection of *E. coli* [64]; (b) PCF core biosensor based on gold-coated side polished hexagonal structure [67].

scattering occurs, which gives rise to magnetic resonance along with the electric resonance. These materials get much attention when the operating frequency is tetra hertz.

Here, we provide an overview of the applications of these optical biosensors for the detection of different analytes and their detection limits in the years 2024–2025.

From here it is observed that with the enhancement of technology, various biosensors have been developed which can utilize more than one type of principle to enhance the limit of detection of various analytes. Still, a few challenges remain, which are summarized below.

5.6 CHALLENGES IN FUTURE GENERATION BIOSENSORS

The following are a few of the future challenges in optical biosensors.

i. The development of lab-on-a-chip technology: This technology will allow the detection of samples with less volume. This will also reduce the detection time along with the minimization of chemical waste.
ii. The minimization of background noise: Background noise is responsible for the reduction of signal strength, giving rise to low sensitivity.
iii. The performance of multitasks at the same time: These tasks include sample injection, mixing, reaction, and detection.
iv. The development of a multiplexing detection facility: This will allow multiple bio-analytes to be detected simultaneously without any interference. This will save time and resources.
v. The integration of wearable and implantable biosensors: This will render them biocompatible for the continuous monitoring of various biomarkers.
vi. The integration of microfluidics with biosensing: This will allow miniaturization of the sensor.
vii. The stability and reproducibility of the sensor over time: This is another big challenge.

TABLE 5.1
Optical Biosensors in Recent Years (2024–2025) for the Detection of Various Analytes

Technology Platform	Materia/methods used in sensing	Analyte	Detection limit	Year	Ref.
Array sensor with 18 different synthetic fluorescent probes	4 different classes of fluorescent probes (i) BODIPY (ii) Naphthylamide (iii) Rhodamine (iv) Phenanthrene	Human saliva with cortisol, dopamine, and adrenaline	1 pM–1 mM	2025	[68]
Pixel diversity Interferometric reflectance imaging sensor	Anti-A29 monoclonal antibody (mAb 69-126-3)	Monkeypox virus	200PFD/ml or 3.3aM	2025	[69]
Dual-mode assay with colorimetric and SERS detection modes	Nanozyme (Fe_3O_4@Au-Pt) and bifunctional nanobody A2.3-SBP	Microcystin-LR	0.26 and 0.032 ng/mL	2025	[70]
Portable colorimetric sensing platform integrated with smartphone	Cryogel encapsulating the DNA hairpins	17β-oestradiol	0.5 nM	2025	[71]
A sensor array with two-dimensional nanoparticles with fluorescently labeled single-stranded DNA	Machine learning models were used	8 pathogenic and spoilage bacteria		2025	[72]
Hybridization chain reaction powered optical fiber-embedded microfluidic biochip		microRNA lethal-7a in serum	100.0 pM	2025	[73]
Bovine serum albumin-coated gold nanocluster immobilized on paper fabric (BSA-AuNC embedded paper fabric)	Direct visualization of fluorescent intensity loss	Glucose and cholesterol	10 µM–5 mM	2025	[74]
Portable lateral flow immune assay integrated with smartphone	ZrFe-MOF@PtNPs	Aflatoxins	0.0062 ng/mL	2025	[75]

(*Continued*)

TABLE 5.1 (CONTINUED)
Optical Biosensors in Recent Years (2024–2025) for the Detection of Various Analytes

Technology Platform	Materia/methods used in sensing	Analyte	Detection limit	Year	Ref.
Liquid crystal-based sensor	Cationic polymer poly-L-lysine	Tau-381 protein	2.77 pg/mL	2025	[76]
Gold nanoparticle-based plasmonic chip	Antipyrin antibody	Pyrin protein	0.24 ng/mL	2025	[77]
Whispering gallery mode optical fiber optofluidic laser	H_2O_2 rhodamine B	Hemoglobin	46.0 pM	2024	[78]
Dual mode fluorometric and colorimetric sensor	UiO-66 (Zr)	H_2O_2	0–600 nM	2024	[79]
Apta sensor with colorimetric detection	Zr-MOF (PCN-224) along with Au nanoparticles	Exosomes	9.1×10^4 particles/mL	2024	[80]
Dual-enhanced enzyme cascade hybrid system (UFD-DEC) hydrogel-based sensor coupled to a smartphone	DNA-controlled nano enzymes (Fe-cdDNA)	Urea	0.12 mmol L^{-1}	2024	[81]

To overcome these challenges, nowadays researchers are focusing on accuracy and signal enhancement by combining plasmonic with fluorescence. The use of nanomaterials for functionalization and detection is also being tried. Machine learning models are also being used to analyze data quickly.

5.7 CONCLUSION

In this chapter, we discuss operating principles, classifications, and advancements in optical biosensors. We also provide a detailed examination of the limitations associated with various biosensor types. Furthermore, we discuss potential modifications that could enhance both sensitivity and selectivity.

REFERENCES

1. Nikhil, B., Pawan, J., Nello, F., & Pedro, E. (2016). Introduction to biosensors. *Essays Biochem*, *60*(1), 1–8.
2. Pushpanjali, P. A., Manjunatha, J. G., & Hareesha, N. (2021). An overview of recent developments of carbon-based sensors for the analysis of drug molecules. *Journal of Electrochemical Science and Engineering*, *11*(3), 161–177.
3. Tigari, G., & Manjunatha, J. G. (2020). Poly (glutamine) film-coated carbon nanotube paste electrode for the determination of curcumin with vanillin: an electroanalytical approach. *Monatshefte für Chemie-Chemical Monthly*, *151*, 1681–1688.
4. Monnappa, A. B., Manjunatha, J. G., & Bhatt, A. S. (2020). Design of a sensitive and selective voltammetric sensor based on a cationic surfactant-modified carbon paste electrode for the determination of alloxan. *ACS Omega*, *5*(36), 23481–23490.
5. Raril, C., Manjunatha, J. G., Ravishankar, D. K., Fattepur, S., Siddaraju, G., & Nanjundaswamy, L. (2020). Validated electrochemical method for simultaneous resolution of tyrosine, uric acid, and ascorbic acid at polymer-modified nano-composite paste electrode. *Surface Engineering and Applied Electrochemistry*, *56*, 415–426.
6. Pushpanjali, P. A., Manjunatha, J. G., Hareesha, N., D'Souza, E. S., Charithra, M. M., & Prinith, N. S. (2021). Voltammetric analysis of antihistamine drug cetirizine and paracetamol at poly (L-Leucine) layered carbon nanotube paste electrode. *Surfaces and Interfaces*, *24*, 101154.
7. Hareesha, N., & Manjunatha, J. G. (2021). Electro-oxidation of formoterol fumarate on the surface of novel poly (thiazole yellow-G) layered multi-walled carbon nanotube paste electrode. *Scientific Reports*, *11*(1), 12797.
8. Damborský, P., Švitel, J., & Katrlík, J. (2016). Optical biosensors. *Essays in Biochemistry*, *60*(1), 91–100.
9. Singh, A. K., Mittal, S., Das, M., Saharia, A., & Tiwari, M. (2023). Optical biosensors: A decade in review. *Alexandria Engineering Journal*, *67*, 673–691.
10. Mostufa, S., Rezaei, B., Ciannella, S., Yari, P., Gómez-Pastora, J., He, R., & Wu, K. (2024). Advancements and perspectives in optical biosensors. *ACS Omega, 9*, 24181–24202.
11. Prasad, R. D., Prasad, R. S., Prasad, R. B., Prasad, S. R., Singha, S. B., Singha, D., ... Navathe, G. J. (2024). A review on modern characterization techniques for analysis of nanomaterials and biomaterials. *ES Energy & Environment*, *23*, 1087.
12. Kaur, B., Kumar, S., Nedoma, J., Martinek, R., & Marques, C. (2024). Advancements in optical biosensing techniques: From fundamentals to future prospects. *APL Photonics*, *9*(9), 091102.

13. Gupta, B. D., & Kant, R. (2018). Recent advances in surface plasmon resonance-based fiber optic chemical and biosensors utilizing bulk and nanostructures. *Optics & Laser Technology, 101*, 144–161.
14. Wu, Y., Feng, J., Hu, G., Zhang, E., & Yu, H. H. (2023). Colorimetric sensors for chemical and biological sensing applications. *Sensors, 23*(5), 2749.
15. Song, Y., Wei, W., & Qu, X. (2011). Colorimetric biosensing using smart materials. *Advanced Materials, 23*(37), 4215–4236.
16. Gu, C. (2017). Quantum dots-based fluorescence resonance energy transfer biosensor for monitoring cell apoptosis. *Luminescence, 32*(7), 1186–1191.
17. Jha, R., Gorai, P., Shrivastav, A., & Pathak, A. (2024). Label-free biochemical sensing using processed optical fiber interferometry: a review. *ACS Omega, 9*(3), 3037–3069.
18. Linh, V. T. N., Ja'farawy, M. S. A., Koh, E. H., Lee, M. Y., Park, S. G., Kim, D. H., & Jung, H. S. (2024). Flexible surface-enhanced Raman scattering substrates toward sampling approaches for on-site sensing and diagnosis applications. *Applied Spectroscopy Reviews, 59*(1), 90–123.
19. Jeanmaire, D. L., & Van Duyne, R. P. (1977). Heterocyclic, aromatic, and aliphatic amines adsorbed on the anodized silver electrode. *Journal of Electroanalytical Chemistry, 84*, 1–20.
20. Gu, C., Zhao, Z., & Shi, P. (2021). Development of monolayer AuNPs decorated on an optical fiber facet for SERS analysis. *Applied Optics, 60*(3), 792–798.
21. Li, J., Chen, X., & Zhu, J. J. (2023). DNA functionalized plasmonic nano assemblies as SERS sensors for environmental analysis. *Aggregate, 4*(2), e271.
22. Vo-Dinh, T. (1995). SERS chemical sensors and biosensors: new tools for environmental and biological analysis. *Sensors and Actuators B: Chemical, 29*(1–3), 183–189.
23. Zong, C., Xu, M., Xu, L. J., Wei, T., Ma, X., Zheng, X. S., ... Ren, B. (2018). Surface-enhanced Raman spectroscopy for bioanalysis: reliability and challenges. *Chemical Reviews, 118*(10), 4946–4980.
24. Liu, J., Zhang, B., Abdullah, A., Muhsin, S. A., Huang, J., & Almasri, M. (2024). A fiber optics-basedsurface-enhanced Raman spectroscopy sensor for chemical and biological sensing. *Sensing and Bio-Sensing Research, 46*, 100686.
25. Chen, C., & Wang, J. (2020). Optical biosensors: An exhaustive and comprehensive review. *Analyst, 145*(5), 1605–1628.
26. Vinogradov, A. P., Dorofeenko, A. V., Pukhov, A. A., & Lisyansky, A. A. (2018). Exciting surface plasmon polaritons in the Kretschmann configuration by a light beam. Physical Review B, *97*, 235407
27. Akimov, Y. (2018). Optical resonances in Kretschmann and Otto configurations. *Optics Letters, 43*(6), 1195–1198.
28. Pol, E., Roos, H., Markey, F., Elwinger, F., Shaw, A., & Karlsson, R. (2016). Evaluation of calibration-free concentration analysis provided by Biacore™ systems. *Analytical Biochemistry, 510*, 88–97.
29. Khan, A. U., Zhao, S., & Liu, G. (2016). Key parameter controlling the sensitivity of plasmonic metal nanoparticles: aspect ratio. *The Journal of Physical Chemistry C, 120*(34), 19353–19364.
30. Lee, S. E., Chen, Q., Bhat, R., Petkiewicz, S., Smith, J. M., Ferry, V. E., ... Bissell, M. J. (2015). Reversible aptamer-Au plasmon rulers for secreted single molecules. *Nano Letters, 15*(7), 4564–4570.
31. Su, J., Wang, D., Nörbel, L., Shen, J., Zhao, Z., Dou, Y., Peng, T., Shi, J., Mathur, S., Fan, C., & Song, S. (2017). Multicolor gold–silver nano-mushrooms as ready-to-use SERS probes for ultrasensitive and multiplex DNA/miRNA detection. Analytical Chemistry, 89, 2531.

32. Pang, Y., Rong, Z., Wang, J., Xiao, R., & Wang, S. (2015). A fluorescent aptasensor for H5N1 influenza virus detection based-on the core–shell nanoparticles metal-enhanced fluorescence (MEF). *Biosensors and Bioelectronics*, *66*, 527–532.
33. Su, L., Yuan, H., Lu, G., Rocha, S., Orrit, M., Hofkens, J., & Uji-i, H. (2016). Super-resolution localization and defocused fluorescence microscopy on resonantly coupled single-molecule, single-nanorod hybrids. *ACS Nano*, *10*(2), 2455–2466.
34. Sugimoto, H., Yashima, S., & Fujii, M. (2018). Hybridized plasmonic gap mode of gold nanorod on mirror nanoantenna for spectrally tailored fluorescence enhancement. *ACS Photonics*, *5*(8), 3421–3427.
35. Khan, M. S., & Chaudhari, V. R. (2014). Morphological effect on fluorescence behavior of silver nanoparticles. *Journal of Fluorescence*, *24*, 751–757.
36. Campu, A., Susu, L., Orzan, F., Maniu, D., Craciun, A. M., Vulpoi, A., ... Astilean, S. (2019). Multimodal biosensing on paper-based platform fabricated by plasmonic calligraphy using gold nanobypiramids ink. *Frontiers in Chemistry*, *7*, 55.
37. Kozma, P., Kehl, F., Ehrentreich-Förster, E., Stamm, C., & Bier, F. F. (2014). Integrated planar optical waveguide interferometer biosensors: A comparative review. *Biosensors and Bioelectronics*, *58*, 287–307.
38. Dell'Olio, F., & Passaro, V. M. (2007). Optical sensing by optimized silicon slot waveguides. *Optics Express*, *15*(8), 4977–4993.
39. Barrios, C. A., Gylfason, K. B., Sánchez, B., Griol, A., Sohlström, H., Holgado, M., & Casquel, R. (2007). Slot-waveguide biochemical sensor. *Optics Letters*, *32*(21), 3080–3082.
40. Gylfason, K. B., Carlborg, C. F., Kaźmierczak, A., Dortu, F., Sohlström, H., Vivien, L., ... Stemme, G. (2010). On-chip temperature compensation in an integrated slot-waveguide ring resonator refractive index sensor array. *Optics Express*, *18*(4), 3226–3237.
41. Khodadad, I., Clarke, N., Khorasaninejad, M., Henneke, D., & Saini, S. S. (2014). Optimization of multiple-slot waveguides for biochemical sensing. *Applied Optics*, *53*(23), 5169–5178.
42. Pan, C., & Rahman, B. M. A. (2016). High-sensitivity polarization-independent biochemical sensor based on silicon-on-insulator cross-slot waveguide. *IEEE Journal of Selected Topics in Quantum Electronics*, *23*(2), 64–71.
43. Sun, X., Dai, D., Thylén, L., & Wosinski, L. (2015). High-sensitivity liquid refractive-index sensor based on a Mach-Zehnder interferometer with a double-slot hybrid plasmonic waveguide. *Optics Express*, *23*(20), 25688–25699.
44. Misiakos, K., Raptis, I., Salapatas, A., Makarona, E., Botsialas, A., Hoekman, M., ... Jobst, G. (2014). Broad-band Mach-Zehnder interferometers as high-performance refractive index sensors: Theory and monolithic implementation. *Optics Express*, *22*(8), 8856–8870.
45. Hartman, N. F., Wyvill, J. C., Campbell, D. P., & Edmonds, P. (1995, January). Rapid response biosensor for detection and identification of common foodborne pathogens. In *Optics in agriculture, forestry, and biological processing* (Vol. 2345, pp. 128–137). SPIE.
46. Schmitt, K., Schirmer, B., & Brandenburg, A. (2005, May). Label-free detection of biomolecules by waveguide interferometry. In *17th International conference on optical fibre sensors* (Vol. 5855, pp. 459–462). SPIE.
47. Foreman, M. R., Swaim, J. D., & Vollmer, F. (2015). Whispering gallery mode sensors. *Advances in Optics and Photonics*, *7*(2), 168–240.
48. Bozzola, A., Perotto, S., & De Angelis, F. (2017). Hybrid plasmonic–photonic whispering gallery mode resonators for sensing: A critical review. *Analyst*, *142*(6), 883–898.
49. Tu, X., Chen, S. L., Song, C., Huang, T., & Guo, L. J. (2019). Ultrahigh Q polymer microring resonators for biosensing applications. *IEEE Photonics Journal*, *11*(2), 1–10.

50. Zhao, X., Guo, Z., Zhou, Y., Guo, J., Liu, Z., Li, Y., ... Wu, X. (2022). Optical whispering-gallery-mode microbubble sensors. *Micromachines*, *13*(4), 592.
51. Mallik, A. K., Farrell, G., Ramakrishnan, M., Kavungal, V., Liu, D., Wu, Q., & Semenova, Y. (2018). Whispering gallery mode microresonators for multi-parameter sensing applications. *Optics Express*, *26*(24), 31829–31838.
52. De Angelis, F., Patrini, M., Das, G., Maksymov, I., Galli, M., Businaro, L., ... Di Fabrizio, E. (2008). A hybrid plasmonic-photonic nanodevice for label-free detection of a few molecules. *Nano Letters*, *8*(8), 2321–2327.
53. Dantham, V. R., Holler, S., Barbre, C., Keng, D., Kolchenko, V., & Arnold, S. (2013). Label-free detection of single protein using a nanoplasmonic-photonic hybrid microcavity. *Nano Letters*, *13*(7), 3347–3351.
54. Zhou, L., Sun, X., Li, X., & Chen, J. (2011). Miniature microring resonator sensor based on a hybrid plasmonic waveguide. *Sensors*, *11*(7), 6856–6867.
55. Zhang, M., Wu, G., & Chen, D. (2015). Silicon hybrid plasmonic microring resonator for sensing applications. *Applied Optics*, *54*(23), 7131–7134.
56. Runnerstrom, E. L., Bergerud, A., Agrawal, A., Johns, R. W., Dahlman, C. J., Singh, A., ... Milliron, D. J. (2016). Defect engineering in plasmonic metal oxide nanocrystals. *Nano Letters*, *16*(5), 3390–3398.
57. Baaske, M. D., Foreman, M. R., & Vollmer, F. (2014). Single-molecule nucleic acid interactions monitored on a label-free microcavity biosensor platform. *Nature Nanotechnology*, *9*(11), 933–939.
58. Hou, J., Li, M., & Song, Y. (2016). Recent advances in photonic crystal sensors. *Scientia Sinica Chimica*, *46*, 1080–1092.
59. Joannopoulos, J. D., Johnson, S. G., Winn, J. N., & Meade, R. D. (2008). *Molding the flow of light*. Princeton University Press.
60. Choi, E., Choi, Y., Nejad, Y. H. P., Shin, K., & Park, J. (2013). Label-free specific detection of immunoglobulin G antibody using nanoporous hydrogel photonic crystals. *Sensors and Actuators B: Chemical*, *180*, 107–113.
61. Inan, H., Poyraz, M., Inci, F., Lifson, M. A., Baday, M., Cunningham, B. T., & Demirci, U. (2017). Photonic crystals: Emerging biosensors and their promise for point-of-care applications. *Chemical Society Reviews*, *46*(2), 366–388.
62. Gallegos, D., Long, K. D., Yu, H., Clark, P. P., Lin, Y., George, S., ... Cunningham, B. T. (2013). Label-free biodetection using a smartphone. *Lab on a Chip*, *13*(11), 2124–2132.
63. Xu, X., Subbaraman, H., Chakravarty, S., Hosseini, A., Covey, J., Yu, Y., ... Chen, R. T. (2014). Flexible single-crystal silicon nanomembrane photonic crystal cavity. *ACS Nano*, *8*(12), 12265–12271.
64. Kaushik, S., Tiwari, U. K., Pal, S. S., & Sinha, R. K. (2019). Rapid detection of Escherichia coli using fiber optic surface plasmon resonance immunosensor based on biofunctionalized Molybdenum disulfide (MoS_2) nanosheets. *Biosensors and Bioelectronics*, *126*, 501–509.
65. Chaudhary, V. S., Kumar, D., & Kumar, S. (2021). Gold-immobilized photonic crystal fiber-based SPR biosensor for detection of malaria disease in human body. *IEEE Sensors Journal*, *21*(16), 17800–17807.
66. Yu, R., Chen, Y., Shui, L., & Xiao, L. (2020). Hollow-core photonic crystal fiber gas sensing. *Sensors*, *20*(10), 2996.
67. Wu, T., Shao, Y., Wang, Y., Cao, S., Cao, W., Zhang, F., Liao, C., He, J., Huang, Y., Hou, M. and Wang, Y., (2017). Surface plasmon resonance biosensor based on gold-coated side-polished hexagonal structure photonic crystal fiber. *Optics Express*, *25*(17), 20313–20322.

68. Santonocito, R., Cavallaro, A., Pappalardo, A., Puglisi, R., Marano, A., Andolina, M., ... Sfrazzetto, G. T. (2025). Detection of human salivary stress biomarkers using an easy-to-use array sensor based on fluorescent organic molecules. *Biosensors and Bioelectronics*, *270*, 116986.
69. Aslan, M., Seymour, E., Brickner, H., Clark, A. E., Celebi, I., Townsend, M. B., ... Ray, P. (2025). A label-free optical biosensor-based point-of-care test for the rapid detection of Monkeypox virus. *Biosensors and Bioelectronics*, *269*, 116932.
70. Wu, L., Jiao, L., Xue, D., Li, Y., Han, Y., Ouyang, W., & Chen, Q. (2025). Nanozyme and bifunctional nanobody-based colorimetric-SERS dual-mode Immunosensor for microcystin-LR detection. *Food Chemistry*, *464*, 141574.
71. Guo, Y., Xiao, J., Ai, F., Yang, X., Zeng, K., Wang, Y., ... Zhao, H. (2025). Highly responsive cryogel-based sensing platform by encapsulating programmed DNA for colorimetric detection of 17β-estradiol. *Analytica Chimica Acta*, *1334*, 343394.
72. Wang, Y., Feng, Y., Xiao, Z., & Luo, Y. (2025). Machine learning supported single-stranded DNA sensor array for multiple foodborne pathogenic and spoilage bacteria identification in milk. *Food Chemistry*, *463*, 141115.
73. Lu, Y., Xu, J., Liu, S., Tian, B., & Long, F. (2025). A facile optical fiber-embedded microfluidic biochip for rapid and sensitive detection of microRNA-let-7a in serum. *Microchimica Acta*, *192*(1), 1–10.
74. Krishnendu, M. R., Bhagat, S., Jain, V., Mehta, D., & Singh, S. (2025). Paper immobilized BSA-decorated gold nanoclusters for single-step optical sensing of glucose and cholesterol without cross-reactivity. *Colloids and Surfaces B: Biointerfaces*, *245*, 114303.
75. Sun, B., Panferov, V., Guo, X., Xiong, J., Zhang, S., Qin, L., ... Jiang, H. (2025). A novel triple-signal biosensor based on ZrFe-MOF@PtNPs for ultrasensitive aflatoxins detection. *Biosensors and Bioelectronics*, *267*, 116797.
76. Ahn, J. S., & Jang, C. H. (2025). Real-time detection of Tau-381 protein using liquid crystal-based sensors for Alzheimer's disease diagnosis. *Colloids and Surfaces B: Biointerfaces*, *245*, 114211.
77. Karaca Acari, I., Kurul, F., Avci, M. B., Yasar, S. D., Topkaya, S. N., Açarı, C., ... Cetin, A. E. (2024). A plasmonic biosensor pre-diagnostic tool for familial mediterranean fever. *Nature Communications*, *15*(1), 8515.
78. Zhang, H., Zhang, Y. N., Wang, M., Zhao, Y., & Han, B. (2024). Quantification measurement of hemoglobin with large dynamic range and low detection limit via an optical fiber optofluidic laser with enzyme-catalyzed reaction. *Sensors and Actuators A: Physical*, *379*, 115936.
79. Sojdeh, S., Bagherzadeh, M., Daneshgar, H., Edrisi, M., Ahmadifar, M., & Rabiei, N. (2024). Dual-mode optical biocompatible-sensor enabled by enzyme-like activity of UiO-66 (Zr) for ultra-sensitive ROS detection in vitro. *Talanta*, *280*, 126742.
80. Kuang, J., Zhao, L., Ruan, S., Sun, Y., Wu, Z., Zhang, H., ... Hu, P. (2024). The integration platform for exosome capture and colorimetric detection: Site occupying effect-modulated MOF-aptamer interaction and aptamer-Au NPs-dopamine interaction. *Analytica Chimica Acta*, *1329*, 343234.
81. Zou, R., Li, H., Shi, J., Sun, C., Lu, G., & Yan, X. (2024). Dual-enhanced enzyme cascade hybrid hydrogel for the construction of optical biosensor. *Biosensors and Bioelectronics*, *263*, 116613.

6 Magnetic Biosensors
Principles and Biomedical Applications

Kalyani Chepuri, Chittepu Pranitha, Kathuroju Harikrishna, Vadakavila Geethikalal, Rishitha Gayathri Botlapally, Pratiksha Pandey and Ramavath Bharath Kumar

6.1 INTRODUCTION

Magnetic biosensors are advanced analytical tools that use laws of magnetism to identify and measure specific compounds in a sample, converting biological interactions into quantifiable magnetic signals for high sensitivity and specificity. The use of magnetic field biosensors is rapidly expanding and plays a crucial role in various industrial and biomedical applications. Its development is driven by new technologies and capabilities, such as flexible and elastic electronics, as well as emerging use cases and demands like the Internet of Things. Magnetic field sensors operate based on different physical principles, resulting in varying specifications for sensitivity, linearity, field range, power consumption, and cost. Several physical phenomena, including the Hall effect, giant magnetoresistance (GMR), anisotropic magnetoresistance (AMR), tunnel magnetoresistance (TMR), giant magnetoimpedance (GMI), and electromagnetic induction, are utilized to construct magnetic sensors. Superconducting quantum interference devices (SQUIDs) are preferred for applications requiring the highest sensitivity (Khan et al. 2021). The remarkable advantages of magnetic biosensors have led to a surge in interest.

Magnetic nanoparticles (MNPs) are at the forefront of magnetic biosensing and biomedical applications. Over the past two decades, the use of magnetic biosensors has surged, driven by advancements in materials science and nanotechnology. Modern biosensing platforms now leverage MNPs for their durability, biocompatibility, and unique magnetic properties, enabling the development of highly sensitive and specific biosensors for food safety, environmental monitoring, and medical diagnostics (Chen et al. 2024). MNPs typically serve two primary roles: as magnetic labels for signal amplification to enhance detection sensitivity, and as magnetic carriers to separate target molecules from complex samples, thereby reducing background interference (Wang and Lin 2020). Magnetic biosensors utilize magnetic

DOI: 10.1201/9781003536239-6

particles, such as paramagnetic or superparamagnetic nanoparticles, to detect biological interactions. These particles, ranging from nanometers to micrometers in size, are extensively used in modern bioanalytical techniques and biomedical treatments (Chen et al. 2017).

The applications of magnetic biosensors in biomedicine are vast, including pathogen detection, cancer biomarker analysis, and drug delivery monitoring. They offer high sensitivity and label-free detection in complex samples. However, challenges persist, such as enhancing sensitivity, integrating with microfluidics, and achieving clinical translation and regulatory compliance. Addressing these challenges requires multidisciplinary efforts to improve functionality, scalability, and ethical implementation.

This chapter offers an in-depth understanding of various types of magnetic biosensors and their detection principles. It also delves into biomedical applications and challenges, emphasizing sensitivity, selectivity, and clinical translation.

6.2 DESIGN OF MAGNETIC BIOSENSORS

A biosensor is an instrument for identifying a specific biological analyte, including a biological component, physicochemical transducer, and detector. It consists of biorecognition, transduction, and interface elements. A biological recognition element, or bioreceptor, is a molecule that latches onto a target molecule, ensuring high specificity and stability under storage conditions. It must also possess high affinity to the transducer surface through immobilization. The physicochemical transducer measures changes in the bioreceptor's physical property during the reaction, converting energy into quantifiable output. This output is amplified and analyzed, and the signal from the transducer is converted to appropriate units and displayed. The biosensor's performance depends on its ability to detect and convert biological molecules (Inamuddin et al. 2019).

6.2.1 Elements of Magnetic Biosensors

Magnetic biosensors comprise several essential components that collaboratively detect and measure biological substances. The key parts of a magnetic biosensor are represented in Figure 6.1.

6.2.1.1 Bioreceptor

Bioreceptors with high binding affinity to specific analytes can capture trace amounts, enhancing biosensor sensitivity and accuracy. High-selectivity bioreceptors ensure targeted analyte detection in complex biofluids, reducing false positives or negatives (Wang, Zhou, and Wang 2024). Bioreceptors vary, including antibodies, nucleic acids, enzymes, proteins, cells, tissues, and molecular imprints, based on the target molecule. There is no perfect bioreceptor; selection should consider the analytical method and application (Pereira, Sales, and Rodrigues 2019).

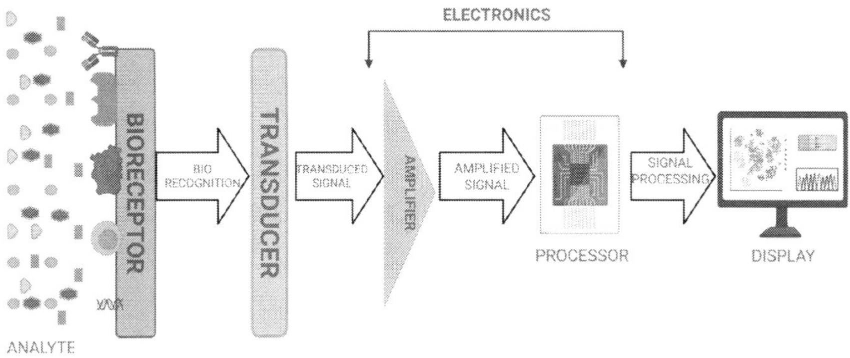

FIGURE 6.1 Components of a magnetic biosensor.

6.2.1.2 Magnetic Labels

Magnetic labels, consisting of magnetic beads and nanoparticles, are used to attach biomolecules in biological samples, as they are biocompatible, uniform in size, and functionalized to minimize non-specific binding (Bhatia et al. 2024). They are typically classified as paramagnetic, ferromagnetic, ferrimagnetic, antiferromagnetic, or superparamagnetic based on their magnetic behavior in and without an applied magnetic field (Chen et al. 2017).

6.2.1.3 Transducer

Signal generation occurs when a biological sensing element binds to a target analyte, and the analyte is then magnetically labeled or tagged with magnetic particles. Magnetic transducers detect changes in magnetic properties due to magnetic labels, including magnetic nanoparticles functionalized with biological sensing elements. Giant magnetoresistive (GMR) sensors detect resistance changes due to magnetic fields, while magnetoresistive sensors measure changes in electrical resistance caused by magnetic fields (Sharma et al. 2021). The transducer converts a detected magnetic signal into an electrical signal, which is then processed and amplified to improve its quality and interpretability (Grieshaber et al. 2008).

6.2.1.4 Signal Detector

The magnetic signal produced by the magnetic particles attached to the target analyte is captured by the detector. Changes in magnetic characteristics, such as inductance, resistance, or magneto-optical properties, frequently cause this signal (Nabaei, Chandrawati, and Heidari 2018).

6.2.1.5 Signal Processor

The refined signals are processed through a signal conditioning unit, which employs a low-pass RC filter to remove noise and ensure data accuracy.

6.2.1.6 Digital Conversion

A processor or microcontroller analyses processed signals, making data analysis and decision-making easier. Analogue signals are converted into digital format, simplifying analysis, processing, and storage.

6.2.1.7 Display

The measured biological quantity is displayed on an LCD screen, and the data is visualized and interpreted to enhance its accessibility and usability (Bhatia et al. 2024).

6.3 TYPES OF MAGNETIC BIOSENSORS AND THEIR PRINCIPLES

There are various types of magnetic biosensors, each operating on different physical principles to detect magnetic fields. The optimal choice of sensor depends on the specific application and the required sensitivity. Some commonly used magnetic biosensors include:

6.3.1 Magnetoresistive Biosensors

Magnetoresistive (MR) biosensors have advanced significantly in mapping biomagnetic fields and detecting magnetic nanoparticles (MNPs) used as biomarkers. These sensors convert magnetic impulses into electrical signals, used in applications like biosensing, hard disk drives, MRAM, and position sensing (Su et al. 2019). MR sensors offer flexibility, low detection limits, and direct electrical signal availability, making them suitable for lab-on-a-chip systems. Applications include giant magnetoresistance (GMR), tunneling magnetoresistance (TMR), and anisotropic magnetoresistance (AMR) (Yang and Zhang 2021). Magnetoresistive (MR) sensors are layered ferromagnetic/nonmagnetic heterostructures exhibiting TMR or GMR effects, used in biosensing due to their sensitivity to small magnetic fields. These sensors have two conduction channels in ferromagnetic metals, divided into spin-up and spin-down bands. Insulating barriers like MgO can enhance the GMR ratio by improving resistance contrast (Xiaolian Sun et al. 2012). Magnetic particles attach via direct (complementary DNA or streptavidin-biotin) or indirect (sandwich immunoassay) labeling (Koh and Josephson 2009). MR sensors are ideal for point-of-care testing (POCT) and can be combined with advanced technologies like microfluidics (Huang et al. 2019).

6.3.1.1 Anisotropic Magnetoresistive Biosensor

In the 1850s, William Thomson discovered that iron and nickel had higher electric resistance when magnetization was parallel to the current direction. This phenomenon, known as anisotropic magnetoresistance (AMR), allows for the detection of magnetic fields by monitoring resistance change upon realignment. AMR sensors are popular due to their resilience and simplicity, and are preferred for alloys of transition metals like Fe, Ni, and Co (Bauer 2021). The AMR effect is a relativistic effect caused by the Zeeman-like interaction between electron spin and magnetic fields. High resistance is observed when magnetization is parallel to current (0° or 180°) and low resistance when perpendicular (θ). The most sensitivity and linearity are achieved by applying magnetization at 45° relative to current (Jander, Smith, and Schneider 2005).

6.3.1.2 Giant Magnetoresistive Biosensors

GMR sensors are effective in detecting low magnetic fields in confined spaces due to the GMR effect, which shifts electrical resistance in multilayer systems (Wu et al. 2022). An external magnetic field causes two conduction channels' resistivity in GMR materials to vary, which lowers the material's electrical resistance overall. Changes in the likelihood of an electron scattering throughout the conduction channel inside the lattice of the GMR material are the cause of the conduction channel's shift in resistance. For example, in the absence of a magnetic field, spin-up and spin-down electrons will undergo the same amount of scattering as they move through the layers of GMR material. The ferromagnetic layer electrons in the GMR material will align themselves parallel to the spin-up conduction electrons when a magnetic field is applied, reducing the likelihood that the electrons would scatter. While the magnetic field decreases the spin-up probability of scattering, the GMR effect is quantified by applying a magnetic field and monitoring the resulting changes in electrical resistance (Cox et al. 2012). GMR biosensors come in two types: multilayer and spin-valve.

6.3.1.2.1 Multilayered GMR Biosensors

Multilayered giant magnetoresistive (GMR) biosensors work by utilizing the giant magnetoresistance effect, which occurs in alternating ferromagnetic and non-magnetic layers in the nanometer range. When an external magnetic field is applied, the magnetization of these layers changes, affecting the electrical resistance of the multilayered structure. A parallel alignment results in low resistance, while an antiparallel alignment results in high resistance (Ennen et al. 2016). When the target analyte binds to the probe, the generated magnetic field affects the magnetization of the ferromagnetic layers, causing a change in resistance, which is measured and correlated to the concentration of the target analyte (Djamal and Ramli 2017).

6.3.1.2.2 Spin-Valve GMR Biosensors

When exposed to an external magnetic field, the layers' resistance in a spin-valve GMR biosensor varies dramatically. It usually comprises a thin ferromagnetic layer, another ferromagnetic layer, and a non-magnetic spacer layer. The main characteristic of the spin-valve is that, in reaction to an external magnetic field, the magnetization of one ferromagnetic layer remains fixed while the magnetization of the other layer moves freely (Hall et al. 2010). The resistance of the two ferromagnetic layers is low when their magnetizations are parallel and high when they are antiparallel. When particular biological molecules are present, this shift in resistance is utilized to identify them (Liang et al. 2022).

6.3.1.3 Tunneling Magnetoresistive Biosensors

Tunnelling magnetoresistance (TMR)-based magnetic sensors are compact, low-power consumption, and highly sensitive. Initially observed in 1975, they were only reproducible at low temperatures (Yan et al. 2022). The discovery of the GMR effect in 1988 marked the beginning of spintronics. In 1995, reproducible room-temperature TMR effects were found in systems using amorphous alumina as the tunneling layer (Ghemes et al. 2023). Subsequent studies on MgO-based MTJs have led to their widespread adoption in various applications, including TMR heads for HDDs, sensors, MRAM, and neuromorphic computing (Matsuura 2021).

Magnetic tunneling junctions (MTJs), which are made up of two ferromagnetic layers spaced apart by a small insulating spacer, exhibit tunneling magnetoresistance (TMR). When a target analyte binds to the probe, a potential difference is applied across the junction, and an electric tunneling current is produced. The output of the sensor depends upon the width of the spacer. The sensitivity of the sensor is maximum when the spacer is sufficiently thin. The relative orientation of the two layer magnetizations determines the tunneling resistance of the junction because of the spin-dependent tunneling effect. Noise measurements (Pnoise/Hz) in magnetic tunnel junctions are crucial for obtaining maximum field resolution at any frequency. Key mechanisms include tunneling across barriers, shot noise at nonzero bias currents, thermal or Johnson-Nyquist noise with zero bias currents, and 1/f noise, which includes both magnetic and non-magnetic noise. Magnetic noise is generated by magnetization oscillations, while 1/f noise appears as resistance fluctuations in the low frequency range. These mechanisms contribute to the noise in MTJs (Yan et al. 2022).

6.3.2 Magneto-Impedance Sensors

The magnetoimpedance effect is a phenomenon that alters the electrical impedance of a soft magnetic material when exposed to an external magnetic field. The sensor element of a MI biosensor is usually a tiny ribbon or wire of this kind. The sensor element creates a magnetic field when an alternating current (AC) passes through it. When an external magnetic field is present, its distribution is changed, which results in modifications to the sensor element's impedance (Sayad, Skafidas, and Kwan 2020). This change in impedance is measured and correlated to the concentration of the target analyte. Magnetoimpedance (MI) biosensors are classified based on their frequency, which affects their performance and sensitivity. Low-frequency MI biosensors, operating below 1 MHz, are suitable for applications requiring high sensitivity and low noise, like detecting low concentrations of biological analytes. High-frequency MI biosensors, operating above 1 MHz, offer faster response times and are used for real-time monitoring and dynamic measurements (Jimenez et al. 2022).

6.3.2.1 Giant Magnetoimpedance Effect–Based Biosensors

The giant magnetoimpedance (GMI) effect is a phenomenon whereby an external magnetic field significantly alters the electrical impedance of a soft magnetic material. Materials like amorphous and nanocrystalline alloys exhibit this effect more strongly. The skin effect, a classical electromagnetic phenomenon, explains why the electromagnetic field associated with an alternating current passing through conductive materials only penetrates a certain amount of them (García-Arribas, Fernández, and Cos 2017).The external magnetic field modifies the magnetic field's distribution, which in turn affects the material's resistance. The skin depth, the depth to which the current penetrates, is inversely related to the square root of the product of the AC current's frequency and the material's permeability. GMI biosensors have high sensitivity and can identify target compounds at extremely low concentrations (Gazda and Szewczyk 2020). Amorphous cobalt-rich ribbons, wires, glass-covered microwires, and multilayers of metallic material encased in ferromagnetic materials are excellent candidates for GMI (Tannous and Gieraltowski 2004).

6.3.3 Hall Effect Sensors

Edwin Herbert Hall discovered the Hall effect in 1879, which is the transverse electric field produced when a current-carrying wire is positioned in a magnetic field perpendicular to it. This creates a Hall voltage, which indicates positive or negative charge carriers (Anderson and Slotkin 1975).The Hall effect and magnetoresistance effect are crucial components of the magnetic sensor realization principle. The Hall effect refers to the movement of carriers influenced by the Lorentz force, shifting when current flows through a semiconductor in the presence of an external magnetic field. This results in a potential difference, known as the Hall voltage. Hall sensors are of two types: linear and switch. Linear Hall effect sensors create a potential difference called Hall voltage when exposed to a perpendicular magnetic field. When magnetic nanoparticles (MNPs) are present, the local magnetic field changes, impacting the sensor's Hall voltage. This change is linked to the concentration of the target analyte. Switch Hall effect sensors, also known as digital sensors, operate similarly to binary switches, activating when a magnetic field strength exceeds a threshold and deactivating when it drops. The Hall voltage, created by the Lorentz force, determines the strength and direction of the magnetic field (Fan et al. 2021).

6.3.4 SQUID-Based Biosensors

The superconducting quantum interference device (SQUID) sensor is a sensitive magnetic sensor with a detection limit as low as 10–14 T, based on the Josephson effect and flux quantization. During measurements, SQUIDs can be set up as gradiometric superconducting flux transformers, which record the gradient of the magnetic flux and inherently eliminate background noise (Faley et al. 2017). The three magnetic properties that SQUID magnetometers can measure are susceptibility, remanence, and relaxation (Chen et al. 2017).

Josephson junctions, also known as superconducting tunnel junctions, are made up of two superconductors encased in a thin layer of insulating material. The Josephson effect is the phenomenon of an electron pair travelling through the barrier tunnel at the Josephson junction and creating a superconducting current when the insulating layer thickness is decreased to 30A.

There is an alternating current distribution in the area where the superconducting ring has a Josephson junction, and the magnetic field controls the alternating current's phase. A slight alteration in the magnetic field will result in a substantial phase shift due to the wave's very high frequency, which causes a huge change in the current. The SQUID sensor can convert weak magnetic signals into significant current signals because of the Josephson junction in its superconducting ring. SQUID sensors are classified as either radio frequency (RF) or direct-current (DC) SQUID sensors based on the bias current. DC SQUIDs create a double-junction superconducting ring by paralleling two superconducting junctions with the same properties. Superconducting junctions are inserted into a superconducting ring to create RF SQUID. DC SQUIDs are classified as high-temperature superconductors (HTS) or low-temperature superconductors (LTS). HTS SQUIDs are superconducting at liquid nitrogen's boiling point, while LTS SQUIDs are more sensitive, stable, and noise-reducing. HTS SQUIDs are more accessible and cost-effective than liquid helium, making them more commonly used (Li et al. 2021).

6.3.4.1 DC SQUID

A superconducting loop with two Josephson junctions makes up a DC superconducting quantum interference device (SQUID). A resistor is used to shunt the junctions in order to avoid hysteresis in the SQUID's I-V behavior. A current significantly greater than double the Josephson junction's critical current is used to bias the SQUID. The current flowing in the superconducting loop will change if an external magnetic flux is inductively connected into the SQUID in this state. A change in the voltage across the junctions is a result of the current. One magnetic flux quantum $\Phi 0 = 2.0678$ Wb is discovered to be the period of this voltage's periodicity with applied flux.

6.3.4.2 rf-SQUID

The rf-SQUID is a superconducting ring with a single Josephson junction, operating by applying an rf current via a tank circuit and measuring the rf voltage. It consists of a single Josephson junction inserted into a superconducting loop, which is inductively coupled to an LC-resonant circuit. The amplitude of the oscillating voltage across the resonant circuit is periodic, allowing for detection of changes in flux of the order of 10. Most SQUIDs are made of low-transition temperature superconductor Nb and operated below the boiling point of liquid helium. The development of high-performance superconductors led to the development of thin films of these materials, resulting in successful technology.

In both varieties, the ring's impedance fluctuates on a regular basis in response to the magnetic flux passing through it due to the characteristics of the Josephson junction. SQUIDs can function as flux-to-voltage converters with the highest magnetic sensitivity of any magnetometer available by measuring the impedance using a lock-in detector (Kleiner et al. 2004).

6.4 BIOMEDICAL APPLICATIONS OF MAGNETIC BIOSENSORS

Magnetic biosensors are emerging as powerful tools in the field of biomedical applications. These devices utilize magnetic fields to detect and measure biological entities, offering several advantages over traditional methods. They are highly sensitive, capable of detecting minute quantities of biological substances, and can be used for a variety of applications, including disease diagnosis, monitoring health conditions, and even guiding targeted therapies (Murzin et al. 2020). Some notable applications include:

6.4.1 Biomolecular Detection

The use of magnetic nanoparticles, which may be engineered to attach selectively to target biomolecules such as DNA, proteins, or pathogens, has made magnetic biosensors an effective tool for biomolecule detection. These systems involve the functionalization of magnetic particles using molecular probes that have a high affinity for particular biomolecules, such as aptamers, nucleic acids, or antibodies. Magnetic biosensors detect biomolecules by measuring changes in magnetic properties, such as magnetic resonance (MR) and giant magnetoresistance (GMR). The sensor can detect a unique magnetic signal produced by the binding of a target molecule to the magnetic particle without the use of extra chemical labels or dyes. Magnetic biosensors are particularly effective for complex biological samples, like tissue or blood, because this label-free detection method reduces interference from sample turbidity or autofluorescence (Pflipsen et al. 2013; Sun et al. 2008). However, high concentrations of biomolecules can lead to particle aggregation, which complicates quantitative measurements. To address this, solutions include modifying the separation distance and particle diameter to prevent dense clustering and ensure a single-layer distribution (Wang et al. 2017).

6.4.2 Pathogen Detection

Infectious diseases cause 13 million deaths annually, with 700,000 deaths from bacteria. Detection techniques are crucial for health and safety. Magnetic biosensors, using magnetic nanoparticles and biological recognition components, help in environmental monitoring and public health (Poltronieri et al. 2014). They are essential tools in pathogen detection, addressing issues in agriculture, public health, and the food sector. They aid in medical diagnosis, treatment, and prevention of infections. Magnetic nanosensors merge biosensing and magnetics to enhance early detection of diseases like SARS-CoV-2. They use magnetic nanoparticles with antibodies or DNA/RNA probes to generate signals proportional to target analyte concentrations (Wu et al. 2022). In the food industry, magnetic biosensors detect pathogens such as *Salmonella* and *E. coli*, enhancing customer safety and reducing the risk of

TABLE 6.1
Comparison of Various Attributes of Different Types of Magnetic Biosensors

Types of Biosensors		MNP Biomarkers	Analytes Detected	Sensor Array Configuration	SNR	Sensitivity	References
Magneto-resistive biosensors	AMR	Fe_3O_4, γ-Fe_2O_3, $CoFe_2O_4$, $NiFe_2O_4CoFe_2O_4$, NWs & NRs (Ni,Co)	IL-8, anti-zika virus IgM and IgG Abs, HBV proteins, HCC biomarkers (alpha-fetoprotein (AFP) and C-reactive protein (CRP)), DNA of *Mycobacterium bovis* BCG (*Bacillus Calmette–Guérin*)	Linear, grid	20–100dB	0.1–10mV/V/Oe	Wang et al. (2024)
	GMR	AuMNPs, C-Coated MNPs, dextran-coated MNPs	HIV-1 p24 Ag, hepatitis B surface antigen (HBsAg), *E. coli* DNA, nucleoprotein (NP) of influenza A virus, RNA of SARS-CoV-2, ovalbumin, protein Arah1	Linear, grid, circular, hexagonal	30–150dB	0.0272–0.0307V/A	Liang et al. (2022)

(*Continued*)

TABLE 6.1 (CONTINUED)

Comparison of Various Attributes of Different Types of Magnetic Biosensors

Types of Biosensors		MNP Biomarkers	Analytes Detected	Sensor Array Configuration	SNR	Sensitivity	References
	TMR	SiO_2MNPs, graphene oxide-coated MNPs, lipid-coated (phospholipids, cholesterol, sphingolipids, lipid-PEG conjugates)	Prostate-specific antigen (PSA), carcinoembryonic antigen (CEA), Troponin, *E. coli* DNA, cholesterol (LDL), estrogen receptor (ER), cytokeratin fragment 21-1 (CYFRA 21-1), Matrix Metallopeptidase 9 (MMP-9)	Linear, grid, circular, hexagonal	40–200dB	0.1–100mV/V/Oe	Parajuli et al. (2021)
Magnetoimpedance biosensors	Low frequency (MI)	AuMNPs, SiO_2MNPs, PEG-coated MNPs, BSA-coated MNPs	Troponin I, HIV Abs, cancer antigen 15-3 (CA15-3), MUC1, osteopontin (OPN), CA-125, prostate cancer antigen 3 (PCA3), acetylcholinesterase (AChE), lactate dehydrogenase (LDH)	Linear, grid	10–30dB	0.01–1mV/V/Oe	Zhao et al. (2022)

(Continued)

TABLE 6.1 (CONTINUED)
Comparison of Various Attributes of Different Types of Magnetic Biosensors

Types of Biosensors		MNP Biomarkers	Analytes Detected	Sensor Array Configuration	SNR	Sensitivity	References
	High frequency (MI)	PEI-coated MNPs, PVA-coated MNPs, PLGA-coated MNPs	Alkaline phosphatase (ALP), glucose oxidase (GOx), Cancer Ag 15-3 (CA15-3), Epithelial cell adhesion molecule (EpCAM), synthetic influenza A DNA and RNA, miR-21, miR-34a, miR-223, dynabeads protein A (DPA), SARS-CoV-2 S1 protein, Myoglobin	Circular, spiral	20–50dB	0.1–10mV/V/Oe	Dey et al. (2023)
Hall effect biosensor	Linear	Silica-coated MNPs, chitosan-coated MNPs,	DNA (*S. aureus*, *S. typhimurium*, *B. subtilis*, cytokeratin 19 fragment (CYFRA 21-1), cytokeratin 18 (CK18),	Linear, grid, circular	10–80dB	0.01–1mV/V/Oe	Ghemes et al. (2023)

(Continued)

TABLE 6.1 (CONTINUED)
Comparison of Various Attributes of Different Types of Magnetic Biosensors

Types of Biosensors	MNP Biomarkers	Analytes Detected	Sensor Array Configuration	SNR	Sensitivity	References
	polymer-coated MNPs, Au-MNPs, GO-coated MNPs	insulin-like growth factor 1 receptor (IGF1R), endotoxins (*E. coli*-lipid A, *S. typhimurium* & *P. aeruginosa-O-*antigen), phospholipase A2				
Switch	Lipid-coated MNPs, carbon-coated MNPs, Amino-silane-coated MNPs	O157:H7 strain *E. coli* DNA, Human Epidermal Growth Factor Receptor 2 (HER2), candidalysin, TR34 gene of *A. fumigatus*, acetylcholinesterase (AChE), soy allergens (Gly m1, m2, m4)	Linear, grid, circular, hexagonal	15–90dB	0.1–10mV/V/Oe	Memon, Niazi, and Qureshi (2024)

(Continued)

TABLE 6.1 (CONTINUED)
Comparison of Various Attributes of Different Types of Magnetic Biosensors

Types of Biosensors		MNP Biomarkers	Analytes Detected	Sensor Array Configuration	SNR	Sensitivity	References
SQUID-based biosensor	DC-SQUID	Ferrites MNPs (CO, Zn, Ni), amino silane-coated MNPs, protein-coated MNPs	Drug targeting (doxorubicin, paclitaxel, gentamicin, vancomycin, ibuprofen, aspirin), B-type natriuretic peptide,	Linear, planar, 3D	100–300dB	1–1000fT/$\sqrt{\text{Hz}}$	Balaban Hanoglu et al. (2023)
	RF-SQUID	PLGA-coated MNPs, phosphatidylcholine-coated MNPs, esterified cholesterols (C oleate, C palmitate), sphingolipids	NP of influenza A virus IL-6, IL-10, TNF-α, fibrinogen, HER2/neu, Fe(II) of myoglobin, N-procalcitonin, AFP receptor, HIV p24 Ag receptor, troponin receptor, cardiac troponin T (cTnT).	Linear, planar, 3D	90–250dB	10–1000fT/$\sqrt{\text{Hz}}$	Thamminana et al. (2024)

foodborne illnesses. In environmental monitoring, they help identify contaminants in water sources, preventing waterborne diseases. With their portability and ease of use, magnetic biosensors are well-suited for on-site applications across a range of industries (Vidic and Manzano 2021). Notable examples of magnetic biosensor used in pathogen detection include the following:

6.4.2.1 Magnetic Biosensor for the Detection of *Yersinia pestis*

Yersinia pestis, a non-motile, Gram-negative bacterium, poses a global health threat due to its potential as a biological weapon. A magnetic biosensor detects the disease using an immunosensory technique based on the binding of the fraction 1 capsule protein to the *Y. pestis* YP19 antibody (Riedel 2005; Meyer et al. 2007). The magnetic detection system for the *Y. pestis* F1 antigen uses PE sintered filters to capture the antigen, with a biotin-streptavidin system linking the biotinylated detection antibody and magnetic bead. It offers high specificity and sensitivity, with optimal results from covalent methods and adsorptive binding. Adsorptive immobilization on PE-sintered filters is simple and effective. The magnetic bead-based sensor performs well with complex samples, efficiently separating beads from problematic matrices (Figure 6.2) (Meyer et al. 2007; Tito et al. 2001).

6.4.2.2 Magnetic Biosensor for the Detection of *Francisella tularensis*

Tularemia is a highly contagious bacterial zoonosis prevalent in the northern hemisphere, affecting various animal hosts, including rodents and carnivores. It is caused by *Francisella tularensis*, a Gram-negative coccobacillus. PCR and ELISA are the primary detection methods for *F. tularensis*. Figure 6.3 illustrates

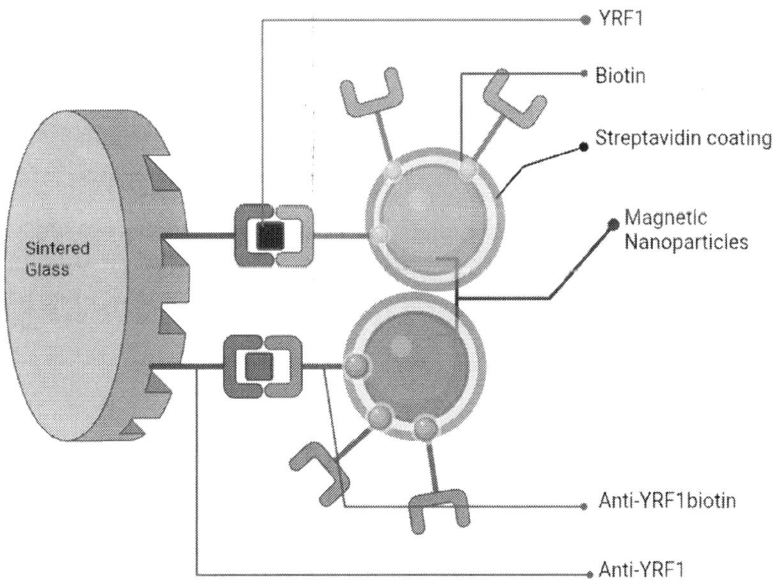

FIGURE 6.2 Detection of *Yersinia pestis* using magnetic biosensors.

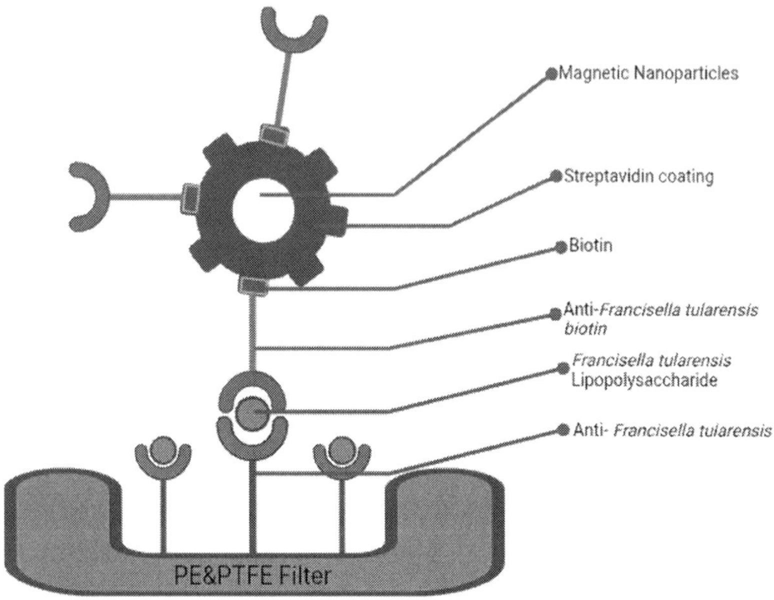

FIGURE 6.3 Detection of *Francisella tularensis* LPS by using anti-Ft antibodies and magnetic beads.

an immunomagnetic bead-based test for detecting *F. tularensis*. The sample first passes through a polyethylene (PE) filter, where anti-Ft antibodies bind to bacterial lipopolysaccharides (LPS), forming antigen–antibody complexes. Biotinylated antibodies then link these complexes to streptavidin-coated magnetic beads. Using an external magnetic field, the beads with the bound antigen-antibody complexes are isolated, enriching the target for further detection or analysis (Meyer et al. 2007) (Figure 6.4).

6.4.3 Cancer Biomarkers

Early detection of biomarkers is essential for clinical diagnosis of cancer. The combination of electrochemical devices with nanomaterials such as gold, magnetic particles, and carbon nanotubes allows for simultaneous and highly sensitive testing of cancer biomarkers, leveraging amplification and multiplexing capabilities (Ravalli and Marrazza 2015).

6.4.3.1 Magnetic Nanoparticles (MNPs)–Based Biosensors for Cancer Biomarkers Detection

Magnetic particles are synthesized by reacting Fe(II) and Fe(III) salts with a base, forming iron oxides like magnetite and maghemite. These particles enhance bioreceptor–analyte interactions, as direct immobilization on electrodes can impede electron transfer. Modified magnetic bead suspensions increase sensitivity and reduce

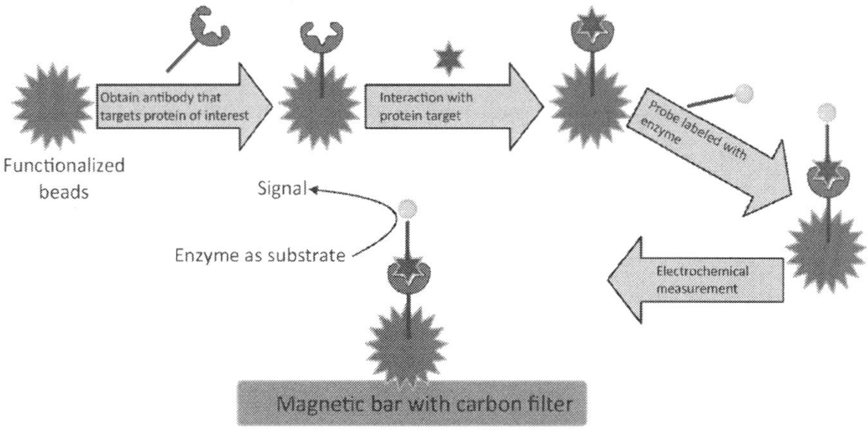

FIGURE 6.4 Schematic representation of magnetic beads–based assay for detection of protein of interest.

assay time by improving bioreceptor–analyte contact. In sandwich assays, functionalized magnetic beads are captured with a magnetic field to separate unbound biomolecules. After the antigen and detection probe reaction, the modified particles are placed on a working electrode for electrochemical analysis.

Surface-functionalized magnetic beads are coated with a capture antibody specific to a target cancer biomarker. When the sample is introduced, the capture antibody binds to the protein biomarker, forming an antigen–antibody complex. An enzyme-labeled tracer antibody is then added, which binds to the biomarker and creates a sandwich-like structure. These modified beads are placed on a magnetic (carbon working) electrode. Upon the enzyme and substrate reaction, an electrochemical signal is generated. The signal's intensity correlates directly with the biomarker's concentration, enabling the detection and quantification of cancer biomarkers. This method supports cancer diagnosis and monitoring by providing precise measurements of biomarker levels (Ravalli and Marrazza 2015).

6.4.3.1.1 Prostate Specific Antigen (PSA) Detection and Carcino Embryonic Antigen (CEA) Detection

Prostate specific antigen (PSA) is a key marker for prostate cancer. Researchers have developed an electrochemical sandwich immunoassay for detecting free PSA (f-PSA) using polyurethane-coated magnetic beads. Reported by Sarkar et al. (2008), anti-PSA antibodies are covalently attached to these beads, which are then washed to remove any remaining molecules. The modified beads are collected on a working electrode to measure hydrogen peroxide in the presence of TMB, enabling f-PSA quantification in human serum samples. Another method involves using protein G-coated magnetic beads to immobilize anti-PSA capture antibodies with PSA antigen, capture antibodies, and a tertiary alkaline phosphatase-labeled antibody. The transducer consists of screen-printed sensor arrays with separate electrodes (Sarkar et al. 2008).

Pan and Yang (2007) developed a biosensor for detecting carcinoembryonic antigen (CEA) using anti-CEA antibody-modified magnetic beads and a flow-injection electrochemical device. This system immobilizes the anti-CEA antibody on core-shell magnetic nanoparticles, streamlining the analytical process. When exposed to CEA, the sensor completes a sandwich assay, achieving a detection limit of 1 pg/ml and showing strong correlation with ELISA reference results (Pan and Yang 2007).

6.4.4 Drug Delivery Monitoring

Magnetic nanoparticles (MNPs) are gaining popularity in drug delivery and sensorics, particularly in the field of theranostics. They can be used in biogenic and chemically synthesized nanoparticles, and magnetic biosensors, which detect chemical interactions with great sensitivity, are becoming popular in drug delivery monitoring. Magnetic biosensors are utilized to monitor magnetic nanoparticles (MNPs), often coupled with drugs or drug carriers like liposomes or polymers. These sensors detect changes in magnetic signals as the medication is released from the carrier at the target site. Additionally, they can be designed to respond to external magnetic fields, allowing for real-time, controlled drug release. This approach is particularly beneficial in cancer treatment, where precise, localized medication delivery is crucial for minimizing side effects and maximizing therapeutic efficacy. Magnetic biosensors also provide quantitative data on drug concentration, release kinetics, and biodistribution, ensuring the treatment remains within the therapeutic window (Zimina et al. 2022).

6.5 CHALLENGES AND FUTURE DIRECTIONS

Magnetic biosensors hold significant promise for a wide range of applications, from medical diagnostics to environmental monitoring. However, several challenges remain in their development and widespread adoption. These include the following:

6.5.1 Sensitivity and Selectivity Improvements

Magnetic biosensors face challenges in detecting low concentrations of target analytes in complex samples due to issues like signal-to-noise ratio, external magnetic field interference, and signal loss. Precise functionalization of magnetic nanoparticles with biorecognition molecules is crucial, but scaling down detection systems to the nanoscale introduces difficulties such as reduced reproducibility and increased sensitivity to environmental changes. Enhancing selectivity in complex sample environments with numerous non-target molecules is particularly challenging, and designing biorecognition elements with high specificity demands significant resources and technical expertise (Roth, Margulis, and Danielli 2022).

6.5.2 Integration with Microfluidics

Combining magnetic biosensors with microfluidics presents technical challenges in design, manufacturing, and operational compatibility. Laminar flow, which dominates fluid dynamics in microchannels, can hinder the effective mixing of reagents or target analytes with magnetic nanoparticles, making precise fluid manipulation at the microscale both essential and difficult. Additionally, advanced materials and methods are required to fabricate microfluidic devices that are highly compatible with magnetic sensing. This ensures seamless integration without compromising the sensitivity or selectivity of the biosensors (Wang et al. 2024).

Maintaining the stability and functionality of biorecognition elements during microfluidic operations poses several challenges. Variations in flow rates, shear stress, or sample viscosity can affect magnetic nanoparticle capture efficiency or disrupt target-binding interactions. Additionally, scaling microfluidic systems for practical applications, such as clinical diagnostics, requires reliable and cost-effective manufacturing processes and user-friendly interfaces, which are still under research and development (Roth, Margulis, and Danielli 2022).

6.5.3 Clinical Translation and Regulatory Issues

Regulatory issues are significant challenges and considerations when adapting magnetic biosensors for clinical use. Key points include the following:

6.5.3.1 Clinical Translation

- *Validation and standardization*: Thorough validation and standardization are necessary to guarantee that magnetic biosensors satisfy clinical standards. In clinical contexts, this entails exhibiting sensitivity, specificity, accuracy, and precision (Kim et al. 2020; Jimenez et al. 2022).
- *Clinical trials*: Regulatory approval requires clinical trials to assess magnetic biosensors' efficacy and safety in real-world situations (Jimenez et al. 2022).
- *Integration with existing systems*: For magnetic biosensors to be widely used in clinical settings, they must work with the current workflows and medical infrastructure (Kim et al. 2020).

6.5.3.2 Regulatory Issues

- *Regulatory approval*: A comprehensive regulatory evaluation process is required for magnetic biosensors, which involves obtaining approval from organizations like the European Medicines Agency (EMA) and the Food and Drug Administration (FDA) of the United States (Karim 2020).
- *Ethical and legal concerns*: Legal and ethical issues include patient consent, data privacy, and the possibility of sensitive health information being misused (Karim 2020).
- *Compliance with standards*: Ensuring the safety and effectiveness of magnetic biosensors requires adherence to international standards and

guidelines, such as those set forth by the International Electrotechnical Commission (IEC) and the International Organization for Standardization (ISO) (Karim 2020; Abdel-Karim 2024).

These challenges underscore the importance of a multidisciplinary approach, involving collaboration among engineers, physicians, regulatory specialists, and ethicists, to effectively transition magnetic biosensors from the laboratory to patient care. Future directions in magnetic biosensor research are focused on enhancing sensor sensitivity and multiplexing capabilities by integrating nanomaterials like magnetic nanoparticles. Significant advancements in miniaturization, portability, and real-time data processing will be crucial for these technologies' success. Additionally, leveraging machine learning and artificial intelligence for enhanced signal processing and pattern recognition can help address existing limitations and enable more efficient, point-of-care diagnostics.

6.6 CONCLUSION

Magnetic biosensors are sophisticated and adaptable tools applicable across the biomedical, industrial, and environmental sectors. Their magnetic interactions allow unparalleled sensitivity and specificity in the detection and measurement of specific biological compounds. They operate based on various physical principles, including magnetoresistance, magneto-impedance, the Hall effect, and superconducting quantum interference devices (SQUIDs). Each type of magnetic biosensor has distinct characteristics, allowing them to be tailored for specific applications depending on factors like sensitivity, field range, and cost. These sensors work by detecting changes in electrical resistance in response to magnetic fields. The integration of magnetic nanoparticles (MNPs) has significantly enhanced these sensors, sparking innovations in applications such as pathogen detection, cancer biomarker identification, and disease diagnostics.

As magnetic biosensor technology continues to evolve, it is likely that we will see even greater integration with emerging technologies like the Internet of Things (IoT), where sensors could be embedded in smart devices for continuous health monitoring or environmental surveillance. These advancements in sensor miniaturization, flexibility, and power efficiency will further expand the scope and applicability of magnetic biosensors, making them even more accessible and versatile across industries. Magnetic biosensors hold significant promise but face challenges in sensitivity, selectivity, and integration with microfluidics. Overcoming regulatory, validation, and ethical hurdles is essential for clinical translation. Despite these challenges, magnetic biosensors are poised to revolutionize health, environmental, and industrial monitoring, making them indispensable in biosensing.

REFERENCES

Abdel-Karim, Randa. 2024. "Nanotechnology-Enabled Biosensors: A Review of Fundamentals, Materials, Applications, Challenges, and Future Scope." *Biomedical Materials & Devices* 2 (2): 759–77. doi:10.1007/s44174-023-00147-z.

Anderson, Thomas R., and Theodore A. Slotkin. 1975. "Maturation of the Adrenal Medulla—IV." *Biochemical Pharmacology* 24 (16): 1469–74. doi:10.1016/0006-2952(75)90020-9.

Balaban Hanoglu, Simge, Duygu Harmanci, Nursima Ucar, Serap Evran, and Suna Timur. 2023. "Recent Approaches in Magnetic Nanoparticle-Based Biosensors of miRNA Detection." *Magnetochemistry* 9 (1): 23. doi:10.3390/magnetochemistry9010023.

Bauer, Gerrit E. W. 2021. "Anisotropic Magnetoresistance: A 170-Year-Old Puzzle Solved." *Science China Physics, Mechanics & Astronomy* 64 (1): 217531. doi:10.1007/s11433-020-1618-7.

Bhatia, Dinesh, Sohini Paul, Tania Acharjee, and Shrimanata Sundar Ramachairy. 2024. "Biosensors and Their Widespread Impact on Human Health." *Sensors International* 5: 100257. doi:10.1016/j.sintl.2023.100257.

Chen, Dongfei, Teresa Zardán Gómez De La Torre, Fuxiang Wei, Bo Tian, and Kai Wu. 2024. "Editorial: Magnetic Particle-Assisted Sensing and Magnetic Biosensors." *Frontiers in Bioengineering and Biotechnology* 12 (November): 1518156. doi:10.3389/fbioe.2024.1518156.

Chen, Yi-Ting, Arati G. Kolhatkar, Oussama Zenasni, Shoujun Xu, and T. Randall Lee. 2017. "Biosensing Using Magnetic Particle Detection Techniques." *Sensors* 17 (10): 2300. doi:10.3390/s17102300.

Cox, Bryan, Don Liu, Chester Wilson, and Despina Davis. 2012. "GMR Sensors: Technologies and Medical Applications." *Recent Patents on Nanomedicine* 1 (2): 130–37. doi:10.2174/1877913111101020130.

Dey, Anusree, Jyoti Prakash, Rituparna Das, Sandeep Shelar, Ajay Saini, Susan Cherian, Sofia C Patel, et al. 2023. "Development of a Rapid and Ultra-Sensitive RNA:DNA Hybrid Immunocapture Based Biosensor for Visual Detection of SARS-CoV-2 RNA." Edited by Richard Stanton. *PNAS Nexus* 2 (3): pgad031. doi:10.1093/pnasnexus/pgad031.

Djamal, Mitra, and Ramli Ramli. 2017. "Giant Magnetoresistance Sensors Based on Ferrite Material and Its Applications." In *Magnetic Sensors - Development Trends and Applications*, edited by Aktham Asfour. InTech. doi:10.5772/intechopen.70548.

Ennen, Inga, Daniel Kappe, Thomas Rempel, Claudia Glenske, and Andreas Hütten. 2016. "Giant Magnetoresistance: Basic Concepts, Microstructure, Magnetic Interactions and Applications." *Sensors* 16 (6): 904. doi:10.3390/s16060904.

Faley, M I, J Dammers, Y V Maslennikov, J F Schneiderman, D Winkler, V P Koshelets, N J Shah, and R E Dunin-Borkowski. 2017. "High- T_c SQUID Biomagnetometers." *Superconductor Science and Technology* 30 (8): 083001. doi:10.1088/1361-6668/aa73ad.

Fan, Hua, Jiangming Wang, Quanyuan Feng, Qiang Hu, Siming Zuo, Vahid Nabaei, and Hadi Heidari. 2021. "Detection Techniques of Biological and Chemical Hall Sensors." *RSC Advances* 11 (13): 7257–70. doi:10.1039/D0RA10027G.

García-Arribas, Alfredo, Eduardo Fernández, and David De Cos. 2017. "Thin-Film Magneto-Impedance Sensors." In *Magnetic Sensors - Development Trends and Applications*, edited by Aktham Asfour. InTech. doi:10.5772/intechopen.70084.

Gazda, P., and R. Szewczyk. 2020. "Influence of Joule-Annealing on Double-Peak GMI Effect in Co-Based Amorphous Ribbons." *Acta Physica Polonica A* 137 (5): 818–20. doi:10.12693/APhysPolA.137.818.

Ghemes, Crina, Oana-Georgiana Dragos-Pinzaru, Mihai Tibu, Mihaela Lostun, Nicoleta Lupu, and Horia Chiriac. 2023. "Tunnel Magnetoresistance-Based Sensor for Biomedical Application: Proof-of-Concept." *Coatings* 13 (2): 227. doi:10.3390/coatings13020227.

Grieshaber, Dorothee, Robert MacKenzie, Janos Vörös, and Erik Reimhult. 2008. "Electrochemical Biosensors - Sensor Principles and Architectures." *Sensors* 8 (3): 1400–58. doi:10.3390/s80314000.

Hall, D. A., R. S. Gaster, T. Lin, S. J. Osterfeld, S. Han, B. Murmann, and S. X. Wang. 2010. "GMR Biosensor Arrays: A System Perspective." *Biosensors and Bioelectronics* 25 (9): 2051–57. doi:10.1016/j.bios.2010.01.038.

Huang, H. T., P. Garu, C. H. Li, W. C. Chang, B. W. Chen, S. Y. Sung, C. M. Lee, et al. 2019. "Magnetoresistive Biosensors for Direct Detection of Magnetic Nanoparticle Conjugated Biomarkers on a Chip." *SPIN* 09 (02): 1940002. doi:10.1142/S2010324719400022.

Inamuddin, Ali Mohammad, Abdullah Mohamed Asiri, and Raju Khan, eds. 2019. *Advanced Biosensors for Health Care Applications*. Amsterdam: Elsevier. doi:10.1016/B978-0-12-815743-5.00017-2.

Jander, Albrecht, Carl Smith, and Robert Schneider. 2005. "Magnetoresistive Sensors for Nondestructive Evaluation (Invited Paper)." In *Proceedings of SPIE—The International Society for Optical Engineering: Advanced Sensor Technologies for Nondestructive Evaluation and Structural Health Monitoring*, edited by Norbert Meyendorf, George Y. Baaklini, and Bernd Michel, 1. San Diego, CA: SPIE. doi:10.1117/12.601826.

Jimenez, Valery Ortiz, Kee Young Hwang, Dang Nguyen, Yasif Rahman, Claire Albrecht, Baylee Senator, Ongard Thiabgoh, et al. 2022. "Magnetoimpedance Biosensors and Real-Time Healthcare Monitors: Progress, Opportunities, and Challenges." *Biosensors* 12 (7). Multidisciplinary Digital Publishing Institute: 517. doi:10.3390/bios12070517.

Karim, Mohammad Ershadul. 2020. "Biosensors: Ethical, Regulatory, and Legal Issues." In *Handbook of Cell Biosensors*, edited by Gerald Thouand, 1–27. Cham: Springer International Publishing. doi:10.1007/978-3-319-47405-2_23-1.

Khan, Mohammed Asadullah, Jian Sun, Bodong Li, Alexander Przybysz, and Jürgen Kosel. 2021. "Magnetic Sensors-A Review and Recent Technologies." *Engineering Research Express* 3 (2): 022005. doi:10.1088/2631-8695/ac0838.

Kim, Seong-Eun, My Van Tieu, Sei Young Hwang, and Min-Ho Lee. 2020. "Magnetic Particles: Their Applications from Sample Preparations to Biosensing Platforms." *Micromachines* 11 (3): 302. doi:10.3390/mi11030302.

Kleiner, R., D. Koelle, F. Ludwig, and J. Clarke. 2004. "Superconducting Quantum Interference Devices: State of the Art and Applications." *Proceedings of the IEEE* 92 (10): 1534–48. doi:10.1109/JPROC.2004.833655.

Koh, Isaac, and Lee Josephson. 2009. "Magnetic Nanoparticle Sensors." *Sensors* 9 (10): 8130–45. doi:10.3390/s91008130.

Li Yongying, Hui Cheng, Zahrah Alhalili, Guiying Xu, and Guo Gao. 2021. "The Progress of Magnetic Sensor Applied in Biomedicine: A Review of Non-invasive Techniques and Sensors." *Journal of the Chinese Chemical Society* 68 (2): 216–27. doi:10.1002/jccs.202000353.

Liang, Shuang, Phanatchakorn Sutham, Kai Wu, Kumar Mallikarjunan, and Jian-Ping Wang. 2022. "Giant Magnetoresistance Biosensors for Food Safety Applications." *Sensors* 22 (15): 5663. doi:10.3390/s22155663.

Matsuura, Yukihito. 2021. "Tunnel Magnetoresistance of a Natural Protein Molecule." *Chemical Physics Letters* 777 (August): 138741. doi:10.1016/j.cplett.2021.138741.

Memon, Roomia, Javed H. Niazi, and Anjum Qureshi. 2024. "Biosensors for Detection of Airborne Pathogenic Fungal Spores: A Review." *Nanoscale* 16 (33): 15419–45. doi:10.1039/D4NR01175A.

Meyer, Martin H.F., Hans-Joachim Krause, Markus Hartmann, Peter Miethe, Jürgen Oster, and Michael Keusgen. 2007. "*Francisella tularensis* Detection Using Magnetic Labels and a Magnetic Biosensor Based on Frequency Mixing." *Journal of Magnetism and Magnetic Materials* 311 (1): 259–63. doi:10.1016/j.jmmm.2006.10.1175.

Murzin, Dmitry, Desmond J. Mapps, Kateryna Levada, Victor Belyaev, Alexander Omelyanchik, Larissa Panina, and Valeria Rodionova. 2020. "Ultrasensitive Magnetic Field Sensors for Biomedical Applications." *Sensors* 20 (6): 1569. doi:10.3390/s20061569.

Nabaei, Vahid, Rona Chandrawati, and Hadi Heidari. 2018. "Magnetic Biosensors: Modelling and Simulation." *Biosensors and Bioelectronics* 103 (April): 69–86. doi:10.1016/j.bios.2017.12.023.

Pan, Jin, and Qingwei Yang. 2007. "Antibody-Functionalized Magnetic Nanoparticles for the Detection of Carcinoembryonic Antigen Using a Flow-Injection Electrochemical Device." *Analytical and Bioanalytical Chemistry* 388 (1): 279–86. doi:10.1007/s00216-007-1224-0.

Parajuli, Ishwari, Gianluigi Ciovati, Jean Delayen, and Alexander Gurevich. 2021. "Evaluation of Anisotropic Magnetoresistive (AMR) Sensors for a Magnetic Field Scanning System for SRF Cavities." PDF. Proceedings of the 12th International Particle Accelerator Conference IPAC2021. JACoW Publishing, Geneva, Switzerland: 4 pages, 1.254 MB. doi:10.18429/JACOW-IPAC2021-TUPAB344.

Pereira, Ac, Mgf Sales, and Lr Rodrigues. 2019. "Biosensors for Rapid Detection of Breast Cancer Biomarkers." In *Advanced Biosensors for Health Care Applications*, 71–103. Elsevier. doi:10.1016/B978-0-12-815743-5.00003-2.

Pflipsen, C., D. Forge, S. Benali, and Y. Gossuin. 2013. "Improved Stability and Relaxivity of a Commercial Magnetic Ferrofluid." *The Journal of Physical Chemistry C* 117 (40): 20919–26. doi:10.1021/jp4050786.

Poltronieri, Palmiro, Valeria Mezzolla, Elisabetta Primiceri, and Giuseppe Maruccio. 2014. "Biosensors for the Detection of Food Pathogens." *Foods* 3 (3): 511–26. doi:10.3390/foods3030511.

Ravalli, Andrea, and Giovanna Marrazza. 2015. "Gold and Magnetic Nanoparticles-Based Electrochemical Biosensors for Cancer Biomarker Determination." *Journal of Nanoscience and Nanotechnology* 15 (5): 3307–19. doi:10.1166/jnn.2015.10038.

Riedel, Stefan. 2005. "Plague: From Natural Disease to Bioterrorism." *Baylor University Medical Center Proceedings* 18 (2): 116–24. doi:10.1080/08998280.2005.11928049.

Roth, Shira, Michael Margulis, and Amos Danielli. 2022. "Recent Advances in Rapid and Highly Sensitive Detection of Proteins and Specific DNA Sequences Using a Magnetic Modulation Biosensing System." *Sensors* 22 (12). Multidisciplinary Digital Publishing Institute: 4497. doi:10.3390/s22124497.

Sarkar, P., D. Ghosh, D. Bhattacharyay, S.J. Setford, and A.P.F. Turner. 2008a. "Electrochemical Immunoassay for Free Prostate Specific Antigen (f-PSA) Using Magnetic Beads." *Electroanalysis* 20 (13): 1414–20. doi:10.1002/elan.200804194.

Sayad, Abkar, Efstratios Skafidas, and Patrick Kwan. 2020. "Magneto-Impedance Biosensor Sensitivity: Effect and Enhancement." *Sensors* 20 (18): 5213. doi:10.3390/s20185213.

Sharma, Apoorva, Ashok D. Chougale, Georgeta Salvan, and Prashant B. Patil. 2021. "Magnetoresistance-Based Biosensors." In *Analytical Applications of Functionalized Magnetic Nanoparticles*, edited by Chaudhery Mustansar Hussain, 369–96. The Royal Society of Chemistry. doi:10.1039/9781839162756-00369.

Su, Diqing, Kai Wu, Renata Saha, Chaoyi Peng, and Jian-Ping Wang. 2019. "Advances in Magnetoresistive Biosensors." *Micromachines* 11 (1): 34. doi:10.3390/mi11010034.

Sun, Ying, Daqian Song, Yuping Bai, Liying Wang, Yuan Tian, and Hanqi Zhang. 2008. "Improvement of Surface Plasmon Resonance Biosensor with Magnetic Beads via Assembled Polyelectrolyte Layers." *Analytica Chimica Acta* 624 (2): 294–300. doi:10.1016/j.aca.2008.06.042.

Tannous, C., and J. Gieraltowski. 2004. "Giant Magneto-Impedance and Its Applications." *Journal of Materials Science: Materials in Electronics* 15 (3): 125–33. doi:10.1023/B:JMSE.0000011350.93694.91.

Thamminana, Bhargava, Mayurkumar Revabhai Patel, Madhura Pradeep Deshpande, Tae Jung Park, and Suresh Kumar Kailasa. 2024. "Fluorescence Turn-off Detection of Myoglobin as a Cardiac Biomarker Using Water-Stable L-Glutamic Acid Functionalized Cesium Lead Bromide Perovskite Quantum Dots." *Microchimica Acta* 191 (11): 674. doi:10.1007/s00604-024-06752-z.

Tito, Mark A., Julie Miller, Kate F. Griffin, E. Diane Williamson, Richard W. Titball, and Carol V. Robinson. 2001. "Macromolecular Organization of the *Yersinia Pestis* Capsular F1 Antigen: Insights from Time-of-flight Mass Spectrometry." *Protein Science* 10 (11): 2408–13. doi:10.1110/ps.11101.

Vidic, Jasmina, and Marisa Manzano. 2021. "Electrochemical Biosensors for Rapid Pathogen Detection." *Current Opinion in Electrochemistry* 29 (October): 100750. doi:10.1016/j.coelec.2021.100750.

Wang, Lei, and Jianhan Lin. 2020. "Recent Advances on Magnetic Nanobead Based Biosensors: From Separation to Detection." *TrAC Trends in Analytical Chemistry* 128 (July): 115915. doi:10.1016/j.trac.2020.115915.

Wang, Miaomiao, Lian Jin, Polly Hang-Mei Leung, Franklin Wang-Ngai Chow, Xiaoni Zhao, Hui Chen, Wenjing Pan, Hongna Liu, and Song Li. 2024. "Advancements in Magnetic Nanoparticle-Based Biosensors for Point-of-Care Testing." *Frontiers in Bioengineering and Biotechnology* 12 (April). Frontiers. doi:10.3389/fbioe.2024.1393789.

Wang, Tao, Yong Zhou, Chong Lei, Jun Luo, Shaorong Xie, and Huayan Pu. 2017. "Magnetic Impedance Biosensor: A Review." *Biosensors and Bioelectronics* 90 (April): 418–35. doi:10.1016/j.bios.2016.10.031.

Wang, Xueqi, Jianhua Zhou, and Han Wang. 2024. "Bioreceptors as the Key Components for Electrochemical Biosensing in Medicine." *Cell Reports Physical Science* 5 (2): 101801. doi:10.1016/j.xcrp.2024.101801.

Wu, Kai, Denis Tonini, Shuang Liang, Renata Saha, Vinit Kumar Chugh, and Jian-Ping Wang. 2022. "Giant Magnetoresistance Biosensors in Biomedical Applications." *ACS Applied Materials & Interfaces* 14 (8): 9945–69. doi:10.1021/acsami.1c20141.

Xiaolian Sun, Don Ho, Lise-Marie Lacroix, J. Q. Xiao, and Shouheng Sun. 2012. "Magnetic Nanoparticles for Magnetoresistance-Based Biodetection." *IEEE Transactions on NanoBioscience* 11 (1): 46–53. doi:10.1109/TNB.2011.2176509.

Yan, Shaohua, Zitong Zhou, Yaodi Yang, Qunwen Leng, and Weisheng Zhao. 2022. "Developments and Applications of Tunneling Magnetoresistance Sensors." *Tsinghua Science and Technology* 27 (3): 443–54. doi:10.26599/TST.2021.9010061.

Yang, Songlin, and Jin Zhang. 2021. "Current Progress of Magnetoresistance Sensors." *Chemosensors* 9 (8): 211. doi:10.3390/chemosensors9080211.

Zhao, Wenlei, Xinchen Tao, Chaofeng Ye, and Yu Tao. 2022. "Tunnel Magnetoresistance Sensor with AC Modulation and Impedance Compensation for Ultra-Weak Magnetic Field Measurement." *Sensors* 22 (3): 1021. doi:10.3390/s22031021.

Zimina, Tatiana M., Nikita O. Sitkov, Kamil G. Gareev, Viacheslav Fedorov, Denis Grouzdev, Veronika Koziaeva, Huile Gao, Stephanie E. Combs, and Maxim Shevtsov. 2022. "Biosensors and Drug Delivery in Oncotheranostics Using Inorganic Synthetic and Biogenic Magnetic Nanoparticles." *Biosensors* 12 (10): 789. doi:10.3390/bios12100789.

7 Point-of-Care Biosensors for Rapid Diagnostics

Priyanka Thakral

7.1 INTRODUCTION TO POINT-OF-CARE (POC) DIAGNOSTICS

The COVID-19 pandemic has brought immense attention towards the need for fast and rapid detection of medical ailments that could be chronic, contagious, or problematic. Accurate and timely diagnosis is crucial to hinder the progression of a disease. Emerging trends in the field of biotechnology, life sciences, electrochemistry, optics, electronics, superconductor materials, and other interdisciplinary areas of engineering have brought a breakthrough transition from bench testing (conventional diagnostic techniques) to point-of-care applications (portable diagnostic methods).

Point-of-care (POC) diagnostics (Gubala et al., 2012) refers to medical assessments done outside of traditional laboratories, typically at or near the site of patient care, such as hospitals, clinics, or even homes. The need for rapid, accurate, and cost-effective diagnostics has fueled the development of various technologies, among which biosensors stand out as a transformative solution. These POC biosensing systems are enabled by a wide range of sensing procedures that can provide accurate, highly selective, and sensitive information in resource-limited environments (Figure 7.1). The expanding demand worldwide for convenient diagnosis, monitoring, and screening tests projects that the global POC diagnostics industry will reach $77.8 billion by 2028.

In general, POC testing encompasses any tests that are performed at or near a patient and at the site where care or treatment is provided. The most common case of POC devices used worldwide is that of glucose home testing kits (glucometers), which depend on potentiometric sensors used for monitoring blood glucose levels for the effective management of diabetes (Zhang et al., 2021). Today, POC diagnostic devices come in an array of forms. They may use basic dipsticks (urinalysis, pregnancy kits), handheld devices (glucose meters), advanced molecular analyzers to detect infectious diseases (COVID-19 testing kits), lab-on-a-chip devices (blood pressure monitoring) or wearable/implantable devices.

The outbreaks of infectious diseases over the last two decades has spread awareness of the need for rapid detection of diseases, thereby exponentially increasing the demand for POC devices. Herein, the current chapter focuses on the basics of biosensor operations: their integration for POC devices, the types of POC biosensor technologies known today, and the role of artificial intelligence (AI) in POC

DOI: 10.1201/9781003536239-7

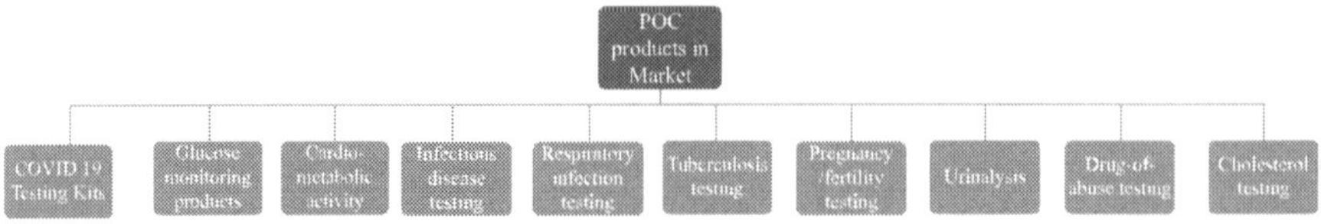

FIGURE 7.1 Some of the POC products available on the market.

biosensing (Cui et al., 2020). Applications of various POC biosensing systems are also presented in detail, followed by the challenges associated with them and the future of next-generation POC devices.

7.2 BASICS OF BIOSENSOR DESIGN AND OPERATIONS

A biosensor (Haleem et al., 2021) is a bioanalytical tool that detects a biological or a biochemical response by converting a biochemical signal into an electrical signal. A biosensor has four main components: the *analyte*, the *recognition element*, the *transducer*, and the *data handling* procedures.

- The *analyte* (small molecules, ions, viruses, bacteria, proteins, nucleic acids, or even whole human cells) is the molecule of interest that is to be detected using a biosensor. Analyte selection is important for the development of suitable and practical biosensor applications. Although there are no limitations on what can act as an analyte, there are four relevant factors that should be considered when deciding on an analyte. Firstly, the selected analyte should have an established link to the disease of interest for which the biosensor is to be designed. Secondly, the chosen analyte should be present at sufficient concentration for its assessment. Thirdly, the concentration variation of the analyte should reflect some physiologically relevant parameter. Fourth and the last, an analyte should have a known, selective *recognition element* to enable its targeting.
- The *recognition element* (antibodies, enzymes etc.) is the molecule or structure that interacts with the analyte for its detection. Biorecognition elements may be catalysis based or affinity based, depending on their ability to react or bind with the analyte of interest. The recognition elements can be classified into three categories: Protein based (enzymes, antibodies), nucleic acid based (DNA, RNA), and lab synthesized (nano-enzymes, molecularly imprinted polymers).
- The *transducer* is the method or the instrument that converts the biochemical interaction between the analyte and the recognition element into a readable signal (electrochemical, optical, colorimetric, electronic, thermal, acoustic, etc.)
- The *data handling* procedures allow for acquisition, analysis, and interpretation (measurement, denoising, deconvolution, correlation, etc.) of the generated data by the transducer.

7.2.1 Types of Biosensors for POC Diagnostics Based on Recognition Elements

Biosensors for POC devices are classified here on the basis of recognition elements as enzyme-based biosensors, immunosensors, DNA/RNA biosensors, and molecularly imprinted biosensors.

7.2.1.1 Enzyme-Based Biosensors

These sensors use enzymes as biological recognition elements. The enzyme catalyzes a reaction with the analyte molecule, generating a measurable signal (such as current, light, or heat) (Turner and Karube, 2011). A good example is that of glucose meters (e.g., Accu-Chek, OneTouch). These devices measure blood glucose levels in diabetic patients by detecting the oxidation of glucose, facilitated by a glucose oxidase enzyme. The glucose oxidase catalyzes the oxidation of glucose to gluconolactone, producing hydrogen peroxide (H_2O_2) as a byproduct, the amount of which directly correlates to glucose concentration and can be measured using an electrochemical or optical sensor (Wang & Chen, 2019). Cholesterol test strips are another example here. The cholesterol oxidase enzyme catalyzes the oxidation of cholesterol to produce cholest-4-*en*-3-one and hydrogen peroxide. The H_2O_2 produced can be measured electrochemically or optically, indicating cholesterol concentration (Huang & Li, 2014). Lactate intolerance can be detected using lactate sensors where a lactate oxidase enzyme catalyzes the oxidation of lactate to pyruvate, generating hydrogen peroxide as a byproduct, the concentration of which correlates with lactate levels and can be measured using electrochemical sensors (Ali et al., 2020). Urine test strips are used to detect urea levels where a urease enzyme catalyzes the hydrolysis of urea into ammonia and carbon dioxide. The ammonia produced can alter the pH of the solution, and this change can be detected electrochemically or optically (Yuan et al., 2016).

Many enzyme-based biosensors for POC applications are being used widely because of their ease of use, portability, cost effectiveness, and real-time result availability. However, enzymes are sensitive to temperature, humidity, and other environmental factors which can potentially affect the sensors' performance and output. Another challenge with enzyme-based biosensors is the interference of other metabolites, which can lead to inaccurate results. It is important to calibrate enzyme based biosensors to ensure accuracy and effectiveness.

7.2.1.2 Immunosensors

Immunosensors rely on the interaction between an antibody and its specific antigen. The binding causes a detectable signal (electrical, optical, etc.) (Wang & Lee, 2014). A popular example in this case is that of rapid antigen tests used for COVID-19 detection. These tests use antibodies to detect viral antigens in a sample, providing a rapid diagnosis at the point of care. Rapid strep throat test immunosensors can detect group A *Streptococcus* bacterium in throat swabs (Patnaik et al., 2019). Prostate-specific antigen (PSA) tests use immunosensors to detect PSA levels in blood, which can be an indicator of prostate cancer (Kumar & Singh, 2020). Pregnancy tests use lateral flow immunoassays for home pregnancy tests to detect the presence of human chorionic gonadotropin (hCG) in urine (Manzano & García, 2018). Allergen-specific IgE tests use immunosensors to measure specific IgE antibodies in response to allergens, which can help diagnose allergies and autoimmune conditions (Verma & Rana, 2021).

The main challenges with immunosensors include sample impurities, cross reactivity, non-specific binding, and the inability to detect low analyte concentrations, which can affect its accuracy. The shelf life and storage conditions of immunosensors, especially those relying on antibodies, can be affected by environmental factors like temperature.

7.2.1.3 DNA/RNA Biosensors

These biosensors use nucleic acid sequences (DNA or RNA) that specifically bind to complementary strands, generating a detectable signal. For example, Cepheid GeneXpert detects genetic material from pathogens (e.g., COVID-19, tuberculosis) in patient samples (Li & Li, 2020). DNA/RNA biosensors are at the forefront of diagnostics for viral infections like SARS-CoV-2 (Koch & Baijal, 2021). Many rapid tests use RNA biosensors to detect the virus's RNA in nasal swabs or saliva samples, enabling quick results. Nucleic acid biosensors have been developed for detecting Mycobacterium tuberculosis (TB) DNA in sputum samples, helping to diagnose TB (Wang & Zhang, 2020). DNA/RNA biosensors can detect mutations or biomarkers related to cancers, such as BRCA1/BRCA2 mutations in breast cancer or KRAS mutations in colon cancer (Huang & Zhang, 2020). These tests can be used for early screening and personalized treatment plans. DNA/RNA biosensors are used in the detection of foodborne pathogens like *Salmonella*, *E. coli*, or *Listeria* (Kumar & Vashist, 2021). These sensors can be integrated into portable devices for onsite testing in food safety applications. Non-invasive prenatal testing (NIPT) using DNA/RNA biosensors can detect fetal DNA in maternal blood to screen for conditions like Down syndrome, trisomy 18, and other genetic disorders (Bianchi & Chiu, 2018).

7.2.1.4 Molecularly Imprinted Polymers (MIPs)

These are lab synthesized materials that have highly selective recognition capability for specific target molecules (Vlatakis & Mo, 2020). These polymers are "imprinted" with molecular templates during their synthesis process, creating cavities in the polymer structure that are complementary in shape, size, and functional groups to the target analyte. The process of creating MIPs involves four main steps:

- *Template molecule*: A specific molecule (the template) that is the target for detection is selected. This could be a drug, pathogen, biomarker, or any other target molecule.
- *Polymerization*: The template molecule is mixed with functional monomers and a cross-linking agent in a solvent. These monomers form a polymer matrix around the template.

- *Template removal*: After polymerization, the template molecule is removed, leaving behind cavities in the polymer that are complementary to the template in terms of size, shape, and functional groups.
- *Binding to target molecule*: The imprinted polymer now has a high affinity for the target molecule, and it can selectively bind to it when exposed to a sample.

MIP-based pregnancy test strips use a synthetic polymer that specifically binds to hCG, which is a hormone released during pregnancy. When urine is applied to the strip, a visible color change occurs if hCG is present, indicating pregnancy (Liu et al., 2013). MIP-based glucose sensors involve interaction between glucose and the MIP, inducing an electrochemical or optical signal, which is measured and used to determine glucose concentration in the blood (Patil et al., 2017). MIP-based drug test strips are designed to recognize drugs like cocaine, marijuana, and opioids. These strips provide a visible signal (color change) when the target drug or its metabolites are present in a sample (Wang & Lin, 2007). MIP-based sensors can be used to detect heavy metals in water samples. These sensors are imprinted to bind specifically to metal ions, and the interaction leads to a detectable signal (e.g., color change) (You et al., 2018). MIPs are designed to bind specifically to the spike protein or RNA of SARS-CoV-2. This sensor generates a detectable signal (optical or electrochemical) when the virus is present, allowing for rapid diagnostics (Zhang & Zhang, 2021).

MIPs exhibit high selectivity towards the target molecules; are stable under a wide range of environmental conditions (temperature, pH, etc.); can be synthesized in bulk, making them cost effective; can be reused for multiple detection cycles; and can work directly with complex biological samples without requiring extensive sample preprocessing. However, there are some challenges associated with MIPs. The process of removing the template molecule can be complex, and any remaining template could affect the performance of the MIP. Complex biological samples may contain molecules that interfere with MIP binding, leading to false positives or negatives. Although MIPs are highly specific, their use may still be limited by the availability of suitable templates for certain analytes. MIP-based POC devices must undergo rigorous validation and regulatory approval, which can be time-consuming and costly (Vlatakis & Mo, 2020).

7.2.2 Types of Biosensors for POC Diagnostics Based on Transduction Methods

Transduction is the most essential component in biosensing systems as it provides a means to convert molecular interactions into signals or information that can be more easily processed and analyzed. Common transduction methods include electrochemical, colorimetric, optical, acoustic, thermal and piezoelectric methods.

7.2.2.1 Electrochemical Biosensors

Electrochemical biosensors (Zhang & Tan, 2020) are among the most common for point-of-care applications. These sensors detect changes in electrical properties (such as current, voltage, or impedance) caused by the biochemical reaction between the target analyte and the biological receptor.

- *Amperometric*: Measures the current generated by the electrochemical reaction.
- *Potentiometric*: Measures the change in voltage generated when the analyte interacts with the receptor.
- *Impedimetric*: Detects changes in the impedance of the sensor when the analyte binds to the receptor.

Devices are used in emergency settings to measure blood oxygen, carbon dioxide, and pH levels. These sensors often rely on electrochemical reactions involving enzymes or ion-selective electrodes. Amperometric glucose biosensors measure the current produced when glucose is oxidized by glucose oxidase at the electrode surface. The FreeStyle Libre system is a continuous glucose monitoring device using electrochemical sensors to monitor glucose levels in interstitial fluid (Potyrailo & Makarov, 2013). A CRISPR/Cas-based electrochemical sensor has been developed for detecting SARS-CoV-2 RNA directly from clinical samples with high sensitivity (Niu & Liao, 2021). MIP-based pregnancy tests combined with electrochemical impedance spectroscopy to detect hCG levels at low concentrations in urine (Liu & Zhao, 2015). Electrochemical biosensors are used for *E. coli* detection in drinking water, offering rapid, on-site pathogen screening (Liu & Huang, 2018).

With advancements in nanotechnology and bioreceptor design, electrochemical biosensors are becoming more sensitive, efficient, and adaptable for real-time diagnostics.

7.2.2.2 Optical Biosensors

Optical biosensors (Li & Zhang, 2021) utilize light to detect the interaction between the target analyte and the biological recognition element. These include:

- *Surface plasmon resonance (SPR)*: Measures the change in the refractive index near the surface of the sensor when the analyte binds to the receptor.
- *Fluorescence*: Involves a change in fluorescence intensity upon interaction of the analyte with a fluorescent probe.
- *Absorption spectroscopy*: Measures changes in light absorption, often using colorimetric methods.

Fluorescent-based biosensors detect glucose using enzymes like glucose oxidase, which causes a fluorescent signal proportional to glucose concentration. The Dexcom G6 continuous glucose monitoring system uses optical sensing to monitor glucose levels in real time (Liao & Lin, 2019). Over-the-counter pregnancy test kits use colorimetric assays to detect hCG levels in urine (Liu & Zhang, 2017). CRISPR-based optical biosensors for rapid detection of SARS-CoV-2 RNA in patient samples, with fluorescence signals indicating the presence of the virus (Zhang & Lu, 2021). SPR-based optical biosensors detect PSA levels in blood or serum based on the refractive index changes caused by PSA binding to immobilized antibodies (Wu & Li, 2018).

Optical biosensors offer a promising approach for point-of-care diagnostics due to their high sensitivity, real-time detection capabilities, and non-invasive nature. However, challenges such as environmental interference, cost, reproducibility, and complexity in interpretation need to be addressed for widespread adoption.

7.2.2.3 Piezoelectric Biosensors

Piezoelectric biosensors (Skladel & Beranek, 2010) work by utilizing piezoelectric materials (e.g., quartz crystal, polymers, ceramics) that convert mechanical energy into electrical signals. When an analyte binds to the sensor's surface (often functionalized with a bioreceptor such as an antibody, enzyme, or DNA), the mass of the sensor changes. This causes a change in its resonant frequency, which is proportional to the amount of mass added to the surface. The frequency shift is measured and correlated with the concentration of the target analyte.

7.2.2.3.1 Working Mechanism

- *Piezoelectric effect*: When mechanical stress is applied to a piezoelectric material, it generates an electrical charge. This effect is reversible, meaning that when an electric field is applied, the material undergoes mechanical deformation.
- *Mass sensing*: When a target analyte interacts with the immobilized receptor, it adds mass to the sensor surface. This mass change causes a shift in the sensor's resonant frequency, which can be measured electronically and correlated with the analyte concentration.
- *Resonance frequency shift*: The amount of mass change is detected by observing shifts in the resonant frequency of the piezoelectric material. A smaller mass change results in a larger frequency shift, making piezoelec tric biosensors highly sensitive.

A quartz crystal microbalance (QCM) sensor functionalized with glucose oxidase has been developed to detect glucose levels in blood or interstitial fluids (Salvo et al., 2020). Piezoelectric biosensors for *E. coli* detection have been employed in food safety and environmental monitoring. These sensors provide rapid detection without the need for complex sample preparation (Lee et al., 2017). A piezoelectric DNA sensor has been developed to detect the RNA of SARS-CoV-2 by targeting specific genetic sequences. The sensor detects binding events and reports real-time changes in frequency (Cheng et al., 2021).

Piezoelectric biosensors are an exciting option for point-of-care diagnostics, offering significant advantages such as high sensitivity, real-time detection, and portability. However, challenges such as environmental interference, surface functionalization, and limited dynamic range must be addressed to improve their reliability, accuracy, and versatility in POC settings.

7.2.2.4 Thermometric Biosensors

Thermometric biosensors (Saito & Nakajima, 2021) are a type of biosensor that measures temperature changes as an indicator of biochemical reactions or interactions. These sensors exploit the heat released or absorbed during biochemical processes such as enzyme–substrate reactions, antigen–antibody binding, or nucleic acid hybridization. Thermometric biosensors are advantageous in POC applications because they offer a simple and direct detection method with minimal or no sample preparation.

A thermometric biosensor using glucose oxidase for glucose monitoring has been developed for non-invasive or minimally invasive testing of glucose levels in blood (Zhang et al., 2019). A thermometric biosensor has been developed for real-time DNA detection, allowing for rapid and direct measurement of nucleic acid sequences (Lazcka et al., 2007). Thermometric biosensors for hCG detection have been developed as a simple and rapid method for pregnancy testing (Chang and Wang, 2019).

These sensors have significant advantages, including high sensitivity, real-time detection, cost-effectiveness, and portability. However, challenges related to temperature interference, small temperature changes, and sample matrix effects need to be addressed.

7.2.3 Modern Biosensing Technologies

In recent years, POC sensing technologies have seen significant development. Lab-on-a-chip platforms, microfluidic devices, paper-based sensors and wearable or implantable sensors are some of the most transformative technologies in the field of biosensing. These devices offer miniaturization, integration, portability, and rapid processing, making them ideal for real-time diagnostics in clinical settings, as well as in resource-limited environments.

7.2.3.1 Lab-on-a-Chip (LOC) and Lab-on-a-Disk (LOD) Devices

LOC devices integrate multiple laboratory functions (such as sample preparation, detection, and analysis) onto a single chip using microfluidics. Microchannels on the chip handle the flow and manipulation of small volumes of fluids (from nanoliters to microliters) to enable highly specific assays. These devices typically use technologies like electrochemical sensing, optical detection (fluorescence, colorimetric), or immunoassays, enabling rapid and accurate biomarker detection. LOC devices can incorporate various features such as pumps, valves, and mixers on-chip to automate sample preparation and detection. LOC devices have been developed to detect SARS-CoV-2 antigens or antibodies, offering a fast and cost-effective solution for COVID-19 testing. PCR-based LOC devices are also being developed for more accurate and sensitive viral RNA detection (Chin et al., 2012). Researchers have also been able to develop LOC devices to detect PSA levels in blood to diagnose prostate cancer (Sharma et al., 2020) and also biomarkers to detect breast cancer (Jaing et al.,

2019) Microfluidic-based glucose monitoring systems for effective diabetes management have been developed (Vashist et al., 2013).

LOD devices (also known as centrifugal microfluidic systems) use a spinning disk to drive fluid through microchannels and chambers. The centrifugal force from rotation controls the movement of fluids, enabling processes such as mixing, separation, and detection. These devices are especially advantageous in POC diagnostics due to their simplicity, cost-effectiveness, and ability to automate complex laboratory workflows on a single platform. LOD systems often use fluorescence, absorbance, or electrochemical detection to analyze samples (Bano et al., 2019; Müller et al., 2020; Jeong et al., 2019).

7.2.3.2 Wearable/Implantable Sensors

Wearable and implantable sensors represent a significant leap forward in healthcare by providing continuous, real-time monitoring of a wide range of physiological parameters. These non-invasive or minimally invasive portable sensors are designed to be worn or implanted in the body, allowing for seamless integration with the human body and enabling constant monitoring of health metrics (Yeo et al., 2016). Many wearable and implantable sensors have wireless capabilities, transmitting data to smartphones, tablets, or cloud-based platforms for real-time monitoring by patients and healthcare providers. Wearable/implantable continuous glucose monitors (CGMs) (Miao et al., 2020), ECG monitors (Yeo et al., 2016), Blood pressure monitors (Kim et al., 2020), temperature monitoring smart patches (Liu et al., 2019), implantation temperature sensors (Zhou et al., 2021), pulse oximeters (Zhang et al., 2020), sweat sensors (Park et al., 2019), and EEG (electroencephalography) wearables (He et al., 2018) are the new-generation POC devices that have enhanced health surveillance, thereby enabling routine health monitoring and improving global health outcomes.

7.2.3.3 Paper-Based Biosensors

Paper-based sensors, being low cost, simple, portable, and disposable, are a revolutionary class of diagnostic tools for POC applications. Colorimetric paper strips use enzymatic reactions to change the color of paper upon glucose interaction for glucose level monitoring. The color change is then analyzed visually or using a smartphone camera to quantify glucose levels (Xie et al. 2019). Lateral flow urine sample strips work by using capillary action to move urine samples across the paper, where they interact with antibodies or other reagents to produce visible results (Ahn et al., 2016). Paper-based sensors can be designed to detect specific bacterial pathogens in water by using immunoassays or biochemical reactions that change the color of the paper (Patel et al., 2015).

7.3 BIOSENSING AT POINT OF CARE AND CHALLENGES

The WHO special program for research and training in tropical diseases (WHO/TDR) in 2003 published criteria for the ideal test that can be used at all levels of the healthcare system to guide treatment and clinical management decisions for infectious tropical diseases and sexually transmitted infections. These criteria, known by the acronym ASSURED (affordable, sensitive, specific, user-friendly, rapid, equipment free and delivered), have become widely accepted as the benchmark for an ideal test that can be used at the POC (Land et al., 2021).

In recent years, the increasing frequency and severity of global health emergencies caused by infectious diseases of epidemic potential such as SARS, coronavirus, Ebola, Zika, or human metapneumovirus (HMPV), have made the global community realize the need to accelerate ASSURED POC test development, validation, manufacturing, and deployment.

Bedside clinical care: POC sensors are now routinely used in bedside clinical applications for rapid diagnostic screening and assessment. The integration of these biosensors into clinical practice has significantly enhanced medical intervention efficiency, and the future development of new biosensing technologies is expected to pave the way for a rapid medical turnaround (Hong et al., 2022; Chang et al., 2021).

At-home care: The concept of at-home POC sensing began in the late 1970s with the introduction of the first pregnancy tests (Ehrenkranz, 2002), which utilized lateral flow assays (LFAs) to detect human chorionic gonadotropin (hCG) in urine. These tests were a game-changer in personal healthcare, allowing individuals to perform diagnostics at home quickly, easily, and privately. To this day, pregnancy tests based on LFA technology remain among the most widely available and commonly used biosensors globally (Bahadır & Sezgintürk, 2015).

Another pioneering biosensor is the personal glucose meter (Clark & Foster, 2018), which has been available since the early 1980s. These devices enable people with diabetes to monitor their blood glucose levels independently, offering rapid results in just a few seconds. The personal glucose meter has become a cornerstone of diabetes management, allowing patients to track their glucose levels at home and make adjustments to their treatment as needed. This self-monitoring capability has not only improved diabetes care by enhancing patient control but has also helped reduce the need for frequent clinical visits, ultimately making diabetes management more efficient and cost-effective (Clarke & Foster, 2018). The deployment of POC biosensors for at-home use has transformed personal health management by ensuring medical monitoring is accessible and convenient.

Public health and environment: The deployment of biosensors in the field represents a critical aspect of POC assessment that can significantly improve environmental and public health monitoring. One notable area where in-field biosensing strategies have made a substantial impact is in the drug-of-abuse testing. Rapid testing kits for substances such as fentanyl have enhanced safety for first responders by enabling quick detection of hazardous substances, thus assisting medical personnel and other emergency responders in identifying dangerous drugs and mitigating exposure risks (Dashtian et al., 2023; Barnett et al., 2023).

Another important application of in-field biosensors is in the monitoring of water quality, particularly for harmful substances like heavy metals and bacterial pathogens. Biosensors have been developed to detect these contaminants, offering a practical solution for areas with limited access to traditional water-testing infrastructure. These sensors can greatly benefit impoverished communities, providing critical tools for monitoring water safety and protecting public health (Maity et al., 2023; Samani et al., 2020). Food safety monitoring is another area where biosensors play a vital role. These devices enable the rapid detection of harmful pathogens, such as *Salmonella* and *E. coli*, helping to prevent foodborne infections and reduce public health risks (Griesche & Baeumner, 2020). Moreover, biosensors are also being deployed for infectious disease monitoring—for example, SARS-CoV-2 detection tools—which enable widespread tracking of disease incidence. The real-time data gathered from these biosensors allows public health officials to make informed decisions quickly, implement effective policies, and respond to evolving health threats more efficiently.

While POC technologies have greatly enhanced the accessibility and rapidity of medical diagnostics, they often fall short of the accuracy and reliability offered by traditional laboratory tests. This limitation is primarily due to the inherent constraints in the design and functionality of POC devices. These devices typically rely on simplified methods and smaller sample sizes, often targeting single analytes, which can lead to less accurate results. Additionally, the transducers used in POC devices tend to produce signals with lower sensitivity, partly due to higher levels of background noise, compared to the more advanced equipment used in laboratory settings. The use of POC devices in uncontrolled environments, such as at home or in the field, where factors like temperature, humidity, and user handling skills can vary, introduces additional variability that may affect the reliability of the results. However, the future of POC diagnostics on integration with artificial intelligence (AI) looks promising and can dramatically improve the performance and reliability of POC diagnostics.

7.4 ARTIFICIAL INTELLIGENCE–ASSISTED POC BIOSENSING

By integrating AI with biosensing technologies, the ability to analyze complex datasets, detect early health abnormalities, and make personalized healthcare decisions has expanded dramatically. This synergy promises to improve patient outcomes, streamline healthcare processes, and make diagnostics more accessible, particularly in remote or resource-limited settings. AI-assisted POC biosensing involves the fundamental steps of identifying a problem, gathering relevant information, training a model using algorithms, and applying the model.

AI approaches can be applied at each stage of the biosensor development process, assisting in the selection of analytes, development of recognition elements, enhancement of signal transduction, and analysis and interpretation of data. Here, various methods by which AI is being employed in the biosensing field are discussed, providing a clear picture of the strength of the computational procedures.

AI can help analyze and interpret the data generated by POC biosensors, which often involve complex signals that may not be immediately obvious to human operators. Machine learning (ML) (Ganjalizadeh et al., 2023) and deep learning (DL) algorithms can process and analyze these signals, identify and extract relevant patterns, classify data, and provide more accurate diagnostics (Thakral et al., 2016). This is especially useful when dealing with large, multi-dimensional data sets, such as those generated from glucose sensors or sensors for infectious diseases or cardiac rhythm tracking. In glucose monitoring, AI can analyze patterns in continuous glucose data to predict hypoglycemic or hyperglycemic events, providing early warnings to patients and healthcare providers (Wang et al., 2019). ML algorithms have been employed for mental fatigue analysis with a predictive accuracy of up to 89% by evaluating different kinds of physiological features using classification algorithms (Zeng et al., 2020). Large volumes of diverse data can be handled with the help of AI technologies supporting more precise diagnostics, tailored treatment plans, and effective health management.

Integrating AI into transduction stage of biosensing focuses on material design, instrument miniaturization, and biorecognition free transduction. Deep learning models have been employed to predict plasmonic nanostructure geometries that produce optimal responses upon interaction with various biomarkers (Malkiel et al., 2018). Enhanced computational abilities have made it possible to develop systems that have less complex instrumentation with comparable resolution and efficacy to laboratory-based instruments (McGlennen, 2020). ML algorithms are capable of extracting patterns and data that make it possible to develop biosensors without the need for analyte specific biological interactions. Nicoliche et al. (2020) were able to extract a signal fingerprint that could detect the p16 protein down to 0.6 pg/mL using ML model.

Incorporation of AI into recognition element identification can revolutionize POC biosensing through enhanced screening methods enabling de novo design of in-silico receptors (Bennett et al., 2023). Numerous AI-based tools have been developed to select recognition molecules which exhibit the desired specificity and sensitivity (Sun et al., 2022). AI procedures allow accurate prediction of biomolecule structures directly from primary sequences, including nucleic acid or amino acid sequences, offering exciting opportunities in the design of recognition elements. One widely recognized AI tool is AlphaFold2 (AF2) by Google DeepMind, which has made predictions of over 200 million protein structures (Jumper et al., 2021). This tool predicts the three-dimensional structure of proteins using a deep neural network assessing distances and angles between amino acid pairs within a protein sequence. The Baker Lab developed RoseTTAFold (RF), a three-track neural network that simultaneously considers the patterns in protein sequences, interactions among protein amino acids, and potential three-dimensional protein structures, achieving structure predictions with accuracies comparable to those of DeepMind (Baek et al., 2021). AI approaches have also made it possible to develop and design catalysts including protein enzymes and nanozymes that can act as recognition elements in biosensors (Zhuang et al., 2024).

AI is known to enhance biosensing through omics-driven prediction of biomolecular pathways, biomarker discovery, and multi-analyte analysis. The prediction of biomolecular pathways with an AI-omics approach proves to provide a better understanding of human health and disease (Siristatidis et al., 2021). AI-assisted discovery of biomarkers can be accomplished by various machine learning avenues (Chung et al., 2021). AI-assisted molecular profiling (the process of assessment of multiple biomarkers) makes it possible to improve both disease diagnosis and prediction of individual treatment response (Malekzadeh et al., 2017). Lyme disease remains a difficult-to-diagnose disease due to its non-specific or non-existent symptoms (Flynn et al., 2020). Through the application of machine learning to identify optimal biomarkers, Joung et al. developed a paper-based biosensor capable of diagnosing Lyme disease in blinded samples with better success than previous POC tests (Joung et al., 2020).

7.5 CONCLUSIONS AND SCOPE OF NEXT-GENERATION POC BIOSENSING DEVICES

The next generation of POC biosensing devices holds great potential to bring in a colossal revolution in healthcare by making POC diagnostic testing ASSURED certified around the globe. One of the key challenges is ensuring that next-generation POC biosensors maintain high accuracy and reliability comparable to laboratory-based tests, with minimal risk of false positives or negatives. For next-generation POC biosensing devices to gain widespread adoption, they must be easy to use, non-invasive, cost-effective, and patient-friendly. As POC biosensors increasingly collect and transmit sensitive health data, data privacy and cybersecurity are another important concern which needs to be dealt with by using highly sophisticated encryption methods to improve data safety. Developing streamlined and adaptable regulatory frameworks for emerging POC technologies will be crucial to accelerating their adoption. Integration of AI and ML with POC biosensing will be pivotal in improving the performance of next-generation POC biosensing devices. AI can enhance data analysis, optimize sensor sensitivity, and interpret complex biomarker data, providing immediate and accurate results. However, the "black box" nature of many AI models poses a challenge in clinical settings. Multiplexed biosensing and multi-analyte detection will further enhance the ability of POC devices to detect a wide range of biomarkers. Sustainable and environmentally friendly materials will be used in the development of biosensors to reduce waste and promote recycling, aligning with the broader trend toward green healthcare technologies.

REFERENCES

Abid Haleem, Mohd Javaid, Ravi Pratap Singh, Rajiv Suman, and Shanay Rab. "Biosensors Applications in Medical Field: A Brief Review." *Sensors International* 2 (2021): 100100.

Ahn, J. H., et al. "Development of Paper-Based Test Strips for Multi-Analyte Detection in Urine for POC Applications." *Lab on a Chip* 16, no. 8 (2016): 1489–1495.

Ali, M. A., et al. "Enzyme-Based Wearable Biosensors for Continuous Monitoring of Lactate and Glucose in Sweat: Challenges and Prospects." *Trends in Analytical Chemistry* 124 (2020): 115803.

Baek, M., F. DiMaio, I. Anishchenko, J. Dauparas, S. Ovchinnikov, G. R. Lee, J. Wang, Q. Cong, L. N. Kinch, R. D. Schaeffer, et al. "Accurate Prediction of Protein Structures and Interactions Using a Three-Track Neural Network." *Science* 373 (2021): 871–876.

Bahadır, E.B., and M.K. Sezgintürk. "Applications of Commercial Biosensors in Clinical, Food, Environmental, and Biothreat/Biowarfare Analyses." *Analytical Biochemistry* 478 (2015): 107–120.

Bano, M., et al. "Lab-on-a-Disc Technology for Point-of-Care Diagnostics: A Review." *Lab on a Chip* 19, no. 11 (2019): 1823–1838.

Barnett, B.S., P.R. Chai, and J. Suzuki. "Scaling up Point-of-Care Fentanyl Testing—A Step Forward." *New England Journal of Medicine* 389 (2023): 1643–1645.

Bennett, N.R., B. Coventry, I. Goreshnik, B. Huang, A. Allen, D. Vafeados, Y.P. Peng, J. Dauparas, M. Baek, L. Stewart, et al. "Improving de Novo Protein Binder Design with Deep Learning." *Nature Communications* 14 (2023): 2625.

Bianchi, D.W., and R.W. Chiu. "Non-Invasive Prenatal Testing: The Advent of Genomics in the Fetal-Maternal Interface." *Nature Reviews Genetics* 19, no. 11 (2018): 649–660.

Chang, D., S. Zakaria, S.E. Samani, Y. Chang, C.D.M. Filipe, L. Soleymani, J.D. Brennan, M. Liu, and Y. Li. "Functional Nucleic Acids for Pathogenic Bacteria Detection." *Accounts of Chemical Research* 54 (2021): 3540–3549.

Chang, J., and Y. Wang. "Label-free Sensors Based on Graphene Field-effect Transistors for the Detection of Human Chorionic Gonadotropin." *Sensors* 18, no. 1 (2018): 229.

Cheng, X., et al. "Development of a Piezoelectric Biosensor for COVID-19 Detection." *Sensors* 21, no. 2 (2021): 395.

Chin, C. D., et al. "Lab-on-a-Chip Devices for Global Health: Past Studies and Future Opportunities." *Lab on a Chip* 12, no. 12 (2012): 2118–2134.

Chung, H., Y. Jo, D. Ryu, C. Jeong, S. Choe, and J. Lee. "Artificial-Intelligence-Driven Discovery of Prognostic Biomarker for Sarcopenia." *Journal of Cachexia, Sarcopenia and Muscle* 12 (2021): 2220–2230.

Clarke, S. F., and J. R. Foster. "A History of Blood Glucose Meters and Their Role in Self-Monitoring of Diabetes Mellitus." *British Journal of Biomedical Science* 69 (2018): 83–93.

Cui, F., Y. Yue, Y. Zhang, Z. Zhang, and H.S. Zhou. "Advancing Biosensors with Machine Learning." *ACS Sensors* 5 (2020): 3346–3364.

Dashtian, K., F. Amourizi, N. Shahbazi, A. Mousavi, B. Saboorizadeh, S.S. Astaraei, and R. Zare-Dorabei. "Biosensors for Drug of Abuse Detection." In *Advanced Sensor Technology*, 125–172. Amsterdam: Elsevier, 2023. ISBN 9780323902229.

Ehrenkranz, J.R.L. "Home and Point-of-Care Pregnancy Tests: A Review of the Technology." *Epidemiology* 13 (2002): S15–S18.

Flynn, C., and A. Ignaszak. "Lyme Disease Biosensors: A Potential Solution to a Diagnostic Dilemma." *Biosensors* 10 (2020): 137.

Ganjalizadeh, V., G.G. Meena, M.A. Stott, A.R. Hawkins, and H. Schmidt. "Machine Learning at the Edge for AI-Enabled Multiplexed Pathogen Detection." *Scientific Reports* 13 (2023): 4744.

Griesche, C., and A.J. Baeumner. "Biosensors to Support Sustainable Agriculture and Food Safety." *TrAC Trends in Analytical Chemistry* 128 (2020): 115906.

Gubala, V., L.F. Harris, A.J. Ricco, M.X. Tan, and D.E. Williams. "Point of Care Diagnostics: Status and Future." *Analytical Chemistry* 84 (2012): 487–515.

He, H., et al. "Wearable Electroencephalography for Brain Activity Monitoring and Neurological Disorder Diagnosis." *Biomedical Engineering Letters* 8, no. 2 (2018): 129–137.

Hong, J.M., H. Lee, N.V. Menon, C.T. Lim, L.P. Lee, and C.W.M. Ong. "Point-of-Care Diagnostic Tests for Tuberculosis Disease." *Science Translational Medicine* 14 (2022): 4124.

Huang, Y., H. Li, and C. Li. "Cholesterol Biosensors: A Review of Development and Application." *Journal of Biomedical Science and Engineering* 7, no. 11 (2014): 1073–1083.

Huang, Y., and X. Zhang. "Biosensors for Cancer Biomarker Detection: Recent Advances and Future Challenges." *Biosensors and Bioelectronics* 166 (2020): 112403.

Jeong, S.H., et al. "Foodborne Pathogen Detection Using Lab-on-a-Disc Technology: Recent Advances and Future Directions." *Sensors and Actuators B: Chemical* 291 (2019): 170–184.

Jiang, Z., et al. "A Portable Lab-on-a-Chip System for Monitoring Cancer Biomarkers in Blood." *Microfluidics and Nanofluidics* 23, no. 7 (2019): 177.

Joung, H.-A., Z.S. Ballard, J. Wu, D.K. Tseng, H. Teshome, L. Zhang, E.J. Horn, P.M. Arnaboldi, R.J. Dattwyler, O.B. Garner, et al. "Point-of-Care Serodiagnostic Test for Early-Stage Lyme Disease Using a Multiplexed Paper-Based Immunoassay and Machine Learning." *ACS Nano* 14 (2020): 229–240.

Jumper, J., R. Evans, A. Pritzel, T. Green, M. Figurnov, O. Ronneberger, K. Tunyasuvunakool, R. Bates, A. Žídek, A. Potapenko, et al. "Highly Accurate Protein Structure Prediction with AlphaFold." *Nature* 596 (2021): 583–589.

Kim, J., et al. "Wearable Blood Pressure Monitoring Systems: Current Status and Future Directions." *Journal of Micromechanics and Microengineering* 30, no. 7 (2020): 073001.

Koch, C. H., and R. Baijal. "Advances in Rapid Molecular Diagnostics with the Cepheid GeneXpert System." *Clinical Infectious Diseases* 72, no. 8 (2021): 1406–1412.

Kumar, P., and M. Singh. "Point-of-Care Immunosensors for PSA: Challenges and Future Perspectives." *Journal of Nanobiotechnology* 18, no. 1 (2020): 61.

Kumar, P., and S. K. Vashist. "Nucleic Acid-Based Biosensors for Detection of Foodborne Pathogens: Current Status and Future Perspectives." *Food Control* 122 (2021): 107688.

Land, K.J., D.I. Boeras, X.-S. Chen, A.R. Ramsay, and R.W. Peeling. "The Application of Molecular Diagnostics for Point-of-Care Detection of Infectious Diseases." *Nature Microbiology* 6, no. 3 (2021): 311–319.

Lazcka, G.A., et al. "Biosensors for Pathogen Detection: an Overview." *Biosensors and Bioelectronics* 22, no. 8 (2007): 1635–1643.

Lee, H.J., et al. "Detection of Foodborne Pathogens Using a Piezoelectric Biosensor." *Food Control* 73 (2017): 351–357.

Li, X., and X. Zhang. "Portable Optical Biosensors for Point-of-Care Diagnostics: From Lab-Based to Field-Based Applications." *Trends in Analytical Chemistry* 134 (2021): 116150.

Li, Y., and X. Li. "Point-of-Care Molecular Diagnostics: The Cepheid GeneXpert Platform." *Journal of Clinical Microbiology* 58, no. 9 (2020): e00945–20.

Liao, S.M., and C.T. Lin. "Glucose Biosensors Based on Optical Techniques." *Journal of Diabetes Science and Technology* 13, no. 1 (2019): 123–129.

Liu, X., et al. "Wearable Temperature Sensors for Healthcare Monitoring: A Review." *Sensors* 19, no. 10 (2019): 2285.

Liu, Q., and C. Huang. "Electrochemical Biosensors for Rapid Detection of Foodborne Pathogens." *Sensors and Actuators B: Chemical* 265 (2018): 120–131.

Liu, Y., Q. Wei, Z. Zhao, et al. "Molecularly Imprinted Polymer-Based Colorimetric Sensor for Detection of Human Chorionic Gonadotropin Hormone (hCG) in Human Urine." *Biosensors and Bioelectronics* 39, no. 1 (2013): 191–196.

Liu, Y., and Z. Zhao. "Molecularly Imprinted Polymer-Based Electrochemical Sensors for hCG Detection." *Sensors and Actuators B: Chemical* 206 (2015): 141–146.

Liu, Y., and H. Zhang. "Colorimetric Biosensors for hCG Detection." *Biosensors and Bioelectronics* 96 (2017): 366–372.

Maity, A., H. Pu, X. Sui, J. Chang, K.J. Bottum, B. Jin, G. Zhou, Y. Wang, G. Lu, and J. Chen. "Scalable Graphene Sensor Array for Real-Time Toxins Monitoring in Flowing Water." *Nature Communications* 14 (2023): 4184.

Malekzadeh, A., H. Twaalfhoven, N. J. Wijnstok, J. Killestein, M. A. Blankenstein, and C. E. Teunissen. "Comparison of Multiplex Platforms for Cytokine Assessments and Their Potential Use for Biomarker Profiling in Multiple Sclerosis." *Cytokine* 91 (2017): 145–152.

Malkiel, I., M. Mrejen, A. Nagler, U. Arieli, L. Wolf, and H. Suchowski. "Plasmonic Nanostructure Design and Characterization via Deep Learning." *Light Science & Applications* 7 (2018): 60.

Manzano, M., and A. García. "Recent Advances in Lateral Flow Immunoassays for Detection of Human Chorionic Gonadotropin (hCG) in Pregnancy Tests." *Biosensors and Bioelectronics* 114 (2018): 111–122.

McGlennen, R.C. "Miniaturization Technologies for Molecular Diagnostics." *Clinical Chemistry* 47 (2020): 393–402.

Miao, W., et al. "Recent Advances in Wearable Non-Invasive Glucose Sensors for Diabetes Monitoring." *Trends in Analytical Chemistry* 127 (2020): 115853.

Müller, M., et al. "Advancements in Lab-on-a-Disc Devices for Non-Invasive Prenatal Testing (NIPT)." *Trends in Biotechnology* 38, no. 4 (2020): 420–430.

Nicoliche, C.Y.N., R.A.G. de Oliveira, G.S. da Silva, L.F. Ferreira, I.L. Rodrigues, R.C. Faria, A. Fazzio, E. Carrilho, L.G. de Pontes, G.R. Schleder, et al. "Converging Multidimensional Sensor and Machine Learning toward High-Throughput and Biorecognition Element-Free Multidetermination of Extracellular Vesicle Biomarkers." *ACS Sensors* 5 (2020): 1864–1871.

Niu, X., and L. Liao. "Electrochemical Biosensors for Detection of COVID-19 Virus and Its Variants." *Sensors and Actuators B: Chemical* 330 (2021): 129437.

Park, J., et al. "Wearable Sweat Sensors for Continuous Health Monitoring." *Nature Biomedical Engineering* 3, no. 3 (2019): 184–191.

Patel, S., et al. "Paper-Based Sensors for Detection of Environmental Pollutants: Current Status and Future Prospects." *Environmental Science and Technology* 49, no. 16 (2015): 9265–9278.

Patil, P., R. Phadke, and M. Ghosh. "Molecularly Imprinted Polymer-Based Electrochemical Glucose Sensor: A Review." *Biosensors and Bioelectronics* 91 (2017): 81–90.

Patnaik, S., M.S. Nair, and V. Ramesh. "Rapid and Cost-Effective Immunosensors for Strep Throat Detection: A Review." *Journal of Medical Systems* 43, no. 12 (2019): 358.

Potyrailo, R.A., and D. Makarov. "Electrochemical Glucose Biosensors for Diabetes Management." *Analytical Chemistry* 85, no. 24 (2013): 1188–1194.

Salvo, C., et al. "Quartz Crystal Microbalance (QCM) Sensor for Glucose Monitoring: Design and Application in Point-of-Care Diagnostics." *Sensors* 20, no. 9 (2020): 2590.

Saito, Y., and M. Nakajima. "Thermometric Biosensors for Point-of-Care Diagnostics: Current Progress and Future Trends." *Sensors and Actuators B: Chemical* 324 (2021): 128747.

Samani, S.E., D. Chang, E.M. McConnell, M. Rothenbroker, C.D.M. Filipe, and Y. Li. "Highly Sensitive RNA-Cleaving DNAzyme Sensors from Surface-to-Surface Product Enrichment." *ChemBioChem* 21 (2020): 632–637.

Sharma, P., et al. "Lab-on-a-Chip Devices for Cancer Biomarker Detection: Advances and Applications." *Biosensors and Bioelectronics* 168 (2020): 112567.

Skladal, P., and M. Beranek. "Piezoelectric Biosensors: From Nanostructure to Application." *Trends in Analytical Chemistry* 29, no. 5 (2010): 400–405.

Siristatidis, C., S. Stavros, A. Drakeley, S. Bettocchi, A. Pouliakis, P. Drakakis, M. Papapanou, and N. Vlahos. "Omics and Artificial Intelligence to Improve in Vitro Fertilization (IVF) Success: A Proposed Protocol." *Diagnostics* 11 (2021): 743.

Sun, D., M. Sun, J. Zhang, X. Lin, Y. Zhang, F. Lin, P. Zhang, C. Yang, and J. Song. "Computational Tools for Aptamer Identification and Optimization." *TrAC Trends in Analytical Chemistry* 157 (2022): 116767.

Thakral, Priyanka, Vimal Rarh, and A.K. Bakhshi. "In-Silico Analysis of Electronic Structures of Model Polypeptide Chains Using Particle Swarm Optimization." *Current Physical Chemistry* 6, no. 4 (2016): 297–306.

Turner, A.P.F., and I. Karube. Biosensors: Applications in Environmental Monitoring. In *Environmental Biotechnology*, edited by S.M.K. Bhatia. Springer, 2011.

Vashist, S.K., et al. "Point-of-Care Diagnostics: Recent Advances in Microfluidic Devices for Glucose Monitoring." *Biosensors and Bioelectronics* 42 (2013): 349–364.

Verma, P., and A. Rana. "Recent Advances in Immunosensors for Detection of Allergen-Specific IgE Antibodies." *Biosensors and Bioelectronics* 175 (2021): 112877.

Vlatakis, G., and S. Mo. "Recent Advances in Molecularly Imprinted Polymers for Biosensing Applications." *Analytica Chimica Acta* 1114 (2020): 90–106.

Wang, H., and L. Zhang. "DNA and RNA Biosensors for the Detection of Tuberculosis and Other Infectious Diseases." *Biosensors and Bioelectronics* 148 (2020): 111813.

Wang, J., and D. Chen. "Enzyme-Based Electrochemical Biosensors for Glucose Monitoring: Advances and Challenges." *Sensors and Actuators B: Chemical* 288 (2019): 98–110.

Wang, J., and S.Y. Lee. "Immunosensors for Point-of-Care Testing: From Lab to Field." *Biosensors and Bioelectronics* 56 (2014): 287–304.

Wang, J., and Y. Lin. "MIP-Based Sensors for Illicit Drug Detection." *Analytical and Bioanalytical Chemistry* 389, no. 4 (2007): 1065–1072.

Wang, Y., et al. "AI for Continuous Glucose Monitoring and Prediction of Hypoglycemia." *Sensors and Actuators B: Chemical* 298 (2019): 127–133.

Wu, P., and Z. Li. "Optical Biosensors for Prostate Cancer Detection." *Biosensors and Bioelectronics* 100 (2018): 404–409.

Xie, Z., et al. "A Paper-Based Glucose Biosensor for Point-of-Care Diagnostics." *Biosensors and Bioelectronics* 137 (2019): 104–111.

Yeo, W.-H., et al. "Wearable and Stretchable Electrochemical Sensors for Health Monitoring." *Advanced Materials* 28, no. 22 (2016): 4328–4334.

Yuan, R., et al. "A Colorimetric Urine Test Strip for Urea Detection Using an Enzyme-Based Method." *Analytical Chemistry* 88, no. 5 (2016): 2874–2880.

You, J., D. Zhang, and H. Choi. "Molecularly Imprinted Polymer-Based Sensor for Heavy Metal Ion Detection: Recent Advances and Challenges." *Environmental Science and Technology* 52, no. 3 (2018): 1149–1159.

Zeng, Z., Z. Huang, K. Leng, W. Han, H. Niu, Y. Yu, Q. Ling, J. Liu, Z. Wu, and J. Zang. "Nonintrusive Monitoring of Mental Fatigue Status Using Epidermal Electronic Systems and Machine-Learning Algorithms." *ACS Sensors* 5 (2020): 1305–1313.

Zhang, S., J. Zeng, C. Wang, L. Feng, Z. Song, W. Zhao, Q. Wang, and C. Liu. "The Application of Wearable Glucose Sensors in Point-of-Care Testing." *Frontiers in Bioengineering and Biotechnology* 9 (2021): 774210.

Zhang, W., and W. Tan. "Electrochemical Biosensors for DNA/RNA Detection: Current Trends and Future Perspectives." *Biosensors and Bioelectronics* 164 (2020): 112285.

Zhang, Y., and J. Lu. "Optical Biosensors for Rapid Detection of COVID-19 Virus." *Sensors and Actuators B: Chemical* 330 (2021): 129249.

Zhang, Y., H. Zhang, and X. Xu. "A Molecularly Imprinted Polymer-Based Biosensor for the Rapid Detection of COVID-19 Virus." *Sensors and Actuators B: Chemical* 333 (2021): 129579.

Zhou, Y., et al. "Implantable Temperature Sensors for Long-Term Monitoring of Body Temperature in Clinical Applications." *Bioengineering* 8, no. 2 (2021): 21.

Zhang, Y., et al. "Recent Advances in Calorimetric Biosensors for Point-of-care Testing." *Biosensors and Bioelectronics* 141 (2019): 111476.

Zhang, Z., et al. "A Wearable Pulse Oximeter for Real-Time Monitoring of Oxygen Saturation Levels in Patients with Respiratory Disorders." *IEEE Access* 8 (2020): 67124–67131.

Zhuang, J., A.C. Midgley, Y. Wei, Q. Liu, D. Kong, and X. Huang. "Machine-Learning-Assisted Nanozyme Design: Lessons from Materials and Engineered Enzymes." *Advanced Materials* 36 (2024): 2210848.

8 Wearable Biosensors for Continuous Health Monitoring

Radha Doddamani

8.1 INTRODUCTION

Continual monitoring of physiological parameters from the body of an individual is crucial in health monitoring, disease diagnosis, and treatment. In spite of the requirements, wearable medical devices relied on either bulky wired systems or battery-powered devices that needed regular recharging (Sattar et al., 2024). However, wearable devices which are lightweight and portable can be worn directly on the body or integrated into the user's clothing or accessories. They are usually connected to the internet and combined with various software applications to monitor the user's physical conditions (Gao et al., 2024). In this conjunction, flexible and wearable biosensors (WBS) have received tremendous traction over the past decade owing to their great potential applications in the field of health and medicine. The capabilities of wearable technology have been steadily expanded with more varied forms and more precise physiological indicators due to advancements in sensor and operating system hardware technologies. As a significant contribution towards improving personalized healthcare, these sensors are getting closer to high precision, continuity, and comfort. Simultaneously, the ubiquitous regulatory capabilities have been launched in light of the rapid advancement of the Internet of Things (Kim et al., 2019; Smith et al., 2023).

Biosensors can provide information about an individual's health status as it directly interacts with the biological components of the body. WBS are devices that can be worn on, inside, or near the body to continuously monitor physiological and biochemical signals. They serve as an ideal platform for real-time and continuous health monitoring, exhibiting unique properties such as being self-powered, being lightweight, low cost, high flexibility, detection convenience, and great conformability (Kim et al., 2019; Song et al., 2023).

Since the modern era, wearable sensors have become widely used in biomedical monitoring systems and healthcare, facilitating the continuous measurement of critical biomarkers for medical diagnostics, evaluation of biological fluids such as blood, sweat and saliva, and monitoring of health conditions. Tracking physiological parameters including heart rate, respiration rate, temperature, blood oxygen

DOI: 10.1201/9781003536239-8

saturation, blood pressure, and ion concentrations has usually required costly and intimidating devices. But over the last few decades, research has advanced to the point where non-invasive, wearable, and soft systems are superseding current practices (Sharma et al., 2021; Zhou et al., 2024).

Over time, as a result of concentrated research that has developed different kinds of WBS such as electrochemical, optical biosensors, thermal sensors, and pressure and strain sensors, along with advances in the non-invasive monitoring of hormones, microorganisms, biomarkers, and so on, numerous apparent problems in the fields of sports, medicine, and health have been resolved by wearable sensors that have built-in capabilities for sensing recognized indicators. Based on different parameters measured, WBSs are classified into motion state, biophysical, and biochemical sensors (Sharma et al., 2021).

Motion state sensors are largely used to measure human physical parameters such as gait, sleep, and tremors for real-time monitoring and gathering of long-term information. *Wearable biochemical sensors* parallelly measure the trace and processing of different samples with the integration of lab-on-a-chip technology. Utilizing wearable biochemical sensors precisely measure biomarkers in biological fluids in order to check metabolism and monitor health conditions. *Wearable biophysical sensors'* unique characteristic is to make contact with skin, which gives information about real-time measurement parameters such as blood pressure, heart rate, and temperature, which provide significant data in healthcare applications. Among these, motion state and biophysical sensors are commercially available and extensively utilized by customers. However, because biological fluids are complex matrices that make it difficult to detect the target analyte, biochemical biosensors currently not commercially available, though they do have the potential to be so.

Wearable technology has steadily advanced with a mix of multiplexed biosensing, microfluidic sampling, and transport systems integrated with body attachment and flexible materials for easier wear and convenience. These wearables have the potential to improve knowledge of the correlations between analyte concentrations in blood or non-invasive biofluids and provide feedback to the patient, which is crucial for immediate medical condition diagnosis, treatment, and management. However, research for cohort validation and performance evaluation of WBS are needed to underpin their clinical acceptance.

Several recent reviews have considered various biosensors for breath and saliva analysis in disease diagnosis. However, most of the reviews focus on a specific type of sensor (electrochemical, optical, piezoelectric, semiconductor, etc.) or sample (breath VOCs, EBC, saliva), lacking a comprehensive overview of the entire field. Additionally, these reviews address biomarker detection for a wide range of diseases rather than focusing solely on continuous monitoring. Multiple perspectives have already been reported. Given their significant relevance, it is imperative to consolidate recent advances in biosensors for detecting early disease. In the present chapter, we discuss the the growing demand, salient features with underlying principles, types of wearables, providing insights particularly into types of materials used and their applications in wearable/implantable biosensors, while carefully examining the challenges encountered that should be addressed for future work to improve the equity and accessibility of quality healthcare, with some guidelines provided to help drive the further research needed and applications of wearable devices for real-time monitoring of human health (Liu et al., 2025; Xiong et al., 2025). Further, the chapter also provides a perspective on the promising potential of the introduction of wearable devices capable of detecting different biomarkers in biological fluids aimed at the monitoring of human health. Providing exciting technological progress with advanced fabrication techniques (namely 3D printing, microfluidics with machine learning and artificial intelligence) are discussed in the developments of wearable/implantable biosensors with the essential branches of physiological monitoring, including biopotential, optical, electrochemical, and multi-signal sensing (P. Zhang et al., 2024; Sharma et al., 2021; Zhou et al., 2024). Lastly, future perspectives and challenges are also discussed in the conclusion to the chapter, and it seeks to provide comprehensive guidance for designing fully functional and stable wearable devices.

8.2 WORKING PRINCIPLE OF BIOSENSORS

A biosensor, also referred to as a biological sensor, is an integrated, self-contained device that uses a biomolecular recognition system in close proximity to a transducer to deliver quantitative or semi-quantitative analytical data. Several essential features, such as high sensitivity, low limits of detection (LOD), selectivity, repeatability, and durability, are expected to exist in biosensors (Baskar et al., 2025)

Typically, biosensors consist of three main components; the sensor element (bio-recognition layer), the transducer, and the electronic system, which work together to generate measurable signals, which are then converted into data. The signal sensor elements plays a significant role in the detection and conversion of the biological signal from the target molecule into an electrical signal. The sensor element can be an enzyme, an antibody, nanostructured approaches, electrochemical methods, antigens, nucleic acids, microorganisms, DNA strands, or more. These elements function as markers of the presence or concentration of the target material by altering the electrical signal when they interact with the target molecule. Any physical changes brought about by this contact, such as variations in conductance, absorbance, emission, resistance, or capacitance, are detected by the sensor. The transducer comprehends a range of sensors, including electrochemical sensors, optical sensors, transistors, and more. The electrical signal generated by the sensor element is often amplified and converted into a format that the system can understand by these interfaces. The electronic system consists of a signal amplifier, processor, and screen or display. In other words, these changes are then electronically converted, stored, and analyzed. Based on detection mechanism, biosensors can be classified into quite a few types, including electrochemical, optical, calorimetric, piezoelectric, and amperometry biosensors. They can also be categorized by the type of analyte they are designed to detect, such as immunosensors, enzymatic biosensors, DNA biosensors, and whole-cell biosensors (Alavian et al., 2025; Baskar et al., 2025). Figure 8.1 is a representation of the working principle of biosensors (Eser et al., 2022).

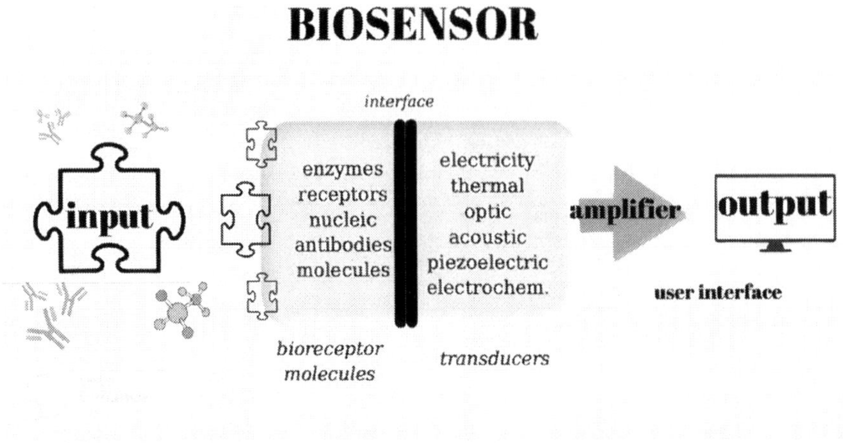

FIGURE 8.1 Representation of the working principle of biosensors (Eser et al., 2022).

8.3 TYPES OF BIOSENSORS

Biosensors are divided mainly into electrochemical and optical categories, based on their detecting method and transducer system.

8.3.1 Electrochemical Biosensors

Electrochemical biosensors and other electroanalytical devices are regarded as a subfield of analytical chemistry that includes both charge transfers and oxidation–reduction reactions. In order to identify analytes, biorecognition components are essential. This component must be used in conjunction with a converter in order to provide a signal based on the analyte's concentration. Measurable signals like current, conductivity, impedance, and potential are obtained in an electrochemical transducer system as a result of sample–bioreceptor interaction. Based on the signals they generate, electrochemical biosensors are divided into different types, such as conductometric, amperometric, impedimetric, and potentiometric.

Electrochemical-based sensors exhibit notable advantages over other types of biosensors, including ease of manufacture, increased sensitivity, rapid response, and low power consumption. Wearable sensors, which are mostly based on the electrochemical approach, depend significantly on sensing electrodes. However, in order to enhance the functionality of wearable biochemical sensors, significant challenges including the functional material and manufacturing methods employed in the preparation of sensing electrodes remain difficult and must be resolved (Sharma et al., 2021).

In this conjunction, metal-based film electrodes are the standard sensing electrode technology used in wearable sensing. The performance of wearable biochemical sensors has improved as a result of numerous developments in the search for novel materials, including hybrid and metallic nanoparticles (Chenthattil Raril et al.,

2020) and nanocomposite (Tigari & Manjunatha, 2020), carbon (Monnappa et al., 2020), and polymeric materials (Pushpanjali et al., 2021; Hareesha & Manjunatha, 2021), to be used as the electrode materials.

However, the design and optimization of the sensing electrodes with a durable and effective micro-manufacturing methods has gained attention. As the importance of identifying the different biomarkers that impact health has increased, considerable efforts have been made in recent years to develop these wearable sensors. .

As the importance of identifying the different biomarkers that impact health has increased, considerable efforts have been made in recent years to develop these wearable sensors. In this way, bioelectronic devices have been designed to translate biological information into electrical signals and *vice versa*, thereby bridging the gap between the living biological world and electronic systems (P. Zhang et al., 2024). Park et al. (Park et al., 2024), in their comprehensive study, focus on promising biomarkers for Alzheimer's disease detection in various biofluids, as well as the accompanying obstacles and advantages. They have explored bioelectronic tongues and bioelectronic noses as innovative, non-invasive tools for detecting biomarkers in Alzheimer's disease, which further facilitates point-of-care applications.

Sattar et al. have introduced flexible wireless wearable bioelectronics to detect electrophysiological signals on the skin via continuous monitoring of physiological signals without recharging batteries, which harvests a high open circuit voltage of 175–180 mV from the human body with long-term stability. Figure 8.2 gives a pictorial representation of a flexible thermoelectric wearable device for physiological monitoring, wireless acquisition of ECG and EMG signals, with the integration of human body heat–charged biosensor system and signal processing. It also utilizes screen printing, laser micromachining, and soft packaging technologies, permitting a multilayered, soft, wearable device to be mounted on any body part. The demonstration of this self-sustainable wearable system for detecting electromyograms and electrocardiograms apprehends the potential of the technology to offer various

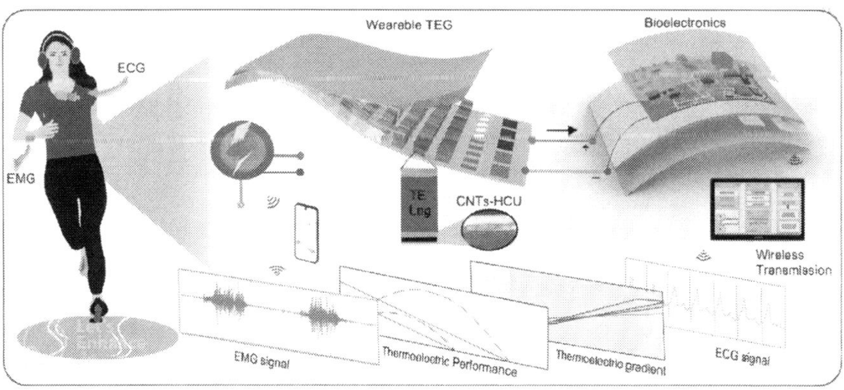

FIGURE 8.2 Representation of a flexible thermoelectric wearable device for physiological monitoring, wireless acquisition of ECG and EMG signals with the integration of human body heat–charged biosensor system and signal processing (Sattar et al., 2024).

opportunities for continuous monitoring of health monitoring health (Sattar et al., 2024).

In a study by Chowdhury et al., research work developed a textile-based sweat glucose sensor, with the comparative analysis of three low-cost conductive textiles, namely Adafruit 1364, 1167, and 4762. For electrochemical glucose detection with glucose-specific enzymes, commercially available *conductive textile* Adafruit 1364 outclassed others in morphological, electrochemical, and wearable properties, avoiding the complexity of traditional fabrication methods. Cyclic voltammetry and differential pulse voltammetry studies highlighted the reproducibility, effect of interferents, and advantageous wearable properties of these sensors (Chowdhury et al., 2024).

8.3.1.1 Optical Biosensors

The most prevalent kind of biosensor is an optical biosensor. Utilizing the interplay between a biorecognition element and the optical field, optical detection is accomplished. Label-free and label-based are the two broad categories into which optical biosensing may be divided. In simpler terms, the interaction between the transducer and the analyzed substance directly produces the detected signal in a label-free mode. Label-based sensing, on the other hand, uses a label and produces the optical signal using a colorimetric, fluorescent, or luminescent technique. Label-assisted sensing can be used to detect simple compounds like glucose through enzymatic oxidation. The most commercially successful (to date) use of a biosensor is for blood glucose analysis; diabetics use a handheld glucose meter for this purpose. However, labeling might alter the features of binding and hence add systematic errors into the biosensor analysis in some instances, such as antibody–antigen interaction where a label is coupled with one of the bioreactants.

An optical biosensor is a device integrated with an optical transducer system containing a biorecognition sensing element. The objective of an optical biosensor is to give a signal which is equivalent to the concentration of a measured substance (analyte). Enzymes, antibodies, antigens, receptors, nucleic acids, entire cells, and tissues are just a few of the biological materials that the optical biosensor can use as biorecognition components.

Surface plasmon resonance (SPR), evanescent wave fluorescence, and optical waveguide interferometry utilize the evanescent field in close proximity to the biosensor to detect the interaction of the biorecognition element with the analyte. The construction of optical biosensors comes in a vast array of varieties. Examples include evanescent wave fluorescence, surface plasmon resonance, localized surface plasmon resonance, and bioluminescent optical fiber biosensors, which are widely used and tend to detect the most physiologically significant compounds. Other optical biosensors are reflectometric interference spectroscopy biosensors, surface-enhanced Raman scattering biosensors, ellipsometric biosensors, and optical waveguide interferometric biosensors.

Recently, psychiatric biosensors mainly based on electrochemical and optical sensing technologies with high sensitivity, selectivity, and reproducibility have been widely developed. The modern biosensors developed for the detection of representative psychiatric biomarkers (e.g., cortisol, dopamine, and serotonin) were comprehensively summarized in their sensitivities, sensing technologies, applicable biological platforms, and integrative readouts, facilitating clinical utility and the commercialization of point-of-care diagnostics in the literature (Marchianò et al., 2025).

8.4 MATERIALS AND INTERFACES FOR BIOSENSORS

8.4.1 Organic Conducting and Semiconducting Materials

Materials for wearable and implantable biosensors include organic materials, conducting polymers (CPs), carbon materials and composites of graphene and carbon nanotubes (CNTs), happen to be the most promising candidates. Their unique electrical, electrochemical, and mechanical properties present new possibilities for bioelectronics by utilizing metal- or silicon-based counterparts. Thus with notable advancements in recent years, the application of organic and carbon-based conductors in the creation of wearable and implantable biosensors has become a fast-expanding research area (P. Zhang et al., 2024).

8.4.2 Smart Polymers

Smart polymers are a type of material that responds (can be deformed and fixed into temporary shape by stimuli) under specific stress conditions. These materials react to external stimuli, including light, magnetic or electric fields, temperature, pH, and analytes of interest (such as ions and bioactive compounds) (Kariduraganavar et al., 2016).

A subclass of smart polymers known as electroresponsive or electroactive polymers is capable of reversibly changing its physicochemical characteristics in response to electrical impulses. Depending on the strength, duration, and frequency of the electric current, they can change shape (swell, shrink, or bend) by transforming electric energy into mechanical energy. The pH level is changed and the hydrogen bonds between the polymer chains are broken by the electric current. The polymer aligns itself parallel to the electric field as the voltage and current rise. These polymers are perfect for converting biological signals into quantifiable electrical signals because of their special electrical characteristics (Karami et al., 2025).

The movement and shape change of a polymer are also inclined by its dielectric properties and stored free energy. Several advantages offered by electroresponsive polymers are high strain capacity, low specific gravity, low cost, easy processability, and downscalability. Polypyrrole (PPY) and polythiophene (PT), both conductive polymers, are examples of electroresponsive polymers. Utilizing such materials improves the precision and authenticity of the information sent while also addressing the issue of mismatched properties between biological tissues and electronic devices (Fattah-alhosseini et al., 2024; Hosseine et al., 2024).

Multiple covalent modification enables a nylon fiber biosensor with robust scrub-resistant and signal-capture ability to be used for multiscenario health monitoring and security warning (Dong et al., 2024).

Among smart wearables, nylon fibers have great potentials in smart textiles. For instance, Dong et al. (Dong et al., 2024) have worked on nylon fibers/di-aldehyde cellulose nanocrystals/polypyrrole biosensors which were fabricated by interfacial multiple covalent reactions. Even after numerous washings and frictions, they demonstrated strong mechanical strength, stretching-releasing dynamic durability, high conductivity (354 S/m), and exceptional sensing capability. Additionally, this biosensor was included into a number of multipurpose health monitoring and security alert systems for deep-sea research, high-altitude activities, and "stress reducing exercise" enthusiasts, showcasing the enormous potential of conductive nylon fiber biosensors in flexible electronics. With the advancement of nanotechnology, fiber-based sensors have been considered the most promising flexible sensor due to their breathability and editable shapes (yarns, fabrics, spacer fabrics, etc.). Among them, nylon fibers have been recognized as an ideal fiber matrix for flexible sensors due to their suitable tensile strength, excellent corrosion resistance, wear resistance, and self-lubricating properties.

8.4.3 Nanomaterials

Nanomaterial-based electrochemical signal amplifications hold considerable promise for enhancing the sensitivity and selectivity of electrochemical biosensors, given the remarkable developments in nanotechnology and nanoscience. First of all, it is commonly recognized that electrode materials are essential to the development of high-performance electrochemical sensing platforms that use a variety of analytical techniques to identify target molecules (Zhu et al., 2015). Nanomaterials have been investigated for their potential use in creating distinctive point-of-care systems, as the sensing mechanism has changed with the recent shift to focus on compacting diagnostic laboratories to produce portable and miniature diagnostic instruments that can perform quick, on-site detection (Karami et al., 2025).

The recent comprehensive review by Chaudhary et al. (2024) provides insights on employing chemiresistive-module nano-biosensors for lung cancer diagnosis through breath analysis by monitoring diversified biomarkers such as volatile organic compounds/gases in exhaled breath.

8.4.4 Edible Materials

Edible biocatalytic biosensors represent a new frontier within healthcare technology available for remote medical diagnosis. Temperature, pH, gaseous substances, gastrointestinal markers, enzymes, hormones, glucose, and levels of drugs are wide range of physiological and biochemical characteristics that edible biosensors may measure to provide real-time data. The development of edible biosensors also faces challenges, namely finding edible materials (which can be mediators for redox reactions, conductive substances, or binders such as enzymes) that satisfy Food and Drug

Administration, European Food Safety Authority, and European Medicines Agency regulations. A further challenge is to develop bioelectronics which are operatable in conditions such as low pH (~pH 1.5 gastric fluids, etc.), body temperature (between 37°C and 40°C), and highly viscous bodily fluids (Marchianò et al., 2025).

8.4.5 Metal-Organic Framework

Continuous sweat nutrient monitoring provides useful information about metabolic cycle and health status. For instance, a skin-mountable flexible biosensor with metal-organic framework (MOF)-derived composites, namely a copper-based MOF attached with poly(3,4-ethylenedioxythiophene) (Cu-MOF/PEDOT), has been developed for real-time monitoring of sweat ascorbic acid levels by Ling et al. The biosensor employs a tiny, highly integrated technology capable of producing an undetectable, stretchable skin patch, ensuring conformal integration with curvilinear skin contours within a detection limit and great sensitivity. For health management, this biosensor represents a viable path for scalable health monitoring employing adaptive and user-friendly technology (Ling et al., 2025).

Because of their efficacy and efficiency in monitoring livestock health, even smart technologies for farming are developing quickly. These include biosensor gadgets and smart wearables for illness detection or health monitors that track the health of cattle (Alipio & Villena, 2023).

Leveraging advancements in flexible electronics, nanomaterials, biosensing, and data analytics, wearable sweat biosensors promise to revolutionize disease management, prevention, and prediction, promoting healthier lifestyles and transforming medical practices globally (Childs et al., 2024).

Another approach which offers motivation for the development of point-of-care testing equipment, indicating significant application potential, where 3,3′,5,5′-tetramethylbenzidine (TMB) which is a common chromogenic probe used in visual detection due to its low toxicity and very sensitive response. The TMB probe and bimetallic MOF nanozymes are encapsulated in a hydrogel patch for reusable vision sensor fabrication. By a utilizing bimetallic MOF (CuFe-MOF), hydrophobic TMB is oxidized to hydrophilic ox-TMB to facilitate diffusion into a hydrogel patch which is hydrophilic agarose, which is reduced back to TMB. As a proof-of-concept application, the color change between TMB and ox-TMB can be quickly read by the naked eye, in this way a visual "On–Off–On" sensor is simply constructed and successfully applied to detect vitamin C in human sweat. T(X. Zhou et al., 2024).

8.4.6 Graphene Oxide

Graphene oxide (GO) is one of the most acclaimed materials for creating new opportunities in the creation of next-generation biosensors (Lee et al., 2016). **A** biosensor with a graphene-based electrode with a polyelectrolyte interlayer to immobilize cholesterol oxidase (ChOx), enabling selective cholesterol detection, happened to be an innovative electrochemical biosensor integrated into a microneedle-based system (non-invasive with sensitive quantification) with cholesterol levels in interstitial fluid

(ISF). By choosing polyaniline as the optimal polyelectrolyte for ChOx immobilization further, graphene oxide is electrochemically reduced to form a conductive layer. The biosensor's potential for real-world interstitial fluid extraction and cholesterol monitoring is demonstrated by its application to skin-mimicking agarose gel and porcine skin. The integrated microneedle-based technology provides a viable approach for non-invasive cholesterol monitoring, with implications for personalized healthcare diagnostics (Kim et al., 2024).

8.4.7 Hydrogels

Hydrogels are a class of biomaterials offer promising alternatives for developing biosensors. Hydrogels, which have properties similar to biological tissues, are three-dimensional and hydrophilic polymers with cross-linked networks. To achieve high sensitivity, specificity, and stability within the device, they can be utilized in various biosensors. Recent advances have been made in the use of hydrogel-based biosensors for cancer detection, and in one of the reviews, their synthesis, applications, and challenges as well as the future directions of this emerging field are discussed (Chaudhary et al., 2024).

Conductive hydrogels exhibit tremendous potential for wearable biosensors for health monitoring applications, yet simultaneously, enhancing their biocompatibility and antimicrobial properties remains a long-standing challenge. In one study, the biocompatibilities of gelatin hydrogel are bolstered with tannic acid (TA), facilitating antibacterial effects through interactions, while borax enhances conductivity (Li et al., 2024). Hydrogel functions as a flexible sensor for monitoring human movement, facial expressions, and speech recognition, paving the way for advanced wearable biosensors with improved safety and multifunctionality like biocompatibility, antimicrobial activity, and conductivity into a single system (Li et al., 2024)

The detection of proteins in sweat might be greatly hindered by the challenges associated in collecting and the existence of contaminants in sweat. Especially the real-time monitoring of low-concentration cytokines such as TNF-α in sweat can aid clinical physicians in assessing the severity of inflammation. For instance, in a study by by Huang et al., this challenge is addressed by incorporating a nanosphere array coupled with a reusable sensor that employs a Nafion/aptamer-modified MoS_2 field-effect transistor specifically designed for automatic sweat transportation. The findings indicate that in clinical scenarios, this device could potentially provide real-time evaluation and management of patients' immune status via sweat testing (Huang et al., 2024).

Li et al. (2024) in their work proposed a polyaniline/platinum nanoparticle (PANI/PtNPs) modified glassy carbon electrode (GCE) as the working electrode for the biosensor. The solutions containing aniline (monomer), phytic acid, and polymerization initiator were mixed and coated onto the GCE. By coating the electrode with the solution, a gelated hydrogel thin film was formed within 3 min. Metabolites such as uric acid, cholesterol, and triglycerides in human serum could give rapid response times when the thin film was put into contact with a solution of chloroplatinic acid and formic acid to form the hydrogel matrix.

The types and characteristics of hydrogels are summarized in a review, where the details of the conductive hydrogels and the role of tough hydrogels with robust mechanical properties are discussed. They can expand sensing applications such as stress detection, which has been challenging to measure utilized as wearable sensing and adhesive layers. Introducing research on double-sided skin-sensitive conductive hydrogel-based adhesive layers that ensure adhesion to the skin and can be utilized as sensors were also discussed, along with their advantages and limitations (Binabaji et al., 2024).

8.4.8 Carbon Polymer Dots

Wearable sweat sensors are revolutionizing healthcare monitoring by offering non-invasive, user-friendly gadgets that provide real-time data on electrolyte and hydration levels. In a recent work, an innovative highly portable two-channel microfluidic wearable system is tailored for monitoring diabetes. The core of this system lies in the use of carbon polymer dots (CPDs) and carbon dots (CDs) derived from monomers such as catechol, resorcinol, *o*-phenylenediamine, urea, and citric acid. These dots are combined into gelatin and poly(vinyl alcohol) hydrogels, resulting in an advanced solid-state fluorometric sensor coated on a cellulose paper substrate. As it is utilized for simultaneous sweat sampling and the real-time detection of biomarkers, by the continuous and non-invasive tracking of acetone and ammonia (which is volatile organic compound (VOC)), it can seamlessly integrate with smartphones for easy data management. These sensors exhibit exceptional performance, offering linear detection ranges of 0.05–0.15 ppm for acetone and 0.25–0.37 ppm for ammonia, with notably low detection limits of 0.01 and 0.08 ppm, respectively (Baek et al., 2024).

8.4.9 Organic Thin-Film Transistor (TFT)

Soft sensors that match the modulus of human skin have shown significant potential for wearable sensing applications by ensuring robust, conformal contact that enables the acquisition of high-quality signals. Soft sensor arrays constructed with organic thin-film transistor (TFT) have been crucial for advanced spatiotemporal measurements of signals, namely arterial pulse waves, using active-matrix pressure and optical sensing arrays, which minimizes signal crosstalk, laying the groundwork for their useful implementation in environmental monitoring, wearable technology, and healthcare. Despite these advancements, challenges such as limited sensitivity, high power consumption, and the need for cost-effective, large-area integration technologies persist, hindering their practical application (Baek et al., 2024; Liu et al., 2024).

Jiang et al. (2024) in their work, report a drug-mediated organic electrochemical transistor (DM-OECT) by implementing the drug gefitinib and an epidermal growth factor receptor (EGFR) as a dual-functional probe and analyte, respectively. The electrostatic interaction of protonated gefitinib with poly(3,4-ethylenedioxythiophene) polystyrene sulfonate (PEDOT:PSS) and the dragging-off binding of gefitinib to the EGFR target endow the device with a unique refresh-in-sensing (RIS) capability. The device displayed an exceptional low limit of detection. It was further tested for non-small-cell lung cancer patients with a sensing performance that can rival commercial enzyme-linked immunosorbent assay (ELISA) test kits.

8.4.10 2D Nanomaterials

Wearable sensors, the capacity of which to gather physiological vital signs is dependent on sensing materials, are becoming more widely used as personal healthcare awareness gradually grows. Over other 2D nanomaterials, MXenes have distinct chemical and physical superiorities for wearable sensors. The latest advancements in MXenes-based materials are being explored recently for wearable physical sensors as they possess versatile functionalities. Perhaps, a detailed description of different sensing mechanisms is included in the review by Yang et al. (2024) to illustrate the contribution of MXenes to sensing performance and its improvement, the challenges involved, promising opportunities are narrated, along with future potential for development.

8.4.11 Silver/Gold Nanowire

Silver/gold nanowire has great potential use as interconnects in electronic, photonic, and optoelectronic devices (Joshi et al., 2022). Liu et al. (2024) in their work have put out a novel wearable patch biosensor that uses a heat-stimulated method to gather perspiration for ongoing, non-invasive diabetes treatment. To encourage sweat gland secretion, an electrothermal film loaded with silver nanowire was created as the heat source. A Prussian blue and glucose oxidase co-modified glucose electrochemical biosensor was then used to continuously and sensitively assess the sweat sample that had been secreted after it had been transferred and enriched through microfluidic channels.

Gong et al. (2023) have reported on a hierarchically resistive (HR) skin offering the ability to disentangle a single resistive signal into multiple types of biometric data. This HR skin was devised to have three resistive skin layers with unique resistivity and stretchability, but sharing common leads to minimize the number of pixels required, which could be realized with low-resistivity cracked platinum film, medium-resistivity vertically aligned gold nanowires (v-AuNWs) and high-resistivity ultrathin gold nanowires (u-AuNWs). Gong et al. discuss how the biosensors are worn on the throat, and proclaim that these fabricated HR skins demonstrated unique linearity and staircase-shaped electrical responses in three distinct strain ranges, which allowed for the identification of five physical/psychological activities (speech, heart and respiration rates, touch, and neck movement). They further infer that these classes of biosensor offer broad applications in future for remote healthcare, soft robotics, and human–machine interactions.

Effective cancer treatment depends on early detection, but existing techniques have drawbacks. In another work, vertically aligned gold nanowire electrodes were utilized in sensors with intrinsic porous sponge with a high electrochemically active surface area of ~67 cm^2. Biosensors work in such a way where, as natural sweat collection and outflow layers, two biocompatible sweat pads will be positioned between the gold sponge biosensing electrodes. This naturally biocompatible and permeable platform can be integrated with wireless communication circuits, leading to a wireless sweat analysis system for the real-time monitoring of glucose, lactate, and pH during exercise (Lin et al., 2024).

8.5 DEVICE FABRICATION

In the past decade, significant studies have been reported on the fabrication of smart biosensors. In one context, they can be used to monitor driver alertness to improve road safety. For instance, a metamaterial biosensor has been designed in a research work, interestingly, that can be directly embroidered onto seat belts, enabling the contactless measurement of drivers' vital signs in moving vehicles. This metamaterial is made of conductive threads stitched in a comb-like pattern, which guides radio waves emitted by a low-power wireless sensor along the surface of the belt. This study claims that this design allows for sensitive detection of subtle cardiopulmonary motions from the torso while tuning down the noise. Utilizing signal processing algorithms to analyze the data recorded by the sensor, the driver's heartbeat and respiration can be measured in real time, even in challenging kinetic environments (Zeng et al., 2024). As another type of smart wearable sensor, functionalized face masks have emerged for multimodal detection of physical parameters and biochemical markers since the COVID-19 pandemic. These sensors are also widely used for environmental pollution monitoring and physiological health management (Song et al., 2024).

Due to technological advances in wearable biosensors, aspects like the miniaturization of sensors (Liu et al., 2020), wireless communication technologies in biosensors (Xu et al., 2024), and power management in wearables (Wittmann et al., 2020) have also gained attention.

For the sensor to accurately receive signals from the human body while preserving mechanical and electrical characteristics, such as self-recovery and dependability, high skin adhesion is essential. It can be difficult to guarantee each of these characteristics in the adhesion layer at the same time. A high signal-to-noise ratio hinders real-time monitoring when there is no conformal contact between the skin and the sensor. In a review, the importance of the adhesion layer for real-time healthcare monitoring was addressed, focusing on reduced noise and the absence of irritation or side effects for long-term usage. Various types of adhesion layers with hydrogels by introducing research on double-sided skin-sensitive conductive hydrogel-based adhesive layers that ensure adhesion to the skin and can be utilized as sensors were also discussed, along with their advantages and limitations. Moreover, due to their sustained adhesion, they can expand sensing applications such as stress detection, which has been challenging to measure (Ahmed et al., 2024).

Recently, advanced printing techniques have revolutionized the design and manufacturing of biosensors. Certain reviews outline research on biosensors, focusing on substrates for printing and discusses various printing methods, highlighting challenges and perspectives along with electrochemical properties such as electrical conductivity and redox potential (Marchianò et al., 2025).

Apporva et al. (2024) have reported on 3D-printed smart biosensors for online health monitoring of canines for patella luxation issues. This study highlights the detailed investigations for the selection of processing parameters to be fixed for the fabrication of a smart biosensor with different meta structures (prepared by 3D printing) to be used as a stifle prosthesis for sensing capability as an Internet of Things (IoT)–based solution. The result of this study suggests meta-structures and fabrication of smart biosensors by 3D printing—for instance, the schematic representation of the ink-jet printing process (3D printing) for the sensor (Figure 8.2) (Song et al., 2023).

Thus, advanced 3D printing techniques have also enabled precise fabrication for flexible sensors. In a recent work, PANI-based biosensors have been developed, having applications in detecting lung, liver, breast, and prostate cancers, highlighting the discussion about detection principles, limits, and specificity for each cancer type. Moreover, the role of PANI in signal amplification through nanocomposite (NC) integration, which progresses diagnostic accuracy, has been explored (Hosseine et al., 2024). By a unique one-step continuous manufacturing process, a 3D printed flexible wearable was fabricated with self-sufficient microfluidic channels and novel single-atom catalyst-based bioassays were introduced by Chen et al. (2024) for monitoring sweat rate and the concentration of three other biomarkers. The direct ink-writing technique of 3D printing is utilized to print the microfluidic device with self-supporting structures to detect human sweat and glucose, lactate, and uric acid concentrations. Contamination and sweat evaporation issues existed in traditional sampling methods, and the current method eliminates and addresses these. A single atom catalyst is utilized in this work for colorimetric bioassays to enhance sensitivity and accuracy (Chen et al., 2024); for instance, a flexible epidermal biomicrofluidic blood glucose continuous monitoring sensor is illustrated in Figure 8.3 (Smith et al., 2023).

In the latest research, the integration of microfluidic technology into wearable devices can be seen. Microfluidics enables the realization of real-time monitoring of body fluids and skin interstitial fluid, which is highly significant in medical diagnosis and even therapeutic interventions. Nevertheless, despite advancements, major challenges remain in the integration of microfluidics into wearable devices at the current technology level. Recent work has focused on applications of microfluidic contact lenses and provided understandings into the escalating intersection between head-worn wearables and microfluidics, offering future prospects (Gao et al., 2024). The wearable microfluidic chip concept is at the forefront of leading-edge mainstream technology. In a work by Lin et al., which was typically based on elastomeric polydimethylsiloxane (PDMS) as wearables for wireless sweat analysis as a promising non-invasive method for real-time health monitoring without the need for invasive blood sampling for management of health issues (Lin et al., 2024). For instance, blood glucose monitoring flexible sensor with epidermal bio-microfluidics chip is illustrated in Figure 8.4.

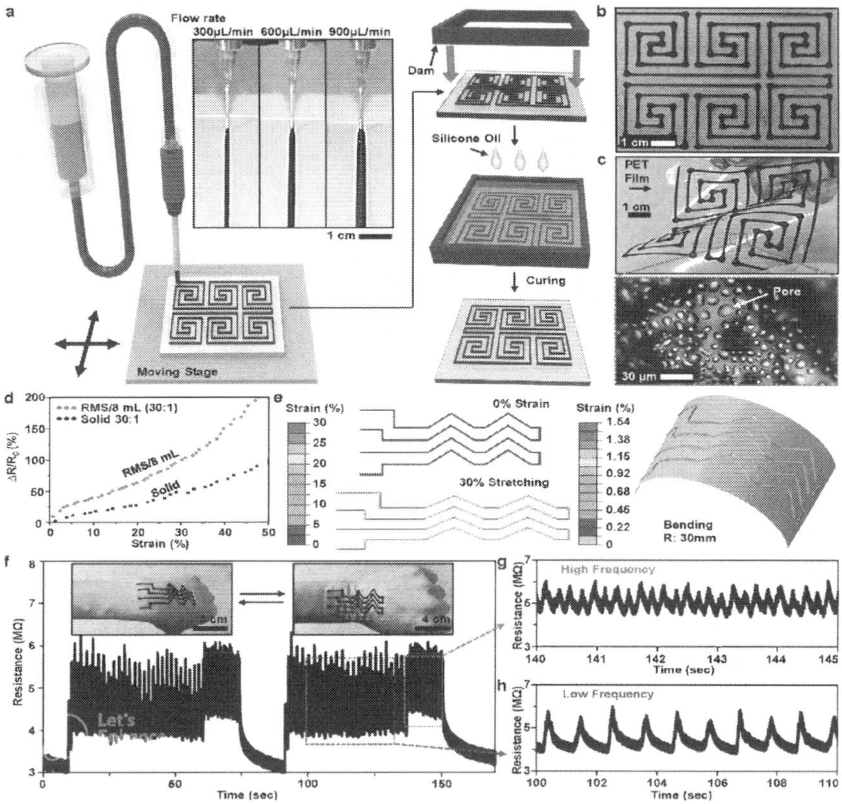

FIGURE 8.3 Schematic representation of the ink-jet printing process (3D printing) for the sensor (Song et al., 2023).

8.6 FUTURE DIRECTIONS AND CHALLENGES

The implementation of machine learning (ML) into biosensors has ushered in a new age of technological advancement in the field of point-of-care testing (PoCT). The integration of data analytics and artificial intelligence (AI) also permits early risk forecast and aids preventive health measures. Despite ongoing hurdles in miniaturization, power economy, and data security, these devices have the potential to democratize healthcare and change patient–doctor interactions (Liu et al., 2020). This viewpoint also covers the breakthrough advances in wearable aptasensor (WA) technology, which combines wearable devices and aptamer-based detection systems for tailored, real-time health monitoring. The devices use aptamers' specificity and sensitivity to target specific molecules, making them useful for a wide range of applications (Rabiee et al., 2024).

For instance, a recent review by Bhaiyya et al. (2024) explores biosensors for PoCT with machine learning (ML). ML algorithms, which can analyze and understand complex biological data, have played a major role in improving the accuracy,

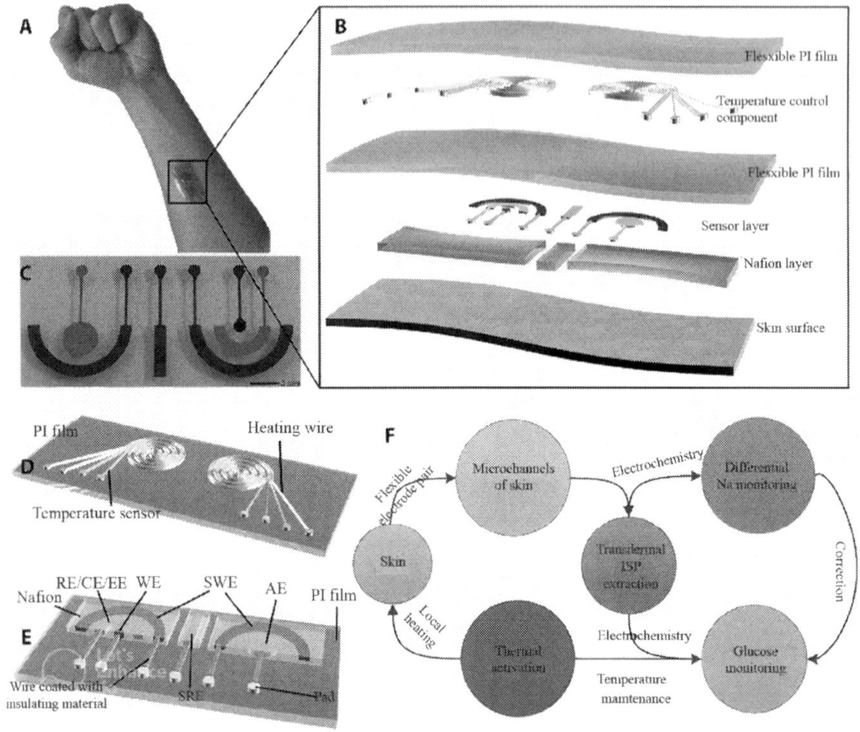

FIGURE 8.4 Blood glucose monitoring flexible sensor with epidermal bio-microfluidics. (i) Proposed device; (ii) integrated epidermal biological microfluidic device, the complete structure; (iii) flexible glucose detection patch; (iv) temperature control component; (v) glucose detection patch structure; (vi) working mechanism of integrated flexible epidermal biological microfluidic device (Smith et al., 2023).

sensitivity, and speed of diagnostic processes in a wide range of healthcare settings. This review explores the multifaceted applications of ML models, demonstrating how they subsidize into improving the diagnostic competences of biosensors. Further, the roles of ML-assisted electrochemical sensors, lab-on-a-chip sensors, electrochemiluminescence/chemiluminescence sensors, colorimetric sensors, and wearable sensors in diagnosis are also discussed in the review.

Moreover, integrating these biosensors with artificial intelligence, machine learning, Internet-of-Things, bioinformatics, and omics technologies provides insights into the prospects of intelligent nose-on-chip nano-biosensors to sniff lung cancer. One review (Chaudhary et al., 2024) consolidates knowledge on breath omic biosensor-based lung cancer screening, its significance and potential applications in improving cutting-edge medical diagnostics to minimize hospital load and save lives.

The advancements in the miniaturization of electronics have allowed great strides toward improving healthcare (L. Zhou et al., 2024a). Sensors can be made small enough to be unobtrusive and comfortable for long-term wear, which is particularly important for the health monitoring of infant patients in intensive care, as testing situations involve a tangle of electrodes, probes, and catheters that keep the infant patient bedridden, which is strenuous for both the patient and their caregiver. Figure 8.5 illustrates a design for all-in-one devices, prioritizing non-invasive monitoring and increased accessibility; wireless communication via Bluetooth, NFC, or radio; and cloud-based data processing, which comprises electronic miniaturization, soft and flexible materials, and gentler adhesion mechanisms (L. Zhou et al., 2024). Emerging technologies (e.g., AI, ML, 3D/4D printing) (Apoorva et al., 2024; Liu et al., 2020) can show potential for personalized medicine. In conjunction with the integration of sensors, the detection of diversified biomarkers in the future of wearable biosensors is a challenge, and it requires continuous developments in sensing devices, with the current market trends and user acceptance. The latest advances in this field are centered on electrochemical and optical biosensors, with significant progress achieved in the non-invasive monitoring of new biomarkers ranging from metabolites to

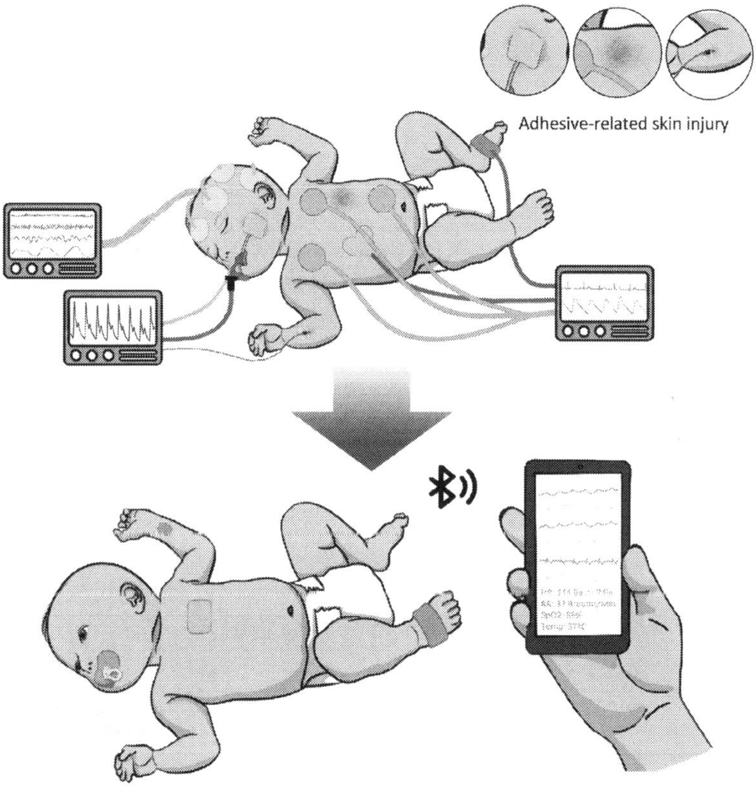

FIGURE 8.5 An illustration for a design for all-in-one devices, aimed at translating conventional monitoring practices to single smart wearables. Methods include electronic miniaturization; soft and flexible materials; milder adhesion mechanisms; non-invasive monitoring and improved accessibility; wireless communication via Bluetooth, NFC, or radio; and cloud-based data processing (L. Zhou et al., 2024a).

microorganisms and hormones. As a whole, data accuracy and calibration are critical for assuring sensor accuracy over time and under varying circumstances (Xu et al., 2024).

In terms of power utilization, most wearable sensors must be energy efficient for prolonged use. Furthermore, in terms of data security and privacy, sensitive health data must be secured while users' privacy is secured. Wearable biosensors' widespread acceptance by the medical and commercial worlds will require comprehensive large-scale validation of their functionality through multidisciplinary collaboration across the technical, biological, and clinical disciplines (Wittmann et al., 2020). Lastly, wearable biosensing technologies offering real-time body sensing and communication of comprehensive physiological information hold great promise for improving performance monitoring and personal healthcare, which could have a wide-ranging effect on daily life. By continuously monitoring physiological data and identifying deviations from healthy baselines, wearable sensors can personalize health management and disease diagnosis (Song et al., 2024). However, a number of factors related to the subpar system-level performance of integrated wearables have caused the commercial adoption of these wearables to remain delayed. Many of the wearables' features are compromised by the wearability consideration. Additionally, wearables' limited power capacity makes it harder to conduct continuous monitoring for prolonged periods of time. Furthermore, thermal problems in the small form factor caused by peak-power operations for demanding computations can quickly impede wearability and sensor functionality. Moreover, wearable technology is always exposed to mechanical, chemical, electrical, and environmental factors that might bias the data it collects. In order to enable proper data interpretation, this creates a requirement for advanced data analytics to contextually identify, include, and exclude data points per multisensor fusion (Y. Zhang et al., 2024). This chapter discuses the challenges surrounding wearable device integration, from aspects in understandings the types of biosensors with materials used along with advanced fabrication techniques employed, with insights of emerging technologies.

REFERENCES

Ahmed, F., Song, J., Masud, Lee, S., & Kim, J. (2024). Advances in the development of hydrogel-based adhesion layers for wearable health monitors: Focusing on biocompatibility, conductivity, and mechanical strength. *ACS Applied Polymer Materials*, *6*(22), 13497–13511. https://doi.org/10.1021/acsapm.4c02727

Alavian, F., Khodabakhshi, F., & Chenary, F. H. (2025). Biosensors for early stroke detection. *Clinica Chimica Acta*, *567*, 120079. https://doi.org/10.1016/j.cca.2024.120079

Alipio, M., & Villena, M. L. (2023). Intelligent wearable devices and biosensors for monitoring cattle health conditions: A review and classification. *Smart Health*, *27*, 100369. https://doi.org/10.1016/j.smhl.2022.100369

Apoorva, S., Nguyen, N.-T., & Sreejith, K. R. (2024). Recent developments and future perspectives of microfluidics and smart technologies in wearable devices. *Lab on a Chip*, *24*(7), 1833–1866. https://doi.org/10.1039/D4LC00089G

Baek, S., Jo, Y., Lee, Y., Kwon, J., & Jung, S. (2024). Design and integration of organic printed thin-film transistor-based soft biosensors for wearable applications. *ACS Applied Electronic Materials*, *6*(11), 7657–7678. https://doi.org/10.1021/acsaelm.4c01632

Baskar, A., Madhivanan, K., Atchudan, R., Arya, S., & Sundramoorthy, A. K. (2025a). Nanoparticle electrochemical biosensors for virus detection. *Clinica Chimica Acta*, *566*, 120054. https://doi.org/10.1016/j.cca.2024.120054

Bhaiyya, M., Panigrahi, D., Rewatkar, P., & Haick, H. (2024). Role of machine learning assisted biosensors in point-of-care-testing for clinical decisions. *ACS Sensors*, *9*(9), 4495–4519. https://doi.org/10.1021/acssensors.4c01582

Binabaji, F., Dashtian, K., Zare-Dorabei, R., Naseri, N., Noroozifar, M., & Kerman, K. (2024). Innovative wearable sweat sensor array for real-time volatile organic compound detection in noninvasive diabetes monitoring. *Analytical Chemistry*, *96*(33), 13522–13532. https://doi.org/10.1021/acs.analchem.4c02034

Chaudhary, V., Taha, B. A., Lucky, Rustagi, S., Khosla, A., Papakonstantinou, P., & Bhalla, N. (2024). Nose-on-chip nanobiosensors for early detection of lung cancer breath biomarkers. *ACS Sensors*, *9*(9), 4469–4494. https://doi.org/10.1021/acssensors.4c01524

Chen, C., Fu, Y., Sparks, S. S., Lyu, Z., Pradhan, A., Ding, S., Boddeti, N., Liu, Y., Lin, Y., Du, D., & Qiu, K. (2024). 3D-printed flexible microfluidic health monitor for *In Situ* sweat analysis and biomarker detection. *ACS Sensors*, *9*(6), 3212–3223. https://doi.org/10.1021/acssensors.4c00528

Chenthattil Raril, Manjunatha, J. G., Ravishankar, D. K., Fattepur, S., Siddaraju, G., & Nanjundaswamy, L. (2020). Validated electrochemical method for simultaneous resolution of tyrosine, uric acid, and ascorbic acid at polymer modified nano-composite paste electrode. *Surface Engineering and Applied Electrochemistry*, *56*(4), 415–426. https://doi.org/10.3103/S1068375520040134

Childs, A., Mayol, B., Lasalde-Ramírez, J. A., Song, Y., Sempionatto, J. R., & Gao, W. (2024). Diving into sweat: Advances, challenges, and future directions in wearable sweat sensing. *ACS Nano*, *18*(36), 24605–24616. https://doi.org/10.1021/acsnano.4c10344

Chowdhury, M.-U.-S., Roy, S., Aryal, K. P., Leung, H., & Pandey, R. (2024). Realizing the potential of commercial E-textiles for wearable glucose biosensing application. *ACS Materials Au*, *4*(6), 592–603. https://doi.org/10.1021/acsmaterialsau.4c00033

Dong, Y., Tian, Y., Yang, Y., Gong, R., Mi, Q., Zou, F., & Yu, H.-Y. (2024). Multiple covalent modification enables nylon fiber biosensor with robust scrub-resistant and signal-capture ability for multiscenario health monitoring and security warning. *International Journal of Biological Macromolecules*, *281*, 136518. https://doi.org/10.1016/j.ijbiomac.2024.136518

Eser, E., Bahar, Ö., Bilgi, İ., Bioe, B., Yin, Y., Supervisor, N., Betül, G., & By, A. (2022). *Gluten detection by using aptamer-based biosensor*. https://www.researchgate.net/publication/364957522

Fattah-alhosseini, A., Chaharmahali, R., Alizad, S., Kaseem, M., & Dikici, B. (2024). A review of smart polymeric materials: Recent developments and prospects for medicine applications. *Hybrid Advances*, *5*, 100178. https://doi.org/10.1016/j.hybadv.2024.100178

Gao, B., Jiang, J., Zhou, S., Li, J., Zhou, Q., & Li, X. (2024). Toward the next generation human–machine interaction: Headworn wearable devices. *Analytical Chemistry*, *96*(26), 10477–10487. https://doi.org/10.1021/acs.analchem.4c01190

Gong, S., Zhang, X., Nguyen, X. A., Shi, Q., Lin, F., Chauhan, S., Ge, Z., & Cheng, W. (2023). Hierarchically resistive skins as specific and multimetric on-throat wearable biosensors. *Nature Nanotechnology*, *18*(8), 889–897. https://doi.org/10.1038/s41565-023-01383-6

Hosseine, M., Bakhshi, A., Naghib, S. M., & Rabiee, N. (2024). Recent advancements in polyaniline-based biosensors for diagnosis of cancers: A comprehensive review. *TrAC Trends in Analytical Chemistry*, *181*, 118040. https://doi.org/10.1016/j.trac.2024.118040

Huang, C., Yang, W., Wang, H., Huang, S., Gao, S., Li, D., Liu, J., Hou, S., Feng, W., Wang, Z., Li, F., Hao, Z., Zhao, X., Hu, P., & Pan, Y. (2024). Flexible/regenerative nanosensor with automatic sweat collection for cytokine storm biomarker detection. *ACS Nano*, *18*(32), 21198–21210. https://doi.org/10.1021/acsnano.4c04456

Jiang, Z., Ye, D., Xiang, L., He, Z., Dai, X., Yang, J., Xiong, Q., Ma, Y., Zhi, D., Zou, Y., Peng, Q., Wang, S., Li, J., Zhang, F., & Di, C. (2024). A drug-mediated organic electrochemical transistor for robustly reusable biosensors. *Nature Materials*, *23*(11), 1547–1555. https://doi.org/10.1038/s41563-024-01970-5

Joshi, F. M., Viar, G. A., Pigino, G., Drechsler, H., & Diez, S. (2022). Fabrication of high aspect ratio gold nanowires within the microtubule lumen. *Nano Letters*, *22*(9), 3659–3667. https://doi.org/10.1021/acs.nanolett.2c00255

Karami, Y., Ehtiati, S., Ghasemi, H., Rafiee, M., Zamani Sani, M., Hosseini, S. E., Moradi Kazerouni, H., Movahedpour, A., Aiiashi, S., & Khatami, S. H. (2025). Non-coding RNA biosensors for early detection of brain cancer. *Clinica Chimica Acta*, *566*, 120041. https://doi.org/10.1016/j.cca.2024.120041

Kariduraganavar, M. (2016). *Shape memory polymers in biomedical applications* (pp. 1–60). https://doi.org/10.21741/9781945291098-1

Kim, J., Campbell, A. S., de Ávila, B. E.-F., & Wang, J. (2019). Wearable biosensors for healthcare monitoring. *Nature Biotechnology*, *37*(4), 389–406. https://doi.org/10.1038/s41587-019-0045-y

Kim, J., Kim, M. Y., Han, Y., Lee, G. Y., Kim, D. H., Heo, Y. J., & Park, M. (2024). Development of an electrochemical biosensor for non-invasive cholesterol monitoring via microneedle-based interstitial fluid extraction. *Talanta*, *280*, 126771. https://doi.org/10.1016/j.talanta.2024.126771

Lee, J., Kim, J., Kim, S., & Min, D.-H. (2016). Biosensors based on graphene oxide and its biomedical application. *Advanced Drug Delivery Reviews*, *105*, 275–287. https://doi.org/10.1016/j.addr.2016.06.001

Li, M., Wang, S., Li, Y., Meng, X., Wei, Y., Wang, Y., Chen, Y., Xiao, Y., & Cheng, Y. (2024). An integrated all-natural conductive supramolecular hydrogel wearable biosensor with enhanced biocompatibility and antibacterial properties. *ACS Applied Materials & Interfaces*, *16*(38), 51618–51629. https://doi.org/10.1021/acsami.4c08690

Lin, F., Vera Anaya, D., Gong, S., Yap, L. W., Lu, Y., Yong, Z., & Cheng, W. (2024). Gold nanowire sponge electrochemistry for permeable wearable sweat analysis comfortably and wirelessly. *ACS Sensors*, *9*(10), 5414–5424. https://doi.org/10.1021/acssensors.4c01635

Ling, W., Shang, X., Liu, J., & Tang, T. (2025). A skin-mountable flexible biosensor based on Cu-MOF/PEDOT composites for sweat ascorbic acid monitoring. *Biosensors and Bioelectronics*, *267*, 116852. https://doi.org/10.1016/j.bios.2024.116852

Liu, C., Liu, Q., Chen, X., Guo, M., Chen, Z., Zhao, J., Chen, H., Guo, S., Cen, H., Yao, G., Chen, L., Wang, Y., Yang, P. C., Wang, L., & Chen, F. (2025). A novel label-free biosensor for myocardial ischemia biomarker detection via CRISPR/12a. *Biosensors and Bioelectronics*, *270*, 116954. https://doi.org/10.1016/j.bios.2024.116954

Liu, D., Wang, J., Wu, L., Huang, Y., Zhang, Y., Zhu, M., Wang, Y., Zhu, Z., & Yang, C. (2020). Trends in miniaturized biosensors for point-of-care testing. *TrAC Trends in Analytical Chemistry*, *122*, 115701. https://doi.org/10.1016/j.trac.2019.115701

Liu, Z., Zhang, C., Xiang, L., Zhang, F., & Di, C. (2024). Organic transistors-driven wearable electronics for smart life. *Wearable Electronics*, *1*, 211–227. https://doi.org/10.1016/j.wees.2024.09.004

Marchianò, V., Tricase, A., Cimino, A., Cassano, B., Catacchio, M., Macchia, E., Torsi, L., & Bollella, P. (2025). Inside out: Exploring edible biocatalytic biosensors for health monitoring. *Bioelectrochemistry, 161*, 108830. https://doi.org/10.1016/j.bioelechem.2024.108830

Monnappa, A. B., Manjunatha, J. G., & Bhatt, A. S. (2020). Design of a sensitive and selective voltammetric sensor based on a cationic surfactant-modified carbon paste electrode for the determination of alloxan. *ACS Omega, 5*(36), 23481–23490. https://doi.org/10.1021/acsomega.0c03517

Park, J. S., Ahmad, W., Choe, K., Ahmad, R., Park, T. J., & Kim, M. O. (2024). Biosensors and biomarkers: A dynamic duo towards Alzheimer's disease detection. *Sensing and Bio-Sensing Research, 46*, 100704. https://doi.org/10.1016/j.sbsr.2024.100704

Pushpanjali, P. A., Manjunatha, J. G., Hareesha, N., Souza, E. S. D', Charithra, M. M., & Prinith, N. S. (2021). Voltammetric analysis of antihistamine drug cetirizine and paracetamol at poly(L-Leucine) layered carbon nanotube paste electrode. *Surfaces and Interfaces, 24*, 101154. https://doi.org/10.1016/j.surfin.2021.101154

Rabiee, N., & Rabiee, M. (2024). Wearable aptasensors. *Analytical Chemistry, 96*(49), 19160–19182. https://doi.org/10.1021/acs.analchem.4c05004

Sattar, M., Lee, Y. J., Kim, H., Adams, M., Guess, M., Kim, J., Soltis, I., Kang, T., Kim, H., Lee, J., Kim, H., Yee, S., & Yeo, W.-H. (2024). Flexible thermoelectric wearable architecture for wireless continuous physiological monitoring. *ACS Applied Materials & Interfaces, 16*(29), 37401–37417. https://doi.org/10.1021/acsami.4c02467

Sharma, A., Badea, M., Tiwari, S., & Marty, J. L. (2021). Wearable biosensors: An alternative and practical approach in healthcare and disease monitoring. *Molecules, 26*(3), 748. https://doi.org/10.3390/molecules26030748

Smith, A. A., Li, R., & Tse, Z. T. H. (2023). Reshaping healthcare with wearable biosensors. *Scientific Reports, 13*(1), 4998. https://doi.org/10.1038/s41598-022-26951-z

Song, Y., Wang, X., Wang, L., Qu, L., & Zhang, X. (2024). Functionalized face masks as smart wearable sensors for multiple sensing. *ACS Sensors, 9*(9), 4520–4535. https://doi.org/10.1021/acssensors.4c01705

Song, Z., Zhou, S., Qin, Y., Xia, X., Sun, Y., Han, G., Shu, T., Hu, L., & Zhang, Q. (2023). Flexible and wearable biosensors for monitoring health conditions. *Biosensors, 13*(6), 630. https://doi.org/10.3390/bios13060630

Tigari, G., & Manjunatha, J. G. (2020). Poly(glutamine) film-coated carbon nanotube paste electrode for the determination of curcumin with vanillin: An electroanalytical approach. *Monatshefte Für Chemie - Chemical Monthly, 151*(11), 1681–1688. https://doi.org/10.1007/s00706-020-02700-8

Wittmann, J., Cannillo, F., Ciomaga, D., Jefremow, M., & Rigoni, F. (2020). Highly Efficient Power management in wearables and IoT devices. In *Next-generation ADCs, high-performance power management, and technology considerations for advanced integrated circuits* (pp. 125–142). Springer International Publishing. https://doi.org/10.1007/978-3-030-25267-0_8

Xiong, H., Zhang, X., Sun, J., Xue, Y., Yu, W., Mou, S., Hsia, K. J., Wan, H., & Wang, P. (2025). Recent advances in biosensors detecting biomarkers from exhaled breath and saliva for respiratory disease diagnosis. *Biosensors and Bioelectronics, 267*, 116820. https://doi.org/10.1016/j.bios.2024.116820

Xu, Z., Hao, Y., Luo, A., & Jiang, Y. (2024). Technologies and applications in wireless biosensors for real-time health monitoring. *Med-X, 2*(1), 24. https://doi.org/10.1007/s44258-024-00041-3

Yang, Y., Yang, S., Xia, X., Hui, S., Wang, B., Zou, B., Zhang, Y., Sun, J., & Xin, J. H. (2024). MXenes for wearable physical sensors toward smart healthcare. *ACS Nano, 18*(36), 24705–24740. https://doi.org/10.1021/acsnano.4c08258

Zeng, Q., Tian, X., Nguyen, D. T., Li, C., Chia, P., Tee, B. C. K., Wu, C., & Ho, J. S. (2024). A digitally embroidered metamaterial biosensor for kinetic environments. *Nature Electronics*, *7*(11), 1025–1034. https://doi.org/10.1038/s41928-024-01263-4

Zhang, P., Zhu, B., Du, P., & Travas-Sejdic, J. (2024). Electrochemical and electrical biosensors for wearable and implantable electronics based on conducting polymers and carbon-based materials. *Chemical Reviews*, *124*(3), 722–767. https://doi.org/10.1021/acs.chemrev.3c00392

Zhang, Y., Zheng, X. T., Zhang, X., Pan, J., & Thean, A. V.-Y. (2024). Hybrid integration of wearable devices for physiological monitoring. *Chemical Reviews*, *124*(18), 10386–10434. https://doi.org/10.1021/acs.chemrev.3c00471

Zhou, L., Guess, M., Kim, K. R., & Yeo, W.-H. (2024). Skin-interfacing wearable biosensors for smart health monitoring of infants and neonates. *Communications Materials*, *5*(1), 72. https://doi.org/10.1038/s43246-024-00511-6

Zhou, X., Shao, Z., Yan, S., Lin, Y., Liu, Y., Feng, X., Sha, J., Ding, L., & Wang, K. (2024). Coencapsulating TMB probes and bimetallic MOF nanozymes in a hydrogel patch for fabricating reusable visual VC sensors. *Analytical Chemistry*, *96*(43), 17310–17318. https://doi.org/10.1021/acs.analchem.4c03665

Zhu, C., Yang, G., Li, H., Du, D., & Lin, Y. (2015). Electrochemical sensors and biosensors based on nanomaterials and nanostructures. *Analytical Chemistry*, *87*(1), 230–249. https://doi.org/10.1021/ac5039863

9 Biosensors for Cancer Diagnosis and Monitoring

Maliana El Aamri, Ghita Yammouri, Hasna Mohammadi and Aziz Amine

9.1 INTRODUCTION

Cancer is one of the leading causes of morbidity and mortality worldwide, with millions of new cases and deaths occurring each year. In 2020, there were an estimated 19.3 million new cancer cases worldwide. Additionally, an estimated 50.6 million people were living with cancer within five years of diagnosis. The global burden of cancer is projected to grow, with new cases expected to rise to 28.4 million by 2040, a 47% increase from 2020, driven by population growth, aging, and changes in cancer risk factors (Sung et al. 2021). Cancer can arise from genetic mutations, lifestyle factors (like smoking and diet), environmental exposures, and infections. Progression varies by cancer type, stage, and patient. Early detection and monitoring are vital for improving outcomes, reducing mortality, and enhancing patients' quality of life (Siegel, Miller, and Jemal 2019). Early diagnosis minimizes treatment intensity and duration, leading to fewer side effects. Early-stage cancers often require less invasive surgeries and lower doses of radiation or chemotherapy, reducing recovery time and side effects. Also, continuous monitoring is crucial for promptly detecting recurrence or progression, enabling quick intervention.

Traditional cancer diagnostics, the gold standard, rely on imaging techniques like X-rays and computed tomography (CT) scans to visualize tumors and assess their spread. A biopsy with histopathological analysis remains the definitive method, allowing tissue examination to confirm malignancy (Edge and Compton 2010). However, these methods have limitations. Imaging techniques may not detect microscopic tumors or early-stage cancers, and biopsies are invasive and can be painful. Moreover, traditional methods often require specialized equipment and trained personnel, which may not be readily available in all healthcare settings, especially in low-resource regions (Ferlay et al. 2019).

In recent years, molecular diagnostics have become vital for early cancer detection due to their high sensitivity and specificity. Techniques like PCR, next-generation sequencing (NGS), and liquid biopsy identify cancer-related genetic mutations, offering insights into cancer status and informing treatment choices (Mardis 2017,

DOI: 10.1201/9781003536239-9

2006–2016). Liquid biopsy, which involves the analysis of various molecules named biomarkers in the blood, saliva, and urine, offers a non-invasive alternative to traditional tissue biopsies. More details about these biomarkers and their specific roles will be discussed in the subsequent sections of this chapter. One device used to detect these molecules (biomarkers) is the biosensor. In the context of cancer diagnosis and monitoring, biosensors play a crucial role by enabling the rapid and precise detection of tumor-associated biomarkers (Tothill 2009).

Biosensors offer distinct advantages in cancer diagnostics and monitoring (El Aamri et al. 2020). They provide high sensitivity and specificity, detecting low biomolecule concentrations for early diagnosis (Lino et al. 2022). Biosensors can be non-invasive, analyzing targets in saliva or urine to reduce patient discomfort (Jiawei Li et al. 2020). They allow real-time monitoring, offering rapid data on biomarker levels for timely intervention (Pohanka 2018). Portable and suitable for point-of-care testing, biosensors increase access in remote areas (Luong, Male, and Glennon 2008). Additionally, they are cost-effective, requiring less equipment and qualified personnel, making diagnostics more accessible (Zong et al. 2018).

9.2 TYPES OF BIOSENSORS FOR CANCER DIAGNOSIS

Biosensors are innovative analytical devices that have revolutionized cancer diagnosis and monitoring. As described in Figure 9.1, these devices combine a biological recognition element with a transducer to detect specific cancer biomarkers with high sensitivity and specificity.

This section will explore the critical biosensors used in cancer diagnosis, including electrochemical, optical, and mass-sensitive biosensors. Additionally, the utilization of nanozymes in the development of biosensors will be discussed.

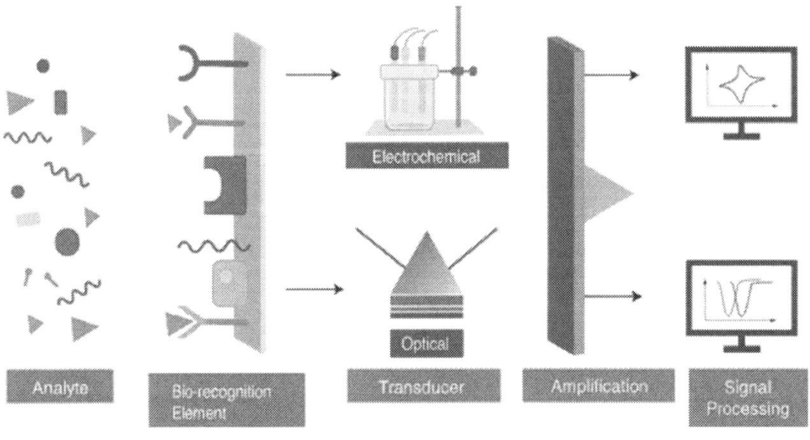

FIGURE 9.1 The basic components and mechanism of a biosensor for application in cancerdiagnosis. Reused with permission from Elsevier publisher (Jain, Chauhan, and Saxena 2023).

9.2.1 Electrochemical Biosensors

Electrochemical biosensors detect cancer biomarkers through specific interactions at the sensor surface. When biomarkers bind to recognition elements, they trigger measurable changes in electrochemical properties (current, voltage, or impedance). The resulting signal correlates directly with the biomarker concentration, enabling quantitative detection (Wang 2006).

9.2.2 Optical Biosensors

Optical biosensors detect cancer biomarkers based on changes in optical properties, such as absorbance, fluorescence, or luminescence, resulting from the interaction between the cancer biomarker and the recognition element. These changes are measured using optical transducers, which convert the optical signals into quantifiable data (Fan et al. 2008).

9.2.3 Mass-Sensitive Biosensors

Mass-sensitive biosensors, also known as piezoelectric biosensors, operate on the principle of detecting changes in mass on the sensor surface. The most common type is the quartz crystal microbalance (QCM). When a target cancer biomarker binds to the recognition element on the QCM surface, it changes the oscillation frequency of the quartz crystal. This change in frequency is directly proportional to the mass of the bound biomarker, allowing for precise quantification (Gronewold 2007).

9.2.4 Thermometric Biosensors

Thermometric (or thermal/calorimetric) biosensors detect cancer biomarkers by measuring temperature changes that occur during biochemical reactions. When biomarkers interact with recognition elements on the sensor surface, the resulting reaction causes heat absorption or release. A thermistor measures these thermal changes, producing signals proportional to biomarker concentration (Ramanathan and Danielsson 2001).

9.3 CANCER BIOMARKERS DETECTED BY BIOSENSORS

Cancer biomarkers are biological molecules found in blood, body fluids, or tissues that indicate the presence or progression of cancer (De Gruttola et al. 1997). Biomarker detection is crucial in oncology, especially for early cancer diagnosis and treatment. Research in this field is rapidly evolving, uncovering new applications, while biomarker detection in oncology is shifting from blood samples to non-invasive matrices like urine, saliva, and sweat (Jiawei Li et al. 2020; Corrie et al. 2015). These methods offer easier and more patient-friendly sample collection. In cancer care, biomarkers are key indicators for detecting stages, treatment responses, and monitoring progression or recurrence (Luo and Davis 2013).

An ideal biomarker must fulfill multiple critical criteria to ensure its effectiveness in clinical applications. It should maintain stability during storage without degradation or artificial induction, while being accurately detectable through specific, sensitive, and reproducible analytical methods. The biomarker should remain unaffected by external factors such as diet, and be readily accessible in tissues or fluids for minimally invasive collection. Additionally, it must be measurable within current analytical detection limits, even in small quantities. These combined characteristics determine a biomarker's suitability for reliable clinical diagnostics and monitoring, ensuring accurate and consistent results across different testing conditions (Vaidya and Bonventre 2010).

9.3.1 Cancer Biomarker Types

Cancer biomarkers encompass a diverse biological entities, including (Figure 9.2) nucleic acids and proteins including cytokines, cells, and exosomes, as well as metabolites and processes such as apoptosis, angiogenesis, and cell proliferation.

9.3.1.1 Proteins

Proteins are essential for cell function, and abnormal levels can signal cancer risk. Detecting these protein markers supports early diagnosis and monitoring, offering

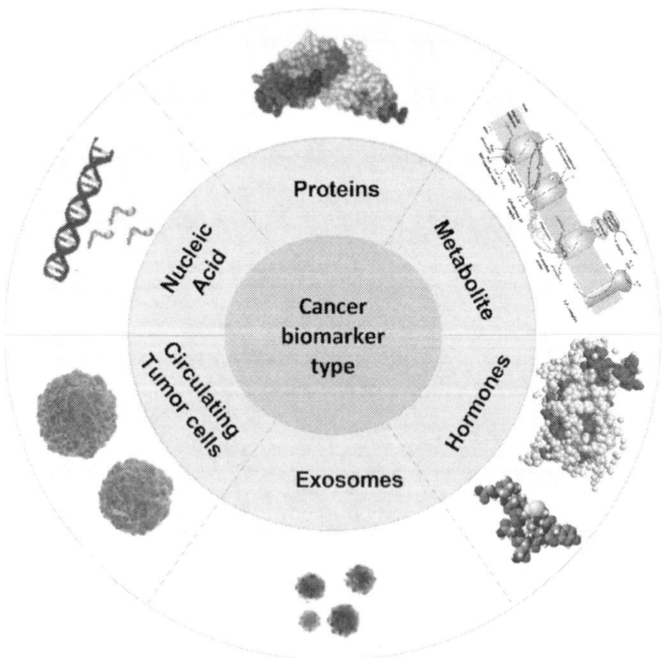

FIGURE 9.2 Different types of cancer biomarkers.

key insights into cancer progression. One of the extensively used protein biomarkers is carcinoembryonic antigen (CEA), particularly in colorectal cancer. CEA levels can indicate tumor presence, metastasis, or recurrence (Jelski and Mroczko 2020). Similarly, alpha-fetoprotein (AFP) and prostate-specific antigen (PSA) are other significant protein biomarkers, used for liver and prostate cancer diagnosis, respectively (Duffy 2020; Singal et al. 2022).

Biosensors detect protein biomarkers by using bioreceptors, like antibodies, to bind target proteins. This interaction generates a measurable signal, enabling precise cancer biomarker detection. Table 9.1 provides an overview of various biosensors used for detecting protein biomarkers in cancer diagnostics.

9.3.1.2 Hormones

As critical regulators of physiological functions, hormones play a significant role as cancer biomarkers. Elevated levels of certain hormones can indicate the presence of malignancies, as normal tissues may start producing hormones abnormally due to cancerous changes. For example, thyroid hormones like thyroxine and triiodothyronine can be markers for thyroid cancer (Prpić et al. 2018). Calcitonine, produced by the thyroid gland, is a biomarker for medullary thyroid carcinoma (Niederle et al. 2020).

Hormone-detecting biosensors use specific bioreceptors—like antibodies—that bind hormones through selective molecular interactions, ensuring accurate detection even among other substances. Table 9.2 provides an overview of various biosensors used for detecting hormones biomarkers in cancer diagnostics

9.3.1.3 Cytokines

Cytokines are signaling molecules that regulate immune responses, inflammation, and cell proliferation. They play a crucial role as biomarkers in cancer research, particularly for understanding the tumor microenvironment and the body's immune response to cancer. Certain cytokines, such as IL-6 and IL-10, have been linked to various cancers, including lymphoma and leukemia. Elevated levels of these cytokines can indicate the presence of malignancy and the body's immune reaction to tumor cells. For example, IL-6 promotes tumor growth and metastasis in certain cancers (Wu et al. 2013). IL-10, on the other hand, can have both tumor-promoting and tumor-inhibiting effects, making its role in cancer complex and context-dependent (Mirlekar 2022).

Cytokine-detecting biosensors use specialized bioreceptors to bind target cytokines, producing signals that allow precise measurement in samples. Table 9.3 provides an overview of various biosensors used for detecting cytokine biomarkers in cancer diagnostics.

9.3.1.4 Exosomes

Exosomes, small extracellular vesicles with a lipid bilayer membrane, range from 50 to 150 nm in size and are released by all cell types (Raposo and Stoorvogel 2013). In cancer research, exosomes have been identified as significant players in tumor progression, facilitating tumorigenesis through the regulation of angiogenesis,

TABLE 9.1
Overview of Various Biosensors for Proteins as Cancer Biomarker Detection

Protein Biomarker	Type of Biosensor	Type of Cancer	Bioreceptor	Limit of Detection	References
PSA	Fluorometric biosensor	Prostate cancer	Aptamer	0.05 nM	R. Zhao et al. (2021)
PSA	Electrochemical biosensor	Prostate cancer	Ab	86 fg/mL	Wang et al. (2021)
PSA	QCM biosensor	Prostate cancer	Ab	0.05 μg/L	Pohanka (2021)
CA-125	Fluorometric biosensor	Ovarian cancer	Magnetic molecularly imprinted polymers	10 pg/mL	Bahari, Babamiri, and Salimi (2021)
CA-125	Electrochemical biosensor	Ovarian cancer	Ab	6 μU/mL	Pakchin et al. (2020)
CA-125	Electrochemical biosensor	Ovarian cancer	Aptamer	0.027 U/mL	Liu et al. (2023)
CEA	Chemiluminescence immunoassay	Lung cancer	Ab	85 pg/ml	Mao et al. (2019)
CEA	Photoelectrochemical	—	Ab	40 fg/ml	Guan et al. (2021)
CEA	Colorimetric biosensor	Colorectal cancer	Aptamer (ssDNA)	1 ng/mL	Shahbazi, Hosseinkhani, and Ranjbar (2017)
AFP	Colorimetric biosensor	—	Ab	7 pg/mL	Zhu et al. (2019)
CA 19-9	Electrochemiluminescence immunosensor	Pancreatic cancer, colorectal cancer, gastric cancer	Ab	31 μU/mL	Mo et al. (2021)
CA 15-3	Fluorometric biosensor	Breast cancer	FAM-aptamer (ssDNA)	25.6 μU/mL	Y. Wu et al. (2021)
Thyroglobulin	Electrochemical biosensor	Thyroid cancer	Sortilin	0.01 fM	Jalalvand et al. (2024)
Beta-2-microglobulin	Surface plasmon resonance	Multiple myeloma, lymphoma	Ab	40 ng/mL	L. Zhang et al. (2022)

(Continued)

TABLE 9.1 (CONTINUED)
Overview of Various Biosensors for Proteins as Cancer Biomarker Detection

Protein Biomarker	Type of Biosensor	Type of Cancer	Bioreceptor	Limit of Detection	References
NSE	Photoelectrochemical biosensors	Small cell lung cancer, neuroblastoma	Ab	70 fg/mL	Y. Zhang et al. (2022)
NSE	Electrochemical biosensor	Lung cancer	Ab	6.1 fg/mL	Aydın, Aydın, and Sezgintürk (2020)
Her2/neu (ErbB-2)	Fluorometric biosensor	Breast cancer	Ab	0.6 ng/mL	Pileri et al. (2024)
S-100 protein	Electrochemical biosensor	Melanoma	Ab	1.9 pg/mL	Huihua Zhang et al. (2022)
Chromogranin A	Electrochemical biosensor	Neuroendocrine tumors	Ab	5.3 fg/mL	X. Liu et al. (2021)
CYFRA-21-1	Electrochemical biosensor	Lung cancer	Ab	4.59 fg/mL	Aydın, Aydın, and Sezgintürk (2023)
SCCA	Electrochemical biosensor	Cervical cancer	Ab	3 fg/mL	Mao et al. (2022)
CD44	Electrochemical biosensor	Gastric, breast, ovarian, colorectal, and pancreatic cancer	Molecularly imprinted polymers	1.41×10^{-5} ng/mL	Cui et al. (2022)

Abbreviations: CA: cancer antigen; SCCA: squamous cell carcinoma antigen.

TABLE 9.2
Overview of Various Biosensors for Hormone Biomarker Detection in Cancer

Hormone Biomarker	Type of Biosensor	Type of Cancer	Bioreceptor	Limit of Detection	Reference
ProGRP	Electrochemiluminescence biosensor	Lung cancer	Aptamer (ssDNA)	1.3 fg/mL	Q. Wu and Zheng (2022)
ProGRP	Electrochemical biosensor	Lung cancer	Ab	77.62 fg/mL	Fu et al. (2024)
HCG	Electrochemical biosensor	Testicular cancer	Aptamer (ssDNA)	0.4 mIU/mL	Xia et al. (2017)
Calcitonin	Luminescence	Thyroid cancer	Ab	0.71pg/mL	Omer et al. (2020)
17β-estradiol	Photoelectrochemical biosensor	Ovarian and breast cancer	Split aptamer	0.02 pg/mL	Qin et al. (2024)
P4	Electrochemical biosensor	Lung and breast cancers	Aptamer (ssDNA)	1.86 pM	Samie and Arvand (2020)
TSH	Electrochemical biosensor	Thyroid cancer	Ab	0.026 mIUL-1	Ozcan and Aydin (2021)
PTH	Electrochemical biosensor	Cancers of the parathyroid glands	Ab	65 fg/ml	Chen, Liu, and Wu (2021)

Abbreviations: ProGRP: progastrin-releasing peptide; HCG: human chorionic gonadotropin; P4: progesterone; TSH: thyroid-stimulating hormone; PTH: parathyroid hormone.

TABLE 9.3
Detection of Cytokine Biomarkers in Cancer Using Various Biosensors

Cytokine Biomarker	Type of Biosensor	Type of Cancer	Bioreceptor	Limit of Detection	References
IL-6	Electrochemical biosensor	Prostate cancer	Ab	6 fg/mL	Aydın (2020)
TNF-α	Electrochemical biosensor	Cancer	Ab	3.1 fg/mL	Aydin (2020)
TNF-α	Fluorescence	Cancer	Aptamer (ssDNA)	97.1 pM	Ghosh et al. (2018)
IL-8	Electrochemical immunosensor	Oral cancer	Ab	1 pg/mL	Bhatia et al. (2021)
IL-10	Electrochemical immunosensor	Cancer	Ab	7.89 fg/mL	Frias et al. (2023)
IFN-γ	Electrochemical aptasensor	Cancer	Aptamer (ssDNA)	0.46 fM	Saputra et al. (2024)
IL-1β	Electrochemical biosensor	Cancer	IL-1β binding peptide	2.4 pM	Yang et al. (2024)
VEGF	Colorimetric biosensor	Cancer	DNA aptamer	0.5 pg/mL	Xu et al. (2017)
AMF	Electrochemical biosensor	Cancer	Phosphoglucose isomerase (enzyme)	0.7 fM	Devillers et al. (2017)

Abbreviations: TNF-α: tumor necrosis factor-alpha; IL-8: interleukin-8; IL-10: interleukin-10; IFN-γ: interferon-gamma; IL-1β: interleukin-1beta; VEGF: vascular endothelial growth factor; AMF: autocrine motility factor.

immunity, and metastasis (Azmi, Bao, and Sarkar 2013). Exosomes present in the tumor microenvironment and circulating in body fluids like serum, plasma, and urine are being explored as non-invasive biomarkers for early cancer detection and diagnosis (Liu et al. 2021). Their stability in body fluids makes them promising candidates for liquid biopsy applications in oncology (Jafari et al. 2023). For instance, exosomes carrying the protein Glypican-1 have been identified as a biomarker for pancreatic cancer (Melo et al. 2015).

Biosensors for detecting exosomes as cancer biomarkers are highly advanced. For instance, antibodies against exosomal markers like CD63, CD81, and CD9 are commonly used in biosensors to isolate and detect exosomes (Kowalczyk et al. 2023). Table 9.4 provides an overview of various biosensors used for detecting exosomes biomarkers in cancer diagnostics

9.3.1.5 Circulating Tumor Cells

Circulating tumor cells (CTCs) are key biomarkers for metastasis, offering real-time cancer insights. Their count helps assess metastasis risk, guide treatments, and track therapy response, especially in breast and prostate cancers.(Min Yu et al. 2011). Biosensors developed recently can detect all these CTCs as cancer biomarkers. Table 9.5 describes various biosensors for CTCs as cancer biomarkers.

9.3.1.6 Nucleic Acid

Nucleic acid-based biomarkers, encompassing DNA and RNA, are pivotal in cancer diagnostics due to their ability to reveal significant genetic and molecular alterations in cells. These changes, such as mutations, gene expression variations, and epigenetic modifications, are integral in understanding the nature and progression of cancer.

9.3.1.6.1 DNA-Based Cancer Biomarker

Changes in circulating DNA levels are linked to tumor growth, with cell-free DNA in blood reflecting tumor-specific genetic changes. This supports developing a minimally invasive cancer diagnostic test using DNA biomarkers.

- Circulating tumor DNA (ctDNA)

Circulating tumor DNA (ctDNA) consists of small fragments of DNA released into the bloodstream from cancer cells. Though present in minute quantities in healthy individuals, its concentration increases in cancer patients. The release of ctDNA is primarily attributed to cell death processes like necrosis or apoptosis.

- Mitochondrial DNA (mtDNA)

Mitochondrial DNA (mtDNA) mutations are commonly observed in various human cancers. Factors contributing to these mutations include the mitochondrial microenvironment and the limited capacity of mitochondria to repair DNA damage. Due to the high number of mitochondria per cell, mtDNA offers an advantage in detecting abnormal cells among a mixture of normal and cancerous cells. Studies have also

TABLE 9.4
Detection of Exosome Biomarkers in Cancer using Various Biosensors

Exosome Biomarker	Type of Biosensor	Type of Cancer	Bioreceptor	Limit of Detection	References
Exosomes from prostate cancer	Plasmon Resonance imaging	Prostate cancer	Aptamer	1.00×10^{5} particles/mL	Chen et al. (2020)
MiRNAs extracted from exosomal RNAs (exomiR-21 and exomiR-451)	Electrochemical biosensor	Prostate cancer	DNA probe	10 aM	Chiou et al. (2022)
Exosomes in serum (CD63)	Electrochemical biosensor	Prostate cancer	Anti-CD63	7.23 ng/5 μL	Su et al. (2022)
MiR-21, miR-27a and miR-375 from exosomes	Fluorescence	Breast cancer	DNA probe	462 pM for miR-21, 301 pM for miR-27a, and 154 pM for miR-375	Wang et al. (2020)
Exosomes (CD9, CD63, and CD81)	Electrochemical biosensor	Breast cancer	Anti-CD9, anti-CD63, and anti-CD81	175 exosomes/μL, 189 exosomes/μL, and 262 exosomes/μL for exosomes derived from MCF-7, MDA-MB231, and SKBr3, respectively.	da Fonseca Alves et al. (2024)

TABLE 9.5

Detection of Circulating Tumor Cell Biomarkers in Cancer using Various Biosensors

CTC Biomarker	Type of Biosensor	Type of Cancer	Bioreceptor	Limit of Detection	References
Hela cells	Electrochemical biosensors	Cancer	Ab	0.43 cell/mL	Chen et al. (2019)
CL1-5 lung cancer cell lines	Surface plasmon resonance	Lung cancer	Ab	13 cells/mL	Mousavi et al. (2015)
MDA MB 231 cells	QCM	Breast cancer	Ab	12 cells/mL	Bakhshpour et al. (2019)
CCRF-CEM Cells, Ramos cells	Electrochemical biosensor	Breast cancer	Aptamer	4 and 3 cells/mL	Dou et al. (2019)
EPCAM	Fluorescence biosensor	Cancer	Aptamer	10 cells/mL	Lulu Xu et al. (2018)

Abbreviation: EPCAM: epithelial cell adhesion molecule.

shown that mtDNA mutations are present in precursor lesions, suggesting their role in the early stages of carcinogenesis (Larman et al. 2012).

- DNA methylation

DNA methylation, a key epigenetic modification, is often one of the first changes observed in detection research (Jones and Baylin 2007). DNA methylation changes are particularly attractive for cancer biomarker development due to their detectability, even in samples with low tumor purity. DNA methylation can be used as a biomarker for cancer due to its characteristic alterations in cancer cells compared to normal cells (Casula et al. 2024).

- Viral DNA biomarkers

In cancers associated with viral infections, circulating viral DNA can be a specific biomarker (Ozcelikay et al. 2022). For example, the presence of Epstein-Barr virus DNA is notably significant in lymphomas and nasopharyngeal carcinoma.

To provide a comprehensive overview of the advancements in biosensing technologies for cancer diagnosis, Table 9.6 summarizes various biosensors developed for detecting specific DNA biomarkers associated with different cancer types.

9.3.1.6.2 RNA-Based Cancer Biomarker

In the context of cancer, non-coding RNAs play a significant role as regulatory molecules, acting both as oncogenes and tumor suppressors. Therefore, their involvement in cancer biology is of great importance (Figure 9.3).

- Messenger RNA (mRNA)

mRNA alterations are a characteristic of many cancers. They result from changes in mRNA metabolism, including aberrant processing, nuclear export, and translation. These alterations can lead to the loss of function in tumor suppressors, affecting cell proliferation, survival, and differentiation. Tumor-related mRNA, especially those involved in these processes, can be used to assess the migration of tumor cells and correlate with tumor burden and malignant progression (Movva et al. 2015). The study of mRNA methylation modifications, particularly m6A (N6-methyladenosine), is gaining attention. Mutations in the promoter region of the TERT gene are associated with high mRNA expression levels in glioblastoma (Yiyang Zheng et al. 2021).

- Long non-coding RNA (lncRNA)

Long non-coding RNAs (lncRNAs) have emerged as significant biomarkers in cancer due to their roles in gene regulation and expression. These RNA molecules, typically longer than 200 nucleotides, are not translated into proteins but can influence various cellular processes related to cancer development and progression. Their altered expression patterns in cancer cells can reflect changes in gene regulation, cellular

TABLE 9.6
Overview of Biosensors for DNA Biomarkers in Cancer Detection

DNA Biomarker	Type of Biosensor	Type of Cancer	Bioreceptor	Limit of Detection	References
KRAS G12DM (ctDNA)	Electrochemical biosensor	Colorectal cancer	DNA probe	2.4 aM	Wang et al. (2018)
TNBC (ctDNA)	Electrochemical biosensor	Breast cancer	DNA probe	0.5 aM	H. Zhao et al. (2021)
PIK3CA (ctDNA)	Electrochemical biosensor	Gastric carcinoma	DNA probe	18 aM	Zhang et al. (2018)
EGFR L858R (ctDNA)	Electrochemical biosensor	Lung cancer	DNA probe	3.3 aM	Zhang et al. (2018)
KRAS and BRAF (ctDNA)	Electrochemical biosensor	Lung cancer	DNA probe	1 fg/μL	Das et al. (2016)
5-mc MGMT (ctDNA)	Electrochemical biosensor	Colorectal cancer	Ab	0.03 nM	Povedano et al. (2019)
DNA TP53 (ctDNA)	Photoelectrochemical biosensor	—	DNA probe	5.83 fM	N. Li et al. (2021)
Ccf-mtDNA	Fluorescence biosensor	Cancer	DNA probe	0.91 ng/μL	Shandilya et al. (2022)
EBV (DNA)	Electrochemical immunosensor	Nasopharyngeal cancer	Ab	0.1 pg/mL	Miao Yu, Liu, and Li (2022)
HCV core antigen	Electrochemical biosensor	Liver cancer	Aptamer	0.16 fg/mL	Rahmati, Roushani, and Hosseini (2021)
HBV (DNA)	Electrochemical aptasensor	Liver cancer	Aptamer	1.4 pg/mL	Mohsin, Mashkour, and Fatemi (2021)
HPV (DNA)	Electrochemical aptasensor	Cervical cancer	Aptamer	1.75 pM	Chekin et al. (2018)

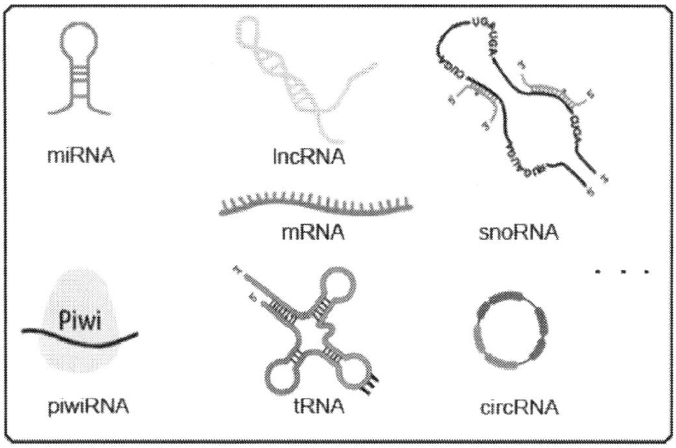

FIGURE 9.3 Different types of RNA cancer biomarkers (Xi et al. 2017).

stress responses, and tumor microenvironment interactions. One notable example is HOTAIR (HOX Transcript Antisense Intergenic RNA). Overexpression of HOTAIR has been linked to metastasis and poor prognosis in several cancers, including breast and colorectal cancers (Hajjari and Salavaty 2015).

- Circular RNA (circRNA)

Circular RNAs (circRNAs) are increasingly recognized as vital biomarkers in cancer research due to their stability and abundance in various biological fluids. These non-coding RNAs, characterized by their covalently closed loop structures, play significant roles in gene regulation. An example is ciRS-7 (circular RNA sponge for miR-7), known for its role in acting as a sponge for miR-7, a microRNA with tumor-suppressive functions. Its upregulation has been linked to a poor prognosis in several cancers, including colorectal and gastric cancer (Weng et al. 2017).

- Small nucleolar RNA (snoRNA)

Small nucleolar RNAs (snoRNAs) are emerging as significant biomarkers in cancer due to their role in the maturation and function of ribosomes. They regulate various aspects of RNA metabolism, including modifications and splicing, which are crucial in cellular processes (Saraiya and Wang 2008). One example is SNORA42, a snoRNA found to be upregulated in lung cancer, and its elevated levels are associated with poor prognosis (Mei et al. 2012).

- MicroRNA (miRNA)

miRNAs regulate gene expression post-transcriptionally and are often dysregulated in cancer. Their presence in body fluids and the ability to reflect the disease state make them ideal non-invasive biomarkers. Changes in miRNA expression can indicate oncogene activation, tumor suppressor gene silencing, or alterations in cellular pathways essential for cancer progression. One example is miR-21, which is frequently upregulated in lung, breast, and pancreatic cancers.

- Piwi-interacting RNA (piRNA)

Piwi-interacting RNAs (piRNAs) are small non-coding RNAs that have proved important in cancer research. They interact with Piwi proteins to play critical roles in gene silencing and epigenetic regulation. Dysregulation of piRNAs can lead to cancer development by impacting gene expression and destabilizing the genome (Ng et al. 2016). One example of a piRNA as a cancer biomarker is piR-651, which is significantly upregulated in gastric cancer and associated with the tumor node metastasis stage (Cheng et al. 2011).

RNA biomarkers like mRNA, lncRNA, circRNA, piRNA, snoRNA, and miRNA are vital in cancer diagnosis and prognosis due to their stability, expression patterns, and roles in gene regulation and cellular communication. Table 9.7 outlines various biosensors for RNA cancer biomarkers.

9.3.1.7 Metabolites

Metabolites are crucial in cellular processes and can serve as important biomarkers for various diseases, including cancer.

Lactate is a key cancer biomarker due to its accumulation in tumor environments. This results from the Warburg effect, where cancer cells rely on glycolysis for energy, increasing lactate production even with oxygen present (Dang 2012). Monitoring lactate levels can thus provide critical insights into tumor presence and progression (Pundir, Narwal, and Batra 2016). Recently, various biosensors have been developed for sensitive lactate detection as a cancer biomarker, including one that involves an amperometric biosensor based on gold nanoparticles anchored on reduced graphene oxide (RGO-AuNPs) (Azzouzi et al. 2015). This biosensor (Figure 9.4 A) incorporates L-lactate dehydrogenase (LDH) to enhance electron transfer and sensitivity, demonstrating high sensitivity and specificity for L-lactate with a detection limit of 0.13 μM and good stability.

On the other hand, altered choline metabolism is a hallmark of cancer. Elevated levels of choline and its derivatives, such as phosphocholine and glycerophosphocholine, are often observed in various cancers, including breast and prostate cancer (Rahimi and Joseph 2019). One significant work in this field involves the development of an optical biosensor using a chalcone derivative, specifically (*E*)-1-(4-aminophenyl)-3-(4-(dimethylamino)phenyl)prop-2-*en*-1-one (4ADP), for the determination of choline in human biofluids. This biosensor demonstrated strong adsorption of choline with a low detection limit of 21 nM, and it was applied to the analysis of choline in serum samples (Lavanya et al. 2023).

Polyamines, such as spermidine and spermine, are another group of metabolites that serve as important biomarkers for cancer. Increased levels of polyamines have been found in various cancers, including colorectal and prostate cancer (Coradduzza,

TABLE 9.7
Detection of RNA Biomarkers in Cancer Using Various Biosensors

RNA Biomarker	Type of Biosensor	Type of Cancer	Bioreceptor	Limit of Detection	References
GPC1 mRNA	Colorimetric biosensor	Pancreatic cancer	Probe	100 fM	H. Li et al. (2021)
m6a mRNA	Electrochemical biosensor	Leukemia, brain tumors, liver, breast and lung cancers	Ab	0.9 pM	Povedano et al. (2021)
PSA mRNA (as a control biomarker)	Electrochemical biosensor	Prostate cancer	Probe	—	Moranova et al. (2022)
MALAT1 (lncRNA)	Electrochemical biosensor	Lung cancer	Probe	42.8 fM	Chen et al. (2021)
HOTAIR (lncRNA)	Photoelectrochemical biosensor	Cancer	Probe	0.12 fM	Li et al. (2024)
LncRNA HULC	Fluorescent biosensor	Liver cancer	Probe	0.06 pM	Yao, Duan, et al. (2023)
LncRNA BC032469	Fluorescent biosensor	Gastric cancer	Probe	—	Yang et al. (2021)
MEG3 lncRNA	Electrochemical biosensor	Gastric cancer	Probe	0.25 fM	Li et al. (2018)
hsa_circ_0137439	Electrochemical biosensor	Bladder cancer	Probe/ CRISPR/ Cas13a	0.09 fM	Cheng et al. (2023)
Circcdyl	Electrochemical biosensor	HCC	Probe	1.0 pM	Zhang et al. (2021)
Circsatb2	Surface Raman Spectroscopy	Lung cancer	Probe	0.77 fM	Luyun Xu et al. (2024)
Circtex2	Fluorescence	Liver cancer	Probe	0.1 pM	Dong et al. (2021)
Circ-MAT2B	Fluorescence	Liver cancer	Probe	1 pM	Yang et al. (2022)
Circmto1	Fluorescence	HCC	Probe	0.265 fM	Yao, Cheng, et al. (2023)
CircRNA AHNAK	Electrochemical biosensor	Breast cancer	Probe	0.22 pM	D. Wu et al. (2021)
Circhipk3	Fluorescence	Breast cancer	Probe	6.6 aM	K. Wang et al. (2023)
Hsa_circ_0061276	Fluorescence	Gastric cancer	Probe	19 aM	He et al. (2024)
Circrobo1	Fluorescence	Breast cancer	Probe/ CRISPR-Cas13a/Cas12a	0.01 pM	Tan et al. (2024)

(*Continued*)

TABLE 9.7 (CONTINUED)
Detection of RNA Biomarkers in Cancer Using Various Biosensors

RNA Biomarker	Type of Biosensor	Type of Cancer	Bioreceptor	Limit of Detection	References
PiRNA–651	Fluorescence	Breast, gastric, and lung cancers	Probe	0.1 fM	El Aamri et al. (2024)
PiRNA-31,143	Photoelectrochemical biosensor	Colorectal cancer	Probe	23 aM	Juan Li et al. (2022)
PiRNA-823	Fluorescence	Colorectal cancer	Probe	100 aM	Wang et al. (2023)
MiRNA-122	Electrochemical biosensor	Liver cancer	Probe	1.73 pM	Kasturi et al. (2021)
MiRNA- Let7a	Electrochemical biosensor	Liver cancer	Probe	1 zM	Azab, Elhakim, and Fekry 2019)
MiRNA-21	Colorimetric biosensor	Breast, prostate, gastric, and lung cancers	Probe	0.5 pM	Aamri, Mohammadi, and Amine (2023)
MiRNA-155	Colorimetric biosensor	Cancer	Probe	22 aM	El Aamri, Mohammadi, and Amine (2024)
MicroRNA-34a	Electrochemical biosensor	Thyroid neoplasms; gastric, colorectal, and breast cancer	Probe	0.87 µg/mL	Erdem and Eksin (2023)
MiR-16	Electrochemical biosensor	HCC	Probe	0.14 nM	Gao et al. (2018)
MiR-126	Electrochemiluminescence biosensor	Lung, breast, and colorectal cancer	Probe	3.17 aM	Liu and Huang (2023)
MiR-141	Electrochemical biosensor	Pancreatic cancer	Probe	0.27 pM	Akin et al. (2022)
MiRNA-210	Electrochemiluminescence biosensor	Breast cancer	Probe	0.3 fM	Shi et al. (2022)

(Continued)

TABLE 9.7 (CONTINUED)
Detection of RNA Biomarkers in Cancer Using Various Biosensors

RNA Biomarker	Type of Biosensor	Type of Cancer	Bioreceptor	Limit of Detection	References
MiR-221	Electrochemical biosensor	Liver and breast cancer, and glioblastoma	Probe	0.6 pM	Hua Zhang et al. (2017)
MiR-18a	Electrochemical biosensor	Pancreas cancer	Probe	16 aM	Rezaei, Hosseini, and Radfar (2021)
MiR-196a	Electrochemical biosensor	Pancreas cancer	Probe	15 aM	Guo et al. (2018)
MiR-492	Electrochemical biosensor	Pancreas cancer	Probe (PNA)	6 nM	Moccia et al. (2020)
MiR-196b	Electrochemical biosensor	Pancreas cancer	Probe	0.26 pM	Chen, Chen, and Jou (2021)
MiRNA-25	Electrochemical biosensor	Lung cancer	Probe	0.25 pM	Asadzadeh-Firouzabadi and Zare (2017)
MiR-223	Electrochemical biosensor	Colorectal cancer	Probe	0.012 aM	Akbari et al. (2023)
MiR-486-5p	Fluorescence	Lung cancer	Probe	300 aM	Djebbi et al. (2022)

Abbreviations: HOTAIR: HOX transcript antisense RNA; HULC: highly upregulated in liver cancer; GPC1: Glypican-1; m6a: N6-methyladenosine; MEG3: maternally expressed gene 3; NEAT1: nuclear paraspeckle assembly transcript 1; HCC: hepatocellular carcinoma.

Arru et al. 2022). A significant study developed a selective turn-on biosensor for detecting spermine and spermidine in urine and blood. Using a boronic acid-coumarin hydrogel matrix, the biosensor shows fluorescence upon binding with spermine, with a detection limit of 6 μM, and performs well in real samples (Nair et al. 2019). This non-invasive biosensor is ideal for early cancer diagnosis and monitoring recovery during chemotherapy.

On the flip side, volatile organic compounds (VOCs), organic chemicals that easily evaporate, serve as valuable cancer biomarkers due to their unique profiles in lung cancer patients' breath, differing from healthy individuals (Saalberg and Wolff 2016). In this context, exhaled VOCs were analyzed non-invasively for early lung cancer detection (Figure 9.4 B). A breath analyzer with advanced gas sensors and a 1D convolutional neural network achieved 97.8% accuracy in distinguishing samples from 107 lung cancer patients and 74 healthy controls, showing promise as a rapid alternative to invasive diagnostics (Lee et al. 2024).

9.3.2 Emerging Trends and Advanced Technologies for Biosensors in Cancer Diagnosis and Monitoring

Emerging trends in biosensor technology for cancer biomarker detection are significantly enhancing diagnostic capabilities. Recent advancements focus on biosensors based on nanozymes, molecularly imprinted polymers (MIPs), and artificial intelligence (AI), leading to improved sensitivity and specificity.

9.3.2.1 Nanozyme-Based Biosensors

Nanozymes, nanomaterials with enzyme-like functions, are a major advancement in biosensor technology. Since their discovery in 2007 for peroxidase-like activity, they have proven more stable, cost-effective, and durable than natural enzymes. Their catalytic capabilities under extreme conditions make them valuable for detecting cancer biomarkers, offering enhanced sensitivity and specificity. Integrated into biosensors, nanozymes enable rapid disease diagnosis through electrochemical and colorimetric methods by generating measurable signals or visible color changes, showcasing their versatility in fields like medical diagnostics and monitoring (Li et al. 2023).

One significant application of nanozymes is highlighted in the development of a highly sensitive colorimetric DNA sensor for detecting microRNA-155, a cancer biomarker. This sensor uses copper nanoparticles (CuNPs) with peroxidase-like activity and the hybridization chain reaction (HCR) to achieve ultra-low detection limits. CuNPs catalyze substrate oxidation, creating a blue color that signals the presence of microRNA-155. With a detection limit of 22 attomolar, this approach is highly sensitive and suitable for real samples, including human serum (El Aamri, Mohammadi, and Amine 2024). In another study, a novel ECL biosensor for PSA detection combines graphitic carbon nitride nanosheets encapsulated in MOF as a nanozyme with peroxidase-like activity, developed for PSA detection (Figure 9.5 A). The system uses a sandwich immunoassay featuring magnetic nanoparticles with capture antibodies and nanozyme-coupled signal antibodies. The biosensor achieves exceptional

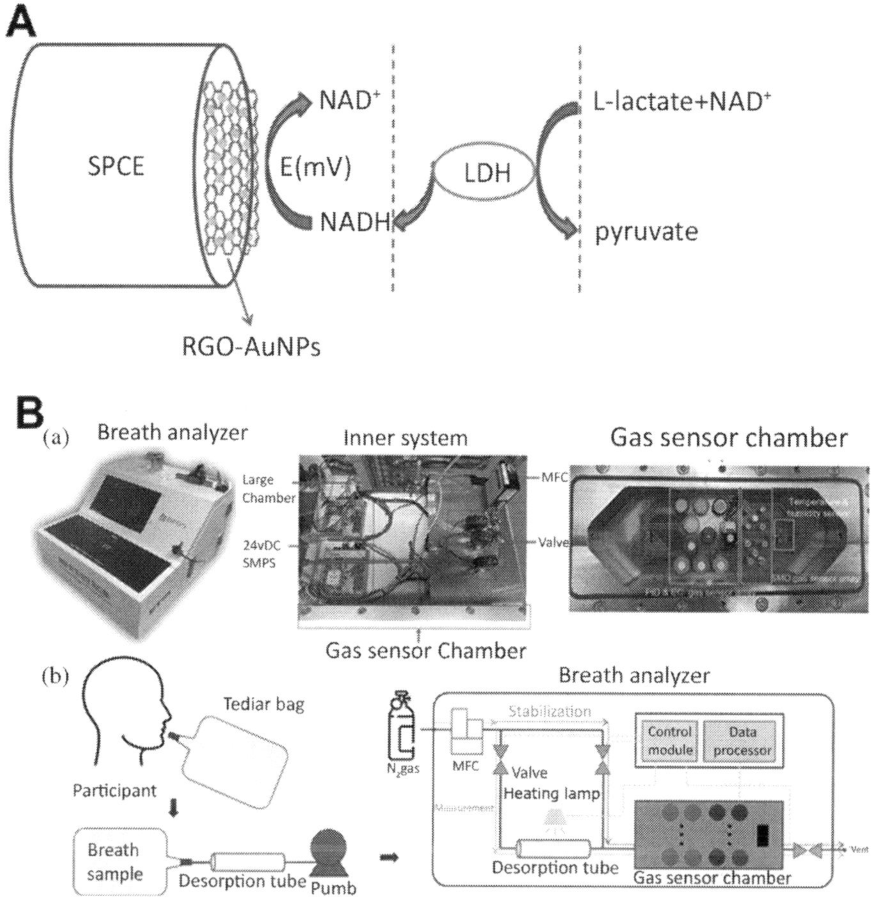

FIGURE 9.4 (A) Schematic representation of working principle of the LDH/RGO-AuNPs/SPCE biosensor. Reused with permission from Elsevier publisher (Azzouzi et al. 2015). (B) Description of the breath analyzer with multimodal sensor array and operation: (a) images of the device and internal systems, (b) diagram of the operation process. Reused with permission from Elsevier publisher (Lee et al. 2024).

sensitivity (0.02 pg/mL PSA) through hydroxyl radical-enhanced ECL signal amplification, showing promise for early cancer detection (Khoshfetrat et al. 2021).

9.3.2.2 Molecularly Imprinted Polymer–Based Biosensors

MIPs are synthetic materials that bind selectively to target molecules, similar to biological antibodies. They are gaining popularity in fields like drug monitoring and separation processes (Keçili et al. 2023). MIPs are fabricated using templates of target molecules to create highly specific binding sites for cancer biomarkers. This technology could greatly enhance cancer detection and treatment, leading to better diagnostic tools and targeted therapies (Suravajhala, Burri, and Malik 2020).

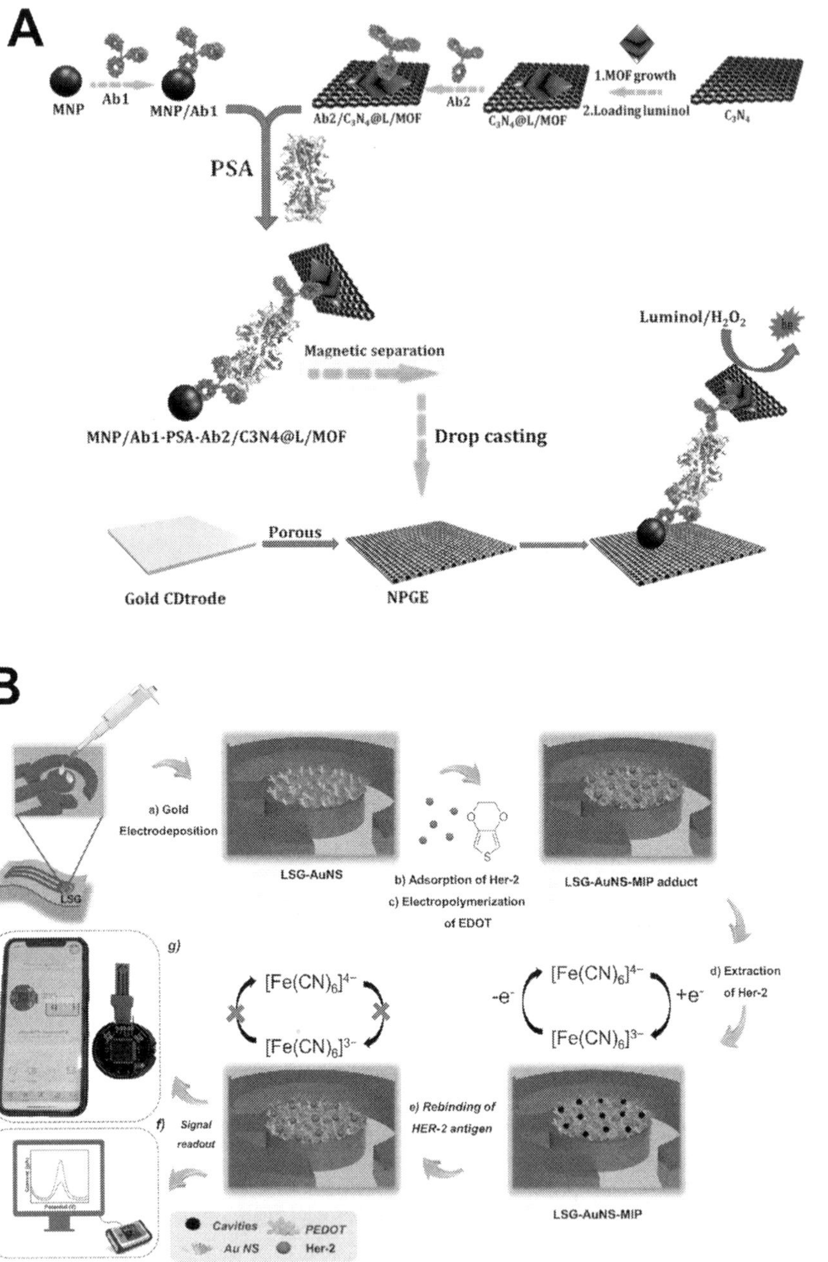

FIGURE 9.5 (A) Schematic representation of the electrochemiluminescence-based biosensor for PSA detection using C3N4@luminol/MOF nanozyme. Reused with permission from Elsevier publisher (Khoshfetrat et al. 2021). (B) Schematic illustration of the LSG-AuNS-MIP based biomimetic sensor. Reused with permission from Elsevier publisher (Lahcen et al. 2021).

MIP-based biosensors are highly selective and sensitive for detecting cancer biomarkers, using MIPs to create specific binding sites for rapid and accurate detection (Kang et al. 2023). In this framework, a novel biomimetic sensor utilizing laser-scribed graphene modified with gold nanostructures and MIPs was developed for the detection of the Her-2 biomarker, which is significant in breast cancer diagnosis (Figure 9.5 B). The biosensor operates on the principle of electrochemical detection, whereby the MIP selectively binds to the Her-2 protein, enhancing sensitivity and specificity due to the tailored cavities formed during the imprinting process. The sensor has a wide detection range (1–200 ng/mL) and a low LOD (0.43 ng/mL), showing high selectivity and effectiveness in clinical applications, even in spiked human serum samples (Lahcen et al. 2021).

9.3.2.3 The Combination of Artificial Intelligence with Biosensors

Integrating AI with biosensors revolutionizes cancer diagnosis by identifying patterns and biomarkers missed by humans, enhancing monitoring, accuracy, and predictive analytics. It automates data processing, reducing healthcare professionals' workload and aiding clinical decisions (Arya et al. 2023). Moreover, AI-powered applications are designed to enhance care coordination by reducing delays in clinical workflows, using AI to generate and send timely alerts to clinicians, including cancer screening results, diagnostics, prevention, and improved cancer management (Kather 2023). The combination of AI and biosensors enhances biomarker discovery, personalized treatments, and cost-effective, portable diagnostic tools. AI techniques like support vector machines and principal component analysis improve diagnostic accuracy and efficiency. This synergy leads to timely cancer diagnoses, personalized treatments, and better patient outcomes, making advanced diagnostics more accessible. In this context, a novel diagnostic system using a dual-gate field-effect transistor (DGFET) and AI was developed to detect prostate cancer biomarkers (PSMA, ENG, ERG, ANXA3) in urine (Figure 9.6 A). This non-invasive method achieves high accuracy through advanced biosensing and machine learning techniques, with random forest algorithms showing 100% accuracy in tests (Kim et al. 2021).

9.3.2.4 CRISPR-based Biosensors for Cancer Diagnosis and Monitoring

CRISPR-based biosensors provide precise, efficient cancer diagnosis by leveraging the CRISPR-Cas12a system to detect specific genetic sequences. Upon binding to a target DNA or RNA, CRISPR-Cas12a triggers a signal amplification mechanism, producing a fluorescent signal. This technology's programmability allows for detecting various biomarkers by adjusting CRISPR components. It offers rapid, sensitive detection, as seen in a study (Figure 9.6 B) where a CRISPR/Cas12a biosensor identified lung-cancer biomarker exosomal miR-21 with a detection limit of 161 fM. This approach supports non-invasive, accurate early cancer detection and monitoring (Lu et al. 2023).

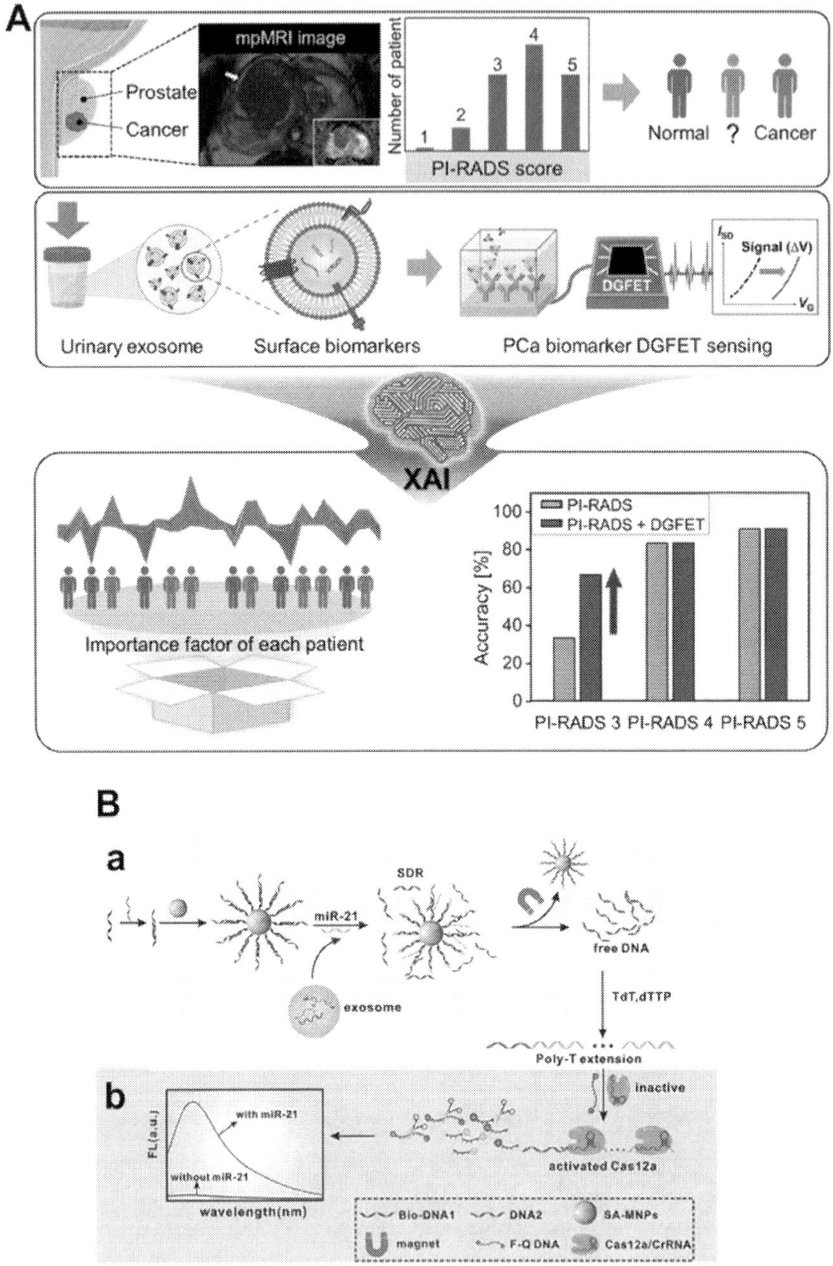

FIGURE 9.6 (A) Schematic illustration of research outline. The distinction between normal and cancer patients remains unclear with the PI-RADS score through mpMRI. Reused with permission from Elsevier publisher (Kim et al. 2021) (B) Schematic diagram of a fluorescence biosensor for detection of exosomal miR-21. (a) Combination of magnetic separation, extraction of exosomal miRNA, SDR and TdT extension; (b) The CRISPR/Cas12a-based fluorescence biosensor for detection exosomal miRNA. Reused with permission from Elsevier publisher (Lu et al. 2023).

9.4 DISCUSSIONS AND PERSPECTIVES

The exploration of cancer biomarkers and the application of biosensors, as presented in this chapter, highlights their critical role in advancing cancer diagnostics and monitoring.

This chapter covers various biosensors—electrochemical, optical, mass-sensitive, and thermometric—each with unique benefits for cancer diagnostics. Electrochemical biosensors are highly sensitive and cost-effective, ideal for low-abundance biomarkers like ctDNA, though they can be limited by electrode stability in clinical settings. Optical biosensors offer high specificity in detecting fluorescence or absorbance changes related to biomarkers such as exosomes but can be costly and complex, though research is reducing these barriers with innovations like smartphone integration. Mass-sensitive biosensors, such as quartz crystal microbalances, detect tiny molecular changes with precision, yet are sensitive to environmental interference. Thermometric biosensors measure temperature changes in biochemical reactions, though they generally require precise temperature control and offer lower sensitivity than other types.

The literature does not recommend universally prescribing or using a single type of biosensor for any particular type of cancer. This is because each cancer type presents unique challenges and biomarkers that vary in their chemical and biological properties. As a result, the selection of a biosensor should be carefully tailored to the specific biomarker being targeted and the clinical requirements of the diagnostic process. For example, while electrochemical biosensors are highly effective for detecting circulating tumor DNA (ctDNA) due to their sensitivity to small changes in the electric signal, they may not be the best choice for detecting exosomes, which are better analyzed using optical biosensors due to their ability to detect size and structure through fluorescence. This emphasizes the importance of matching the biosensor type with the characteristics of the biomarker to ensure accurate and efficient cancer diagnosis and monitoring. Ultimately, the choice of biosensor depends on the clinical need, whether for rapid diagnosis, cost-effectiveness, or long-term monitoring.

Cancer biomarkers, including proteins, nucleic acids, and exosomes, are key to early cancer detection and monitoring. Found in fluids like blood, urine, and saliva, they can be accessed non-invasively, often indicating cancer presence before symptoms arise. These biomarkers aid in diagnosis and tracking tumor progression with real-time data. The portability of many biosensors allows point-of-care diagnostics, reducing reliance on lab infrastructure. This enables faster, accessible testing in clinics, remote areas, and patients' homes, supporting timely diagnosis and monitoring without complex facilities. Each type of cancer biomarker has unique advantages and disadvantages. Proteins are abundant but can be complex to analyze due to their variability. Hormones offer specificity and quantifiable measures but can vary due to non-cancer factors and may degrade quickly. Exosomes protect their cargo and provide comprehensive information but are challenging to isolate. Cytokines enable early detection and sensitive monitoring but are highly variable and short-lived. Circulating tumor cells provide real-time insights but are rare and technically challenging to detect. DNAs as cancer biomarkers offer specific genetic information and are non-invasive but are present in low concentrations and can degrade. RNAs reflect gene expression and allow dynamic monitoring but are unstable and technically demanding to measure.

The future of biosensors in cancer diagnostics is promising, driven by innovations like nanozyme-based biosensors, which offer enhanced stability and sensitivity. Integrating AI-driven analysis further boosts predictive accuracy by identifying subtle biomarker patterns. Wearable biosensors, providing continuous real-time monitoring, are set to revolutionize patient care by reducing hospital visits and enabling personalized, adaptive treatment plans, ultimately improving patient outcomes and quality of life.

9.5 CONCLUSION

This chapter highlights the essential role of biosensors in cancer diagnosis and monitoring. With cancer as a leading global health challenge, early detection and continuous monitoring are crucial for better outcomes. Biosensors offer advantages over traditional methods through improved sensitivity, specificity, and non-invasive detection of biomarkers like proteins, exosomes, and circulating tumor cells. However, challenges remain, such as maintaining reliability in complex environments, managing scalability, and achieving regulatory approval. Addressing these technical and logistical hurdles is key to advancing biosensors for widespread, accessible cancer diagnostics and monitoring.

REFERENCES

Aamri, M.E., H. Mohammadi, and A. Amine. 2023. Based Colorimetric Detection of miRNA-21 Using Pre-Activated Nylon Membrane and Peroxidase-Mimetic Activity of Cysteamine-Capped Gold Nanoparticles. *Biosensors* 13, no. 1: 74.

Akbari, A., H. Hashemzadeh, Z.S. Eshkiki, M. Masoodi, S.P. Tabaeian, H. Naderi-Manesh, A.A. Zare, and S. Agah. 2023. Detection of Plasma miR-223 by a Novel Label-Free Graphene Oxide/Gold Nanocomposite Immunosensor in Colorectal Cancer Patients: An Electrochemical Biosensor Approach. *Biosensors and Bioelectronics: X* 14: 100331.

Akin, M., M. Bekmezci, R. Bayat, Z.K. Coguplugil, F. Sen, F. Karimi, and H. Karimi-Maleh. 2022. Mobile Device Integrated Graphene Oxide Quantum Dots Based Electrochemical Biosensor Design for Detection of miR-141 as a Pancreatic Cancer Biomarker. *Electrochimica Acta* 435: 141390.

Arya, S.S., S.B. Dias, H.F. Jelinek, L.J. Hadjileontiadis, and A.-M. Pappa. 2023. The Convergence of Traditional and Digital Biomarkers through AI-Assisted Biosensing: A New Era in Translational Diagnostics? *Biosensors and Bioelectronics* 235: 115387.

Asadzadeh-Firouzabadi, A., and H.R. Zare. 2017. Application of Cysteamine-Capped Gold Nanoparticles for Early Detection of Lung Cancer-Specific miRNA (miR-25) in Human Blood Plasma. *Analytical Methods* 9, no. 25: 3852–3861.

Aydin, E.B. 2020. A Label-Free and Sensitive Impedimetric Immunosensor for TNF? Biomarker Detection Based on Epoxysilane-Modified Disposable ITO-PET Electrode. *International Journal of Environmental Analytical Chemistry* 100, no. 4: 363–377.

Aydın, E.B. 2020. Highly Sensitive Impedimetric Immunosensor for Determination of Interleukin 6 as a Cancer Biomarker by Using Conjugated Polymer Containing Epoxy Side Groups Modified Disposable ITO Electrode. *Talanta* 215: 120909.

Aydın, E.B., M. Aydın, and M.K. Sezgintürk. 2020. Selective and Ultrasensitive Electrochemical Immunosensing of NSE Cancer Biomarker in Human Serum Using Epoxy-Substituted Poly (Pyrrole) Polymer Modified Disposable ITO Electrode. *Sensors and Actuators B: Chemical* 306: 127613.

Aydın, E.B., M. Aydın, and M.K. Sezgintürk. 2023. Novel Electrochemical Biosensing Platform Based on Conductive Multilayer for Sensitive and Selective Detection of CYFRA 21-1. *Sensors and Actuators B: Chemical* 378: 133208.

Azab, S.M., H.K. Elhakim, and A.M. Fekry. 2019. The Strategy of Nanoparticles and the Flavone Chrysin to Quantify miRNA-Let 7a in Zepto-Molar Level: Its Application as Tumor Marker. *Journal of Molecular Structure* 1196: 647–652.

Azmi, A.S., B. Bao, and F.H. Sarkar. 2013. Exosomes in Cancer Development, Metastasis, and Drug Resistance: A Comprehensive Review. *Cancer and Metastasis Reviews* 32: 623–642.

Azzouzi, S., L. Rotariu, A.M. Benito, W.K. Maser, M.B. Ali, and C. Bala. 2015. A Novel Amperometric Biosensor Based on Gold Nanoparticles Anchored on Reduced Graphene Oxide for Sensitive Detection of L-Lactate Tumor Biomarker. *Biosensors and Bioelectronics* 69: 280–286.

Bahari, D., B. Babamiri, and A. Salimi. 2021. Ultrasensitive Molecularly Imprinted Fluorescence Sensor for Simultaneous Determination of CA125 and CA15–3 in Human Serum and OVCAR-3 and MCF-7 Cells Lines Using Cd and Ni Nanoclusters as New Emitters. *Analytical and Bioanalytical Chemistry* 413, no. 15 (June): 4049–4061.

Bakhshpour, M., A.K. Piskin, H. Yavuz, and A. Denizli. 2019. Quartz Crystal Microbalance Biosensor for Label-Free MDA MB 231 Cancer Cell Detection via Notch-4 Receptor. *Talanta* 204: 840–845.

Bhatia, A., H.S. Na, P. Nandhakumar, B. Yu, S. Jon, J. Chung, and H. Yang. 2021. Electrochemical Detection of Interleukin-8 in Human Saliva Using a Polyenzyme Label Based on Diaphorase and Neutravidin. *Sensors and Actuators B: Chemical* 326: 128979.

Casula, G., S. Lai, E. Loi, L. Moi, P. Zavattari, and A. Bonfiglio. 2024. An Innovative PCR-Free Approach for DNA Methylation Measure: An Application for Early Colorectal Cancer Detection by Means of an Organic Biosensor. *Sensors and Actuators B: Chemical* 398: 134698.

Chekin, F., K. Bagga, P. Subramanian, R. Jijie, S.K. Singh, S. Kurungot, R. Boukherroub, and S. Szunerits. 2018. Nucleic Aptamer Modified Porous Reduced Graphene Oxide/MoS2 Based Electrodes for Viral Detection: Application to Human Papillomavirus (HPV). *Sensors and Actuators B: Chemical* 262: 991–1000.

Chen, G.-C., C.-H. Liu, and W.-C. Wu. 2021. Electrochemical Immunosensor for Serum Parathyroid Hormone Using Voltammetric Techniques and a Portable Simulator. *Analytica Chimica Acta* 1143: 84–92.

Chen, M., D. Wu, S. Tu, C. Yang, D. Chen, and Y. Xu. 2021. A Novel Biosensor for the Ultrasensitive Detection of the lncRNA Biomarker MALAT1 in Non-Small Cell Lung Cancer. *Scientific Reports* 11, no. 1: 3666.

Chen, S.-C., K.-T. Chen, and A.F.-J. Jou. 2021. Polydopamine-Gold Composite-Based Electrochemical Biosensor Using Dual-Amplification Strategy for Detecting Pancreatic Cancer-Associated microRNA. *Biosensors and Bioelectronics* 173: 112815.

Chen, W., J. Li, X. Wei, Y. Fan, H. Qian, S. Li, Y. Xiang, and S. Ding. 2020. Surface Plasmon Resonance Biosensor Using Hydrogel-AuNP Supramolecular Spheres for Determination of Prostate Cancer-Derived Exosomes. *Microchimica Acta* 187, no. 11 (November): 590.

Chen, Y., J. Peng, Y. Lai, B. Wu, L. Sun, and J. Weng. 2019. Ultrasensitive Label-Free Detection of Circulating Tumor Cells Using Conductivity Matching of Two-Dimensional Semiconductor with Cancer Cell. *Biosensors and Bioelectronics* 142: 111520.

Cheng, J., J.-M. Guo, B.-X. Xiao, Y. Miao, Z. Jiang, H. Zhou, and Q.-N. Li. 2011. piRNA, the New Non-Coding RNA, Is Aberrantly Expressed in Human Cancer Cells. *Clinica Chimica Acta* 412, no. 17 (August 17): 1621–1625.

Cheng, L., F. Yang, Y. Zhao, Z. Liu, X. Yao, and J. Zhang. 2023. Tetrahedron Supported CRISPR/Cas13a Cleavage for Electrochemical Detection of Circular RNA in Bladder Cancer. *Biosensors and Bioelectronics* 222: 114982.

Cheng, N., D. Du, X. Wang, D. Liu, W. Xu, Y. Luo, and Y. Lin. 2019. Recent Advances in Biosensors for Detecting Cancer-Derived Exosomes. *Trends in Biotechnology* 37, no. 11: 1236–1254.

Chiou, Y.-E., K.-J. Yu, S.-N. Pang, Y.-L. Yang, S.-T. Pang, and W.-H. Weng. 2022. A Highly Sensitive Urinary Exosomal miRNAs Biosensor Applied to Evaluation of Prostate Cancer Progression. *Bioengineering* 9, no. 12: 803.

Coradduzza, D., C. Arru, N. Culeddu, A. Congiargiu, E.G. Azara, A.M. Scanu, A. Zinellu, M.R. Muroni, V. Rallo, and S. Medici. 2022. Quantitative Metabolomics to Explore the Role of Plasma Polyamines in Colorectal Cancer. *International Journal of Molecular Sciences* 24, no. 1: 101.

Coradduzza, D., T. Solinas, E. Azara, N. Culeddu, S. Cruciani, A. Zinellu, S. Medici, M. Maioli, M. Madonia, and C. Carru. 2022. Plasma Polyamine Biomarker Panels: Agmatine in Support of Prostate Cancer Diagnosis. *Biomolecules* 12, no. 4: 514.

Corrie, S.R., J.W. Coffey, J. Islam, K.A. Markey, and M.A.F. Kendall. 2015. Blood, Sweat, and Tears: Developing Clinically Relevant Protein Biosensors for Integrated Body Fluid Analysis. *Analyst* 140, no. 13: 4350–4364.

Cui, M., X. Sun, R. Liu, M. Du, X. Song, S. Wang, W. Hu, and X. Luo. 2022. A Dual-Responsive Electrochemical Biosensor Based on Artificial Protein Imprinted Polymers and Natural Hyaluronic Acid for Sensitive Recognition towards Biomarker CD44. *Sensors and Actuators B: Chemical* 371: 132554.

Dang, C.V. 2012. Links between Metabolism and Cancer. *Genes & Development* 26, no. 9: 877–890.

Das, J., I. Ivanov, E.H. Sargent, and S.O. Kelley. 2016. DNA Clutch Probes for Circulating Tumor DNA Analysis. *Journal of the American Chemical Society* 138, no. 34 (August 31): 11009–11016.

De Gruttola, V., T. Fleming, D.Y. Lin, and R. Coombs. 1997. Perspective: Validating Surrogate Markers—Are We Being Naive? *Journal of Infectious Diseases* 175, no. 2: 237–246.

Devillers, M., L. Ahmad, H. Korri-Youssoufi, and L. Salmon. 2017. Carbohydrate-Based Electrochemical Biosensor for Detection of a Cancer Biomarker in Human Plasma. *Biosensors and Bioelectronics* 96: 178–185.

Djebbi, K., J. Xing, T. Weng, M. Bahri, M.A. Elaguech, C. Du, B. Shi, L. Hu, S. He, and P. Liao. 2022. Highly Sensitive Fluorescence Multiplexed miRNAs Biosensors for Accurate Clinically Diagnosis Lung Cancer Disease Using LNA-Modified DNA Probe and DSN Enzyme. *Analytica Chimica Acta* 1208: 339778.

Dong, J., Z. Zeng, R. Sun, X. Zhang, Z. Cheng, C. Chen, and Q. Zhu. 2021. Specific and Sensitive Detection of CircRNA Based on Netlike Hybridization Chain Reaction. *Biosensors and Bioelectronics* 192: 113508.

Dou, B., L. Xu, B. Jiang, R. Yuan, and Y. Xiang. 2019. Aptamer-Functionalized and Gold Nanoparticle Array-Decorated Magnetic Graphene Nanosheets Enable Multiplexed and Sensitive Electrochemical Detection of Rare Circulating Tumor Cells in Whole Blood. *Analytical Chemistry* 91, no. 16 (August 20): 10792–10799.

Duffy, M.J. 2001. Carcinoembryonic Antigen as a Marker for Colorectal Cancer: Is It Clinically Useful? *Clinical Chemistry* 47, no. 4: 624–630.

Duffy, M.J. 2020. Biomarkers for Prostate Cancer: Prostate-Specific Antigen and Beyond. *Clinical Chemistry and Laboratory Medicine (CCLM)* 58, no. 3: 326–339.

Edge, S.B., and C.C. Compton. 2010. The American Joint Committee on Cancer: The 7th Edition of the AJCC Cancer Staging Manual and the Future of TNM. *Annals of Surgical Oncology* 17, no. 6 (June): 1471–1474.

El Aamri, M., H. Mohammadi, and A. Amine. 2024. A Highly Sensitive Colorimetric DNA Sensor for MicroRNA-155 Detection: Leveraging the Peroxidase-like Activity of Copper Nanoparticles in a Double Amplification Strategy. *Microchimica Acta* 191, no. 1: 32.

El Aamri, M., G. Yammouri, H. Mohammadi, A. Amine, and H. Korri-Youssoufi. 2020. Electrochemical Biosensors for Detection of microRNA as a Cancer Biomarker: Pros and Cons. *Biosensors* 10, no. 11: 186.

El Aamri, M., R. Zayani, S. Baachaoui, H. Mohammadi, A. Amine, and N. Raouafi. 2024. An Ingenious Double-Amplified Fluorometric Genosensor Based on Hybridization Chain Reaction and Enzymatic Signal for the Sensitive Detection of piRNA–651. *Sensors and Actuators B: Chemical* 399: 134749.

Erdem, A., and E. Eksin. 2023. Zip Nucleic Acid-Based Genomagnetic Assay for Electrochemical Detection of microRNA-34a. *Biosensors* 13, no. 1: 144.

Fan, X., I.M. White, S.I. Shopova, H. Zhu, J.D. Suter, and Y. Sun. 2008. Sensitive Optical Biosensors for Unlabeled Targets: A Review. *Analytica Chimica Acta* 620, no. 1–2: 8–26.

Ferlay, J., M. Colombet, I. Soerjomataram, C. Mathers, D.M. Parkin, M. Piñeros, A. Znaor, and F. Bray. 2019. Estimating the Global Cancer Incidence and Mortality in 2018: GLOBOCAN Sources and Methods. *International Journal of Cancer* 144, no. 8 (April 15): 1941–1953.

da Fonseca Alves, R., A. Pallarès-Rusiñol, R. Rossi, M. Martí, E.R. Vaz, T.G. de Araújo, M.D.P.T. Sotomayor, and M.I. Pividori. 2024. Peptide-Based Biosensing Approaches for Targeting Breast Cancer-Derived Exosomes. *Biosensors and Bioelectronics* 255: 116211.

Frias, I.A., N. Zine, M. Sigaud, P. Lozano-Sanchez, M. Caffio, and A. Errachid. 2023. Non-Covalent π–π Functionalized Gii-sense® Graphene Foam for Interleukin 10 Impedimetric Detection. *Biosensors and Bioelectronics* 222: 114954.

Fu, X., K. Gao, N. Liu, B. Guo, M. He, N. Lai, X. Li, S. Ding, X. He, and L. Wu. 2024. Au/PANI@PtCu-Based Electrochemical Immunosensor for Ultrasensitive Determination of pro-Gastrin-Releasing Peptide. *Microchimica Acta* 191, no. 3 (March): 126.

Gao, T., J. Zhi, C. Mu, S. Gu, J. Xiao, J. Yang, Z. Wang, and Y. Xiang. 2018. One-Step Detection for Two Serological Biomarker Species to Improve the Diagnostic Accuracy of Hepatocellular Carcinoma. *Talanta* 178: 89–93.

Ghosh, S., D. Datta, S. Chaudhry, M. Dutta, and M.A. Stroscio. 2018. Rapid Detection of Tumor Necrosis Factor-Alpha Using Quantum Dot-Based Optical Aptasensor. *IEEE Transactions on Nanobioscience* 17, no. 4: 417–423.

Gronewold, T.M. 2007. Surface Acoustic Wave Sensors in the Bioanalytical Field: Recent Trends and Challenges. *Analytica Chimica Acta* 603, no. 2: 119–128.

Guan, X., X. Deng, J. Song, X. Wang, and S. Wu. 2021. Polydopamine with Tailorable Photoelectrochemical Activities for the Highly Sensitive Immunoassay of Tumor Markers. *Analytical Chemistry* 93, no. 17 (May 4): 6763–6769.

Guo, J., C. Yuan, Q. Yan, Q. Duan, X. Li, and G. Yi. 2018. An Electrochemical Biosensor for microRNA-196a Detection Based on Cyclic Enzymatic Signal Amplification and Template-Free DNA Extension Reaction with the Adsorption of Methylene Blue. *Biosensors and Bioelectronics* 105: 103–108.

Hajjari, M., and A. Salavaty. 2015. HOTAIR: An Oncogenic Long Non-Coding RNA in Different Cancers. *Cancer Biology & Medicine* 12, no. 1: 1.

He, S., L. Chen, Z. Chen, G. Zhang, Y. Huang, H. Zheng, Q. Yang, Z. Mo, X. Lin, and J. Wen. 2024. The Sensing of circRNA with Tetrahedral DNA Nanostructure Modified Microfluidic Chip. *Analytica Chimica Acta* 1319: 342951.

Heikenfeld, J. 2016. Technological Leap for Sweat Sensing. *Nature* 529, no. 7587: 475–476.

Jafari, A., K. Karimabadi, A. Rahimi, G. Rostaminasab, M. Khazaei, L. Rezakhani, and T. Ahmadi jouybari. 2023. The Emerging Role of Exosomal miRNAs as Biomarkers for Early Cancer Detection: A Comprehensive Literature Review. *Technology in Cancer Research & Treatment* 22: 15330338231205999.

Jain, U., N. Chauhan, and K. Saxena. 2023. Fundamentals of Sensors and Biosensors: An Overview. *Multifaceted Bio-Sensing Technology*: 31–44.

Jalalvand, A.R., F. Majidi, S. Mohammadi, S. Fakhri, C. Jalili, and T. Farshadnia. 2024. A Novel and Computer Assisted Electrochemical Biosensor for Determination of Thyroglobulin to Detect Thyroid Cancer Treatment Success. *Sensing and Bio-Sensing Research* 44: 100660.

Jelski, W., and B. Mroczko. 2020. Biochemical Markers of Colorectal Cancer—Present and Future. *Cancer Management and Research*: 4789–4797.

Jones, P.A., and S.B. Baylin. 2007. The Epigenomics of Cancer. *Cell* 128, no. 4: 683–692.

Kang, M.S., E. Cho, H.E. Choi, C. Amri, J.-H. Lee, and K.S. Kim. 2023. Molecularly Imprinted Polymers (MIPs): Emerging Biomaterials for Cancer Theragnostic Applications. *Biomaterials Research* 27, no. 1 (February 9): 45.

Kasturi, S., Y. Eom, S.R. Torati, and C. Kim. 2021. Highly Sensitive Electrochemical Biosensor Based on Naturally Reduced rGO/Au Nanocomposite for the Detection of miRNA-122 Biomarker. *Journal of Industrial and Engineering Chemistry* 93: 186–195.

Kather, J.N. 2023. Artificial Intelligence in Oncology: Chances and Pitfalls. *Journal of Cancer Research and Clinical Oncology* 149, no. 10 (August): 7995–7996.

Keçili, R., Ö.B. Ünlüer, A. Ersöz, and R. Say. 2023. Molecularly Imprinted Polymers (MIPs) for Biomedical Applications. In *Advances in Biomedical Polymers and Composites*, 745–768. Elsevier. https://www.sciencedirect.com/science/article/pii/B9780323885249000085.

Khoshfetrat, S.M., P. Hashemi, A. Afkhami, A. Hajian, and H. Bagheri. 2021. Cascade Electrochemiluminescence-Based Integrated Graphitic Carbon Nitride-Encapsulated Metal-Organic Framework Nanozyme for Prostate-Specific Antigen Biosensing. *Sensors and Actuators B: Chemical* 348: 130658.

Kim, H., S. Park, I.G. Jeong, S.H. Song, Y. Jeong, C.-S. Kim, and K.H. Lee. 2021. Noninvasive Precision Screening of Prostate Cancer by Urinary Multimarker Sensor and Artificial Intelligence Analysis. *ACS Nano* 15, no. 3 (March 23): 4054–4065.

Kowalczyk, A., A. Gajda-Walczak, M. Ruzycka-Ayoush, A. Targonska, G. Mosieniak, M. Glogowski, A. Szumera-Cieckiewicz, et al. 2023. Parallel SPR and QCM-D Quantitative Analysis of CD9, CD63, and CD81 Tetraspanins: A Simple and Sensitive Way to Determine the Concentration of Extracellular Vesicles Isolated from Human Lung Cancer Cells. *Analytical Chemistry* 95, no. 25 (June 27): 9520–9530.

Lahcen, A.A., S. Rauf, A. Aljedaibi, J.I. de Oliveira Filho, T. Beduk, V. Mani, H.N. Alshareef, and K.N. Salama. 2021. Laser-Scribed Graphene Sensor Based on Gold Nanostructures and Molecularly Imprinted Polymers: Application for Her-2 Cancer Biomarker Detection. *Sensors and Actuators B: Chemical* 347: 130556.

Larman, T.C., S.R. DePalma, A.G. Hadjipanayis, C.G.A.R. Network, A. Protopopov, J. Zhang, S.B. Gabriel, L. Chin, C.E. Seidman, and R. Kucherlapati. 2012. Spectrum of Somatic Mitochondrial Mutations in Five Cancers. *Proceedings of the National Academy of Sciences* 109, no. 35: 14087–14091.

Lavanya, R., V. Srinivasadesikan, M. Lin, and V. Padmini. 2023. Development of an Optical Biosensor for the Determination of Choline in Human Biofluids. *Journal of Molecular Structure* 1280: 135040.

Lee, B., J. Lee, J.-O. Lee, Y. Hwang, H.-K. Bahn, I. Park, S. Jheon, and D.-S. Lee. 2024. Breath Analysis System with Convolutional Neural Network (CNN) for Early Detection of Lung Cancer. *Sensors and Actuators B: Chemical* 409: 135578.

Lee, Y.R., K.-Y. An, J. Jeon, N.K. Kim, J.W. Lee, J. Hong, and B.C. Chung. 2020. Untargeted Metabolomics and Polyamine Profiling in Serum before and after Surgery in Colorectal Cancer Patients. *Metabolites* 10, no. 12: 487.

Li, H., A.R. Warden, W. Su, J. He, X. Zhi, K. Wang, L. Zhu, G. Shen, and X. Ding. 2021. Highly Sensitive and Portable mRNA Detection Platform for Early Cancer Detection. *Journal of Nanobiotechnology* 19, no. 1 (December). https://jnanobiotechnology.biomedcentral.com/articles/10.1186/s12951-021-01039-4.

Li, Jiawei, X. Guan, Z. Fan, L.-M. Ching, Y. Li, X. Wang, W.-M. Cao, and D.-X. Liu. 2020. Non-Invasive Biomarkers for Early Detection of Breast Cancer. *Cancers* 12, no. 10: 2767.

Li, Juan, Y. Tong, Z. Sun, Y. Chen, Y. Wang, L. Zhou, Y. Jiang, P. Li, C. Wang, and L. Du. 2022. A Duplex-Specific Nuclease Assisted Photoelectrochemical Biosensor Based on MoS2@ ReS2/Ti3C2 Hybrid for Ultrasensitive Detection of Colorectal Cancer-Related piRNA-31,143. *Acta Biomaterialia* 149: 287–296.

Li, N., Y. Sun, F. Wang, C. Huang, C. Fu, L. Zhang, Y. Liu, S. Ge, and J. Yu. 2021. Target Dual-Recycling-Induced Bipedal DNA Walker and Bi2WO6/Bi2S3 Cascade Amplification Strategy in Photoelectrochemical Biosensor for TP53 Detection. *Sensors and Actuators B: Chemical* 345: 130386.

Li, R., H. Fan, H. Zhou, Y. Chen, Q. Yu, W. Hu, G.L. Liu, and L. Huang. 2023. Nanozyme-Catalyzed Metasurface Plasmon Sensor-Based Portable Ultrasensitive Optical Quantification Platform for Cancer Biomarker Screening. *Advanced Science* 10, no. 24 (August). https://onlinelibrary.wiley.com/doi/10.1002/advs.202301658.

Li, X., G. Peng, F. Cui, Q. Qiu, X. Chen, and H. Huang. 2018. Double Determination of Long Noncoding RNAs from Lung Cancer via Multi-Amplified Electrochemical Genosensor at Sub-Femtomole Level. *Biosensors and Bioelectronics* 113: 116–123.

Li, Y., L. Liu, H. Han, X. Yuan, J. Ji, L. Xue, Y. Wu, and R. Yang. 2024. A Signal-Switchable Photoelectrochemical Biosensor for Ultrasensitive Detection of Long Non-Coding RNA in Cancer Cells. *Talanta* 273: 125878.

Lin, S.-H., C.-Y. Lin, N.-T. Hsu, Y.-H. Yen, K.-M. Kee, J.-H. Wang, T.-H. Hu, C.-H. Chen, C.-H. Hung, and J.-T. Hu. 2022. Reappraisal of the Roles of Alpha-Fetoprotein in Hepatocellular Carcinoma Surveillance Using Large-Scale Nationwide Database and Hospital-Based Information. *Journal of the Formosan Medical Association* 121, no. 10: 2085–2092.

Lino, C., S. Barrias, R. Chaves, F. Adega, P. Martins-Lopes, and J.R. Fernandes. 2022. Biosensors as Diagnostic Tools in Clinical Applications. *Biochimica et Biophysica Acta (BBA)-Reviews on Cancer* 1877, no. 3: 188726.

Liu, J., L. Ren, S. Li, W. Li, X. Zheng, Y. Yang, W. Fu, J. Yi, J. Wang, and G. Du. 2021. The Biology, Function, and Applications of Exosomes in Cancer. *Acta Pharmaceutica Sinica B* 11, no. 9: 2783–2797.

Liu, X., Y. Li, L. He, Y. Feng, H. Tan, X. Chen, and W. Yang. 2021. Simultaneous Detection of Multiple Neuroendocrine Tumor Markers in Patient Serum with an Ultrasensitive and Antifouling Electrochemical Immunosensor. *Biosensors and Bioelectronics* 194: 113603.

Liu, Y., and Y. Huang. 2023. Highly Sensitive Sensing Detection of Micro RNA-126 in Urine Using POCT-Based Electrochemiluminescence Biosensor. *International Journal of Electrochemical Science* 18, no. 4: 100098.

Liu, Z., Q. Huang, Y. Yan, J. Yao, F. Zhong, S. Xie, M. Zhang, H. Zhang, M. Jin, and L. Shui. 2023. A Multi-Unit Integrated Electrochemical Biosensor Array for Synergistic Signal Enhancing Carbohydrate Antigen 125 Detection. *Sensors and Actuators B: Chemical* 393: 134224.

Lu, Z., W. Ni, N. Liu, D. Jin, T. Li, K. Li, Y. Zhang, Q. Yao, and G.-J. Zhang. 2023. CRISPR/Cas12a-Based Fluorescence Biosensor for Detection of Exosomal miR-21 Derived from Lung Cancer. *Microchemical Journal* 187: 108370.

Luo, X., and J.J. Davis. 2013. Electrical Biosensors and the Label Free Detection of Protein Disease Biomarkers. *Chemical Society Reviews* 42, no. 13: 5944–5962.

Luong, J.H., K.B. Male, and J.D. Glennon. 2008. Biosensor Technology: Technology Push versus Market Pull. *Biotechnology Advances* 26, no. 5: 492–500.

Mahmudunnabi, R.G., F.Z. Farhana, N. Kashaninejad, S.H. Firoz, Y.-B. Shim, and M.J. Shiddiky. 2020. Nanozyme-Based Electrochemical Biosensors for Disease Biomarker Detection. *Analyst* 145, no. 13: 4398–4420.

Mao, Y., N. Wang, F. Yu, S. Yu, L. Liu, Y. Tian, J. Wang, Y. Wang, L. He, and Y. Wu. 2019. Simultaneous Detection of Carcinoembryonic Antigen and Neuron-Specific Enolase in Human Serum Based on Time-Resolved Chemiluminescence Immunoassay. *Analyst* 144, no. 16: 4813–4819.

Mao, Y.-W., J.-X. Zhang, D.-N. Chen, A.-J. Wang, and J.-J. Feng. 2022. Bimetallic PtFe Alloyed Nanoparticles Decorated on 3D Hollow N-Doped Carbon Nanoflowers as Efficient Electrochemical Biosensing Interfaces for Ultrasensitive Detection of SCCA. *Sensors and Actuators B: Chemical* 370: 132416.

Mardis, E.R. 2017. DNA Sequencing Technologies: 2006–2016. *Nature Protocols* 12, no. 2: 213–218.

Mei, Y.P., J.P. Liao, J. Shen, L. Yu, B.L. Liu, L. Liu, R.Y. Li, L. Ji, S.G. Dorsey, and Z.R. Jiang. 2012. Small Nucleolar RNA 42 Acts as an Oncogene in Lung Tumorigenesis. *Oncogene* 31, no. 22: 2794–2804.

Melo, S.A., L.B. Luecke, C. Kahlert, A.F. Fernandez, S.T. Gammon, J. Kaye, V.S. LeBleu, E.A. Mittendorf, J. Weitz, and N. Rahbari. 2015. Glypican-1 Identifies Cancer Exosomes and Detects Early Pancreatic Cancer. *Nature* 523, no. 7559: 177–182.

Mirlekar, B. 2022. Tumor Promoting Roles of IL-10, TGF-β, IL-4, and IL-35: Its Implications in Cancer Immunotherapy. *SAGE Open Medicine* 10: 20503121211069012.

Mo, G., X. He, D. Qin, S. Meng, Y. Wu, and B. Deng. 2021. Spatially-Resolved Dual-Potential Sandwich Electrochemiluminescence Immunosensor for the Simultaneous Determination of Carbohydrate Antigen 19–9 and Carbohydrate Antigen 24-2. *Biosensors and Bioelectronics* 178: 113024.

Moccia, M., V. Caratelli, S. Cinti, B. Pede, C. Avitabile, M. Saviano, A.L. Imbriani, D. Moscone, and F. Arduini. 2020. Based Electrochemical Peptide Nucleic Acid (PNA) Biosensor for Detection of miRNA-492: A Pancreatic Ductal Adenocarcinoma Biomarker. *Biosensors and Bioelectronics* 165: 112371.

Mohsin, D.H., M.S. Mashkour, and F. Fatemi. 2021. Design of Aptamer-Based Sensing Platform Using Gold Nanoparticles Functionalized Reduced Graphene Oxide for Ultrasensitive Detection of Hepatitis B Virus. *Chemical Papers* 75, no. 1 (January): 279–295.

Moranova, L., M. Stanik, R. Hrstka, S. Campuzano, and M. Bartosik. 2022. Electrochemical LAMP-Based Assay for Detection of RNA Biomarkers in Prostate Cancer. *Talanta* 238: 123064.

Mousavi, M.Z., H.-Y. Chen, H.-S. Hou, C.-Y.-Y. Chang, S. Roffler, P.-K. Wei, and J.-Y. Cheng. 2015. Label-Free Detection of Rare Cell in Human Blood Using Gold Nano Slit Surface Plasmon Resonance. *Biosensors* 5, no. 1: 98–117.

Movva, S., W. Wen, W. Chen, S.Z. Millis, Z. Gatalica, S. Reddy, M. von Mehren, and B.A. Van Tine. 2015. Multi-Platform Profiling of over 2000 Sarcomas: Identification of Biomarkers and Novel Therapeutic Targets. *Oncotarget* 6, no. 14: 12234.

Nair, R.R., S. Debnath, S. Das, P. Wakchaure, B. Ganguly, and P.B. Chatterjee. 2019. A Highly Selective Turn-On Biosensor for Measuring Spermine/Spermidine in Human Urine and Blood. *ACS Applied Bio Materials* 2, no. 6 (June 17): 2374–2387.

Ng, K.W., C. Anderson, E.A. Marshall, B.C. Minatel, K.S. Enfield, H.L. Saprunoff, W.L. Lam, and V.D. Martinez. 2016. Piwi-Interacting RNAs in Cancer: Emerging Functions and Clinical Utility. *Molecular Cancer* 15, no. 1: 1–13.

Niederle, M.B., C. Scheuba, P. Riss, A. Selberherr, O. Koperek, and B. Niederle. 2020. Early Diagnosis of Medullary Thyroid Cancer: Are Calcitonin Stimulation Tests Still Indicated in the Era of Highly Sensitive Calcitonin Immunoassays? *Thyroid* 30, no. 7 (July 1): 974–984.

Omer, W.E., M.A. El-Kemary, M.M. Elsaady, M.N. Abou-Omar, A.O. Youssef, A.A. Sayqal, A.A. Gouda, and M.S. Attia. 2020. Highly Efficient Gold Nano-Flower Optical Biosensor Doped in a Sol–Gel/PEG Matrix for the Determination of a Calcitonin Biomarker in Different Serum Samples. *ACS Omega* 5, no. 11 (March 24): 5629–5637.

Ozcan, H.M., and U.D. Aydin. 2021. A Simple Immunosensor for Thyroid Stimulating Hormone. *Artificial Cells, Nanomedicine, and Biotechnology* 49, no. 1 (January): 61–70.

Ozcelikay, G., S.I. Kaya, A. Cetinkaya, and S.A. Ozkan. 2022. Recent Advances in Electrochemical Biosensors for Detection of Oncoviruses. *Biosensors and Bioelectronics: X* 12: 100260.

Pakchin, P.S., M. Fathi, H. Ghanbari, R. Saber, and Y. Omidi. 2020. A Novel Electrochemical Immunosensor for Ultrasensitive Detection of CA125 in Ovarian Cancer. *Biosensors and Bioelectronics* 153: 112029.

Parra-Cabrera, C., J. Samitier, and A. Homs-Corbera. 2016. Multiple Biomarkers Biosensor with Just-in-Time Functionalization: Application to Prostate Cancer Detection. *Biosensors and Bioelectronics* 77: 1192–1200.

Pileri, T., A. Sinibaldi, A. Occhicone, N. Danz, E. Giordani, M. Allegretti, F. Sonntag, P. Munzert, P. Giacomini, and F. Michelotti. 2024. Direct Competitive Assay for HER2 Detection in Human Plasma Using Bloch Surface Wave-Based Biosensors. *Analytical Biochemistry* 684: 115374.

Pohanka, M. 2018. Overview of Piezoelectric Biosensors, Immunosensors and DNA Sensors and Their Applications. *Materials* 11, no. 3: 448.

Pohanka, M. 2021. QCM Biosensor for Prostate Specific Antigen Assay Using Antibody–Gold Particle Conjugate. *International Journal of Electrochemical Science* 16, no. 5: 21051.

Povedano, E., M. Gamella, R.M. Torrente-Rodríguez, A. Montero-Calle, M. Pedrero, G. Solís-Fernández, F. Navarro-Villoslada, R. Barderas, S. Campuzano, and J.M. Pingarrón. 2021. Magnetic Microbeads-Based Amperometric Immunoplatform for the Rapid and Sensitive Detection of N6-Methyladenosine to Assist in Metastatic Cancer Cells Discrimination. *Biosensors and Bioelectronics* 171: 112708.

Povedano, E., V.R.-V. Montiel, A. Valverde, F. Navarro-Villoslada, P. Yáñez-Sedeño, M. Pedrero, A. Montero-Calle, et al. 2019. Versatile Electroanalytical Bioplatforms for Simultaneous Determination of Cancer-Related DNA 5-Methyl- and 5-Hydroxymethyl-Cytosines at Global and Gene-Specific Levels in Human Serum and Tissues. *ACS Sensors* 4, no. 1 (January 25): 227–234.

Prpić, M., M. Franceschi, M. Romić, T. Jukić, and Z. Kusić. 2018. Thyroglobulin as a Tumor Marker in Differentiated Thyroid Cancer–Clinical Considerations. *Acta Clinica Croatica* 57, no. 3: 518.

Pundir, C.S., V. Narwal, and B. Batra. 2016. Determination of Lactic Acid with Special Emphasis on Biosensing Methods: A Review. *Biosensors and Bioelectronics* 86: 777–790.

Qin, Y., S. Liu, S. Meng, D. Liu, and T. You. 2024. Split Aptamer-Based Sandwich-Type Ratiometric Biosensor for Dual-Modal Photoelectrochemical and Electrochemical Detection of 17β-Estradiol. *Analytica Chimica Acta* 1285: 342030.

Rahimi, P., and Y. Joseph. 2019. Enzyme-Based Biosensors for Choline Analysis: A Review. *TrAC Trends in Analytical Chemistry* 110: 367–374.

Rahmati, Z., M. Roushani, and H. Hosseini. 2021. Three-Dimensional NiCo2O4 Nanowires Encapsulated in Nitrogen-Doped Carbon Networks as a High-Performance Aptamer Stabilizer for Impedimetric Ultrasensitive Detection of Hepatitis C Virus Core Antigen. *Surfaces and Interfaces* 22: 100813.

Ramanathan, K., and B. Danielsson. 2001. Principles and Applications of Thermal Biosensors. *Biosensors and Bioelectronics* 16, no. 6: 417–423.

Raposo, G., and W. Stoorvogel. 2013. Extracellular Vesicles: Exosomes, Microvesicles, and Friends. *Journal of Cell Biology* 200, no. 4: 373–383.

Rezaei, H., M. Hosseini, and S. Radfar. 2021. A Dual-Signaling Electrochemical Ratiometric Strategy Combining "Signal-off" and "Signal-on" Approaches for Detection of MicroRNAs. *Analytical Biochemistry* 632: 114356.

Saalberg, Y., and M. Wolff. 2016. VOC Breath Biomarkers in Lung Cancer. *Clinica Chimica Acta* 459: 5–9.

Samie, H.A., and M. Arvand. 2020. Label-Free Electrochemical Aptasensor for Progesterone Detection in Biological Fluids. *Bioelectrochemistry* 133: 107489.

Saputra, H.A., J.H. Chung, R.J. Kwon, K.A. Jannath, D.-S. Park, and Y.-B. Shim. 2024. Ultrasensitive Interferon-Gamma Sensor with a Bifunctionalized Conducting Polymer Nanobioconjugate for Therapy-Progress Monitoring of Cancer Patients. *Sensors and Actuators B: Chemical* 398: 134739.

Saraiya, A.A., and C.C. Wang. 2008. snoRNA, a Novel Precursor of microRNA in Giardia Lamblia. *PLoS Pathogens* 4, no. 11: e1000224.

Seitz, H., and S. Schumacher. 2015. *Biomarker Validation: Technological, Clinical and Commercial Aspects*. John Wiley & Sons.

Shahbazi, N., S. Hosseinkhani, and B. Ranjbar. 2017. A Facile and Rapid Aptasensor Based on Split Peroxidase DNAzyme for Visual Detection of Carcinoembryonic Antigen in Saliva. *Sensors and Actuators B: Chemical* 253: 794–803.

Shandilya, R., R. Kumari, N. Bunkar, A. Bhargava, K. Chaudhury, I.Y. Goryacheva, and P.K. Mishra. 2022. A Photonic Dual Nano-Hybrid Assay for Detection of Cell-Free Circulating Mitochondrial DNA. *Journal of Pharmaceutical and Biomedical Analysis* 208: 114441.

Shi, J., Y. Zhang, P. Wang, Y. Nie, and Q. Ma. 2022. Luminous MoS2 Nanosheet-Based Electrochemiluminescence Biosensor with Biomimetic Vesicle for miRNA-210 Detection. *Talanta* 237: 122969.

Siegel, R.L., K.D. Miller, and A. Jemal. 2019. Cancer Statistics, 2019. *CA: A Cancer Journal for Clinicians* 69, no. 1 (January): 7–34.

Singal, A.G., B. Haaland, N.D. Parikh, A.B. Ozbay, C. Kirshner, S. Chakankar, K. Porter, J. Chhatwal, and T. Ayer. 2022. Comparison of a Multitarget Blood Test to Ultrasound and Alpha-fetoprotein for Hepatocellular Carcinoma Surveillance: Results of a Network Meta-analysis. *Hepatology Communications* 6, no. 10: 2925–2936.

Su, J., S. Chen, Y. Dou, Z. Zhao, X. Jia, X. Ding, and S. Song. 2022. Smartphone-Based Electrochemical Biosensors for Directly Detecting Serum-Derived Exosomes and Monitoring Their Secretion. *Analytical Chemistry* 94, no. 7 (February 22): 3235–3244.

Sung, H., J. Ferlay, R.L. Siegel, M. Laversanne, I. Soerjomataram, A. Jemal, and F. Bray. 2021. Global Cancer Statistics 2020: GLOBOCAN Estimates of Incidence and Mortality Worldwide for 36 Cancers in 185 Countries. *CA: A Cancer Journal for Clinicians* 71, no. 3 (May): 209–249.

Suravajhala, R., H.R. Burri, and B. Malik. 2020. Selective Targeted Drug Delivery Mechanism via Molecular Imprinted Polymers in Cancer Therapeutics. *Current Topics in Medicinal Chemistry* 20, no. 22: 1993–1998.

Tan, C., G. Xie, S. Wu, C. Song, J. Zhang, X. Yi, J. Wang, and H. Tang. 2024. Simultaneous Detection of Breast Cancer Biomarkers circROBO1 and BRCA1 Based on a CRISPR-Cas13a/Cas12a System. *Biosensors and Bioelectronics* 258: 116373.

Tothill, I.E. 2009. Biosensors for Cancer Markers Diagnosis. In *Seminars in Cell & Developmental Biology*, 20: 55–62. Elsevier. https://www.sciencedirect.com/science/article/pii/S1084952109000196?casa_token=Nck5ErOnLvIAAAAA:QANmbtFLdNnM5v_3aTcZbQ-izEUk904oj3ihgeMv13wgRnZChxQkhuMCGoUMUctoUJ-xs-IYfw.

Vaidya, V.S., and J.V. Bonventre. 2010. *Biomarkers: In Medicine, Drug Discovery, and Environmental Health*. John Wiley & Sons.

Van Poppel, H., M.J. Roobol, C.R. Chapple, J.W. Catto, J. N'Dow, J. Sønksen, A. Stenzl, and M. Wirth. 2021. Prostate-Specific Antigen Testing as Part of a Risk-Adapted Early Detection Strategy for Prostate Cancer: European Association of Urology Position and Recommendations for 2021. *European Urology* 80, no. 6: 703–711.

Wang, H., D. He, K. Wan, X. Sheng, H. Cheng, J. Huang, X. Zhou, X. He, and K. Wang. 2020. In Situ Multiplex Detection of Serum Exosomal microRNAs Using an All-in-One Biosensor for Breast Cancer Diagnosis. *Analyst* 145, no. 9: 3289–3296.

Wang, H.-F., R.-N. Ma, F. Sun, L.-P. Jia, W. Zhang, L. Shang, Q.-W. Xue, W.-L. Jia, and H.-S. Wang. 2018. A Versatile Label-Free Electrochemical Biosensor for Circulating Tumor DNA Based on Dual Enzyme Assisted Multiple Amplification Strategy. *Biosensors and Bioelectronics* 122: 224–230.

Wang, J. 2006. Electrochemical Biosensors: Towards Point-of-Care Cancer Diagnostics. *Biosensors and Bioelectronics* 21, no. 10: 1887–1892.

Wang, K., X. Bai, Y. Xue, X. Luo, J. Dong, G. Yang, K. Ma, L. Huang, J. Zhou, and J. Wang. 2023. Absolute Quantification of circRNA Using Digital Reverse Transcription-Hyperbranched Rolling Circle Amplification. *Sensors and Actuators B: Chemical* 375: 132893.

Wang, N., J. Wang, X. Zhao, H. Chen, H. Xu, L. Bai, W. Wang, H. Yang, D. Wei, and B. Yuan. 2021. Highly Sensitive Electrochemical Immunosensor for the Simultaneous Detection of Multiple Tumor Markers for Signal Amplification. *Talanta* 226: 122133.

Wang, S., H. Li, K. Dong, W. Shu, J. Zhang, J. Zhang, R. Zhao, S. Wei, D. Feng, and X. Xiao. 2023. A Universal and Specific RNA Biosensor via DNA Circuit-Mediated PAM-Independent CRISPR/Cas12a and PolyA-Rolling Circle Amplification. *Biosensors and Bioelectronics* 226: 115139.

Wasilewski, T., W. Kamysz, and J. Gębicki. 2024. AI-Assisted Detection of Biomarkers by Sensors and Biosensors for Early Diagnosis and Monitoring. *Biosensors* 14, no. 7. https://www.ncbi.nlm.nih.gov/pmc/articles/PMC11274923/.

Weng, W., Q. Wei, S. Toden, K. Yoshida, T. Nagasaka, T. Fujiwara, S. Cai, H. Qin, Y. Ma, and A. Goel. 2017. Circular RNA ciRS-7—a Promising Prognostic Biomarker and a Potential Therapeutic Target in Colorectal Cancer. *Clinical Cancer Research* 23, no. 14: 3918–3928.

Wu, C.-T., M.-F. Chen, W.-C. Chen, and C.-C. Hsieh. 2013. The Role of IL-6 in the Radiation Response of Prostate Cancer. *Radiation Oncology* 8: 1–11.

Wu, D., H. Lu, J. Wang, L. Wu, L. Liu, X. Yi, J. Wang, and H. Tang. 2021. Amplified Electrochemical Detection of Circular RNA in Breast Cancer Patients Using Ferrocene-Capped Gold Nanoparticle/Streptavidin Conjugates. *Microchemical Journal* 164: 106066.

Wu, Q., and W. Zheng. 2022. A Highly Sensitive Electrochemiluminescence Sensor Based on Antimony Tin Oxide/TiO2 for Detection of Gastrin-Releasing Peptide (ProGRP). *International Journal of Electrochemical Science* 17, no. 11: 221188.

Wu, Y., X. Chen, X. Wang, M. Yang, F. Xu, C. Hou, and D. Huo. 2021. A Fluorescent Biosensor Based on Prismatic Hollow Metal-Polydopamine Frameworks and 6-Carboxyfluorescein (FAM)-Labeled Protein Aptamer for CA15-3 Detection. *Sensors and Actuators B: Chemical* 329: 129249.

Xi, X., T. Li, Y. Huang, J. Sun, Y. Zhu, Y. Yang, and Z.J. Lu. 2017. RNA Biomarkers: Frontier of Precision Medicine for Cancer. *Non-Coding RNA* 3, no. 1: 9.

Xia, N., Z. Chen, Y. Liu, H. Ren, and L. Liu. 2017. Peptide Aptamer-Based Biosensor for the Detection of Human Chorionic Gonadotropin by Converting Silver Nanoparticles-Based Colorimetric Assay into Sensitive Electrochemical Analysis. *Sensors and Actuators B: Chemical* 243: 784–791.

Xu, H., F. Kou, H. Ye, Z. Wang, S. Huang, X. Liu, X. Zhu, Z. Lin, and G. Chen. 2017. Highly Sensitive Antibody-Aptamer Sensor for Vascular Endothelial Growth Factor Based on Hybridization Chain Reaction and pH Meter/Indicator. *Talanta* 175: 177–182.

Xu, Lulu, Z. Jiang, Y. Mu, Y. Zhang, Q. Zhan, J. Cui, W. Cheng, and S. Ding. 2018. Colorimetric Assay of Rare Disseminated Tumor Cells in Real Sample by Aptamer-Induced Rolling Circle Amplification on Cell Surface. *Sensors and Actuators B: Chemical* 259: 596–603.

Xu, Luyun, Y. Chen, J. Ye, M. Fan, G. Weng, Y. Shen, Z. Lin, D. Lin, Y. Xu, and S. Feng. 2024. Optical Nanobiosensor Based on Surface-Enhanced Raman Spectroscopy and Catalytic Hairpin Assembly for Early-Stage Lung Cancer Detection via Blood Circular RNA. *ACS Sensors* 9, no. 4 (April 26): 2020–2030.

Yang, H.J., C.V. Raju, C.-H. Choi, and J.P. Park. 2024. Electrochemical Peptide-Based Biosensor for the Detection of the Inflammatory Disease Biomarker, Interleukin-1beta. *Analytica Chimica Acta* 1295: 342287.

Yang, M., H. Li, X. Li, K. Huang, W. Xu, and L. Zhu. 2022. Catalytic Hairpin Self-Assembly Regulated Chameleon Silver Nanoclusters for the Ratiometric Detection of CircRNA. *Biosensors and Bioelectronics* 209: 114258.

Yang, Z., E. Liu, S.M. Wang, Y.F. Xiao, S. Zeng, S.M. Yang, X.Y. Zhao, and Y. Huang. 2021. Development of A Long Noncoding RNA *BC032469* -Dependent Gold Nanoparticle Molecular Beacon for The Detection of Gastric Cancer Cells. *Nanomedicine* 16, no. 25 (October): 2255–2267.

Yao, Y., W. Cheng, T. Li, Y. Chen, C. Duan, Z. Wang, and Y. Xiang. 2023. Triple Amplification Strategy of CHA-PER-DNAzyme for Ultrasensitive Detection of circRNA Associated with Hepatocellular Carcinoma. *Analytical Chemistry* 95, no. 35 (September 5): 13149–13155.

Yao, Y., C. Duan, Y. Chen, Z. Hou, W. Cheng, D. Li, Z. Wang, and Y. Xiang. 2023. Long Non-Coding RNA Detection Based on Multi-Probe-Induced Rolling Circle Amplification for Hepatocellular Carcinoma Early Diagnosis. *Analytical Chemistry* (January 4). https://pubs.acs.org/doi/10.1021/acs.analchem.2c04594.

Yu, Miao, M. Liu, and Y. Li. 2022. Point-of-Care Based Electrochemical Immunoassay for Epstein-Barr Virus Detection. Ed. Yuhong Zheng. *Journal of Analytical Methods in Chemistry* 2022 (May 30): 1–8.

Yu, Min, S. Stott, M. Toner, S. Maheswaran, and D.A. Haber. 2011. Circulating Tumor Cells: Approaches to Isolation and Characterization. *Journal of Cell Biology* 192, no. 3: 373–382.

Zhang, B., Y. Liang, L. Bo, M. Chen, B. Huang, Q. Cao, J. Wei, T. Li, X. Cai, and X. Ye. 2021. A PCR-Free Screen-Printed Magnetic Electrode for the Detection of Circular RNA from Hepatocellular Cancer Based on a Back-Splice Junction. *RSC Advances* 11, no. 29: 17769–17774.

Zhang, Hua, Q. Wang, X. Yang, K. Wang, Q. Li, Z. Li, L. Gao, W. Nie, and Y. Zheng. 2017. An Isothermal Electrochemical Biosensor for the Sensitive Detection of microRNA Based on a Catalytic Hairpin Assembly and Supersandwich Amplification. *Analyst* 142, no. 2: 389–396.

Zhang, Huihua, H. Qu, J. Cui, and L. Duan. 2022. A Simple Electrochemical Immunosensor Based on a Chitosan/Reduced Graphene Oxide Nanocomposite for Sensitive Detection of Biomarkers of Malignant Melanoma. *RSC Advances* 12, no. 40: 25844–25851.

Zhang, L., H. Wang, H. Zhang, N. Zhang, X. Zheng, W. Li, X. Qiu, and D. Yu. 2022. Development of a Portable Multiplexed Instrument for Multi-Proteins Detection in Human Urine Using Surface Plasmon Resonance. *Sensors and Actuators B: Chemical* 369: 132272.

Zhang, W., Z. Dai, X. Liu, and J. Yang. 2018. High-Performance Electrochemical Sensing of Circulating Tumor DNA in Peripheral Blood Based on Poly-Xanthurenic Acid Functionalized MoS2 Nanosheets. *Biosensors and Bioelectronics* 105: 116–120.

Zhang, Y., T. Wu, D. Liu, R. Xu, H. Ma, Q. Wei, and Y. Zhang. 2022. Photoelectrochemical Immunosensor for the Sensitive Detection of Neuron-Specific Enolase Based on the Effect of Z-Scheme WO3/NiCo2O4 Nanoarrays Pn Heterojunction. *Biosensors and Bioelectronics* 213: 114452.

Zhao, F., L. Wang, M. Li, M. Wang, G. Liu, and J. Ping. 2023. Nanozyme-Based Biosensor for Organophosphorus Pesticide Monitoring: Functional Design, Biosensing Strategy, and Detection Application. *TrAC Trends in Analytical Chemistry*: 117152.

Zhao, H., Z. Niu, K. Chen, L. Chen, Z. Wang, M. Lan, J. Shi, and W. Huang. 2021. A Novel Sandwich-Type Electrochemical Biosensor Enabling Sensitive Detection of Circulating Tumor DNA. *Microchemical Journal* 171: 106783.

Zhao, R., L. Zhao, H. Feng, X. Chen, H. Zhang, Y. Bai, F. Feng, and S. Shuang. 2021. A Label-Free Fluorescent Aptasensor Based on HCR and G-Quadruplex DNAzymes for the Detection of Prostate-Specific Antigen. *Analyst* 146, no. 4: 1340–1345.

Zheng, Yiyang, Y. Luo, X. Chen, H. Li, B. Huang, B. Zhou, L. Zhu, X. Kang, and W. Geng. 2021. The Role of mRNA in the Development, Diagnosis, Treatment and Prognosis of Neural Tumors. *Molecular Cancer* 20, no. 1: 1–16.

Zhu, D., Y. Hu, X.-J. Zhang, X.-T. Yang, and Y.-Y. Tang. 2019. Colorimetric and Fluorometric Dual-Channel Detection of α-Fetoprotein Based on the Use of ZnS-CdTe Hierarchical Porous Nanospheres. *Microchimica Acta* 186, no. 2 (February): 124.

Zong, C., M. Xu, L.-J. Xu, T. Wei, X. Ma, X.-S. Zheng, R. Hu, and B. Ren. 2018. Surface-Enhanced Raman Spectroscopy for Bioanalysis: Reliability and Challenges. *Chemical Reviews* 118, no. 10 (May 23): 4946–4980.

10 Electrochemical Tools for Disease Detection in Plants

Ishita Ghosh

10.1 INTRODUCTION

Utilizing the principles of electrochemistry to identify pathogens, biomarkers, or physiological changes in plants associated with disease onset, electrochemical tools have emerged as highly sensitive, cost-effective, and portable alternatives for detecting plant diseases. The application of electrochemical tools for plant disease detection represents a significant advancement in agricultural diagnostics, addressing the growing need for quick, accurate, and economical ways to protect crops. Additionally, early and accurate detection of plant diseases can minimize the use of agrochemicals, reduce crop losses, and support sustainable agriculture. These instruments use the concepts of electrochemistry to identify certain biomarkers linked to plant diseases, such as proteins, nucleic acids, or metabolites. Through the integration of cutting-edge materials like conductive polymers and nanomaterials, electrochemical sensors have attained exceptional sensitivity and specificity, which makes them appropriate for early illness detection even before outward symptoms manifest [1].

A wide range of pathogens, including bacteria, fungi, and viruses, can be identified using electrochemical biosensors, especially those that use DNA, antibodies, or enzymes. Additionally, real-time monitoring has been made possible by the integration of portable and on-field electrochemical systems, enabling farmers to take prompt preventive action. These technologies can greatly lessen the need for labour-intensive, time-consuming, and skill-intensive laboratory-based assays and other traditional diagnostic techniques. A wide range of pathogens, including bacteria, fungi, and viruses, can be identified with electrochemical biosensors, especially those that use DNA, antibodies, or enzymes. Additionally, the use of portable and on-field electrochemical platforms has allowed for real-time monitoring, enabling farmers to take timely preventive action. These tools can greatly reduce the reliance on traditional diagnostic methods, such as laboratory-based assays, which are frequently labor-intensive and time-consuming and require specialized skills.

DOI: 10.1201/9781003536239-10

10.2 PRINCIPLES OF ELECTROCHEMICAL DETECTION

Electrochemical detection is based on the measurement of electrical signals generated during chemical reactions [2]. The core components of an electrochemical system include:

1. *Electrodes*: Typically composed of materials like carbon, gold, or platinum, electrodes act as the interface for electron transfer.
2. *Electrolytes*: These provide the medium for ion movement, facilitating the electrochemical reaction.
3. *Redox reactions*: The basis for signal generation, involving the oxidation or reduction of analytes.
4. *Signal readout*: The electrical signals, such as current, voltage, or impedance, are measured and correlated to the concentration of the target analyte.

10.3 TYPES OF ELECTROCHEMICAL TECHNIQUES

- *Amperometry*: Measures current at a fixed potential, commonly used for enzymatic biosensors.
- *Voltammetry*: Monitors current as a function of applied potential, enabling detailed analyte characterization.
- *Impedance spectroscopy*: Measures the resistance to current flow, useful for label-free detection.
- *Conductometry*: Measures changes in electrical conductivity, often used for ionic analytes (Figure 10.1).

10.4 DETAILS OF ELECTROCHEMICAL TOOLS FOR DISEASE DETECTION IN PLANTS

10.4.1 Amperometric Sensors

Amperometric sensors measure the current produced during a redox reaction at a constant potential. These sensors are particularly useful for detecting enzymatic reactions associated with plant pathogens [3]. For example, glucose oxidase–based sensors can monitor changes in glucose levels, indicating stress or infection in plants.

10.4.2 Voltammetric Sensors

Voltammetry involves varying the potential applied to an electrode and measuring the resulting current. This technique is versatile and allows for the detection of a wide range of plant metabolites and pathogens. Differential pulse voltammetry (DPV) and cyclic voltammetry (CV) are commonly used variants. For instance, DPV has been applied to detect fungal DNA in plant extracts.

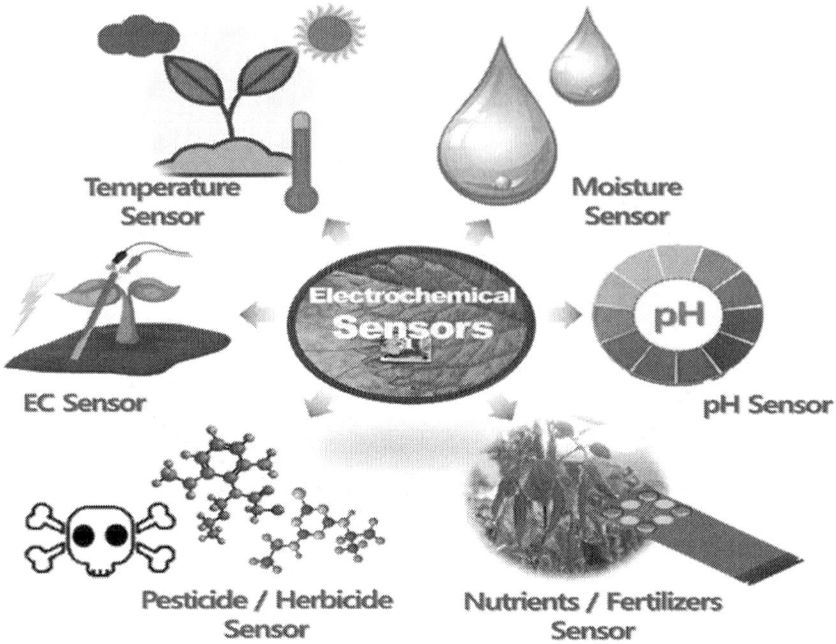

FIGURE 10.1 Types of electrochemical sensors.

Source: https://doi: 10.3389/fchem.2022.848320.

10.4.3 Impedance Spectroscopy

Impedance spectroscopy measures the opposition to electrical flow in a system. It is a label-free technique suitable for detecting pathogens directly on the electrode surface. Impedance-based sensors can identify bacterial pathogens like *Xanthomonas* by measuring changes in the impedance of the electrode surface caused by bacterial adhesion or biofilm formation.

10.4.4 Conductometric Sensors

Conductometric sensors monitor changes in the electrical conductivity of the sample. These sensors are effective for detecting ionic changes in plant sap, which may occur due to infection or stress. For example, conductometric sensors have been employed to monitor ion leakage in leaves, an early indicator of disease.

10.4.5 Potentiometric Sensors

Potentiometric sensors measure changes in the potential of an electrode without drawing significant current. These sensors are used to detect pH changes or ion concentrations in plant tissues. For example, they can monitor changes in soil pH caused by root infections or fungal activity.

10.4.6 Nano-Enabled Electrochemical Sensors

Integration of nanomaterials such as gold nanoparticles, graphene, and carbon nanotubes enhances the performance of electrochemical sensors. These materials increase the electrode surface area and improve signal-to-noise ratios, enabling ultra-sensitive detection of pathogens and biomarkers in plants.

10.5 ELECTROCHEMICAL BIOSENSORS FOR PLANT DISEASE DETECTION

Electrochemical biosensors are analytical devices combining a biorecognition element with an electrochemical transducer [4]. These sensors have been extensively employed for detecting plant pathogens and disease biomarkers (Figures 10.2 and 10.3).

10.6 KEY COMPONENTS

1. *Biorecognition elements*:
   - *Enzymes*: Catalyze specific reactions to generate measurable signals.

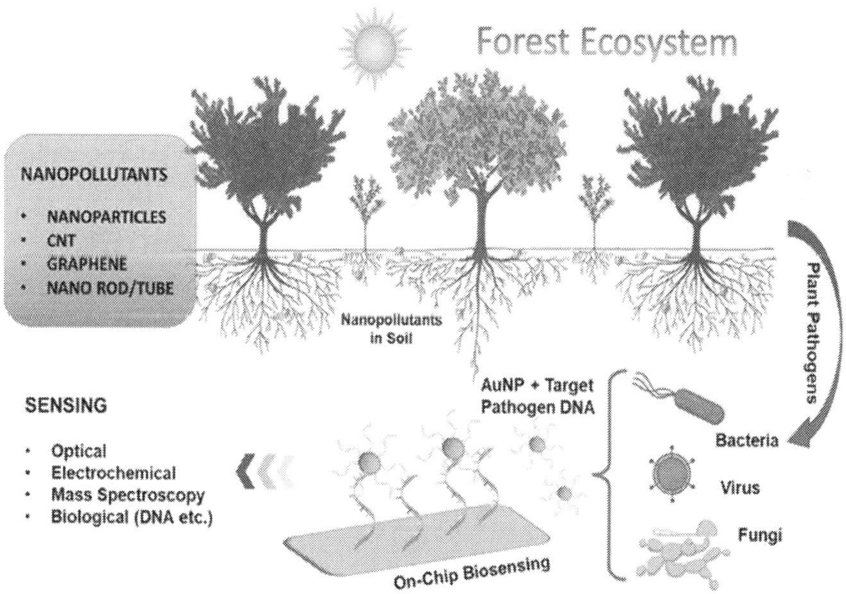

FIGURE 10.2 The entire cycle in brief.

Source: https://doi.org/10.3390/s22031259

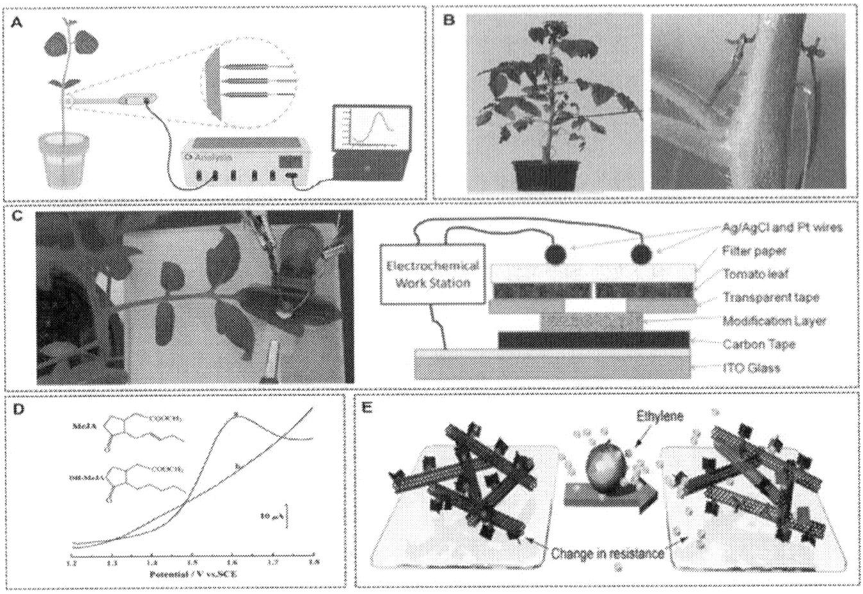

FIGURE 10.3 The process used in electrochemical sensors explained.

Source: https://doi.org/10.3389/fchem.2022.848320

   - *Antibodies*: Provide specificity for pathogen detection.
   - *Nucleic acids*: Detect DNA or RNA sequences of pathogens.
   - *Aptamers*: Synthetic oligonucleotides with high binding affinity.
2. *Transducers*:
   - Convert biological interactions into measurable electrical signals.

10.7 DETECTION STRATEGIES

- *Pathogen identification*: Electrochemical sensors can detect microbial pathogens like fungi, bacteria, and viruses by recognizing specific antigens or nucleic acids.
- *Biomarker monitoring*: These sensors identify stress markers or plant metabolites altered during infection, such as phytohormones, reactive oxygen species, or volatile organic compounds (Figure 10.4).

10.8 CASE STUDIES

- *Fungal pathogen detection*

Electrochemical sensors have been used to identify fungal pathogens like *Fusarium* and *Phytophthora* [5]. For instance, DNA-based biosensors employing voltammetry can detect fungal DNA within hours, offering a rapid alternative to traditional culture methods.

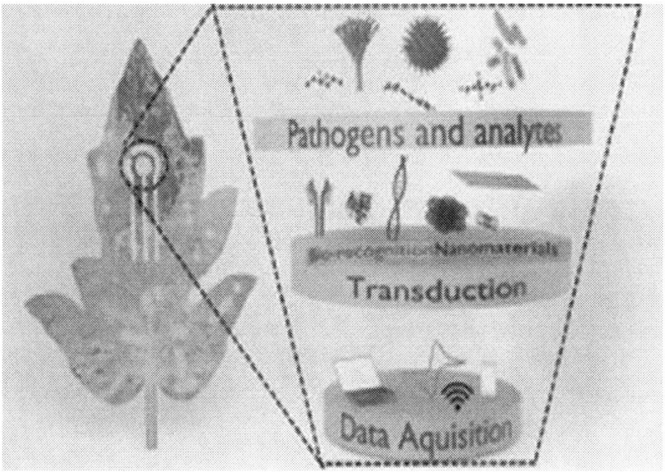

FIGURE 10.4 Key components.

Source: https://doi.org/10.1016/j.snr.2021.100068

- *Viral disease detection*

Plant viral diseases, such as those caused by *Tobacco mosaic virus* (TMV), can be detected using electrochemical immunosensors. These sensors employ virus-specific antibodies immobilized on electrode surfaces, enabling selective detection.

- *Bacterial pathogen detection*

Biosensors targeting bacterial DNA or lipopolysaccharides have shown high specificity in identifying bacterial pathogens like *Xanthomonas* and *Pseudomonas*. Impedance-based sensors have been particularly effective for label-free detection.

- *Detection of Phytophthora infestans using voltammetric DNA sensors*

Phytophthora infestans, the causative agent of late blight in potatoes, poses a significant threat to crop yields. Researchers developed a voltammetric DNA sensor using gold nanoparticles on a carbon electrode. The sensor was functionalized with complementary DNA probes specific to the pathogen's DNA. Differential pulse voltammetry (DPV) was employed to measure hybridization events. This approach demonstrated high sensitivity, detecting *P. infestans* DNA at concentrations as low as 10 fM [6]. The rapid detection capability of this sensor highlights its potential for field applications in managing late blight outbreaks.

- *Impedance-based detection of Xanthomonas campestris*

Xanthomonas campestris, a bacterial pathogen affecting cruciferous vegetables, was detected using an impedance biosensor. The sensor utilized antibodies specific to the pathogen's surface proteins, immobilized on a gold electrode. Upon bacterial binding, the impedance of the electrode surface changed significantly, correlating with the pathogen concentration. The sensor achieved a detection limit of 100 CFU/mL and provided results within 30 min. This rapid and label-free detection method is advantageous for real-time monitoring of bacterial infections in agricultural settings.

10.9 ADVANTAGES OF ELECTROCHEMICAL TOOLS

- *Sensitivity*: Capable of detecting low concentrations of analytes, making them suitable for early diagnosis.
- *Specificity*: Use of selective biorecognition elements ensures accurate identification of pathogens.
- *Portability*: Compact and lightweight designs facilitate field use.
- *Cost-effectiveness*: Require minimal reagents and infrastructure compared to molecular techniques.
- *Rapid response*: Provide results in real-time or within a few hours.

10.10 CHALLENGES AND FUTURE DIRECTIONS

Despite their potential, electrochemical tools face several challenges:

- *Stability of biorecognition elements*: Ensuring long-term stability under field conditions remains a hurdle.
- *Interference from complex matrices*: Plant sap and soil extracts may introduce interfering substances, affecting sensor performance.
- *Standardization*: Establishing standardized protocols and regulatory frameworks is essential for widespread adoption [7].

10.11 INNOVATIONS IN THE FIELD

- *Nanotechnology*: Integration of nanomaterials like graphene, gold nanoparticles, and carbon nanotubes enhances sensor sensitivity and stability.
- *Multiplexing*: Development of platforms capable of detecting multiple pathogens simultaneously.
- *Integration with IoT*: Linking electrochemical sensors with Internet of Things (IoT) devices for real-time monitoring and data analysis.
- *Wearable sensors*: Design of flexible, wearable sensors for continuous plant health monitoring (Figure 10.5).

10.12 CONCLUSION

Despite these developments, there are still obstacles to fully utilizing the potential of electrochemical tools in agricultural settings, including the need to address issues like the scalability of sensor fabrication, stability under variable field conditions, and deployment costs. Additionally, the development of multiplexed sensors that can

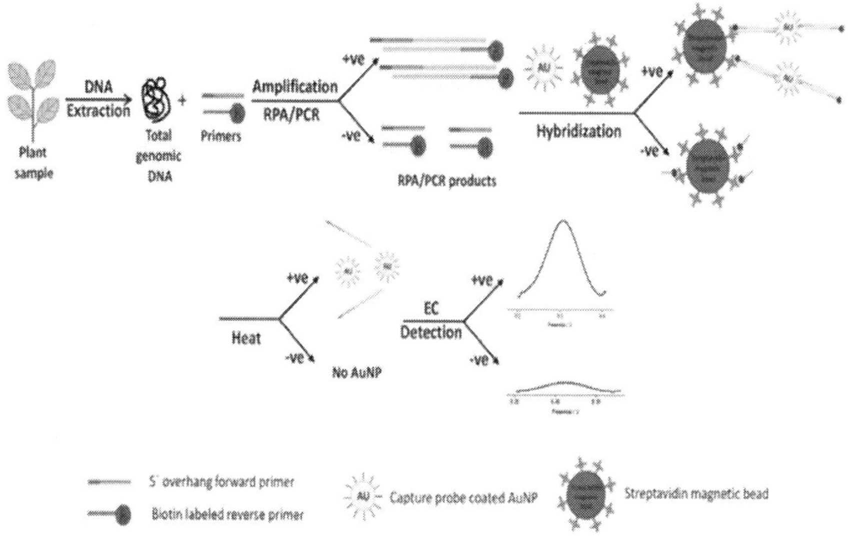

FIGURE 10.5 Case study.

Source: https://www.nature.com/scientificreports

detect multiple pathogens at once could further enhance their utility in integrated pest management strategies. Electrochemical tools offer a promising approach for the early and accurate detection of plant diseases. Their sensitivity, specificity, and adaptability to field conditions make them indispensable for modern agricultural practices.

By addressing current challenges and leveraging advancements in nanotechnology, material science, and IoT, these tools can revolutionize plant disease diagnostics and support global efforts toward sustainable agriculture. Despite these advancements, challenges remain in fully realizing the potential of electrochemical tools in agricultural settings. Issues such as the scalability of sensor fabrication, stability under variable field conditions, and the cost of deployment need to be addressed. Moreover, the development of multiplexed sensors capable of detecting multiple pathogens simultaneously could further enhance their utility in integrated pest management strategies.

Finally, by bridging the gap between laboratory research and field application, electrochemical tools can play a crucial role in ensuring global food security in the face of emerging agricultural challenges. Ongoing research and innovation in sensor technology, along with efforts to make these tools more accessible and affordable, are essential for their widespread adoption. Electrochemical tools for plant disease detection offer a promising path toward precision agriculture, contributing to sustainable farming practices by improving disease management and reducing crop losses (Figure 10.6).

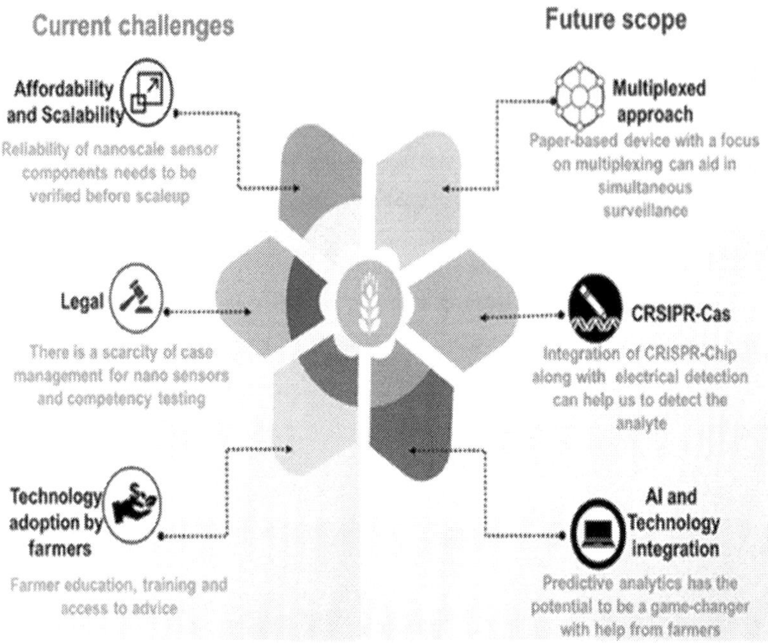

FIGURE 10.6 Current challenges and future scope.

Source: https://pubs.acs.org/doi/10.1021/acsagscitech.2c00150

REFERENCES

1. Arugula, M. A., & Simonian, A. L. (2015). "Biosensors for Plant Pathogen Detection: Electrochemical Approaches." *Analytica Chimica Acta*, 893, 1–15.
2. Chikkaveeraiah, B. V., et al. (2014). "Electrochemical Immunosensors for Pathogen Detection in Plants." *Trends in Biotechnology*, 32(10), 515–529.
3. Cinti, S., et al. (2017). "Electrochemical Biosensors for Pathogen Detection in Agriculture." *Biosensors and Bioelectronics*, 98, 255–268.
4. Hasan, M. R., et al. (2019). "Electrochemical DNA Sensors for Pathogen Detection in Agriculture." *Talanta*, 200, 181–191.
5. Monnappa, A. B., Manjunatha, J. G., & Bhatt, A. S. (2020). "Bhatt, Design of a Sensitive and Selective Voltammetric Sensor Based on a Cationic Surfactant-Modified Carbon Paste Electrode for the determination of Alloxan." *ACS Omega*, 5(36), 23481–23490.
6. Moulick, A., et al. (2020). "Nanomaterial-Based Electrochemical Sensors for Phytopathogen Detection." *Sensors and Actuators B: Chemical*, 309, 127761.
7. Rarila, C., Manjunatha, J. G., Ravishankar, D. K., Fattepur, S., Siddaraju, G., & Nanjundaswamy, L. (2020). "Validated Electrochemical Method for Simultaneous Resolution of Tyrosine, Uric Acid, and Ascorbic Acid at Polymer Modified Nano-Composite Paste Electrode." *Surface Engineering and Applied Electrochemistry*, 56(4), 415–426.

11 Biosensors for Neurological Disorders

Kalyani Chepuri, Vadakavila Geethikalal, Chittepu Pranitha, Kathuroju Harikrishna, Volukula Vaishnavi, Sanchaita Tewary and Ranabothu Madhurima

11.1 INTRODUCTION

Neurological degenerative diseases like Creutzfeldt-Jakob, Alzheimer's, traumatic brain injury (TBI), and Parkinson's significantly impact health and place a major physical and financial strain on society. Alzheimer's cases in the United States are projected to rise from 5.2 million in 2016 to 23.8 million by 2050, with care costs reaching $236 billion in 2016 alone. Additionally, over 40% of caregivers report high emotional stress. TBI affects 100,000 Americans annually, contributing to long-term disability with an annual cost of $48.3 billion. However, standardized TBI diagnostics are still lacking (Dai, Molazemhosseini, and Liu 2017). Biosensors offer rapid, sensitive detection of neurological disease markers. For example, enzyme activity in phenylketonuria or genetic markers in Huntington's disease can aid diagnosis. While single biomarkers may lead to false positives or negatives, using a multitarget disease "signature" can improve reliability. For instance, a phage peptide panel has shown high accuracy in diagnosing prostate cancer, a model potentially adaptable for neurological cancers. A sensor array detecting multiple neurological markers in fluids like CSF could further enhance diagnostic precision (Bell and Korguth 2007).

Healthcare relies on efficient diagnostics, molecular techniques, and pharmacogenomics to support personalized medicine. DNA damage, particularly in non-dividing cells like brain tissue, can lead to neurodegenerative diseases through impaired DNA repair. Enzymes such as TOP and TDP play critical roles in repair pathways, and mutations in TDP2 are linked to neurological issues. Dysregulation of TOP1, essential for synaptic function, contributes to diseases like frontotemporal dementia and ALS, suggesting TOP1 and TDP2 may serve as neurodegenerative disease biomarkers (Shawky et al. 2018). Despite decades of biosensor research, achieving small, affordable, durable, and real-time biosensors with easy device integration remains challenging. Nevertheless, biosensors are showing promise in delivering patient-centered diagnostics for complex neurological illnesses by enabling multiplexed biomarker analysis and advancing fundamental neuroscience.

DOI: 10.1201/9781003536239-11

Chemical signaling is crucial for neuronal function, but directly measuring extracellular neurotransmitter and modulator concentrations is uncommon. Biosensors with spatial resolution between 100 nm and 10 mm and millisecond precision can detect neurotransmitter release, including pH, glutamate, dopamine, and nitric oxide. Advanced microamperometric electrodes enable in vivo neurotransmitter tracking with minimal invasiveness. In vivo biosensors like CNiFERs use fluorescence-engineered reporters to monitor neurotransmitter and receptor activity. However, metabolites alone cannot fully characterize neurological disorders, necessitating multiparametric approaches integrating psychoanalytic, genomic, proteomic, and metabolomic markers for better diagnosis. Examples include Ciphergen's SELDI-TOF MS, identifying Alzheimer's biomarkers, and the biobarcode system for enhanced biomarker detection, including Alzheimer's amyloid-β and tau proteins (Nguyen et al. 2010).

Surface plasmon resonance (SPR) technology is widely used for identifying and quantifying neurodegenerative biomarkers, such as amyloid-β and polyglutamine tracts linked to Alzheimer's disease. Grating-coupled SPR systems enable label-free, flexible assays and allow for simultaneous genomic and proteomic analysis with high adaptability. Simplified point-of-care (POC) diagnostics, like reflection holograms, offer portable solutions. These holograms use analyte-selective polymers and diffraction gratings to detect biomolecule interactions, with potential applications in diagnosing neurological and psychiatric conditions. Additionally, microfluidic devices, such as flexible-film devices with thread-based microfluidics, enable low-cost, accessible diagnostics for non-specialists, advancing the accuracy and accessibility of neurological and psychological disorder diagnostics (Lowe 2011).

11.2 TYPES OF BIOSENSORS USED IN NEUROLOGICAL DISORDERS

Biosensors for neurological disorders are devices designed to detect biological signals and convert them into measurable data, aiding in diagnosis, monitoring, and treatment. Table 11.1 represents the different types of biosensors used in neurology, along with their advantages and disadvantages. The main types of biosensors employed in neurological applications include:

11.2.1 Electrochemical Biosensors

In recent decades, numerous electrochemical biosensors have been developed for biomarker and biomolecule detection, enhancing clinical diagnostics and other fields. The working electrode plays a critical role in sensor performance. Traditional electrodes, such as glassy carbon electrodes (GCE) and screen-printed carbon electrodes (SPE), have been modified with surface treatments and materials to improve sensor effectiveness. Electrodes are often immobilized with chemical groups like amines, carboxyls (e.g., EDC), aldehydes (hydrazide), and thiols (maleimide) to create stable platforms for biomarker labeling (e.g., enzymes, antibodies, nucleic acids).

TABLE 11.1
Advantages and Disadvantages of Different Biosensors in Neurology

Type of Biosensor	Sensor Probe	Advantages	Disadvantages	References
Electrochemical biosensor	Gold Platinum Carbon Electrode ssDNA	High selectivity, stability, sensitivity, and suitability for long-term biosensing	Susceptible to sample matrix effects, lower sensitivity than traditional methods, and limited shelf life	Chen, Ding, and Zhang (2024)
SPR biosensors (surface plasmon resonance)	Metal layer, gold	Uses fewer biomolecules than traditional SPR, enables real-time imaging with a high-resolution CCD camera, and supports arrays with hundreds of active spots	Limited sensitivity to low molecular weight analytes, prone to non-specific binding, and affected by refractive index changes in the sample solution	Singh (2016)
Optical biosensors	Antibodies Aptamers Enzymes Receptors Nucleic acids	Highly sensitive, enables real-time, label-free detection, is versatile, and suitable for miniaturization	Requires complex instrumentation, skilled operators, and is sensitive to environmental factors, with a limited dynamic range and susceptibility to interference	Damborský, Švitel, and Katrlík (2016)
Aptamer-based biosensors	Aptamer	High specificity and affinity, chemically and thermally stable, versatile, easy to produce, low immunogenicity, and suitable for multiplexing	Susceptible to degradation, limited dynamic range, complex signal transduction, and prone to non-specific binding	Kudłak and Wieczerzak (2020)

(Continued)

TABLE 11.1 (CONTINUED)
Advantages and Disadvantages of Different Biosensors in Neurology

Type of Biosensor	Sensor Probe	Advantages	Disadvantages	References
Magnetic biosensors	Magnetic nanoparticle	Highly sensitive, enables label-free, real-time detection, versatile, robust, and easy to integrate	Requires complex instrumentation, susceptible to magnetic interference, has a limited dynamic range, and needs specialized magnetic particles	Balaban Hanoglu et al. (2023)
Nano biosensors	Nano particles made of gold silver of magnetic materials	Enhanced sensitivity, rapid response, real-time monitoring, increased surface area, efficient mass transport, multiple sensing capabilities, and potential for miniaturization and portability	Nanotechnology challenges, characterization complexity, surface reactivity, and biocompatibility concerns	Kulkarni, Ayachit, and Aminabhavi (2022)
Microfluid biosensors	Electrodes Optical fibers Piezoelectric crystals	Reduced sample and reagent use, enhanced sensitivity, rapid analysis, portability, multi-function integration, automation, and high throughput	Complex fabrication, surface effects, bubble formation, limited sample volume, and challenges in scaling up	Choi et al. (2011)

While these modifications improve selectivity, sensitivity, stability, and resistance to fouling, improper immobilization can reduce activity, specificity, and biocompatibility. Recently, high-sensitivity electrochemical biosensors have utilized carbon-based and 2D materials for electrode modifications, showing promise for improved functionality. However, challenges like surface fouling and instability of modified materials remain significant obstacles.

Label-free electrochemical biosensors are increasingly popular for detecting biomarkers, DNA, and proteins due to their simplicity, bypassing complex labeling procedures. Neurotransmitters like dopamine are critical biomarkers for neurological diseases, including Parkinson's, Alzheimer's, and depression. Given dopamine's low concentration in biological fluids (10^{-8} to 10^{-6} M), there is a demand for ultrasensitive and highly selective electrochemical biosensors for real-time clinical diagnostics. Recent research suggests perovskite materials, such as $NaNbO_3$, $SrPdO_3$, $LaFeO_3$, $LaCoO_3$, and β-$NaFeO_2$, as functional materials for electrochemical biosensors targeting dopamine. Though not traditionally known for electrocatalytic activity, these perovskites possess active sites that facilitate the electrooxidation of dopamine to dopamine-o-quinone, making them promising candidates for sensitive dopamine detection (Durai and Badhulika 2022).

11.2.2 Microfluid Nanobiosensors

Microfluidic technology has advanced rapidly in recent years, offering significant advantages like high throughput, minimal sample requirements, and integration of multiple functions. These benefits have made microfluidic-based biosensors essential in fields such as clinical care, environmental monitoring, and food safety. Compact and portable, these biosensors enable precise control of incoming liquids, minimizing interference between reaction systems and allowing for semi-automatic or fully automatic quantification, ensuring highly accurate and reliable measurements. These features are especially valuable for clinical diagnostics.

Lateral flow immunoassay (LFIA) sensors are among the most common microfluidic sensors used today. LFIA sensors are favored for their low cost, high sensitivity, specificity, portability, flexibility, and excellent biocompatibility. Driven by capillary forces, they detect fluids without requiring additional tools or trained personnel. LFIA sensors can differentiate targets by arranging multiple test lines on a single device for spatial resolution. Affordable, self-contained, and easy to modify, paper-based LFIA sensors are among the most popular and widely adopted applications in this category.

Microfluidic chips, frequently employed for disease detection, utilize precisely designed channels fabricated through technologies like hot pressing, injection molding, photolithography, and laser etching, tailored to the properties of polymeric substrates. These channels enable automated multistep reactions on a single chip, enhancing detection capabilities. For instance, Sun et al. 2023 developed a dual-detection microfluidic device for p-tau181 and Aβ1–42 proteins using polystyrene/Au nanoparticle (PS/Au) substrates, leveraging LSPR to amplify SERS signals for ultrasensitive detection (<100 fg/mL). This device allows quick, simultaneous biomarker detection and is suitable for home-based Alzheimer's disease (AD) screening. Similarly, Chou et al. 2008 created a nanofluidic biosensor that trapped gold nanoparticles and Aβ1–40 at nanochannel openings, forming dense clusters to enhance detection sensitivity and repeatability. Additionally, droplet-based microfluidics offers a promising approach for high-throughput biomarker screening, using isolated droplets as independent bioreactors to enable controlled and parallel reactions. Although limited research exists on droplet-based microfluidics for AD biomarker detection, its potential remains significant (Gao et al. 2023).

11.2.3 Optical Biosensors

Optical biosensors detect changes in optical properties, such as wavelength, fluorescence, refractive index, and absorption, that occur during biological reactions. Localized surface plasmon resonance (LSPR) is one of the most sensitive, real-time, label-free optical techniques, detecting changes in the surface plasmon resonance (SPR) angle influenced by the refractive index and other factors. LSPR's high sensitivity makes it ideal for monitoring biological interactions in real time.

Recent advancements in optical biosensors have contributed to the diagnosis and treatment of Alzheimer's disease (AD). For example, Palladino, Aura, and Spoto (2016) immobilized antibodies on biosensor chips and used SPR to study the development of amyloid-β (Aβ) plaques in real time, offering an advantage over methods that rely on fluorophores, which could alter aggregation processes. This method aids in creating therapeutic drugs and provides a novel tool for researching aggregation inhibitors. Additionally, Yi et al. (2018) developed an SPR-based biosensor to measure and genotype apolipoprotein E (ApoE), a primary genetic risk factor for AD. They used biotinylated DNA probes that complement the ApoE ε4 allele sequence, modifying gold (Au) films. After hybridization, streptavidin was added following enzymatic cleavage, binding to the biotinylated probes to detect single-base mismatched sequences.

The biotin–streptavidin binding was inhibited for the complementary sequence in the SPR-based biosensor, resulting in distinct SPR signals. This approach allows DNA sequences to be selectively detected at femtomolar concentrations, demonstrating high sensitivity. Doong, Lee, and Anitha (2010) developed an enzymatic optical biosensor capable of analyzing multiple Alzheimer's disease (AD) biomarkers simultaneously. In this sensor, fluorescent dyes are bound to sol–gel substrates and are excited to produce a signal due to pH shifts caused by enzyme–analyte interactions. The biosensor demonstrated low limits of detection (LODs) and high performance for analytes like glutamate (Glu), Aβ, and acetylcholine (ACh). Additionally, Zhang et al. 2021 created a photoelectrical immunosensor based on SnO_2/$CdCo_3$/CdS nanocomposites, which achieved a high photocurrent intensity for Aβ detection. This immunosensor showed a low detection limit of 50 fg mL^{-1} and a broad linear range from 0.1 pg mL^{-1} to 100 ng mL^{-1}, demonstrating excellent sensitivity and a wide detection range for Aβ.

11.2.4 Nanobiosensors

Nanobiosensors are transforming clinical diagnostics by offering exceptional selectivity and sensitivity in detecting electrical signals generated by specific molecules. These devices integrate biological detection elements, sensor systems, and transducers, providing a significant advantage over traditional diagnostic methods. The ability to detect electrical signals with high precision allows for more accurate and reliable clinical diagnoses.

In Parkinson's disease (PD) diagnostics, nanobiosensors have proven to be particularly effective. These sensors focus on biomarkers such as dopamine, a key neurotransmitter associated with the disease's progression. Recent innovations in nanofabrication have led to advances in detection accuracy. For example, carbon paste electrodes reinforced with carbon nanotubes and silver nanoparticles allow for simultaneous detection of ascorbic acid and dopamine, offering high sensitivity and micromolar detection limits (Kalinke et al. 2023). Additionally, 3D zinc oxide (ZnO) nanowire arrays integrated with graphene foam have been developed to enhance electrical conductivity, enabling the detection of dopamine and uric acid with nanomolar precision. Since Parkinson's patients often exhibit reduced uric acid levels, this biomarker has significant diagnostic potential (Zhang et al. 2021).

The development of aptasensors for alpha-synuclein oligomers marks another breakthrough in PD diagnostics. These sensors, which use colorimetry and electrochemical impedance spectroscopy (EIS), achieve exceptional sensitivity, with detection limits as low as 0.64 femtomolar. They are capable of identifying minute biomarkers in human serum, making them useful for early-stage PD diagnosis. Additionally, molecular fingerprint polymer-based biosensors and immunosensors leveraging EIS and anti-PARK7/DJ-1 antibodies have shown promise for detecting alpha-synuclein and PARK7/DJ-1 biomarkers. These label-free electrochemical sensors offer real-time, non-invasive, and interference-resistant diagnostics, providing a robust method for identifying biomarkers associated with Parkinson's and other neurological conditions (Dhahi et al. 2024).

11.3 PRINCIPLES OF BIOSENSING IN NEUROLOGY

A biosensor is "a self-contained integrated device which is capable of providing specific quantitative or semi-quantitative analytical instrumentation using a biological recognition element (biochemical receptor) which is in direct spatial contact with a transducer element," according to the International Union of Pure and Applied Chemistry (IUPAC) (Bhalla et al. 2016).

There are three primary parts in a biosensor. A biosensor's bioreceptor—which is made up of biomolecules like enzymes, proteins, nucleic acids, aptamers, antibodies, organelles, microorganisms, or cell receptors—is what gives it selectivity towards the target analyte. The transducer—which can be optical, electrochemical, physicochemical, piezoelectric, mechanical, or thermal—converts the biorecognition event proportionate to the concentration of the target analyte into a measurable electrical signal (Chircov et al. 2020). The electronic system is made up of a processor, an amplifier, and a display unit. High ligand specificity and selectivity, high throughput capacity, dynamic range, quick detection, ease of engineering and operation, cost effectiveness, and low power needs are typically included in their design.

The basic idea of biosensor operation is molecular recognition, in which the biorecognition component interacts with the target analyte in a selective manner. The foundation of the biosensor's capacity to recognize and detect particular molecules is this exact and targeted interaction. Because of their high level of molecular specificity, biosensors can detect and capture specific molecules in the midst of a complex sample with remarkable accuracy. The biosensors use signal transduction mechanisms to convert the biological contact into a detectable signal after molecular recognition. In order to transform the binding event into a format appropriate for quantitative analysis, this procedure is necessary. For example, enzymatic processes are frequently used in biosensors to accomplish this signal transmission. In these cases, a specific biochemical reaction is started when the analyte attaches to the enzyme reaction that results in an electrochemical signal. The presence of the analyte is quantitatively represented by this quantifiable signal, enabling accurate detection. Therefore, signal transduction processes are essential for connecting the biological recognition event with a concrete output, making it easier to interpret biosensor results in a useful and significant manner. Understanding the interactions between the analyte and sensor is essential to comprehending how biosensors work. The analyte triggers a sequence of reactions in the sensor when it attaches to the biorecognition component. The foundation for the careful design and optimization of biosensors for a variety of applications is a thorough understanding of these analytesensor interactions (Hemdan et al. 2024).

Bioreceptors are components that interact with the target analyte or biomarker associated with a neurological disorder. These bioreceptors may consist of nucleic acids, enzymes, antibodies, or even whole cells. The transducer then converts this biological interaction into a measurable signal, which can be thermal, optical, or electrical, and this signal is then analyzed by a signal processor to produce a quantitative or qualitative output (Evtugyn et al. 2017).

The method involves three key stages: recognition, transduction, and output, as shown in Figure 11.1. In the recognition phase, the bioreceptor binds to the target biomarker, which could be a specific protein or metabolite associated with a neurological condition. This is followed by transduction, where the transducer detects a change in the bioreceptor triggered by the binding event; for example, if an electrochemical transducer is used, this binding may alter the voltage or current. Lastly, in the output phase, the signal is processed and displayed, often immediately, providing information on the presence and concentration of the biomarker.

11.4 APPLICATIONS OF BIOSENSORS IN SPECIFIC NEUROLOGICAL CONDITIONS

Biosensors are increasingly valuable in neurology, particularly for diagnosing, monitoring, and treating neurological disorders. By enabling real-time, non-invasive, and

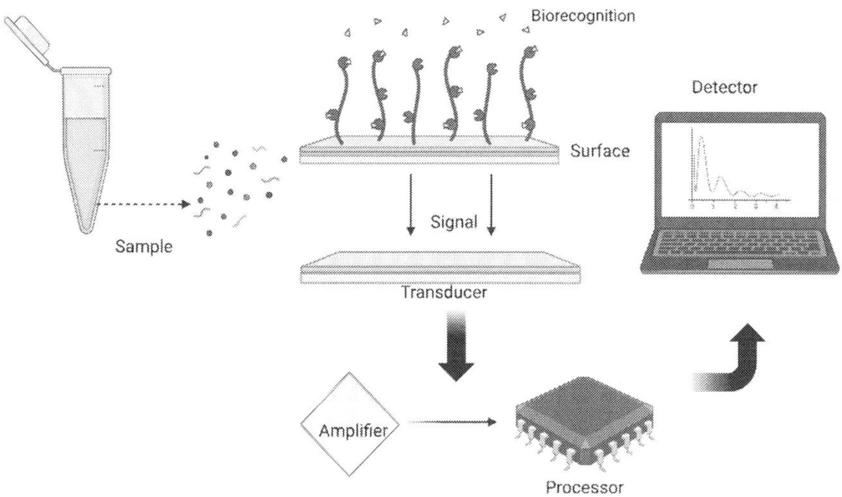

FIGURE 11.1 Principle of biosensor in neurological disorders.

continuous monitoring, biosensors are transforming neurological care. These capabilities support more effective symptom management, facilitate early diagnosis, and enable personalized treatment plans for improved patient outcomes. Table 11.2 provides an overview of neurology-related biosensors, highlighting their applications and examples.

11.4.1 Biosensors for Parkinson's Disease Diagnostics

Parkinson's disease (PD) is a prevalent neurodegenerative disorder, primarily affecting older adults. It has a complex pathogenesis, and diagnosis remains challenging due to the lack of specific laboratory tests, with early symptoms often overlapping with those of other conditions, complicating differential diagnosis (Mobed et al. 2021). As the global population ages, PD cases are expected to increase by over 50% by 2030, posing a significant public health concern. The hallmark symptoms of PD, including bradykinesia, rigidity, and tremor, arise from the accumulation of α-synuclein deposits and the degeneration of dopaminergic neurons in the substantia nigra pars compacta.

PD diagnosis typically occurs after the onset of synucleinopathy, but early biomarker-based detection is crucial for improving patient outcomes. The degeneration of dopaminergic neurons and striatal terminals in the nigrostriatal pathway leads to a depletion of dopamine (DA), resulting in the motor dysfunction characteristic of PD, such as bradykinesia and tremor. Monitoring DA levels holds promise for the early detection of dopaminergic denervation, potentially allowing for the development of treatments that could delay or prevent the progression of PD. However, accurately measuring DA is challenging due to the low concentrations of the neurotransmitter in biological fluids, including plasma (0.01–1 μM) and cerebrospinal

TABLE 11.2
Types of Neurology-Related Biosensors and their Applications

Type of Biosensor	Example	Materials required	Application	References
Electrochemical biosensors	Glucose biosensors	Carbon nanotubes, gold electrodes, conductive polymers	Parkinson's and Alzheimer's monitoring	Choi, Choi, and Yoon (2023)
Optical biosensors	Surface plasmon resonance (SPR) biosensors	Plasmonic nanomaterials, gold nanoparticles	Detection of amyloid-beta proteins	Lee et al. (2018)
Magnetic biosensors	Magnetic immunoassays	Magnetic nanoparticles, iron oxide	Detection of neuron-specific enolase (NSE) in traumatic brain injury	Salehirozveh, Dehghani, and Mijakovic (2024)
Enzyme-based biosensors	Dopamine and acetylcholine sensors	Carbon electrodes, enzymes (e.g., acetylcholinesterase, tyrosinase)	Alzheimer's and Parkinson's monitoring	Fredj et al. (2023)
Immunosensors	Antibody-based biosensors	Antibodies, gold nanoparticles, graphene	Detection of amyloid-beta in Alzheimer's	Ranjan and Khan (2022)
Field-effective transistor (FET) biosensors	FET-based dopamine biosensors	Graphene, silicon, carbon nanotubes	Detection of Parkinson's disease	Choi, Choi, and Yoon (2023)
Piezoelectric biosensors	Resonant mass sensors	Piezoelectric materials (e.g., PZT), quartz crystals	Early detection of Alzheimer's	Cai et al. (2024)
Electroencephalography (EEG) sensors	Portable EEG headbands	Silver/silver chloride, dry or wet electrodes	Detection of seizure in epilepsy	Biondi et al. (2022)
Wearable biosensors	EEG headsets	Conductive fabrics, flexible polymers, electrodes	Detection of Parkinson's disease	Song et al. (2023)
Fluorescence biosensors	Fluorescent probes	Fluorescent dyes, quantum dots, optical fibers	Amyloid-beta detection in Alzheimer's disease	Liu, Liu, and Liu (2023)

fluid (CSF) (0.02–0.07 nM). Advances in electrochemical sensing platforms based on oxidation processes have shown promise for DA detection, offering high sensitivity, selectivity, and cost-effectiveness for in-situ monitoring in biological environments. For instance, Asif et al. (2017) used a Zn-NiAl LDH/rGO superlattice electrode to monitor DA release from SH-SY-5Y cells, and Li et al. (2013) developed a nanoelectronic biosensor for tracking DA in PC12 cells. The miniaturization of sensors has enabled the detection of DA in small sample volumes such as CSF, though DA sensing remains largely a research tool for studying complex biological systems.

Graphene-based sensors, particularly those incorporating nanomaterial-modified electrodes like graphene oxide (GO) and reduced graphene oxide (rGO), have emerged as powerful tools for electrochemical DA detection. Graphene's unique structure enhances its conductivity and provides abundant active sites for the detection of analytes, making it ideal for DA sensing. Butler et al. (2021) developed a graphene ink-based electrochemical sensor with a detection limit as low as 1 nM. Surface modifications to graphene improve sensor sensitivity and selectivity by enhancing its electrocatalytic activity, surface area, and conductivity

Organic electrochemical transistors (OECTs), such as those developed by Wang et al. (2018), also provide highly precise DA sensing with a 1 nM detection limit. These devices, when used with a lock-in amplifier, offer stable, noise-resistant signals over a wide frequency range. Additionally, flexible DA sensors, like the Pt–Au/LIG/PDMS sensor introduced by Hui et al. (2019), have been designed for continuous DA monitoring in self-care applications. This sensor, with a detection limit of 75 nM, holds promise for real-time monitoring in practical settings.

Fluorescence biosensors are another promising approach for DA detection due to their high sensitivity and rapid response times. Techniques such as fluorescence resonance energy transfer (FRET), chemiluminescence, and fluorescent probes are being employed to detect DA. Graphene oxide (GO) derivatives, with their conjugated structure, quench the fluorescence of adsorbed dyes, aiding in DA detection. For example, Teniou, rhouati, and Catanante (2021) developed a GO-based fluorescent aptasensor for DA detection, demonstrating a linear relationship between fluorescence recovery and DA concentrations from 3 to 1680 nM, with a detection limit of 0.031 nM. Fluorescence-based methods offer a cost-effective, sensitive, and simple approach for DA detection in living cells, with applications in real-time monitoring of human blood samples.

These advances in biosensor technology, including electrochemical and fluorescence-based methods, offer significant potential for early detection and monitoring of Parkinson's disease, contributing to improved diagnosis, treatment, and patient care.

11.4.2 Biosensors for Alzheimer's Disease Diagnostics

Alzheimer's disease (AD) primarily affects individuals over the age of 65, evolving from early cognitive impairment and personality changes to severe dementia and eventual mortality. The hallmark symptoms—memory loss, cognitive dysfunction, and a gradual decline in daily functioning—make AD a significant global public health challenge, especially given the aging population. Early diagnosis is crucial, and numerous methods have been proposed to detect AD in its early stages. Among these, biosensors have emerged as a promising tool for real-time identification of AD biomarkers, enabling early intervention.

Biosensors are designed to detect key AD biomarkers such as tau protein, amyloid β peptides (Aβ), and apolipoprotein E4 (APOE4). These biomarkers are typically measured in blood and cerebrospinal fluid (CSF), offering crucial insight into the presence and progression of the disease. Tau, a microtubule-associated protein (MAP), plays an essential role in stabilizing microtubules for axonal transport. However, when tau becomes hyperphosphorylated, it forms neurofibrillary tangles (NFTs), disrupting cellular function and contributing to neurodegeneration. Similarly, extracellular Aβ plaques accumulate in the brain, further impairing neuronal communication and function (Šimić et al. 2016). Both tau tangles and Aβ plaques are therefore considered key pathological markers for AD diagnosis and monitoring.

Advancements in biosensor technology have significantly improved the sensitivity and specificity of AD biomarker detection, allowing for more accurate and early diagnosis. Optical, electrochemical, and piezoelectric biosensors have been developed for the detection of tau protein, particularly phosphorylated tau (P-Tau), which is found in elevated levels in cerebrospinal fluid (CSF) samples of AD patients. These biosensors offer a promising method for monitoring disease progression and facilitating early interventions (Ameri et al. 2020).

The tau protein is encoded by the MAPT gene located on chromosome 17q21, and its splicing is regulated as neurons mature (Wang and Mandelkow 2016). The role of the MAPT gene in tau's structural and functional variability highlights its potential as a diagnostic marker for AD. With the continuous development of biosensors, the real-time monitoring of tau and Aβ concentrations could play a transformative role in the early detection and intervention of AD, offering the possibility to delay the progression of this devastating disease.

11.5 FUTURE DIRECTIONS AND POTENTIAL IMPACT OF BIOSENSORS IN NEUROLOGY

Biosensors are increasingly valued for their potential in medical diagnostics due to their ease of use, scalability, and cost-effectiveness, positioning them as transformative tools in healthcare. Their integration into wearable devices enables continuous, non-invasive monitoring, particularly benefiting the elderly by bridging the gap between healthcare providers and self-care. This shift reduces hospital visits, enhances patient convenience, and creates new opportunities in wellness, fitness, and athletics.

In healthcare, biosensors have revolutionized personalized medicine, enabling early diagnosis of diseases like Parkinson's and Alzheimer's by detecting subtle biomarkers. This facilitates early intervention, improving patient outcomes and quality of life. Additionally, they enable precision medicine, tailoring treatments to individuals to reduce side effects and enhance therapeutic efficacy. Through continuous monitoring, biosensors provide real-time data on vital signs, biomarkers, and disease progression, shifting healthcare from reactive to proactive and remote, ultimately lowering costs.

Early detection of neurological diseases (NDDs) relies on various diagnostic techniques, including biochemical testing, brain imaging, clinical evaluations, and genetic testing. Among these, biochemical tests that measure changes in proteins, RNA, small molecules, and metabolic products in blood and CSF are gaining prominence. Electrochemical sensors, particularly those using nanomaterials and electroactive polymers, have emerged as highly sensitive, stable, and cost-effective tools for detecting biomarkers. These sensors can be miniaturized and integrated into portable devices, facilitating more efficient and accessible monitoring of neurological diseases (Chircov et al. 2020; Gautam 2022).

The ongoing innovation in biosensor technology, particularly when integrated with the Internet of Things (IoT), is shaping the future of healthcare. This combination enables continuous, personalized care and could potentially detect biochemical and cellular changes in the brain at early stages, paving the way for more effective interventions in neurodegenerative diseases.

11.5.1 Advancements in Biosensor Technology

Biosensors are becoming more flexible, compact, wearable, and bioadaptive thanks to advancements in technology, enabling them to meet the needs of clinical medicine and address persistent health challenges (Ramesh et al. 2022). Patients with neurological diseases often face mood changes, mobility difficulties, cognitive impairments, and even fatal outcomes due to abnormalities in the central or peripheral nervous systems. Advanced biosensing devices have been approved for the treatment and monitoring of various nervous system disorders.

11.5.1.1 Brain–Computer Interfaces

Brain–computer interfaces (BCIs) enable direct communication between the brain and external devices, facilitating the exchange of information. This technology is crucial in clinical settings, allowing patients with severe neurological, muscular, psychiatric, and other disorders to regain lost physiological functions. BCIs also assist in monitoring various medical conditions. They have evolved into a multidisciplinary research area, integrating clinical medicine, computer science, neurology, and microelectronics. Current advanced BCI devices are used in clinical practice, with 81% of neurologists believing BCIs could significantly improve patients' quality of life, and 70% recognizing their benefits in clinical settings (Awuah et al. 2024).

11.5.1.2 Wearable Biosensors

Motion biosensors provide valuable insights into the motor functions of patients with clinical diseases, aiding in disease progression monitoring and offering movement data that is difficult to capture in traditional clinical studies. Wearable biosensors have gained popularity for motion monitoring due to their high sensitivity to variations in human activity and their potential for long-term use with enhanced comfort (Asif et al. 2017). These sensors are useful for prognosis evaluation, efficacy assessment, and movement control. Beyond illness-related movement, wearable sensors can assess motor potential in paralyzed patients, supporting neuromuscular activity. Despite challenges like biocompatibility, comfort, customization, and data processing efficiency, wearable biosensors hold promise for individuals with mobility disorders.

In neuroscience, sophisticated biosensors are widely used with limitations based on detection targets and objectives. These devices require clinical applications and ethical approvals. Brain–computer interfaces (BCIs), a product of the fourth industrial revolution, have gained importance as both therapeutic techniques and supplementary tools. Brain pacemakers (DBS) are crucial in neurosurgery, helping patients with severe brain injuries regain mobility and slowing the progression of neurodegenerative diseases. With the advancements of artificial intelligence and precision medicine, inverse neural reactions may enhance treatment efficacy and optimize personalized therapies (Meng et al. 2024).

11.5.1.3 Miniaturization and Portability

The miniaturization of biosensors, driven by advancements in microfabrication techniques, is a significant trend in biosensor technology. This is particularly impactful in diagnostics and point-of-care applications, resulting in smaller, more portable devices (Kulkarni, Ayachit, and Aminabhavi 2023). Microfabrication allows for the precise manufacturing of small devices, with microfluidics and microelectromechanical systems (MEMS) playing a crucial role in shaping biosensors at the micro and nanoscale. MEMS enables the integration of mechanical components, actuators, sensors, and electronics at a microscopic scale, while microfluidics controls small fluid volumes inside microchannels to optimize reaction kinetics. Miniaturized biosensors benefit from improved reaction kinetics, enabling rapid response times with minimal sample amounts. This is especially important in point-of-care settings, where small sample sizes and quick results are critical, making these compact devices ideal for facilitating timely medical decisions.

11.5.1.4 Nanotechnology in Biosensors

Nanotechnology has greatly enhanced biosensor performance and design, introducing nanomaterials that improve sensitivity and broaden detection capabilities. Nanowires, nanoparticles, and nanotubes have transformed biosensing, especially in medical applications (Tang et al. 2022). The integration of materials like gold nanoparticles has notably boosted the electrochemical performance of biosensors by offering a high surface-to-volume ratio. This facilitates better biomolecule immobilization and enhanced interactions, resulting in increased sensitivity and improved binding efficiency for target molecules. The use of nanomaterials in biosensors has advanced healthcare technologies, enabling precise biomarker detection and reliable drug monitoring platforms.

11.5.2 Non-Invasive Monitoring

Non-invasive monitoring has become a preferred alternative to invasive procedures, which often involve painful methods like lancets or finger pricks for blood or interstitial fluid collection. Non-invasive techniques, however, use biofluids already outside the skin barrier, such as sweat, saliva, tears, and urine. Wearable wireless biosensors (WWBs), powered by energy-harvesting technologies, eliminate the need for external power sources, unlike traditional battery-powered devices. These biosensors are cost-effective, energy-efficient, and flexible, making them ideal for continuous monitoring and improving disease management (Salim and Lim 2019).

Low-modulus flexible substrates, such as elastomers and polymer composites, allow for direct skin attachment and seamless movement with the body. Conformable biosensors, designed to enhance the reliability of non-invasive sample collection, are useful for self-health management and point-of-care testing. The development of wearable biosensors requires careful selection of substrates, biocompatibility, active materials, and cost-effective production methods. Stretchability is a key focus for skin-compatible electronics, with elastomers and composites providing mechanical resilience and scalability. Research into self-healing electronic materials, which rely on dynamic intermolecular interactions in polymeric materials, has also gained attention.

11.5.2.1 EEG (Electroencephalography)

EEG sensors measure the brain's electrical activity using external equipment. While traditional EEG is non-invasive, it is limited in frequency range and spatial resolution. Advanced technologies like BCIs, electrocorticography (ECoG), and cortical microelectrodes offer higher spatial and temporal resolution, but still require electrode arrays placed on the cortical surface or within the brain. EEG is essential for diagnosing neurological disorders such as epilepsy, which is marked by abnormal brain firing patterns. Symptoms include dyskinesia, mental disturbances, sensory changes, and loss of consciousness. In epilepsy, EEG readings typically show spike waves, sharp waves, or spike-slow waves, which are key to confirming the diagnosis (Supriya et al. 2023).

In neurodegenerative diseases like Alzheimer's disease (AD), EEG provides valuable insights. AD is characterized by aphasia, memory loss, and cognitive decline. In AD patients, EEG often shows a reduction in the coherence of rapid brainwave activity. EEG changes are also seen in Parkinson's disease (PD) and Lewy body disease (DLB), aiding in disease progression monitoring. As a traditional and essential tool, EEG remains vital for diagnosing and understanding complex neurological conditions (Meng et al. 2024).

11.5.2.2 Salivary Biomarkers

Salivary analysis is emerging as a promising method for detecting neurodegenerative and demyelinating disorders, especially those linked to aging. The advantages of saliva collection include ease of use, non-invasive sampling, reproducibility, and stress-free procedures, making it an attractive option for early disease detection and intervention. Recent innovations include mouthguard-based devices for saliva testing, which allow for sample collection without invasive methods. Another development is a wearable, biocompatible biosensor made from a conformal radiofrequency (RF) construct using modified PNIPAM hydrogel and porous silk film. This biosensor monitors fluid parameters like alcohol content, salinity, pH, and temperature during swallowing, with the hydrogel enclosed between two split ring resonators fixed on a tooth (Tseng et al. 2018).

11.5.3 Early Diagnosis and Intervention

Parkinson's disease (PD) is a progressive neurodegenerative disorder that affects the motor system, leading to symptoms like bradykinesia, tremors, stiffness, and postural instability. It impacts over six million people worldwide and is marked by the degeneration of dopaminergic neurons in the substantia nigra (Mobed et al. 2021). Early diagnosis is vital for effective management, but current methods are limited by low sensitivity, high costs, and complex technologies. Electrochemical biosensors have emerged as promising tools for early PD detection, converting biological reactions into measurable electrical signals to detect specific PD biomarkers.

Advances in material science and nanotechnology have led to novel sensing platforms with improved sensitivity and selectivity for PD biomarkers. These platforms target key biomarkers such as microRNAs (miRNAs), dopamine (DA), and alpha-synuclein (α-syn). Graphene-based biosensors have shown high sensitivity for detecting dopamine, while gold nanoparticle-enhanced graphene immunosensors offer ultra-sensitive detection of alpha-synuclein. Electrochemical aptasensors and nanobiosensors for miR-195 are also showing potential for early PD diagnosis. The integration of nanomaterials like carbon nanotubes, quantum dots, and gold nanoparticles has further boosted the sensitivity and electrochemical activity of these biosensors. These innovations promise more reliable, cost-effective, and rapid methods for early PD diagnosis, improving disease management and patient outcomes. Ongoing research is essential for the commercialization and clinical integration of these biosensors, which could revolutionize PD diagnosis and treatment (Khatami et al. 2025)

REFERENCES

Ameri, Mehrdad, Zahra Shabaninejad, Ahmad Movahedpour, Amirhossein Sahebkar, Soheila Mohammadi, Saereh Hosseindoost, Mohammad Saeid Ebrahimi, Amir Savardashtaki, Mohammad Karimipour, and Hamed Mirzaei. 2020. "Biosensors for Detection of Tau Protein as an Alzheimer's Disease Marker." *International Journal of Biological Macromolecules* 162 (November): 1100–1108. doi: 10.1016/j.ijbiomac.2020.06.239.

Asif, Muhammad, Hongwei Liu, Ayesha Aziz, Haitao Wang, Zhengyun Wang, Muhammad Ajmal, Fei Xiao, and Hongfang Liu. 2017. "Core-Shell Iron Oxide-Layered Double Hydroxide: High Electrochemical Sensing Performance of H_2O_2 Biomarker in Live Cancer Cells with Plasma Therapeutics." *Biosensors and Bioelectronics* 97 (November): 352–59. doi: 10.1016/j.bios.2017.05.057.

Awuah, Wireko Andrew, Arjun Ahluwalia, Kwadwo Darko, Vivek Sanker, Joecelyn Kirani Tan, Pearl Ohenewaa Tenkorang, Adam Ben-Jaafar, et al. 2024. "Bridging Minds and Machines: The Recent Advances of Brain-Computer Interfaces in Neurological and Neurosurgical Applications." *World Neurosurgery* 189 (September): 138–53. doi: 10.1016/j.wneu.2024.05.104.

Balaban Hanoglu, Simge, Duygu Harmanci, Nursima Ucar, Serap Evran, and Suna Timur. 2023. "Recent Approaches in Magnetic Nanoparticle-Based Biosensors of MiRNA Detection." *Magnetochemistry* 9 (1): 23. doi: 10.3390/magnetochemistry9010023.

Bell, Kathryn M., and Steven E. Kornguth. 2007. "Biosensors for Neurological Disease." In *Handbook of Biosensors and Biochips*. Wiley. doi:10.1002/9780470061565.hbb139.

Bhalla, Nikhil, Pawan Jolly, Nello Formisano, and Pedro Estrela. 2016. "Introduction to Biosensors." *Essays in Biochemistry* 60 (1): 1–8. doi:10.1042/EBC20150001.

Biondi, Andrea, Viviana Santoro, Pedro F. Viana, Petroula Laiou, Deb K. Pal, Elisa Bruno, and Mark P. Richardson. 2022. "Noninvasive Mobile EEG as a Tool for Seizure Monitoring and Management: A Systematic Review." *Epilepsia* 63 (5): 1041–63. doi:10.1111/epi.17220.

Butler, Derrick, David Moore, Nicholas R. Glavin, Joshua A. Robinson, and Aida Ebrahimi. 2021. "Facile Post-Deposition Annealing of Graphene Ink Enables Ultrasensitive Electrochemical Detection of Dopamine." *ACS Applied Materials & Interfaces* 13 (9): 11185–94. doi:10.1021/acsami.0c21302.

Cai, Xianfa, Yiqin Wang, Yunqi Cao, Wenyu Yang, Tian Xia, and Wei Li. 2024. "Flexural-Mode Piezoelectric Resonators: Structure, Performance, and Emerging Applications in Physical Sensing Technology, Micropower Systems, and Biomedicine." *Sensors* 24 (11): 3625. doi:10.3390/s24113625.

Chen, Jiatao, Xiuting Ding, and Dongdong Zhang. 2024. "Challenges and Strategies Faced in the Electrochemical Biosensing Analysis of Neurochemicals in Vivo: A Review." *Talanta* 266 (Pt 1): 124933. doi: 10.1016/j.talanta.2023.124933.

Chircov, Cristina, Alexandra Cătălina Bîrcă, Alexandru Mihai Grumezescu, and Ecaterina Andronescu. 2020. "Biosensors-on-Chip: An Up-to-Date Review." *Molecules* 25 (24): 6013. doi:10.3390/molecules25246013.

Choi, Hye Kyu, Jin-Ha Choi, and Jinho Yoon. 2023. "An Updated Review on Electrochemical Nanobiosensors for Neurotransmitter Detection." *Biosensors* 13 (9): 892. doi:10.3390/bios13090892.

Choi, Seokheun, Michael Goryll, Lai Yi Mandy Sin, Pak Kin Wong, and Junseok Chae. 2011. "Microfluidic-Based Biosensors toward Point-of-Care Detection of Nucleic Acids and Proteins." *Microfluidics and Nanofluidics* 10 (2): 231–47. doi:10.1007/s10404-010-0638-8.

Chou, I-Hsien, Melodie Benford, Hope T. Beier, Gerard L. Coté, Miao Wang, Nan Jing, Jun Kameoka, and Theresa A. Good. 2008. "Nanofluidic Biosensing for β-Amyloid Detection Using Surface Enhanced Raman Spectroscopy." *Nano Letters* 8 (6): 1729–35. doi:10.1021/nl0808132.

Dai, Yifan, Alireza Molazemhosseini, and Chung Liu. 2017. "A Single-Use, In Vitro Biosensor for the Detection of T-Tau Protein, A Biomarker of Neuro-Degenerative Disorders, in PBS and Human Serum Using Differential Pulse Voltammetry (DPV)." *Biosensors* 7 (1): 10. doi:10.3390/bios7010010.

Damborský, Pavel, Juraj Švitel, and Jaroslav Katrlík. 2016. "Optical Biosensors." *Essays in Biochemistry* 60 (1): 91–100. doi:10.1042/EBC20150010.

Dhahi, Thikra S., Alaa Kamal Yousif Dafhalla, A. Wesam Al-Mufti, Mohamed Elshaikh Elobaid, Tijjani Adam, and Subash C.B. Gopinath. 2024. "Application of Nanobiosensor Engineering in the Diagnosis of Neurodegenerative Disorders." *Results in Engineering* 24 (December): 102790. doi: 10.1016/j.rineng.2024.102790.

Doong, Ruey-an, Pin-Shang Lee, and Kowthalam Anitha. 2010. "Simultaneous Determination of Biomarkers for Alzheimer's Disease Using Sol–Gel-Derived Optical Array Biosensor." *Biosensors and Bioelectronics* 25 (11): 2464–69. doi: 10.1016/j.bios.2010.04.005.

Durai, Lignesh, and Sushmee Badhulika. 2022. "Current Challenges and Developments in Perovskite-Based Electrochemical Biosensors for Effective Theragnostics of Neurological Disorders." *ACS Omega* 7 (44): 39491–97. doi:10.1021/acsomega.2c05591.

Evtugyn, Gennady A., Tibor Hianik, Georgia-Paraskevi Nikoleli, and Dimitrios P. Nikolelis. 2017. "Biosensors for Detection of Neurotransmitters and Neurodegenerative Related Diseases." In *Frontiers in Clinical Drug Research – CNS and Neurological Disorders*, 184–210. doi:10.2174/9781681085852117050007.

Fredj, Zina, Baljit Singh, Mohamed Bahri, Peiwu Qin, and Mohamad Sawan. 2023. "Enzymatic Electrochemical Biosensors for Neurotransmitters Detection: Recent Achievements and Trends." *Chemosensors* 11 (7): 388. doi:10.3390/chemosensors11070388.

Gao, Feng, Fang Li, Jianhao Wang, Hang Yu, Xiang Li, Hongyu Chen, Jiabei Wang, et al. 2023. "SERS-Based Optical Nanobiosensors for the Detection of Alzheimer's Disease." *Biosensors* 13 (9): 880. doi:10.3390/bios13090880.

Gautam, Akash. 2022. "Towards Modern-Age Advanced Sensors for the Management of Neurodegenerative Disorders: Current Status, Challenges and Prospects." *ECS Sensors Plus* 1 (4): 042401. doi:10.1149/2754-2726/ac973e.

Hemdan, Mohamed, Mohamed A. Ali, Ahmed S. Doghish, Sherif S. Abdel Mageed, Ibrahim M. Elazab, Magdy M. Khalil, Mostafa Mabrouk, Diganta B. Das, and Alaa S. Amin. 2024. "Innovations in Biosensor Technologies for Healthcare Diagnostics and Therapeutic Drug Monitoring: Applications, Recent Progress, and Future Research Challenges." *Sensors* 24 (16): 5143. doi:10.3390/s24165143.

Hui, Xue, Xing Xuan, Jiyoung Kim, and Jae Yeong Park. 2019. "A Highly Flexible and Selective Dopamine Sensor Based on Pt-Au Nanoparticle-Modified Laser-Induced Graphene." *Electrochimica Acta* 328 (December): 135066. doi: 10.1016/j.electacta.2019.135066.

Kalinke, Cristiane, Paulo Roberto De Oliveira, Craig E. Banks, Bruno Campos Janegitz, and Juliano Alves Bonacin. 2023. "3D-Printed Immunosensor for the Diagnosis of Parkinson's Disease." *Sensors and Actuators B: Chemical* 381 (April): 133353. doi: 10.1016/j.snb.2023.133353.

Khatami, Seyyed Hossein, Hamed Khanifar, Ahmad Movahedpour, Mortaza Taheri-Anganeh, Sajad Ehtiati, Hadi Khanifar, and Amir Asadi. 2025. "Electrochemical Biosensors in Early Detection of Parkinson Disease." *Clinica Chimica Acta* 565 (January): 120001. doi: 10.1016/j.cca.2024.120001.

Kudłak, Błażej, and Monika Wieczerzak. 2020. "Aptamer Based Tools for Environmental and Therapeutic Monitoring: A Review of Developments, Applications, Future Perspectives." *Critical Reviews in Environmental Science and Technology* 50 (8): 816–67. doi:10.1080/10643389.2019.1634457.

Kulkarni, Madhusudan B., Narasimha H. Ayachit, and Tejraj M. Aminabhavi. 2022. "Recent Advancements in Nanobiosensors: Current Trends, Challenges, Applications, and Future Scope." *Biosensors* 12 (10): 892. doi:10.3390/bios12100892.

Kulkarni, Madhusudan B., Narasimha H. Ayachit, and Tejraj M. Aminabhavi. 2023. "A Short Review on Miniaturized Biosensors for the Detection of Nucleic Acid Biomarkers." *Biosensors* 13 (3): 412. doi:10.3390/bios13030412.

Lee, Jin-Ho, Hyeon-Yeol Cho, Hye Kyu Choi, Ji-Young Lee, and Jeong-Woo Choi. 2018. "Application of Gold Nanoparticle to Plasmonic Biosensors." *International Journal of Molecular Sciences* 19 (7): 2021. doi:10.3390/ijms19072021.

Li, Bor-Ran, Ying-Jhu Hsieh, Yan-Xi Chen, Ya-Ting Chung, Chien-Yuan Pan, and Yit-Tsong Chen. 2013. "An Ultrasensitive Nanowire-Transistor Biosensor for Detecting Dopamine Release from Living PC12 Cells under Hypoxic Stimulation." *Journal of the American Chemical Society* 135 (43): 16034–37. doi:10.1021/ja408485m.

Liu, Xingyun, Yibiao Liu, and Qiong Liu. 2023. "Fluorescent Sensing Platforms for Detecting and Imaging the Biomarkers of Alzheimer's Disease." *Biosensors* 13 (5): 515. doi:10.3390/bios13050515.

Lowe, Christopher R. 2011. "The Future." In *International Review of Neurobiology*, 375–400. doi:10.1016/B978-0-12-387718-5.00015-8.

Meng, Zexuan, Yuchan Zhang, Lu Yang, Feng Yuan, Jian Wang, Jiajia Chen, Jun Liu, Guixue Wang, and Guangchao Zang. 2024. "Application of Advanced Biosensors in Nervous System Diseases." *Interdisciplinary Medicine* 2 (4). doi:10.1002/INMD.20240024.

Mobed, Ahmad, Siamak Razavi, Ali. Ahmadalipour, Seyed Kazem Shakouri, and Ghazal Koohkan. 2021. "Biosensors in Parkinson's Disease." *Clinica Chimica Acta* 518 (July): 51–58. doi: 10.1016/j.cca.2021.03.009.

Nguyen, Quoc-Thang, Lee F Schroeder, Marco Mank, Arnaud Muller, Palmer Taylor, Oliver Griesbeck, and David Kleinfeld. 2010. "An in Vivo Biosensor for Neurotransmitter Release and in Situ Receptor Activity." *Nature Neuroscience* 13 (1): 127–32. doi:10.1038/nn.2469.

Palladino, Pasquale, Angela M. Aura, and Giuseppe Spoto. 2016. "Surface Plasmon Resonance for the Label-Free Detection of Alzheimer's β-Amyloid Peptide Aggregation." *Analytical and Bioanalytical Chemistry* 408 (3): 849–54. doi:10.1007/s00216-015-9172-6.

Ramesh, Manickam, Ravichandran Janani, Chinnaiyan Deepa, and Lakshminarasimhan Rajeshkumar. 2022. "Nanotechnology-Enabled Biosensors: A Review of Fundamentals, Design Principles, Materials, and Applications." *Biosensors* 13 (1): 40. doi:10.3390/bios13010040.

Ranjan, Pushpesh, and Raju Khan. 2022. "Electrochemical Immunosensor for Early Detection of β-Amyloid Alzheimer's Disease Biomarker Based on Aligned Carbon Nanotubes Gold Nanocomposites." *Biosensors* 12 (11): 1059. doi:10.3390/bios12111059.

Salehirozveh, Mostafa, Parisa Dehghani, and Ivan Mijakovic. 2024. "Synthesis, Functionalization, and Biomedical Applications of Iron Oxide Nanoparticles (IONPs)." *Journal of Functional Biomaterials* 15 (11): 340. doi:10.3390/jfb15110340.

Salim, Ahmed, and Sungjoon Lim. 2019. "Recent Advances in Noninvasive Flexible and Wearable Wireless Biosensors." *Biosensors and Bioelectronics* 141 (September): 111422. doi: 10.1016/j.bios.2019.111422.

Shawky, Sherif M., Ahmed M Awad, Arwa A. Abugable, and Sherif F. El-Khamisy. 2018. "Gold Nanoparticles – an Optical Biosensor for RNA Quantification for Cancer and Neurologic Disorders Diagnosis." *International Journal of Nanomedicine* 13 (November): 8137–51. doi:10.2147/IJN.S181732.

Šimić, Goran, Mirjana Babić Leko, Selina Wray, Charles Harrington, Ivana Delalle, Nataša Jovanov-Milošević, Danira Bažadona, et al. 2016. "Tau Protein Hyperphosphorylation and Aggregation in Alzheimer's Disease and Other Tauopathies, and Possible Neuroprotective Strategies." *Biomolecules* 6 (1): 6. doi:10.3390/biom6010006.

Singh, Pranveer. 2016. "SPR Biosensors: Historical Perspectives and Current Challenges." *Sensors and Actuators B: Chemical* 229 (June): 110–30. doi: 10.1016/j.snb.2016.01.118.

Song, Zhimin, Shu Zhou, Yanxia Qin, Xiangjiao Xia, Yanping Sun, Guanghong Han, Tong Shu, Liang Hu, and Qiang Zhang. 2023. "Flexible and Wearable Biosensors for Monitoring Health Conditions." *Biosensors* 13 (6): 630. doi:10.3390/bios13060630.

Sun, Jianli, Zengliang Shi, Li Wang, Xinyi Zhang, Chunshan Luo, Jianyu Hua, Muyu Feng, Zaozao Chen, Mingliang Wang, and Chunxiang Xu. 2023. "Construction of a Microcavity-Based Microfluidic Chip with Simultaneous SERS Quantification of Dual Biomarkers for Early Diagnosis of Alzheimer's Disease." *Talanta* 261 (August): 124677. doi: 10.1016/j.talanta.2023.124677.

Supriya, Supriya, Siuly Siuly, Hua Wang, and Yanchun Zhang. 2023. "Epilepsy Detection from EEG Using Complex Network Techniques: A Review." *IEEE Reviews in Biomedical Engineering* 16: 292–306. doi:10.1109/RBME.2021.3055956.

Tang, Xiaofang, Yaping Zhu, Weijiang Guan, Wenjuan Zhou, and Peifa Wei. 2022. "Advances in Nanosensors for Cardiovascular Disease Detection." *Life Sciences* 305 (September): 120733. doi: 10.1016/j.lfs.2022.120733.

Tseng, Peter, Bradley Napier, Logan Garbarini, David L. Kaplan, and Fiorenzo G. Omenetto. 2018. "Functional, RF-Trilayer Sensors for Tooth-Mounted, Wireless Monitoring of the Oral Cavity and Food Consumption." *Advanced Materials* 30 (18). doi:10.1002/adma.201703257.

Wang, Naixiang, Yuzhe Liu, Ying Fu, and Feng Yan. 2018. "AC Measurements Using Organic Electrochemical Transistors for Accurate Sensing." *ACS Applied Materials & Interfaces* 10 (31): 25834–40. doi:10.1021/acsami.7b07668.

Wang, Yipeng, and Eckhard Mandelkow. 2016. "Tau in Physiology and Pathology." *Nature Reviews Neuroscience* 17 (1): 22–35. doi:10.1038/nrn.2015.1.

Yi, Xinyao, Yonghong Xia, Binrong Ding, Ling Wu, Shengqiang Hu, Zixiao Wang, Minghui Yang, and Jianxiu Wang. 2018. "Dual-Channel Surface Plasmon Resonance for Quantification of ApoE Gene and Genotype Discrimination in Unamplified Genomic DNA Extracts." *ACS Sensors* 3 (11): 2402–7. doi:10.1021/acssensors.8b00845.

Zhang, Jun, Shuai Shao, Desen Zhou, Quanlong Xu, and Tielin Wang. 2021. "ZnO Nanowire Arrays Decorated 3D N-Doped Reduced Graphene Oxide Nanotube Framework for Enhanced Photocatalytic CO2 Reduction Performance." *Journal of CO2 Utilization* 50 (August): 101584. doi: 10.1016/j.jcou.2021.101584.

12 Biosensors for Environmental Monitoring and Pollution Detection

Nutan Rani, Swati Rani, Sapna Yadav, Kalawati Saini and Dipak Maity

12.1 INTRODUCTION TO BIOSENSORS IN ENVIRONMENTAL MONITORING

Environmental contamination is one of the major issues affecting the planet, and it is the primary cause of human health issues and has detrimental economic effects [1]. Hazardous chemicals and contaminants are discharged into our environment daily from various sources. The three phases, air, water, and soil, are in equilibrium with one another, and therefore contamination of one will eventually affect the others. Different pollutants, such as inorganic, organic, and biological materials, can contaminate air, land, and water resources. The growing number of potentially hazardous ecological chemicals necessitates adopting quick and affordable analytical methods in comprehensive monitoring activities. To monitor the ever-increasing number of analytes of environmental significance as fast and economically as possible, as well as the potential to permit on-site field evaluation, the demand for reusable systems or tools for environmental monitoring has prompted the advancement of innovative technologies and appropriate approaches.

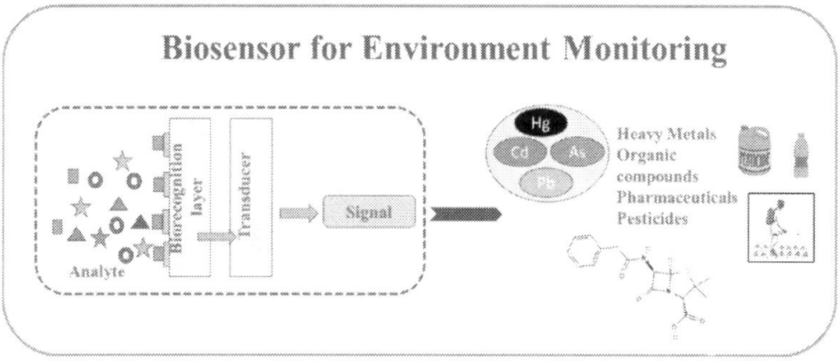

Typically, pollutants are detected using the traditional chromatographic and spectroscopic techniques, which are known for their excellent sensitivity and selectivity [2–4]. Most conventional approaches to analysis (such as chromatographic methods) have costly and time-consuming limitations that frequently make them unsuitable for routine use. Biosensors are an excellent substitute or supplementary analytical tool in this instance. Recently, biosensors have earned a lot of interest due to their use in detecting environmental pollution. The potential for portability, miniaturization, on-site operation, and the capacity to quantify contaminants in intricate matrices with minimum sample preparation are the primary benefits biosensors provide over traditional analytical methods. However, several of the established techniques are not as accurate or reproducible as conventional analytical methods; industry and regulatory organizations can use them to give sufficient details for routine sample screening and testing [5]. Currently, spot water sample collection, extraction, and laboratory-based instrumental analysis are the main methods for monitoring water quality. However, the spatiotemporal fluctuations in water parameters indicate the condition at the moment of sampling and do not provide more actual information [6]. In this situation, biosensors may help keep track of a polluted environment.

Biosensors are essential tools for detecting chemical and biological components because of their performance characteristics, which include high specificity and sensitivity, fast response, inexpensiveness, comparatively small size, and easy functioning. In addition to identifying particular chemicals, biosensors may also identify biological implications such as genotoxicity, toxicity, and cytotoxicity. With their high sensitivity and selectivity, low LOD values, quick analysis, affordability, and potential for miniaturization, biosensors are frequently used in analytical investigations of various substances, including heavy metals, drugs, and environmental contaminants in biomedical and environmental applications.

A biosensor is a discrete integrated device comprised of a transduction element and a biological recognition element close to one another. When the biorecognition element interacts with the target analyte, certain physicochemical changes are induced. The transduction element transforms the observed changes/events into an output signal that can be used. Most catalytic biosensors depend on electrochemical techniques, while affinity biosensors have often shown better compatibility for optical detection techniques [7].

Biosensors are categorized into different types based on the biorecognition principle or the signal transduction technique. Therefore, biosensors can be classified as enzymatic, non-enzymatic receptor, immunochemical, whole-cell, and DNA biosensors based on the biorecognition principle; and as optical, piezoelectric, electrochemical, and thermal sensors based on the transducing element.

Initially, biosensors were catalytic devices that combined transducers, transforming the biological response into a digital electronic signal combined with enzymes, tissues, cellular organelles, or whole microbes. Mostly optical, thermometric, or electrochemical transducers are employed in biosensors. Affinity biosensors (next-generation biosensors) are based on an identical set of measurement principles coupled with magnetic and piezoelectric transducers. Real-time data regarding the binding of cell receptors to their ligands, antibodies to antigens, and DNA and RNA to nucleic acids with complementary sequences is provided by affinity biosensors. Several literature studies are reported on various biosensors including optical, electrochemical, thermal, immuno, whole-cell, and DNA [8–11]. The biosensor technology has advanced to detect heavy metals, phenols, biological oxygen demand (BOD), organic compounds, and various pesticides [12, 13]. It is possible to successfully assess the bioavailable and hazardous fractions of metals and organic contaminants like Cd and toluene in groundwater and river water samples by employing biosensors [14, 15]. Additionally, biosensors were used to determine the bioavailability of pollutants including mercury and phenanthrene [16, 17], indicating the potential utility of biosensors for assessing the availability of contaminants in environmental samples.

There is an increasing demand for employing biosensors to analyze new and emerging contaminants, including hormones, antibiotics, and surfactants. Not only can biosensors identify certain compounds, but they may also identify biological impacts, such as toxicity or endocrine disruption, which are increasingly sought-after details about the sample. The unique sensitivity and specificity of enzymatic biosensors have made them widely utilized for identifying environmental contaminants. The biosensors monitored the pollutants in two different ways, that is, direct and indirect monitoring. For instance, several enzymatic biosensors have been developed for organophosphorus and carbamate pesticides, depending on the activity of choline oxidase and the inhibition of acetylcholinesterase (AChE) [18–20].

Numerous biosensors have been produced and are currently being developed in response to the various effects of contaminants on the biological system. The detection of mercury, heavy metals, phenolic compounds, organophosphorus, and carbamate insecticides is the main concern among hazardous chemicals because of their significant role in elevated pollutant levels. Presently, biosensors that detect heavy metals, pesticides, or pharmaceuticals demonstrate the benefits of biosensors, including their portability, simplicity of use, and time-saving capabilities [21]. Recent research has emphasized advancements in selectivity, stability, and sensitivity as well as their significance in environmental monitoring [22]. There are currently very few commercial biosensors that are used primarily for biochemical oxygen demand biosensing in the environmental context [23].

This chapter presents the latest developments, trends, and uses of biosensors in determining the quality of the environment and identifying different pollutants. It focuses on the use of biosensors in offering rapid, reliable, and affordable methods to monitor environmental contamination while addressing environmental issues and maintaining ecosystem sustainability.

12.2 RECENT PROGRESS IN BIOSENSORS FOR ENVIRONMENTAL MONITORING

12.2.1 Overview of Recent Advancements

The development of extremely reliable and efficient methods for detecting and screening environmental contaminants is currently gaining attention. Since controlling contaminants is a prerequisite for pollution restoration, researchers are searching for long-term solutions to environmental monitoring. Environmental pollution monitoring demands rapid, reliable, affordable, and compact technologies. This requirement can explain the recent developments in biosensing techniques for pollutant detection. In contrast to other biosensor types, such as biomedical ones, environmental monitoring biosensors have a nonaged phase because of the intricacy of the analysis, including the intricate ecological matrix that obstructs pollutant recognition. Heavy metals, pesticides, herbicides, nitrogenous compounds, biochemical oxygen demand (BOD), polychlorinated biphenyls, and several pathogens (including viruses and bacteria) are the main pollutants that may be gradually identified and eliminated utilizing biosensors.

12.2.2 Analytical Performances and Applications of Recent Biosensors

Numerous analytical applications for biosensors in environmental monitoring exist, such as:

- Detection of pollutants: Biosensors can detect various pollutants, such as toxins, pesticides, pathogens, and endocrine-disrupting chemical compounds, in soil, water, and air samples. In addition to identifying particular chemicals, biosensors may also identify biological impacts, such as toxicity or endocrine disruption, which are increasingly sought-after details about the sample [24].
- In situ monitoring: Biosensors can be utilized for in situ monitoring, eliminating the need to carry samples to a lab to monitor the environment.
- Response linearity: Biosensors give an accurate signal proportional to the concentration of the analyte.
- High selectivity: Biosensors exhibit high selectivity and need minimal sample preparation.

The usage of fluorescence tags, nanomaterials, aptamers, nucleotides, peptide arrays, and molecule-imprinted polymers are examples of recent developments in biosensor technology [25]. Figure 12.1 depicts the various types of biosensors utilized in environment monitoring.

The potential uses for different biosensors, which include microorganisms, polymers, and nanomaterials, are more extensive. Table 12.1 depicts the utilization of various biosensors for environment monitoring.

12.3 TYPES OF BIOSENSORS USED IN ENVIRONMENTAL MONITORING

12.3.1 Electrochemical Biosensors

Electrochemical biosensors provide high accuracy, sensitivity, and speed for on-site environmental investigations. They are popular due to their affordability, portability, and compact size. An electrochemical sensor's primary function is to deliver

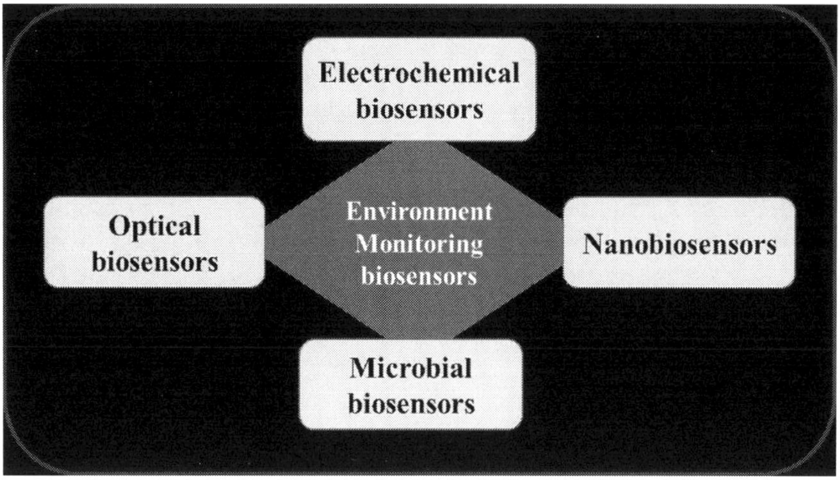

FIGURE 12.1 Various types of biosensors utilized in environment monitoring.

accurate, real-time data regarding the chemical constituents of its immediate vicinity. This is a crucial feature for the on-site monitoring of priority pollutants and other health risks. An electrochemical biosensor is an analytical tool that detects a target compound (analyte) in solution by integrating a particular biological recognition element (bioreceptor) into or closely relating to an electrode (transducer) that transforms an identifiable event into an electrical signal that can be detected. Real-time quantitative and qualitative assessments of an analyte are rendered possible by the produced signal. Miniaturized electrochemical cells often have a small analyte volume, which is essential when working with potentially hazardous substances and assists in making these devices perfect for environmental monitoring. In addition to facilitating on-site analysis, this method eliminates the possibility of sample manipulation and time delays that may arise during transportation to a central laboratory.

Electrochemical biosensors are classified into one of two categories: 1) affinity devices, which rely on selective binding between the biological sensing element and the target analyte, such as DNA or antibody-based biosensors; or 2) biocatalytic devices, which utilize enzymes, whole cells, or tissue-based biosensors to produce electroactive species upon interaction with a target analyte [46].

In recent years, there has been a significant increase in interest in developing electrochemical biosensors for detecting environmental pollutants [47, 48]. The use of electrochemical biosensors for monitoring water pollution was reported in the literature [49]. To quickly measure Hg^{2+}, Cu^{2+}, Pb^{2+}, and Cd^{2+}, Ayenimo et al. reported a sensitive and efficient amperometric glucose biosensor [50]. A stronger suppression of GOx activity on subsequent exposure to trace metals resulted in a decreased glucose amperometric response. Rengaraj et al. reported a novel, easy-to-use, and cost-effective paper-based impedimetric sensor for the detection of bacteria in water with a dynamic range of 10^3–10^6 CFU mL^{-1} [51]. Furthermore, the detection of the ultralow concentration of NO_2 in a complex environment setup was

TABLE 12.1
Utilization of Various Biosensors for Environment Monitoring

S. No.	Biosensor Used	Evaluated Target	Limit of Detection	Pollutant/Analyte Detection	References
1.	Electrochemical immunosensor	Paraoxon	Detection limit 12 μg l^{-1}	Pesticide	[26]
2.	Electrochemical sensor	Capsaicin	0.10 μM	Amide alkaloid	[27]
3.	Electrochemical sensor	Paraoxon methyl	1.6 ng/ml	Pesticide	[28]
4.	Electrochemical sensor	Paraoxon ethyl	1.35 nM	Pesticide	[29]
5.	Electrochemical sensor	Paraoxon ethyl	0.42 nM	Pesticide	[30]
6.	Whole-cell biosensor	Atrazine	80 ng/mL	Herbicide	[31]
7.	Optical biosensor	Methyl-parathion	63.17 ng/L	Pesticide	[32]
8.	SPR immunosensor	Ampicillin	0.3 g/L	Pharmaceuticals	[33]
9.	SPR immunosensor	Ciprofloxacin	0.08 μg/L	Pharmaceuticals	[34]
10.	SPR immunosensor	Naphthalene	0.76 ng/mL	Organic compound	[35]
11.	SPR biosensor	Bisphenol A	5.2 pg/mL	Organic compound	[36]
12.	Fluorescent biosensor	Mercury (II)	0.012 ppb	Heavy metal	[37]
13.	DNA biosensor	Mercury (II)	2.1 nM	Heavy metal	[38]
14.	Electrochemical biosensor	Silver (I)	0.5 fM	Heavy metal	[39]
15.	Fluorescent biosensor	2,4-Dichlorophenoxyacetic acid (2,4-D)	7.53 μg/L	Pesticide	[40]
16.	Optical biosensor	Atrazine Paraquat	0.04 ng/mL 0.05 ng/mL	Pesticides	[41]
17.	Optical biosensor	Diclofenac	1 μg/L	Pharmaceutical	[42]
18.	Fluorescence biosensor	Sulfadimidine	0.06 ppb	Pharmaceutical	[43]
19.	Optical biosensor	Ibuprofen	1000 ppb	Pharmaceutical	[44]
20.	Whole-cell-based biosensor	Pb^{2+}, Hg^{2+}, Cd^{2+}, Cr^{2+}, Cu^{2+}	1 μM–100 μM	Heavy metals	[45]

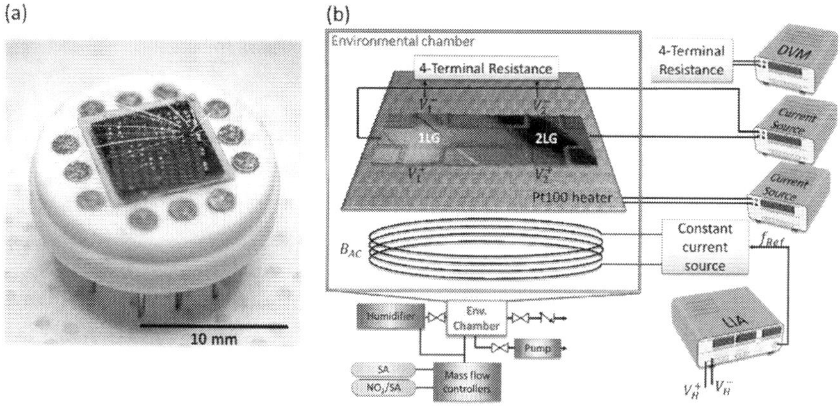

FIGURE 12.2 (a) Illustration of a developed epitaxial graphene chip with 25 sensor devices mounted on a ceramic TO-8 header and a Pt-100 heater. (b) Experimental setup layout for measuring transport characteristics in the environmental chamber with a digital voltmeter (DVM), current source, and lock-in amplifier (LIA) [52]. Copyright © 2018, American Chemical Society, copied with permission.

reported by Melios et al. using a graphene-based sensor [52]. Figure 12.2 illustrates the graphene-based sensor and experimental setup layout for measuring transport characteristics in the environmental chamber. At 150 °C, the sensor's sensitivity to NO_2 increased considerably while its cross-selectivity to carbon monoxide, sulfur dioxide, and water was lowered.

12.3.2 Nanobiosensors

Increasing literature articles, patents, and technology demonstrate the remarkable applications of nanomaterials in biosensors to produce nanobiosensors and improve their characteristics. Furthermore, it has been noted that biosensors produced with nanoparticles typically exhibit improved capabilities. The unique characteristics and cost-effectiveness of nanobiosensors make them extremely popular sensors with enormous applications in nearly every field of science.

They depend on biological detection and identification. Nanobiosensors have been utilized extensively over time because they provide a large platform for using biomolecules and nanoparticles to detect a variety of analytes. Nanobiosensors exhibit significantly better sensitivity, specificity, and biocompatibility than most traditional sensing devices. Nanobiosensors are fabricated using a variety of nanomaterials, such as quantum dots, metal and its oxide nanoparticles, carbon nanotubes, magnetic nanoparticles, and so on, to improve their characteristics [53]. The biomolecules and nanomaterials interact through a bio-interference that makes it simple to identify even the lowest concentration of the target molecules, facilitating the rapid detection of different analytes. Today, one of the most essential demands to prevent environmental pollution is the immediate and precise detection of hazardous metal ions. In this context, gold nanoparticles are vital since they are used to produce intelligent sensors and detection agents. Researchers have been focusing on the production of gold nanoparticles (GNPs)–based biosensors for the detection of heavy metal ions in water samples. Hong et al. reported a poly(allylamine hydrochloride) (PAH) conjugated GNP for the detection of arsenic [54]. The technique exhibits strong arsenate selectivity when other environmentally significant anions, such as $H_2PO_4^-$, SO_4^{2-}, NO_3^-, and Cl^-, were also present. Moreover, the CN^- (cyanide ions) sensing was reported by Dong et al. using an Ag@Au core-shell nanoparticle (NP)/iridium(III) complex-based nano biosensor [55]. Figure 12.3 depicts the reported suggested cyanide sensing platform's mechanism. The designed biosensor demonstrated great selectivity for CN^- over ten-fold excess quantities of other anions, as well as good sensitivity towards CN^- with a 0.036 μM detection limit.

When nanobiosensors are integrated into the water system, users will be able to a) remotely monitor and identify pollution issues; and b) adhere to laws and policies on water quality and management. Technological advancements have also led to the development of wearable, lab-on-a-chip, point-of-care, and Internet of Things (IoT) devices that use nanobiosensors to detect pollutants in real time and on-site. Notwithstanding their many benefits, nanobiosensors have several drawbacks. For example, the presence of certain components may impair their performance in the real world, and their high cost limits their applicability in a variety of disciplines.

12.3.3 Optical Biosensors

Optical biosensors are unique in their ability to reveal important details about a sample (such as kinetic behavior, molecular interaction, concentration, etc.) without being affected by electrical or magnetic interference. Additionally, optical biosensors are highly sensitive and can identify analytes at both the attomolar and femtomolar levels [56, 57]. Because optical devices are insensitive to outside signals, they are more versatile than other transducers in terms of setup and target sample types, making optical biosensors extremely promising for environmental analysis [58]. Device miniaturization, biological recognition element stability, and device

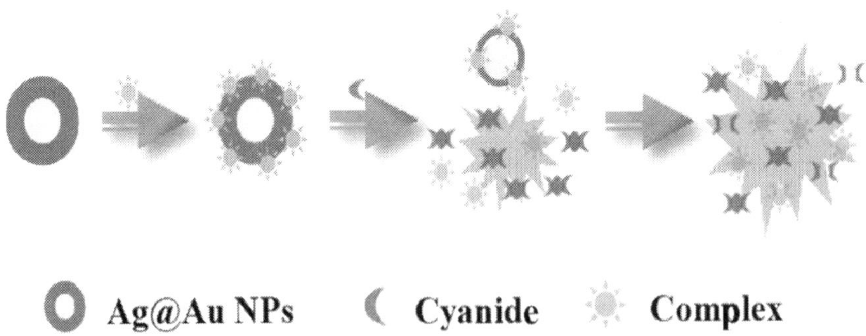

FIGURE 12.3 The suggested cyanide sensing platform's mechanism [55]. Copyright © 2017, American Chemical Society, copied with permission.

reusability are the primary challenges in developing portable commercial optical biosensors. Among the methods for identifying pollutants in the water examined here, interferometric biosensors are particularly significant. Antibiotics, pesticides, and toxic microbes can be detected with these biosensors, the majority of which have an LOD of less than 1 μg/L. Arjmand et al.'s tapered fiber optic for pesticide detection and Lamarca et al.'s U-bent for antibiotic detection are two examples of fiber-optic biosensors utilized in water monitoring [59, 60].

An SPR sensor based on antibodies immobilized on the gold sensor surface was used by Koubaka et al. to detect *Salmonella enteriditis* and *Listeria monocytogenes* in real time at concentrations as low as 10^6 cells/mL [61]. Pathirana et al. demonstrated that an immunosensor based on the acoustic wave concept could detect *Salmonella typhimurium* in liquid samples in less than 2 min, with detection limits of about 100 cells/mL [62]. An immunochemical potentiometric alternating biosensor was reported by Ercole et al. for identifying *Escherichia coli* in water samples [63]. Many lung disorders are linked to exposure to volatile organic compounds (VOCs), which are thought to be harmful to human health. Using aerogel attached to polydiacetylene (PDA), Dolai et al. reported a novel VOC detection technology [64]. Figure 12.4 depicts the diagrammatic representation of the PDA–aerogel formation and different color indications for various VOCs. The PDA–aerogel combination responded promptly to low VOC concentrations in terms of color and fluorescence, and its signals were increased.

12.3.4 Microbial Biosensors

Nowadays, microbial biosensors—which use microorganisms like bacteria, fungi, viruses, etc.—have drawn lots of interest in the environmental field since they can detect a wide range of analytes. Initially, the biological oxygen demand (BOD) was

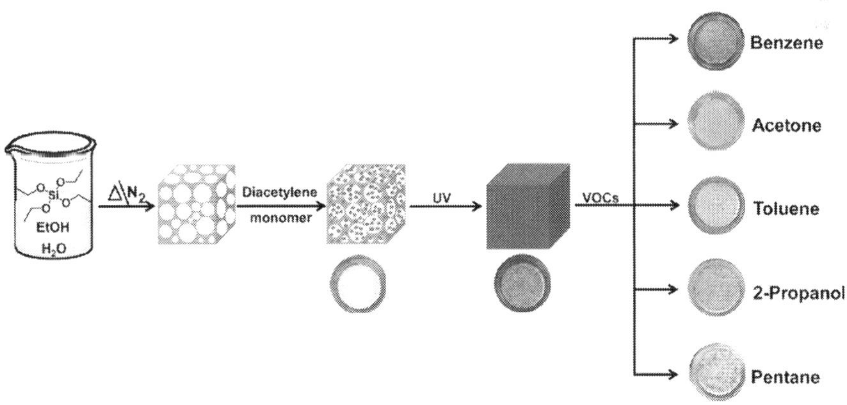

FIGURE 12.4 Diagrammatic representation of PDA–aerogel formation and different color indications for various VOCs [64]. Copyright © 2017, American Chemical Society, reprinted with permission, copied with permission.

investigated and general toxicity was monitored using microbial biosensors. The traditional BOD methods are expensive and time-consuming, and may result in the dumping of wastewater into river bodies before the reporting of test findings. As a result, several microbial biosensors were fabricated to reduce the drawn-out procedure while minimizing the expense of this monitoring procedure. Numerous BOD biosensors have been produced to detect high BOD levels, including those observed in industrial wastewater. Conversely, Chee et al. have designed an optical fiber biosensor to assess low BOD levels in river waters [65]. The authors found that the sensor performed optimally at pH 7.0 and 30°C. BOD was measured at a minimum of 0.5 mgL^{-1}. Although river water has large levels of chloride and other heavy metal ions, the optical fiber biosensor appeared to be appropriate for assessing real samples without significantly altering these ions.

In a whole-cell biosensor (WCB), microorganisms with a rapid growth rate and a short period of generation are used to provide a sufficiently measurable signal [66]. Another significant benefit of using WCBs is their capacity to analyze samples that involve several enzymes. Algae, bacteria, yeast, plant cells, and fungi are among the several microorganisms that are employed in WCBs [67–69].

Kharkova et al. designed a unique highly sensitive microorganism–mediator(s) receptor biosensor system for the fast BOD determination of the water sample [70]. They demonstrated that the yeast *Blasto botrytis adeninivorans* effectively established the association between graphite paste–immobilized ferrocene and phenazine mediators, which aid in the oxidation of bulky materials by accelerating the electron transfer rate. The results obtained from the designed biosensor were closely aligned with the results of the conventional BOD analysis method (R = 0.9693). The BOD biosensors based on microbial membranes and photocatalysis were also reported [71, 72].

The application of microbial biosensors as an alternative method of environmental monitoring has been the subject of extensive research; nevertheless, transfer from laboratory to field has proven challenging. This is partially because naturally changing circumstances, such as reactions to the local microbial communities, pH changes, and nutrient availability, are unpredictable. Whole cells serve an essential role in detecting heavy metal ions as the carrier to adsorb, metabolize, or precipitate heavy metal ions. Green microalgae (*Tetraselmis chuii*) were employed by Alpat et al. for the simple, affordable, sensitive, and efficient biosorption, and measurement of Cu^{2+} [73]. Carbon paste was mixed with green microalgae to produce the functioning electrode. Similarly, for Pb^{2+} detection, a *Phormidium* sp. modified voltammetric sensor was reported by Yuce et al. in the aqueous solutions [74]. Hnaien et al. reported a microbial biosensor for the detection of trichloroethylene [75]. The microbial biosensor exhibited a value of 20 $\mu g\ L^{-1}$ as the limit of detection. Aromatic compounds such as phenols, benzene, and toluene are examples of carcinogenic xenobiotics that contaminate water supplies/resources. Roy et al. developed a whole-cell biosensor (WCB) array for the detection of these harmful aromatic compounds as observed in Figure 12.5 [76].

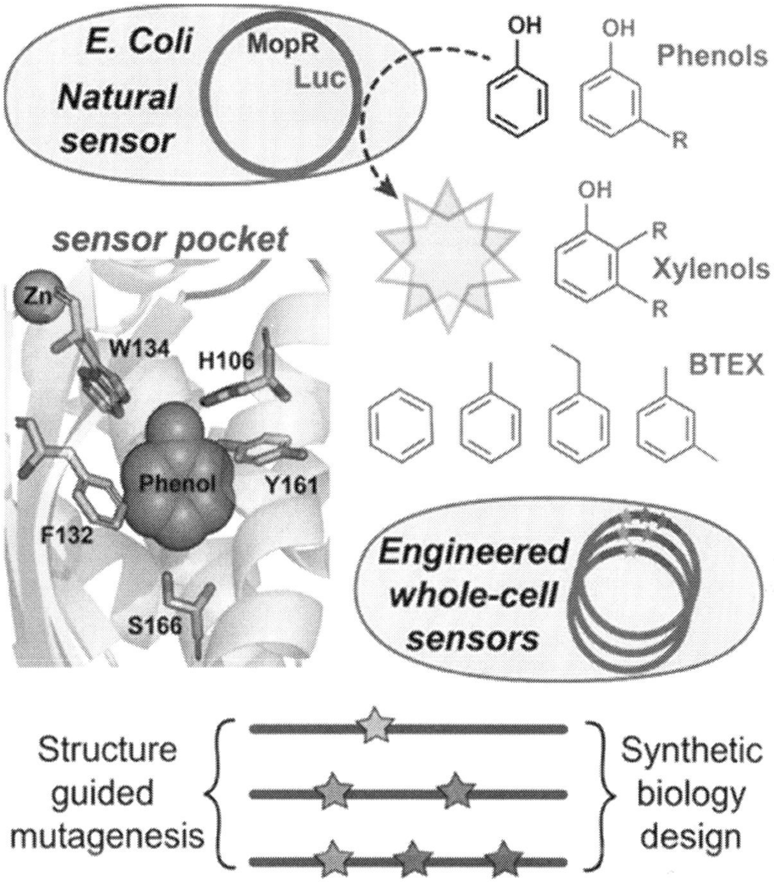

FIGURE 12.5 Design approach for an array of programmable genetic biosensors [76]. Copyright © 2021, American Chemical Society, copied with permission.

12.4 APPLICATIONS OF BIOSENSORS IN DETECTING ENVIRONMENTAL POLLUTANTS

12.4.1 Heavy Metals Detection

One of the biggest environmental issues impacting global sustainability is heavy metal contamination. As industrial activity advances, heavy metals find their way into the environment. The primary factors responsible for the increased concentrations of heavy metals in soil are urbanization, agricultural modernization, and industrialization [77].

Even at trace levels, heavy metals including mercury, arsenic, lead, and cadmium are extremely carcinogenic and poisonous [78]. These are extremely hazardous to human health and the environment because they accumulate in the food chain. Increased levels of lead (Pb) exceeding 0.025 mg/kg body weight (b.w.) have been related to early fetal development, cardiovascular disease, neurological disorders, and renal failure. However, Cd intake exceeding 0.007 mg/kg bw damages the kidneys and liver [79]. Similarly, mercury concentrations greater than 0.0016 mg/kg (b.w.) damage the growing fetus and result in neurological problems in both adults and infants [80].

Thus, monitoring heavy metals in food, drinking water, biological fluids, and the environment is crucial. The tremendous advancement of nanotechnology has opened up new avenues for developing nanostructured sensors that can detect heavy metals. Sensors show great promise for on-site, accurate detection of particular heavy metals.

According to several investigations, biosensors are used to detect heavy metals in food products. To detect $Cr_2O_7^{2-}$ in citrus fruits quickly and accurately, Li et al. used electrochemical polymerization to create a new fluorescence sensor based on polyfluorene (CPs), namely P(Fmoc-GABA-OH) [81]. The detection limit of the reported sensor was 16.6 femtomolar (fM) in various samples, namely soil, farmland water, tangerine, orange, and grapefruit. Metal ions can also be detected using antibody-based biosensors, which rely on forming an antibody–metal ion combination complex to function properly. Even in very small quantities, antibodies can detect antigens [82]. Wu et al. reported DNAzyme and quantum dot (QD)-based nanosensors that are extremely sensitive and selective for the multiplexed detection of heavy metal ions in liquid samples [83]. After 25 min, the detection limits for Pb^{2+} and Cu^{2+} were 0.2 and 0.5 nM, respectively. Moreover, multiplexed detection was achieved using two different QD colors, showing little cross-talk between the Pb^{2+} and Cu^{2+} detection. Ilangovan et al. designed a conductometric biosensor based on the urease enzyme to find heavy metal ions in liquid samples [84]. The sensor could detect urea concentrations between 1 mM and 15 mM, and the concentration range for heavy metals reported was 0.1–10 mM, with Cd exhibiting the highest level of inhibition, followed by Pb and Cu.

The approach known as a lateral flow assay (LFA) is very adaptable and may be modified to detect various targets without the need for equipment. Several investigations have reported LFAs to identify heavy metals. They are suitable for on-site sample analysis due to their compact size, affordable price, simple storage, and easy-to-use layout. The results of LFAs are rapid and simple for both experts and non-experts to understand [85, 86]. An aptamer-based fluorescence-quenching lateral flow strip (LFS) was specially designed by Wu et al. to detect Hg^{2+} in river water samples with high sensitivity [87]. High specificity was shown by this test, which produced results in 15 min. The limit of detection (LOD) was 0.13 ng mL^{-1}, and for Hg^{2+}, the linear detection range of the LFS was 0.13 ng mL^{-1} to 4 ng mL^{-1}. The findings conformed to observations from inductively coupled plasma-mass spectrometry for Hg^{2+} detection in river water.

Srinivasan et al., for example, developed a rapid and inexpensive lateral flow assay with a detection limit of 7.4 μM for thallium and 6.6 nM for Pb [88]. Medina-Sánchez et al. presented an electrochemical lab-on-paper approach with detection limits of 7 and 11 ppb for Pb and Cd, respectively [89]. Furthermore, for the detection of Hg^{2+} with a 1 μM detection limit, Chao et al. successfully designed an easy, affordable, high-throughput, rapid, and quantitative sensor based on the LFA system [90]. All of these research study results indicate that LFAs have the potential to detect heavy metals, although with various degrees of sensitivity.

Arsenic is regarded as one of the most dangerous substances in nature. The two most common inorganic forms of arsenic are arsenite and arsenate. Arsenic levels in drinking water should not exceed 10 parts per billion. Electrochemical sensors with electrodes improved with carbon nanoparticles offer a fast response with low detection limits. It has been reported that thiolated multi-walled carbon nanotube-modified Au electrodes work effectively for measuring arsenic. To detect As^{3+} in water, Kato et al. reported Au nanoparticle (AuNP)-embedded carbon sheets using an unbalanced magnetron (UBM) co-sputtering method in a single, repeatable step [91]. A linear dynamic range of 1–100 ppb and a detection limit of 0.55 ppb were attained with the synthesized electrode using anodic stripping voltammetry (ASV) for As^{3+} detection. With the addition of EDTA, the Au-embedded UBM carbon film electrode additionally obtained exceptional selectivity of As^{3+} detection in interference (Cu^{2+}) even with tap water samples.

12.4.2 Pesticides and Herbicides Detection

Pesticides are among the most harmful environmental contaminants because of their extensive prevalence in the environment. The toxicity of organophosphorus insecticides, for instance, makes them a class of pesticides of significant ecological concern despite their widespread usage in agriculture. Herbicides, fungicides, plant growth regulators and repellents, and insecticides are among the pesticides frequently used in the industry and agriculture sectors. These can be classified chemically as carbamates, pyrethrins, pyrethroids, organochlorines, and organophosphates [92]. Pesticides can potentially leach through the soil into groundwater, resulting in drinking water contamination. Additionally, these can enter surface water through runoff or drift, impacting ecosystems and human health [93].

Although they have been associated with numerous ecological and health issues, pesticides are a highly hazardous substance widely utilized in industry, home care, agriculture, and medicine. Advanced techniques are necessary to detect pesticides at the concentration limits established by the appropriate authorities. The traditional methods of gas chromatography and liquid chromatography combined with mass spectrometry (GC–MS and LC–MS) are employed to analyze pesticide residues in water in a sensitive, repetitive, and selective way [94, 95]. While these methods work well for measuring pesticides, they are not appropriate for survey analyses of flowing water that need to be monitored at an appropriate frequency to identify an eventual pollution peak in real time and to act quickly. This is because the samples under study must be pre-treated before being analyzed in advanced labs.

The development of real-time detection and warning systems, like pesticide biosensors, has attracted significant research attention due to the health risks connected with pesticides. Organophosphate and carbamate insecticides are potent neurotoxins that cause serious health problems by blocking the cholinesterase enzyme's activity. Many biosensing methods based on the concept of inhibition by cholinesterase enzymes have been developed to detect organophosphate and carbamate pesticides [96–99]. The enzyme acetylcholinesterase (AChE), often referred to as serine esterase, catalyzes the breakdown of acetylcholine into choline and acetic acid, which is essential to cholinergic transmission in mammals and insects. Recently, acetylcholinesterase-containing biosensors have received a lot of interest since they can be used to detect anticholinesterase chemicals or inhibitors, such as carbamate and organophosphate insecticides. Biosensor response signals are reduced when organophosphate or carbamate pesticides are present because they change the usual AChE activity. Therefore, pesticide concentration may be related to this decrease in the response signal of the biosensor. A polyaniline-mercaptobenzothiazole-based sensor system was reported by Somerset et al. for the detection of carbamate and organophosphate insecticides [100]. For diazinon and carbofuran, the developed biosensor revealed lower detection limits of 0.07 ppb and 0.06 ppb, respectively. Arduini et al. developed an acetylcholinesterase (AChE) amperometric biosensor for the analysis of organophosphorus insecticides [101]. The biosensor demonstrated a linear range of up to 40 ppb when using paraoxon as the model analyte with a detection limit of 2 ppb. Similarly, piezoelectric biosensors have been designed to identify the pesticidal effects of carbamate and organophosphate [102].

One intriguing sign that quickly conveys the harmful effects of some contaminants is the inhibition of photosynthesis. Using this characteristic, it has been reported that certain biosensors based on Photosystem II (PSII) may identify environmental herbicides [103]. For instance, Koblizek et al.'s amperometric biosensor demonstrated selective sensitivity to triazine, polyphenols, and urea herbicides [104].

Additionally, immunological biosensors for pesticide detection have also been reported. To identify pesticides including atrazine, simazine, isoproturon, 2,4-D, alachlor, and paraquat in natural waterways, Mallat et al. used the River Analyser (RIANA) immunosensor [105, 106]. Using an amperometric immunosensor, Wilmer et al. identified 2,4-dichlorophenoxyacetic (2,4-D) acid in water with a detection limit of 0.1 μg/L [107]. Due to its slow degradation, atrazine is one of the most widespread pesticides and has been found in water bodies all over the world [108]. The simultaneous detection of chlorpyrifos, thiabendazole, and imazalil in water and wine samples was made possible by an immunosensor that a Greek research team designed with a straightforward setup to measure interference. The LODs that were found ranged from 30 to 40 pg/mL [109].

12.4.3 Organic Pollutants Detection

The three main phenolic chemicals that contribute to water contamination are bisphenol-A, hydroquinone, and catechol. Phenolic compounds are widely used for manufacturing dyes, chemical analysis, disinfectants, synthetic resins, petrochemicals, medicines, textiles, paints, and medical devices [110]. Even at low concentrations, phenolic chemicals are extremely hazardous and are considered major pollutants. The toxicity limit level of phenolic compounds in both humans and marine life falls between 9 and 25 mg/L [111]. The adverse health effects of phenol can be both acute and chronic/long-lasting. Long-term exposure can cause fatal levels of respiratory arrest, coma, tremors, muscle weakness, and irregular breathing in humans. Eyes, skin, and mucous membranes get irritated when humans are exposed to phenol. Even from a composite medium like wastewater, biosensors can detect very small amounts of contaminants and are expected to accurately measure the overall phenolic content.

Polyphenols have been identified through a variety of biorecognition components, including enzymes, DNA, whole cells, microbes, and antibodies [112–115]. The most successful enzymatic biosensors for detecting phenolic compounds are those based on horseradish peroxidase, laccase (Lac), and tyrosinase [116–118]. These enzymes can directly catalyze electron transfer processes/reactions without adding cofactors to the reaction media. Hydroquinone (Hq), bisphenol-A, and catechol are the main phenolic chemicals that lead to water contamination. Hydroquinone is most widely used as a raw substance in chemical industries. Research indicates that increased levels of hydroquinone can cause skin irritation and damage to soil microbes. Upan et al. reported a simple and cost-effective amperometric sensor for accurate and sensitive determination of hydroquinone [119]. Furthermore, the detection of hydroquinone in real-water samples was reported by Lou et al. using a polyaniline/magnetic graphene composite [120]. The simultaneous determination of hydroquinone and catechol was reported by Huang et al using a Ce-MOF/CNT composite-based electrochemical biosensor [121].

Bisphenol A (BPA) is a widespread industrial raw chemical that causes a risk to the environment due to its broad utilization and discharge into the environment. BPA can cause endocrine disruption and is utilized in the production of plastics. A GCE modified with a hydroxyl group functionalized MWCNTs was employed by Cosio et al. to fabricate an electrochemical sensor for the detection of BPA [122]. The suggested technique worked effectively to determine the amount of BPA in real water samples (water samples from plastic and infant bottles) with good findings that were in the best possible accordance with those from a separate HPLC approach. Furthermore, Yu et al. utilized the Au NPs deposited rGO-MWCNTs modified glassy carbon electrode for BPA determination [123]. The modified electrode exhibited better electrochemical oxidation for BPA. Using differential pulse voltammetry, the linear range for determining BPA in 0.10 M phosphate buffer (PBS, pH 7.0) was between 5.0×10^{-9}–1.0×10^{-7} M and 1.0×10^{-7}–2.0×10^{-5} M. Also, the modified electrode was used for BPA detection in river water samples exhibiting recovery ranges of 97%–110% and 98%–104%, respectively.

Farre et al. presented an economical method for quickly determining the toxicity of wastewater, utilizing an amperometric biosensor containing *E. coli* [124]. This study examined the reactions of *E. coli* to frequently found contaminants in wastewater from textile and tanneries, including phenolic compounds, non-ionic surfactants, and benzene sulfonate compounds, utilizing an integrated bacterial biosensor system. To develop an efficient, easy, and precise biosensor for environmental pollutants, a novel rapid biosensor technique has been developed that uses the dye resazurin to indicate bacterial respiration [125]. The developed whole-cell biosensor was used to detect toluene analyte using naturally occurring toluene degrading bacteria as the biological element. Another study evaluated the detection of halogenated substances using a less complex fluorescent array. Five distinct chemicals were tested in this investigation, and the LODs for each ranged from 12.1 to 1.4 mg/L [126].

12.4.4 Pharmaceuticals

Pharmaceutical chemicals contaminate the environment through excretion or improper disposal of their leftovers. These substances, like pesticides, are stable and accumulate in the environment; some of them also have oestrogenic effects on particular organisms and can even change how some algae and microalgae species develop. A sandwich-format immunoassay was reported by Tomasseti et al. to quantify ampicillin; it had good selectivity but limited sensitivity, with a limit of detection of 0.3 g/L [127]. The ciprofloxacin antibiotic detection was accomplished using a polymer-based SPR biosensor. This biosensor was calculated to have a detection limit of 0.08 μg/L [128]. A more recent investigation was performed to find out if ciprofloxacin was present in water treatment plant effluent. Utilizing an optical fiber immunosensor, this study yielded a limit of detection of 3.3×10^{-3} ng/L [60].

12.5 CHALLENGES AND FUTURE PERSPECTIVES

12.5.1 Challenges in Biosensor Technology

Biosensor technology has emerged as an essential tool for environmental monitoring and pollution detection [129]. It provides methods that are quick, accurate, and cost-effective for identifying contaminants such as those found in the environment. On the other hand, because of various obstacles, its widespread adoption and effectiveness in environmental applications are restricted. One of the most significant challenges in terms of technology is the achievement of high sensitivity and specificity. A complex matrix of substances is frequently found in environmental samples. Many of these substances have the potential to interfere with the performance of biosensors, which can result in false positives or false negatives [130].

To detect pollutants at extremely low concentrations (roughly 10^{-18} M), biosensors with extremely low detection limits are required. This is a goal that continues to be challenging to accomplish. Furthermore, cross-sensitivity resulting from the overlapping signals of non-target substances complicates the challenge of accurate and precise pollutant detection. Moreover, another important challenge in biosensor technology is mainly the durability of biosensors, which can be affected due to severe environmental conditions that include extreme temperatures, pH fluctuations, and humidity [131]. These factors can compromise biosensor components, especially the biorecognition elements, diminishing their durability and dependability. Additionally, the assurance of reusability and stability of these biosensors under these conditions introduces an additional complexity to their development.

Furthermore, material challenges are another significant factor that contributes to the limitations of biosensors in terms of their scalability and sustainability. Graphene and other nanomaterials are examples of advanced materials that improve sensitivity and performance [132, 133]. However, the incorporation of these materials into mass production is still difficult due to the compatibility issues with the manufacturing technologies that are currently in use. In addition, a significant number of biosensors are made of materials that are not biodegradable, which raises concerns about the environment. Recently, ongoing efforts have been made to transition towards sustainable materials, such as biopolymers; however, significant advancements in material science are required for these efforts to be successful. The high cost of raw materials, particularly for high-performance biosensors, further restricts the accessibility of these sensors. When it comes to their operation, biosensors need to undergo routine maintenance and calibration to guarantee their accuracy over time. It is common for environmental samples to require extensive preprocessing to eliminate impurities that could potentially interfere with the performance of sensors. This significantly increases operational complexity and reduces the real-time monitoring applications.

The widespread adoption of biosensors for environmental monitoring is also hampered by challenges posed by regulatory agencies and competition in the market. The implementation of substantial investments in testing and certification is necessary to satisfy stringent regulatory standards. The high cost of biosensors, in addition to the difficulties associated with striking a balance between cost and performance, is another factor that reduces the widespread market adoption of these devices. As a result, concerns regarding sustainability bring to light the important need for a paradigm shift in the design and manufacturing of biosensors. There are a lot of designs that are only used once, which contributes to waste and has an impact on the environment.

Although these benefits exist, the utilization of biosensors in the environmental sector is still very limited as compared to pharmacy or healthcare applications, where the vast majority of research and development has occurred. However, the vast majority of the systems established are still prototypes that require validation before they can be widely utilized or commercialized.

High-density arrays of biomolecular sensors are anticipated in the future to challenge microelectronics in size and capability, provide the ability to recognize and distinguish intricate analytical instruments, and provide researchers with physiologically significant data.

12.5.2 Emerging Trends and Future Directions

Biosensors are increasingly recognized as revolutionary tools in environmental monitoring and pollution detection, providing rapid, sensitive, precise, and sustainable solutions. Recent advancements in biosensing technologies have incorporated nanomaterials and life cycle analysis (LCA) methodologies to enhance detection efficiency and environmental sustainability [134]. Paper-based and biodegradable plastic biosensors are being investigated for their recyclability and reduced ecological impact, by green chemistry principles [135]. Additionally, single-molecule electrical biosensors employing methods such as scanning tunneling microscopy-assisted molecular break junctions (STM-BJ) have exhibited remarkable sensitivity and specificity, facilitating the identification of pollutants and pathogens at very small concentrations [135]. These methods eliminate the need for resource-intensive procedures such as amplification, which results in a reduction in both waste and resource consumption. To enhance the efficiency of real-time environmental monitoring, future directions emphasize the combination of biosensors with the Internet of Things (IoT) and machine learning. The incorporation of artificial intelligence will improve data processing and predictive analytics, thereby resulting in increased precision in the monitoring of pollution. Furthermore, compact, automated, and portable biosensors utilizing sustainable materials are expected to diminish issues related to environmental impact, scalability, and cost. Researchers are investigating innovative biodegradable materials such as cellulose-based substrates and environmentally friendly nanocomposites to further diminish the carbon footprint of biosensor manufacturing. Biosensors that were developed with life cycle assessment considerations present a viable approach to sustainable pollution management and ecosystem health monitoring. This is particularly important because climate change is increasing environmental crises. These innovations have the potential to revolutionize environmental management by making it easier to identify pollution-related problems at an earlier stage, to take corrective action, and to prevent them.

12.5.3 Role of Biosensors in Sustainable Environmental Management

Biosensors have become essential instruments for sustainable environmental management, providing innovative monitoring methods and detecting and reducing pollution while reducing ecological effects [136]. These analytical instruments combine components of biological recognition with transducers to convert biochemical reactions into quantifiable signals. As a result, they produce real-time data with high sensitivity. Biosensors primarily contribute by detecting pollutants, including heavy metals, pesticides, and organic toxins, in air, water, and soil. Conventional pollutant detection techniques frequently require laborious and resource-intensive laboratory analyses; however, biosensors facilitate immediate, precise, and on-site monitoring. Enzyme-based biosensors can identify specific contaminants in water sources, enabling quick interference to avoid extensive environmental damage [137]. Microbial biosensors are utilized to assess soil health by measuring nutrient concentrations and identifying harmful substances, thus enhancing agricultural practices and minimizing resource wastage [138]. Integrating sustainability principles into biosensor development is essential. Recent advancements emphasize the utilization of biodegradable materials, including cellulose-based substrates and paper, to develop environmentally sustainable biosensors [139, 140]. These materials not only diminish production expenses but also mitigate the environmental issues related to the disposal of single-use devices. Furthermore, nanotechnology is improving the efficacy of biosensors. Nanomaterials such as graphene, metal oxide, and carbon nanotubes enhance sensitivity and selectivity, facilitating the detection of pollutants at trace concentrations while maintaining compactness and energy efficiency [141]. The incorporation of life cycle analysis (LCA) into the design and manufacturing of biosensors represents notable progress in sustainable biosensing [142]. Life cycle assessment (LCA) evaluates the environmental impact of materials, manufacturing processes, and disposal methods to enhance the sustainability of these devices.

This method shows the potential for biosensors to help in pollution reduction without unintentionally intensifying environmental harm through waste production or resource-heavy manufacturing processes. Moreover, biosensors play a pivotal role in data-driven environmental management. Biosensors, when combined with the IoT and machine learning technologies, can establish comprehensive networks for the real-time monitoring of ecosystems [143, 144]. By facilitating predictive analytics, these networks help authorities identify areas of high pollution and take preventative action.

Biosensors utilized in urban environments can assess air quality and identify pollutants such as nitrogen oxides and particulate matter, supplying data to inform sustainable urban planning and policy formulation [145, 146]. Biosensors significantly contribute to the reduction of industrial pollution. Industries can utilize biosensors to monitor effluents and ensure adherence to environmental regulations. This not only reduces their ecological footprint but also encourages the implementation of cleaner production technologies [147]. Furthermore, communities can monitor the environment with the help of easily transportable biosensors, raising public awareness and encouraging group action for sustainable practices. These data show that biosensors are essential for attaining sustainable environmental management. Their capacity to deliver immediate, precise, and environmentally sustainable solutions meets the urgent demand for effective pollution management and resource preservation. Biosensors, through the integration of advanced technologies and sustainability principles, possess the capacity to transform environmental monitoring, facilitating a healthier and more sustainable planet.

12.6 CASE STUDIES AND APPLICATIONS

12.6.1 Case Studies Demonstrating the Effectiveness of Biosensors

Numerous studies demonstrate the efficacy of biosensors in environmental monitoring and pollution detection. An exemplary instance is the application of enzyme-based biosensors for the detection of pesticide residues in water sources within agricultural areas of India [148]. This study demonstrated the effective use of biosensors incorporating acetylcholinesterase enzymes for the detection of organophosphate pesticides at nanomolar concentrations, facilitating prompt interventions and minimizing contamination levels. Another study illustrated the efficacy of microbial biosensors in detecting heavy metals in industrial wastewater [149]. Researchers developed a biosensor employing genetically modified *E. coli* to identify heavy metal ions with elevated specificity and sensitivity [150]. This biosensor facilitated real-time monitoring, allowing industries to adjust effluent treatment processes promptly and comply with environmental regulations. Biosensors integrated with IoT platforms have been utilized to monitor air quality in polluted urban areas such as Beijing, China [151]. These systems, utilizing nanomaterial-based electrochemical sensors, provided real-time data on particulate matter and volatile organic compounds, leading to the implementation of targeted air quality improvement measures. The study demonstrated a significant reduction in pollutant levels within a year of deployment.

A particularly innovative application involved the use of paper-based biosensors for detecting nitrates and phosphates in river water [152]. These portable, disposable sensors allowed people to monitor water quality independently, fostering greater public participation in environmental conservation efforts.

12.6.2 Real-World Applications in Environmental Monitoring and Pollution Detection

The effectiveness of biosensors in addressing ecological challenges has been demonstrated by the fact that they have found extensive real-world applications in environmental monitoring and pollution detection. These applications, which cover a wide range of environments, demonstrate the versatility and impact of biosensor technologies. One of the most important applications is the monitoring of water quality. The detection of organophosphate pesticides in water bodies, particularly in agricultural regions, has been accomplished through the utilization of enzyme-based biosensors. One example of such a biosensor employs acetylcholinesterase enzymes [153]. Similarly, whole-cell microbial biosensors that incorporate genetically engineered microorganisms have been utilized to detect heavy metals such as cadmium and lead in industrial effluents [149]. This has made it easier to comply with environmental regulations. One particularly noteworthy example is the utilization of an optical biosensor by Durrieu et al. to detect zinc and cadmium by inhibiting the activity of alkaline phosphatase [154]. Monitoring the quality of the air is yet another important domain in which biosensors play a pivotal role. To detect airborne pollutants such as nitrogen oxides, carbon monoxide, and particulate matter, electrochemical biosensors that are embedded with nanomaterials such as graphene are utilized [155]. These systems, when combined with platforms that are connected to the IoT, deliver data in real-time that are specifically targeted to urban environments. For instance, cities have implemented such biosensor networks to significantly lower the levels of air pollution [156]. The monitoring of the health of the soil is another area of impact. Microbial biosensors are utilized in agricultural soils to measure nutrient levels and identify harmful compounds [157]. This allows for the optimization of fertilizer use and the reduction of environmental degradation. Another study has shown the real-time monitoring of total bioaerosol by using a localized surface plasmon resonance (LSPR) biosensor developed by scientists using succinimidyl-ester-functionalized gold nanoislands (SEF–AuNIs) [158]. With detection limits of 1.69 cells/mL for *Bacillus subtilis* and 0.5119 cells/mL for *E. coli*, respectively, this sensor showed exceptional sensitivity. The researchers carried out a thorough investigation using aerosol samples gathered from three different locations in Switzerland—Bern, an urban setting; Basel, a suburban area; and Rigi Mountain, a rural, high-altitude site—to evaluate the sensor's practicality. Their research showed notable variations in bioaerosol concentrations among these sites, which were impacted by atmospheric conditions and urbanization. The study also investigated relationships between PM10, endotoxin concentrations, and bioaerosol levels, offering important new information on how microorganisms and air quality parameters interact. In addition, paper-based biosensors have been introduced for community-driven monitoring of soil and water quality [159]. This has led to the empowerment of local stakeholders to actively participate in the conservation of the environment. For instance, Nakamura and Karube have developed microbial biosensors to measure biochemical oxygen demand (BOD) in polluted waters. These biosensors have significantly reduced detection times in comparison to traditional methods. The monitoring and reduction of industrial pollution have also been significantly aided by the utilization of biosensors. To monitor the quality of effluent and identify potentially harmful substances in wastewater, industries make use of biosensors. This not only prevents violations of regulations but also encourages the development of less polluting production methods. Biosensors have been utilized in marine ecosystems to monitor toxic algal blooms and identify oil spills from the environment [160, 161]. For example, biosensors have been utilized to monitor pollutants in coastal waters, which has contributed to the preservation of marine biodiversity [162]. As demonstrated by these examples, biosensors are extremely versatile when it comes to addressing difficult environmental challenges. As a result of their ability to detect pollution in a timely, accurate, and cost-effective manner, biosensors have evolved into indispensable tools for the management of the environment in a sustainable manner.

12.7 CONCLUSION AND FUTURE PROSPECTS

With ongoing advances in nanotechnology, genetic engineering, and miniaturization, a new trend in the effort to develop biosensors is growing. The current environmental situation necessitates the development of more accurate and dependable instruments that can analyze required pollutants efficiently. These are compact, simple devices that offer accuracy, speed, reliability, excellent sensitivity, and bioselectivity for specific analytes. In their design, several user-related issues must also be taken into account, including ease of use and manipulation, safety features, suitability for in situ detection (which eliminates the need for complicated sample preparation), cost-effectiveness, real-time detection, and environmental friendliness.

This chapter intended to demonstrate how biosensors can meet the demand for fast, dependable, and stable instruments for the detection of environmental contaminants. However, when applied to complicated and unreliable environmental samples with variable compositions, these ought to meet the requirements of sensitivity and selectivity.

- In analytical chemistry, biosensors have certain benefits. They speed up the conventional laboratory and analytical monitoring operations, which include extracting different analytes from different samples.
- Furthermore, biosensors improve environmental research efficiency by reducing the amount of time needed for acute and chronic toxicity studies.
- Biosensors facilitate simple, eco-friendly evaluation of specific substances (such as heavy metals) or their mixtures in a variety of samples that are continuously monitored in situ.

Recently, nanomaterials have been used in biosensors more effectively, and numerous nanomaterials have been developed to expand the range of possible uses for biosensors. The sensitivity limit of biosensors has been raised by recent developments in biological approaches and devices that use fluorescent tags on nanomaterials. Future research could concentrate on developing novel detection systems in the fields of micro- and nanoelectronics, enhancing multianalyte detection, and increasing automation and integration. Moreover, emerging synthetic receptors like aptamers and affibodies will offer better alternatives to antibody-based tests/assays. The development of sensitive and selective biosensors with significant regenerative potential was improved by integrated techniques. For the emergence of biosensors, all of these phenomena are significant and ought to happen simultaneously.

REFERENCES

1. Koedrith, P.; Thasiphu, T.; Weon, J.I.; Boonprasert, R.; Tuitemwong, K.; Tuitemwong, P. Recent trends in rapid environmental monitoring of pathogens and toxicants: potential of nanoparticle-based biosensor and applications. *Sci. World J.* **2015**, *2015*(1), 510982.
2. Dhote, S.S.; Deshmukh, L.; Paliwal, L. Miceller chromatographic method for the separation of heavy metal ions and spectrophotometric estimation of UO^{22+} on bismuth silicate layer. *Int. J. Chem. Anal. Sci.* **2013**, *4*(2), 85–90.
3. Chullasat, K.; Huang, Z.; Bunkoed, O.; Kanatharana, P.; Lee, H.K. Bubble-in-drop microextraction of carbamate pesticides followed by gas chromatography-mass spectrometric analysis. *Microchem. J.* **2020**, *155*, 104666.
4. Murzyn, C.M.; Allen, D.J.; Baca, A.N.; Ching, M.L.; Marinis, R.T. Tunable infrared laser absorption spectroscopy of aluminum monoxide A2Πi– X2Σ+. *J. Quant. Spectrosc. Radiat. Transf.* **2022**, *279*, 108029.
5. Parellada, J.; Narvaez, A.; Lopez, M.A.; Domınguez, E.; Fernandez, J.J.; Pavlov, V.; Katakis, I. Amperometric immunosensors and enzyme electrodes for environmental applications. *Anal. Chim. Acta* **1998**, *362*(1), 47–57.
6. Allan, I.J.; Vrana, B.; Greenwood, R.; Mills, G.A.; Roig, B.; Gonzalez, C. A "toolbox" for biological and chemical monitoring requirements for the European Union's Water Framework Directive. *Talanta* **2006**, *69*(2), 302–322.
7. Turner, A.P. Biosensors—sense and sensitivity. *Science* **2000**, *290*(5495), 1315–1317.
8. Tschmelak, J.; Proll, G.; Gauglitz, G. Optical biosensor for pharmaceuticals, antibiotics, hormones, endocrine disrupting chemicals and pesticides in water: Assay optimization process for estrone as example. *Talanta* **2005**, *65*(2), 313–323.
9. Zhou, Y.; Zheng, S.; Qin, W. Electrochemical biochemical oxygen demand biosensors and their applications in aquatic environmental monitoring. *Sens.Bio-Sens. Res.* **2024**, *44*, 100642.
10. Mauriz, E.; Calle, A.; Manclús, J.J.; Montoya, A.; Hildebrandt, A.; Barceló, D.; Lechuga, L.M. Optical immunosensor for fast and sensitive detection of DDT and related compounds in river water samples. *Biosens. Bioelectron.* **2007**, *22*(7), 1410–1418.
11. Daunert, S.; Barrett, G.; Feliciano, J.S.; Shetty, R.S.; Shrestha, S.; Smith-Spencer, W. Genetically engineered whole-cell sensing systems: coupling biological recognition with reporter genes. *Chem. Rev.* **2000**, *100*(7), 2705–2738.
12. Voon, C.H.; Yusop, N.M.; Khor, S.M. The state-of-the-art in bioluminescent whole-cell biosensor technology for detecting various organic compounds in oil and grease content in wastewater: From the lab to the field. *Talanta* **2022**, *241*, 123271.

13. Zhan, S.; Wu, Y.; Wang, L.; Zhan, X.; Zhou, P. A mini-review on functional nucleic acids-based heavy metal ion detection. *Biosens. Bioelectron.* **2016**, *86*, 353–368.
14. Wang, G.H.; Tsai, T.H.; Kui, C.C.; Cheng, C.Y.; Huang, T.L.; Chung, Y.C. Analysis of bioavailable toluene by using recombinant luminescent bacterial biosensors with different promoters. *J. Biol. Eng.* **2021**, *15*, 1–9.
15. Hui, C.Y.; Guo, Y.; Gao, C.X.; Li, H.; Lin, Y.R.; Yun, J.P.; Chen, Y.T.; Yi, J. A tailored indigoidine-based whole-cell biosensor for detecting toxic cadmium in environmental water samples. *Environ. Technol. Innov.* **2022**, *27*, 102511.
16. He, W.; Hu, Z.H.; Yuan, S.; Zhong, W.H.; Mei, Y.Z.; Dai, C.C. Bacterial bioreporter-based mercury and phenanthrene assessment in Yangtze river delta soils of China. *J. environ. Qual.* **2018**, *47*(3), 562–570.
17. Brányiková, I.; Lucáková, S.; Kuncová, G.; Trögl, J.; Synek, V.; Rohovec, J.; Navrátil, T. Estimation of Hg (II) in soil samples by bioluminescent bacterial bioreporter *E. coli* ARL1, and the effect of humic acids and metal ions on the biosensor performance. *Sens.* **2020**, *20*(11), 3138.
18. Andreou, V.G.; Clonis, Y.D. A portable fiber-optic pesticide biosensor based on immobilized cholinesterase and sol–gel entrapped bromcresol purple for in-field use. *Biosens. Bioelectron.* **2002**, *17*(1–2), 61–69.
19. Choi, J.W.; Kim, Y.K.; Lee, I.H.; Min, J.; Lee, W.H. Optical organophosphorus biosensor consisting of acetylcholinesterase/viologen hetero Langmuir–Blodgett film. *Biosens. Bioelectron.* **2001**, *16*(9–12), 937–943.
20. Ivanov, A.N.; Evtugyn, G.A.; Gyurcsányi, R.E.; Toth, K.; Budnikov, H.C. Comparative investigation of electrochemical cholinesterase biosensors for pesticide determination. *Anal. Chim. Acta.* **2000**, *404*(1), 55–65.
21. Sohrabi, H.; Hemmati, A.; Majidi, M.R.; Eyvazi, S.; Jahanban-Esfahlan, A.; Baradaran, B.; Adlpour-Azar, R.; Mokhtarzadeh, A.; de la Guardia, M. Recent advances on portable sensing and biosensing assays applied for detection of main chemical and biological pollutant agents in water samples: A critical review. *TrAC Trends Anal. Chem.* **2021**, *143*, 116344.
22. Justino, C.I.; Duarte, A.C.; Rocha-Santos, T.A. Recent progress in biosensors for environmental monitoring: A review. *Sens.* **2017**, *17*(12), 2918.
23. Bahadır, E.B.; Sezgintürk, M.K. Applications of commercial biosensors in clinical, food, environmental, and biothreat/biowarfare analyses. *Anal. Biochem.* **2015**, *478*, 107–120.
24. Rodriguez-Mozaz, S.; de Alda, M.J.L.; Marco, M.P.; Barceló, D. Biosensors for environmental monitoring: A global perspective. *Talanta*, **2005**, *65*(2), 291–297.
25. Sharma, K.; Sharma, M. Optical biosensors for environmental monitoring: Recent advances and future perspectives in bacterial detection. *Environ. Res.* **2023**, *236*, 116826.
26. Hu, S.Q.; Xie, J.W.; Xu, Q.H.; Rong, K.T.; Shen, G.L.; Yu, R.Q. A label-free electrochemical immunosensor based on gold nanoparticles for detection of paraoxon. *Talanta* **2003**, *61*(6), 769–777.
27. Zhong, F.; Liu, Z.; Han, Y.; Guo, Y. Electrochemical sensor for sensitive determination of capsaicin using Pd decorated reduced graphene oxide. *Electroanalysis* **2019**, *31*(6), 1182–1188.
28. Parab, A.E.; Mohanapriya, K.; Jha, N. Non enzymatic electrochemical detection of paraoxon methyl using zinc oxide graphene nanocomposite in water and food samples. *Mater. Today: Proc.* **2021**, *42*, 710–717.

29. Adiraju, A.; Brahem, A.; Lu, T.; Al-Hamry, A.; Zhou, Y.; Wei, L.; Jalasutram, A.; Tegenkamp, C.; Halouani, K.; Kanoun, O. Electrochemical enrichment of biocharcoal modified on carbon electrodes for the detection of nitrite and paraxon ethyl pesticide. *J. Compos. Sci.* **2024**, *8*(6), 217.
30. Paneru, S.; Sweety and Kumar, D. CuO@ PEDOT: PSS-grafted paper-based electrochemical biosensor for paraoxon-ethyl detection. *J. Appl. Electrochem.* **2023**, *53*(11), 2229–2238.
31. Scognamiglio, V.; Antonacci, A.; Arduini, F.; Moscone, D.; Campos, E.V.; Fraceto, L.F.; Palleschi, G. An eco-designed paper-based algal biosensor for nanoformulated herbicide optical detection. *J. Hazard. Mater.* **2019**, *373*, 483–492.
32. Arjmand, M.; Saghafifar, H.; Alijanianzadeh, M.; Soltanolkotabi, M. A sensitive tapered-fiber optic biosensor for the label-free detection of organophosphate pesticides. *Sensor Actuat. B: Chem.* **2017**, *249*, 523–532.
33. Tomassetti, M.; Conta, G.; Campanella, L.; Favero, G.; Sanzò, G.; Mazzei, F.; Antiochia, R. A flow SPR immunosensor based on a sandwich direct method. *Biosens.* **2016**, *6*(2), 22.
34. Luo, Q.; Yu, N.; Shi, C.; Wang, X.; Wu, J. Surface plasmon resonance sensor for antibiotics detection based on photo-initiated polymerization molecularly imprinted array. *Talanta*, **2016**, *161*, 797–803.
35. Cennamo, N.; Zeni, L.; Ricca, E.; Isticato, R.; Marzullo, V.M.; Capo, A.; Staiano, M.; D'Auria, S.; Varriale, A. Detection of naphthalene in sea-water by a label-free plasmonic optical fiber biosensor. *Talanta* **2019**, *194*, 289–297.
36. Xue, C.S.; Erika, G.; Jiří, H. Surface plasmon resonance biosensor for the ultrasensitive detection of bisphenol A. *Anal. Bioanal. Chem.* **2019**, *411*, 5655–5658.
37. Chen, J.; Zheng, A.; Chen, A.; Gao, Y.; He, C.; Kai, X.; Wu, G.; Chen, Y. A functionalized gold nanoparticles and Rhodamine 6G based fluorescent sensor for high sensitive and selective detection of mercury (II) in environmental water samples. *Anal. Chim. Acta.* **2007**, *599*(1), 134–142.
38. Long, F.; Gao, C.; Shi, H.C.; He, M.; Zhu, A.N.; Klibanov, A.M.; Gu, A.Z. Reusable evanescent wave DNA biosensor for rapid, highly sensitive, and selective detection of mercury ions. *Biosens. Bioelectron.* **2011**, *26*(10), 4018–4023.
39. Zhang, Y.; Li, H.; Chen, M.; Fang, X.; Pang, P.; Wang, H.; Wu, Z.; Yang, W. Ultrasensitive electrochemical biosensor for silver ion based on magnetic nanoparticles labeling with hybridization chain reaction amplification strategy. *Sensor Actuat. B: Chem.* **2017**, *249*, 431–438.
40. Liu, L.; Zhou, X.; Lu, M.; Zhang, M.; Yang, C.; Ma, R.; Memon, A.G.; Shi, H.; Qian, Y. An array fluorescent biosensor based on planar waveguide for multi-analyte determination in water samples. *Sensor. Actuat. B: Chem.* **2017**, *240*, 107–113.
41. Stavra, E.; Petrou, P.S.; Koukouvinos, G.; Kiritsis, C.; Pirmettis, I.; Papadopoulos, M.; Goustouridis, D.; Economou, A.; Misiakos, K.; Raptis, I.; Kakabakos, S.E. Simultaneous determination of paraquat and atrazine in water samples with a white light reflectance spectroscopy biosensor. *J. Hazard. Mater.* **2018**, *359*, 67–75.
42. Steinke, N.; Döring, S.; Wuchrer, R.; Kroh, C.; Gerlach, G.; Härtling, T. Plasmonic sensor for on-site detection of diclofenac molecules. *Sensor. Actuat. B: Chem.* **2019**, *288*, 594–600.
43. Liu, L.H.; Zhou, X.H.; Xu, W.Q.; Song, B.D.; Shi, H.C. Highly sensitive detection of sulfadimidine in water and dairy products by means of an evanescent wave optical biosensor. *RSC Adv.* **2014**, *4*(104), 60227–60233.
44. Sciacca, B.; Secret, E.; Pace, S.; Gonzalez, P.; Geobaldo, F.; Quignard, F.; Cunin, F. Chitosan-functionalized porous silicon optical transducer for the detection of carboxylic acid-containing drugs in water. *J. Mater. Chem.* **2011**, *21*(7), 2294–2302.

45. Wong, L.S.; Lee, Y.H.; Surif, S. Whole cell biosensor using Anabaena torulosa with optical transduction for environmental toxicity evaluation. *J. Sens.* **2013**, *2013*(1), 567272.
46. Ronkainen, N.J.; Halsall, H.B.; Heineman, W.R. Electrochemical biosensors. *Chem. Soc. Rev.* **2010**, *39*(5), 1747–1763.
47. Belkhamssa, N.; da Costa, J.P.; Justino, C.I.; Santos, P.S.; Cardoso, S.; Duarte, A.C.; Rocha-Santos, T.; Ksibi, M. Development of an electrochemical biosensor for alkylphenol detection. *Talanta* **2016**, *158*, 30–34.
48. Zheng, H.; Yan, Z.; Wang, M.; Chen, J.; Zhang, X. Biosensor based on polyaniline-polyacrylonitrile-graphene hybrid assemblies for the determination of phenolic compounds in water samples. *J. Hazard. Mater.* **2019**, *378*, 120714.
49. Hui, Y.; Huang, Z.; Alahi, M.E.E.; Nag, A.; Feng, S.; Mukhopadhyay, S.C. Recent advancements in electrochemical biosensors for monitoring the water quality. *Biosens.* **2022**, *12*(7), 551.
50. Ayenimo, J.G.; Adeloju, S.B. Rapid amperometric detection of trace metals by inhibition of an ultrathin polypyrrole-based glucose biosensor. *Talanta* **2016**, *148*, 502–510.
51. Rengaraj, S.; Cruz-Izquierdo, Á.; Scott, J.L.; Di Lorenzo, M. Impedimetric paper-based biosensor for the detection of bacterial contamination in water. *Sens. Actuat. B: Chem.* **2018**, *265*, 50–58.
52. Melios, C.; Panchal, V.; Edmonds, K.; Lartsev, A.; Yakimova, R.; Kazakova, O. Detection of ultralow concentration NO_2 in complex environment using epitaxial graphene sensors. *ACS Sens.* **2018**, *3*(9), 1666–1674.
53. Maduraiveeran, G.; Jin, W. Nanomaterials based electrochemical sensor and biosensor platforms for environmental applications. *Trends Environ. Anal. Chem.* **2017**, *13*, 10–23.
54. Hong, S.; Park, S.; Lee, S.; Yang, Y.I.; Song, H.D.; Yi, J. The sensitive, anion-selective detection of arsenate with poly (allylamine hydrochloride) by single particle plasmon-based spectroscopy. *Anal. Chim. Acta.* **2011**, *694*(1–2), 136–141.
55. Dong, Z.Z.; Yang, C.; Vellaisamy, K.; Li, G.; Leung, C.H.; Ma, D.L. Construction of a nano biosensor for cyanide anion detection and its application in environmental and biological systems. *ACS Sens.* **2017**, *2*(10), 1517–1522.
56. Xiong, Y.; Huang, Q.; Canady, T.D.; Barya, P.; Liu, S.; Arogundade, O.H.; Race, C.M.; Che, C.; Wang, X.; Zhou, L.; Wang, X. Photonic crystal enhanced fluorescence emission and blinking suppression for single quantum dot digital resolution biosensing. *Nat. Commun.* **2022**, *13*(1), 4647.
57. Sansone, L.; Macchia, E.; Taddei, C.; Torsi, L.; Giordano, M. Label-free optical biosensing at femtomolar detection limit. *Sens. Actuat. B: Chem.* **2018**, *255*, 1097–1104.
58. Nehra, M.; Dilbaghi, N.; Kumar, R.; Kumar, S. Trends in point-of-care optical biosensors for antibiotics detection in aqueous media. *Mater. Lett.* **2022**, *308*, 131235.
59. Arjmand, M.; Saghafifar, H.; Alijanianzadeh, M.; Soltanolkotabi, M. A sensitive tapered-fiber optic biosensor for the label-free detection of organophosphate pesticides. *Sens. Actuat. B: Chem.* **2017**, *249*, 523–532.
60. Lamarca, R.S.; Franco, D.F.; Nalin, M.; de Lima Gomes, P.C.F.; Messaddeq, Y. Label-free ultrasensitive and environment-friendly immunosensor based on a silica optical fiber for the determination of ciprofloxacin in wastewater samples. *Anal. Chem.* **2020**, *92*(21), 14415–14422.
61. Koubova, V.; Brynda, E.; Karasova, L.; Škvor, J.; Homola, J.; Dostalek, J.; Tobiška, P.; Rošický, J. Detection of foodborne pathogens using surface plasmon resonance biosensors. *Sens. Actuat. B: Chem.* **2001**, *74*(1–3), 100–105.

62. Pathirana, S.T.; Barbaree, J.; Chin, B.A.; Hartell, M.G.; Neely, W.C.; Vodyanoy, V. Rapid and sensitive biosensor for Salmonella. *Biosens. Bioelectron.* **2000**, *15*(3–4), 135–141.
63. Ercole, C.; Del Gallo, M.; Pantalone, M.; Santucci, S.; Mosiello, L.; Laconi, C.; Lepidi, A. A biosensor for *Escherichia coli* based on a potentiometric alternating biosensing (PAB) transducer. *Sens. Actuat. B: Chem.* **2002**, *83*(1–3), 48–52.
64. Dolai, S.; Bhunia, S.K.; Beglaryan, S.S.; Kolusheva, S.; Zeiri, L.; Jelinek, R. Colorimetric polydiacetylene–aerogel detector for volatile organic compounds (VOCs). *ACS Appl. Mater. Interfaces.* **2017**, 9(3), 2891–2898.
65. Chee, G.J.; Nomura, Y.; Ikebukuro, K.; Karube, I. Optical fiber biosensor for the determination of low biochemical oxygen demand. *Biosens. Bioelectron.* **2000**, *15*(7–8), 371–376.
66. Gutiérrez, J.C.; Amaro, F.; Martín-González, A. Heavy metal whole-cell biosensors using eukaryotic microorganisms: an updated critical review. *Front. Microbial.* **2015**, *6*, 48.
67. Yilmazer, M. Recent Studies in Yeast-Based Whole-Cell Biosensors. *Biotechnology and Application Areas*, **2022**, 105.
68. Kannappan, S.; Ramisetty, B.C.M. Engineered whole-cell-based biosensors: Sensing environmental heavy metal pollutants in water—a review. *Appl. Biochem. Biotechnol.* **2022**, *194*, 1814–1840.
69. Oluwaseun Adetunji, C.; Mathew, J.T.; Inobeme, A.; Olaniyan, O.T.; RB Singh, K.; Abimbola, O.F.; Nayak, V.; Singh, J.; Singh, R.P. Microbial and plant cell biosensors for environmental monitoring. In *Nanobiosensors for Environmental Monitoring: Fundamentals and Application*. Cham: Springer International Publishing, **2022**, 175–190.
70. Kharkova, A.S.; Arlyapov, V.A.; Ilyukhina, A.S.; Ponamoreva, O.N.; Alferov, V.A.; Reshetilov, A.N. A kinetic approach to the formation of two-mediator systems for developing microbial biosensors as exemplified by a rapid biochemical oxygen demand assay. *3 Biotech* **2021**, *11*, 1–13.
71. Jang, J.D.; Barford, J.P.; Renneberg, R. Application of biochemical oxygen demand (BOD) biosensor for optimization of biological carbon and nitrogen removal from synthetic wastewater in a sequencing batch reactor system. *Biosens. Bioelectron.* **2004**, *19*(8), 805–812.
72. Chee, G.J.; Nomura, Y.; Ikebukuro, K.; Karube, I. Development of photocatalytic biosensor for the evaluation of biochemical oxygen demand. *Biosens. Bioelectron.* **2005**, *21*(1), 67–73.
73. Alpat, Ş.; Alpat, S.K.; Çadırcı, B.H.; Yaşa, İ.; Telefoncu, A. A novel microbial biosensor based on Circinella sp. modified carbon paste electrode and its voltammetric application. *Sens. Actuat. B: Chem.* **2008**, *134*(1), 175–181.
74. Yüce, M.; Nazır, H.; Dönmez, G. An advanced investigation on a new algal sensor determining Pb (II) ions from aqueous media. *Biosens. Bioelectron.* **2010**, *26*(2), 321–326.
75. Hnaien, M.; Bourigua, S.; Bessueille, F.; Bausells, J.; Errachid, A.; Lagarde, F.; Jaffrezic-Renault, N. Impedimetric microbial biosensor based on single wall carbon nanotube modified microelectrodes for trichloroethylene detection. *Electrochim. Acta.* **2011**, *56*(28), 10353–10358.
76. Roy, R.; Ray, S.; Chowdhury, A.; Anand, R. Tunable multiplexed whole-cell biosensors as environmental diagnostics for ppb-level detection of aromatic pollutants. *ACS Sens.* **2021**, *6*(5), 1933–1939.

77. Chen, W.X.; Li, Q.; Wang, Z.; Sun, Z.J. Spatial distribution characteristics and pollution evaluation of heavy metals in arable land soil of China. *Huan Jing ke Xue= Huanjing Kexue* **2020**, *41*(6), 2822–2833.
78. Duffus J. "Heavy metals" a meaningless term? (IUPAC Technical Report). *Pure Appl. Chem.* **2002**, *74*(5), 793–807.
79. Genchi, G.; Sinicropi, M.S.; Lauria, G.; Carocci, A.; Catalano, A. The effects of cadmium toxicity. *Int. J. Environ. Res. Public Health* **2020**, *17*(11), 3782.
80. Balali-Mood, M.; Naseri, K.; Tahergorabi, Z.; Khazdair, M.R.; Sadeghi, M. Toxic mechanisms of five heavy metals: Mercury, lead, chromium, cadmium, and arsenic. *Front. Pharmacol.* **2021**, *12*, 643972.
81. Li, F.; Duan, X.; Li, H.; Zou, L.; Liu, G.; Liu, F.; Zhang, G.; Xu, J. Dual effect of aminobutyric acid group and "molecular wire effect" of conjugated polymer enables ultratrace detection of $Cr_2O_7^{2-}$ in fruits. *Microchem. J.* **2022**, *178*, 107426.
82. Dzantiev, B.B.; Zherdev, A.V. Antibody-based biosensors. In *Portable Biosensing of Food Toxicants and Environmental Pollutants*. Boca Raton, FL: CRC Press Book, **2013**, 161–196.
83. Wu, C.S.; Khaing Oo, M.K.; Fan, X. Highly sensitive multiplexed heavy metal detection using quantum-dot-labeled DNAzymes. *ACS Nano* **2010**, *4*(10), 5897–5904.
84. Ilangovan, R.; Daniel, D.; Krastanov, A.; Zachariah, C.; Elizabeth, R. Enzyme based biosensor for heavy metal ions determination. *Biotechnol. Biotechnol. Equip.* **2006**, *20*(1), 184–189.
85. Li, D.; Wei, S.; Yang, H.; Li, Y.; Deng, A. A sensitive immunochromatographic assay using colloidal gold–antibody probe for rapid detection of pharmaceutical indomethacin in water samples. *Biosens. Bioelectron.* **2009**, *24*(7), 2277–2280.
86. Lin, Y.Y.; Wang, J.; Liu, G.; Wu, H.; Wai, C.M.; Lin, Y. A nanoparticle label/immunochromatographic electrochemical biosensor for rapid and sensitive detection of prostate-specific antigen. *Biosens. Bioelectron.* **2008**, *23*(11), 1659–1665.
87. Wu, Z.; Shen, H.; Hu, J.; Fu, Q.; Yao, C.; Yu, S.; Xiao, W.; Tang, Y. Aptamer-based fluorescence-quenching lateral flow strip for rapid detection of mercury (II) ion in water samples. *Anal Bioanal. Chem.* **2017**, *409*, 5209–5216.
88. Srinivasan, S.; Ranganathan, V.; McConnell, E.M.; Murari, B.M.; DeRosa, M.C. Aptamer-based colorimetric and lateral flow assay approaches for the detection of toxic metal ions, thallium (i) and lead (ii). *RSC Adv.* **2023**, *13*(29), 20040–20049.
89. Medina-Sánchez, M.; Cadevall, M.; Ros, J.; Merkoçi, A. Eco-friendly electrochemical lab-on-paper for heavy metal detection. *Anal. Bioanal. Chem.* **2015**, *407*, 8445–8449.
90. Chao, C.H.; Wu, C.S.; Huang, C.C.; Liang, J.C.; Wang, H.T.; Tang, P.T.; Lin, L.Y.; Ko, F.H. A rapid and portable sensor based on protein-modified gold nanoparticle probes and lateral flow assay for naked eye detection of mercury ion. *Microelectron. Eng.* **2012**, *97*, 294–296.
91. Kato, D.; Kamata, T.; Kato, D.; Yanagisawa, H.; Niwa, O. Au nanoparticle-embedded carbon films for electrochemical As3+ detection with high sensitivity and stability. *Anal. Chem.* **2016**, *88*(5), 2944–2951.
92. Yadav, I.C.; Devi, N.L. Pesticides classification and its impact on human and the environment. *Environ. Eng. Sci.* **2017**, *6*(7), 140–158.
93. Sjerps, R.M.; Kooij, P.J.; van Loon, A.; Van Wezel, A.P. Occurrence of pesticides in Dutch drinking water sources. *Chemosphere*, **2019**, *235*, 510–518.
94. Gamón, M.; Lleó, C.; Ten, A.; Mocholí, F. Multiresidue determination of pesticides in fruit and vegetables by gas chromatography/tandem mass spectrometry. *J. AOAC Int.* **2001**, *84*(4), 1209–1216.

95. Lehotay, S.J.; Kok, A.D.; Hiemstra, M.; Bodegraven, P.V. Validation of a fast and easy method for the determination of residues from 229 pesticides in fruits and vegetables using gas and liquid chromatography and mass spectrometric detection. *J. AOAC Int.* **2005**, *88*(2), 595–614.
96. Singh, A.K.; Flounders, A.W.; Volponi, J.V.; Ashley, C.S.; Wally, K.; Schoeniger, J.S. Development of sensors for direct detection of organophosphates. Part I: immobilization, characterization and stabilization of acetylcholinesterase and organophosphate hydrolase on silica supports. *Biosens. Bioelectron.* **1999**, *14*(8–9), 703–713.
97. Vakurov, A.; Simpson, C.E.; Daly, C.L.; Gibson, T.D.; Millner, P.A. Acetylecholinesterase-based biosensor electrodes for organophosphate pesticide detection: II. Immobilization and stabilization of acetylecholinesterase. *Biosens. Bioelectron.* **2005**, *20*(11), 2324–2329.
98. Montesinos, T.; Pérez-Munguia, S.; Valdez, F.; Marty, J.L., Disposable cholinesterase biosensor for the detection of pesticides in water-miscible organic solvents. *Anal. Chim. Acta* **2001**, *431*(2), 231–237.
99. Andreescu, S.; Noguer, T.; Magearu, V.; Marty, J.L. Screen-printed electrode based on AChE for the detection of pesticides in presence of organic solvents. *Talanta* **2002**, *57*(1), 169–176.
100. Somerset, V.S.; Klink, M.J.; Sekota, M.M.; Baker, P.G.; Iwuoha, E.I. Polyaniline-mercaptobenzothiazole biosensor for organophosphate and carbamate pesticides. *Anal. Lett.* **2006**, *39*(8), 1683–1698.
101. Arduini, F.; Guidone, S.; Amine, A.; Palleschi, G.; Moscone, D. Acetylcholinesterase biosensor based on self-assembled monolayer-modified gold-screen printed electrodes for organophosphorus insecticide detection. *Sens. Actuat. B: Chem.* **2013**, *179*, 201–208.
102. Marrazza, G. Piezoelectric biosensors for organophosphate and carbamate pesticides: A review. *Biosens.* **2014**, *4*(3), 301–317.
103. Giardi, M.T.; Koblízek, M.; Masojí dek J. Photosystem II-based biosensors for the detection of pollutants. *Biosens. Bioelectron.* **2001**, *16*(9–12), 1027–1033.
104. Koblizek, M.; Masojidek, J.; Komenda, J.; Kucera, T.; Pilloton, R.; Mattoo, A.K.; Giardi, M.T. A sensitive photosystem II-based biosensor for detection of a class of herbicides. *Biotechnol. Bioeng.* **1998**, *60*(6), 664–669.
105. Mallat, E.; Barzen, C.; Abuknesha, R.; Gauglitz, G.; Barcelo, D. Fast determination of paraquat residues in water by an optical immunosensor and validation using capillary electrophoresis-ultraviolet detection. *Anal. Chim. Acta.* **2001**, *427*(2), 165–171.
106. Mallat, E.; Barzen, C.; Klotz, A.; Brecht, A.; Gauglitz, G.; Barcelo, D. River analyzer for chlorotriazines with a direct optical immunosensor. *Environ. Sci. Technol.* **1999**, *33*(6), 965–971.
107. Wilmer, M.; Trau, D.; Renneberg, R.; Spener, F. Amperometric immunosensor for the detection of 2, 4-dichlorophenoxyacetic acid (2, 4-D) in water. *Synth. React. Inorg. Met.Org. Chem.* **1997**, *30*(3), 515–525.
108. Scognamiglio, V.; Antonacci, A.; Arduini, F.; Moscone, D.; Campos, E.V.; Fraceto, L.F.; Palleschi, G. An eco-designed paper-based algal biosensor for nanoformulated herbicide optical detection. *J. Hazard. Mater.* **2019**, *373*, 483–492.
109. Koukouvinos, G.; Tsialla, Z.; Petrou, P.S.; Misiakos, K.; Goustouridis, D.; Moreno, A.U.; Fernandez-Alba, A.R.; Raptis, I.; Kakabakos, S.E. Fast simultaneous detection of three pesticides by a white light reflectance spectroscopy sensing platform. *Sens. Actuat. B: Chem.* **2017**, *238*, 1214–1223.
110. Villegas, L.G.C.; Mashhadi, N.; Chen, M.; Mukherjee, D.; Taylor, K.E.; Biswas, N. A short review of techniques for phenol removal from wastewater. *Curr. Pollut. Rep.* **2016**, *2*, 157–167.

111. Kulkarni, S.J.; Kaware, J.P. Review on research for removal of phenol from wastewater. *Int. J. Sci. Res. Publ.* 2013, *3*(4), 1–5.
112. Makarewicz, M.; Drożdż, I.; Tarko, T.; Duda-Chodak, A. The interactions between polyphenols and microorganisms, especially gut microbiota. *Antioxidants* **2021**, *10*(2), 188.
113. Ye, Y.; Guo, H.; Sun, X. Recent progress on cell-based biosensors for analysis of food safety and quality control. *Biosens. Bioelectron.* **2019**, *126*, 389–404.
114. Zheng, Y.; Yang, C.; Pu, W.; Zhang, J. Carbon nanotube-based DNA biosensor for monitoring phenolic pollutants. *Microchim. Acta.* **2009**, *166*, 21–26.
115. Othman, A.; Karimi, A.; Andreescu, S. Functional nanostructures for enzyme based biosensors: Properties, fabrication and applications. *J. Mater. Chem. B* **2016**, *4*(45), 7178–7203.
116. Sekretaryova, A.N.; Volkov, A.V.; Zozoulenko, I.V.; Turner, A.P.; Vagin, M.Y.; Eriksson, M. Total phenol analysis of weakly supported water using a laccase-based microband biosensor. *Anal. Chim. Acta.* **2016**, *907*, 45–53.
117. Ji, X.; Su, Z.; Xu, M.; Ma, G.; Zhang, S. TiO_2–horseradish peroxidase hybrid catalyst based on hollow nanofibers for simultaneous photochemical–enzymatic degradation of 2, 4-dichlorophenol. *ACS Sustain. Chem. Eng.* **2016**, *4*(7), 3634–3640.
118. Abd Manan, F.A.; Hong, W.W.; Abdullah, J.; Yusof, N.A.; Ahmad, I. Nanocrystalline cellulose decorated quantum dots based tyrosinase biosensor for phenol determination. *Mater. Sci. Eng. C* **2019**, *99*, 37–46.
119. Upan, J.; Reanpang, P.; Chailapakul, O.; Jakmunee, J. Flow injection amperometric sensor with a carbon nanotube modified screen printed electrode for determination of hydroquinone. *Talanta* **2016**, *146*, 766–771.
120. Lou, C.; Jing, T.; Zhou, J.; Tian, J.; Zheng, Y.; Wang, C.; Zhao, Z.; Lin, J.; Liu, H.; Zhao, C.; Guo, Z. Laccase immobilized polyaniline/magnetic graphene composite electrode for detecting hydroquinone. *Int. J. Biol. Macromol.* **2020**, *149*, 1130–1138.
121. Huang, H.; Chen, Y.; Chen, Z.; Chen, J.; Hu, Y.; Zhu, J.J. Electrochemical sensor based on Ce-MOF/carbon nanotube composite for the simultaneous discrimination of hydroquinone and catechol. *J. Hazard. Mater.* **2021**, *416*, 125895.
122. Cosio, M.S.; Pellicano, A.; Brunetti, B.; Fuenmayor, C.A. A simple hydroxylated multi-walled carbon nanotubes modified glassy carbon electrode for rapid amperometric detection of bisphenol A. *Sens. Actuat. B: Chem.* **2017**, *246*, 673–679.
123. Hao, Y.U.; Xiao, F.E.N.G.; Xiao-Xia, C.H.E.N.; Jin-Li, Q.I.A.O.; Xiao-Ling, G.A.O.; Na, X.U.; Lou-Jun, G.A.O. Electrochemical determination of bisphenol A on a glassy carbon electrode modified with gold nanoparticles loaded on reduced graphene oxide-multi walled carbon nanotubes composite. *Chin. J. Anal. Chem.* **2017**, *45*(5), 713–720.
124. Farré, M.; Pasini, O.; Alonso, M.C.; Castillo, M.; Barceló, D. Toxicity assessment of organic pollution in wastewaters using a bacterial biosensor. *Anal. Chim. Acta.* **2001**, *426*(2), 155–165.
125. Tizzard, A.C.; Bergsma, J.H.; Lloyd-Jones, G. A resazurin-based biosensor for organic pollutants. *Biosens. Bioelectron.* **2006**, *22*(5), 759–763.
126. Bidmanova, S.; Kotlanova, M.; Rataj, T.; Damborsky, J.; Trtilek, M.; Prokop, Z. Fluorescence-based biosensor for monitoring of environmental pollutants: from concept to field application. *Biosens. Bioelectron.* **2016**, *84*, 97–105.
127. Tomassetti, M.; Conta, G.; Campanella, L.; Favero, G.; Sanzò, G.; Mazzei, F.; Antiochia, R. A flow SPR immunosensor based on a sandwich direct method. *Biosens.* **2016**, *6*(2), 22.
128. Luo, Q.; Yu, N.; Shi, C.; Wang, X.; Wu, J. Surface plasmon resonance sensor for antibiotics detection based on photo-initiated polymerization molecularly imprinted array. *Talanta*, **2016**, *161*, 797–803.

129. Rodriguez-Mozaz, S.; Lopez de Alda, M. J.; Barceló, D. Biosensors as useful tools for environmental analysis and monitoring. *Anal. Bioanal. Chem.* **2006**, *386*, 1025–1041.
130. Chadha, U.; Bhardwaj, P.; Agarwal, R.; Rawat, P.; Agarwal, R.; Gupta, I.; Panjwani, M.; Singh, S.; Ahuja, C.; Selvaraj, S. K. Recent progress and growth in biosensors technology: A critical review. *J. Ind. Eng. Chem.* **2022**, *109*, 21–51.
131. Naresh, V.; Lee, N. A review on biosensors and recent development of nanostructured materials-enabled biosensors. *Sensors* **2021**, *21*(4), 1109.
132. Yu, X.; Cheng, H.; Zhang, M.; Zhao, Y.; Qu, L.; Shi, G. Graphene-based smart materials. *Nat. Rev. Mater.* **2017**, *2*(9), 1–13.
133. Cataldi, P.; Athanassiou, A.; Bayer, I. S. Graphene nanoplatelets-based advanced materials and recent progress in sustainable applications. *Appl. Sci.* **2018**, *8*(9), 1438.
134. Kang, S.; Rahman, A.; McGinnis, S.; Vikesland, P. Toward environmentally favorable nano-sensing by production of reusable gold nanoparticles from gold nano-waste: life cycle and nanocircular economy implications. *Environ. Sci. Nano,* **2024**, *11*(4), 1499–1507.
135. Silva-Neto, H. A.; de Lima, L. F.; Rocha, D. S.; Ataide, V. N.; Meloni, G. N.; Moro, G.; Raucci, A.; Cinti, S.; Paixão, T. R.; de Araujo, W. R. Recent achievements of greenness metrics on paper-based electrochemical (bio) sensors for environmental and clinical analysis. *TrAC Trends Anal. Chem.* **2024**, *174*, 117675.
136. Thakur, A.; Kumar, A. Recent advances on rapid detection and remediation of environmental pollutants utilizing nanomaterials-based (bio) sensors. *Sci. Total Environ.* **2022**, *834*, 155219.
137. BS, M. N. D.; Babu, L. R.; Paul, A. E.; Murugan, S.; Periakaruppan, R. Nanozymes as catalytic marvels for biomedical and environmental concerns: A chemical engineering approach. *J. Clust. Sci.* **2024**, *35*(3), 715–740.
138. Hassan, S. H.; Van Ginkel, S. W.; Hussein, M. A.; Abskharon, R.; Oh, S.-E. Toxicity assessment using different bioassays and microbial biosensors. *Environ. Int.* **2016**, *92*, 106–118.
139. Gabrielli, V.; Frasconi, M. Cellulose-based functional materials for sensing. *Chemosensors* **2022**, *10*(9), 352.
140. Kummari, S.; Panicker, L. R.; Rao Bommi, J.; Karingula, S.; Sunil Kumar, V.; Mahato, K.; Goud, K. Y. Trends in paper-based sensing devices for clinical and environmental monitoring. *Biosens.* **2023**, *13*(4), 420.
141. Guerra, F. D.; Attia, M. F.; Whitehead, D. C.; Alexis, F. Nanotechnology for environmental remediation: materials and applications. *Molecules* **2018**, *23*(7), 1760.
142. Ahamed, A.; Ge, L.; Zhao, K.; Veksha, A.; Bobacka, J.; Lisak, G. Environmental footprint of voltammetric sensors based on screen-printed electrodes: An assessment towards "green" sensor manufacturing. *Chemosphere* **2021**, *278*, 130462.
143. Pratama, A.; Rahayu, S. Biosensors enabled by nanotechnology: Advancing precision agriculture through IoT monitoring. *J. Intell. Connect Emerg. Technol.* **2024**, *9*(1), 25–39.
144. Ali, K.; Nishtar, Z.; Sabir, R. M.; Safdar, M. Biosensors for environmental monitoring in the smart agriculture sector. In Alex Khang (eds.), *Agriculture and Aquaculture Applications of Biosensors and Bioelectronics.* IGI Global, **2024**, 135–154.
145. Mohamed, E. F.; Awad, G. Development of nano-sensor and biosensor as an air pollution detection technique for the foreseeable future. In Nouha Bakaraki Turan, Guleda Onkal Engin, and Mehmet Sinan Bilgili (eds.), *Comprehensive Analytical Chemistry,* **2022**, *99*, 163–188.
146. Seesaard, T.; Kamjornkittikoon, K.; Wongchoosuk, C. A comprehensive review on advancements in sensors for air pollution applications. *Sci. Total Environ.* **2024**, 175696.

147. Gupta, M.; Saini, S.; Sahoo, M. Determinants of ecological footprint and PM2. 5: Role of urbanization, natural resources and technological innovation. *Env. Chall.* **2022**, *7*, 100467.
148. Kaur, J.; Singh, P. K. Enzyme-based optical biosensors for organophosphate class of pesticide detection. *Phys. Chem. Chem. Phys.* **2020**, *22*(27), 15105–15119.
149. Liu, C.; Yu, H.; Zhang, B.; Liu, S.; Liu, C.-g.; Li, F.; Song, H. Engineering whole-cell microbial biosensors: Design principles and applications in monitoring and treatment of heavy metals and organic pollutants. *Biotechnol. Adv.* **2022**, *60*, 108019.
150. Souiri, M.; Gammoudi, I.; Ouada, H. B.; Mora, L.; Jouenne, T.; Jaffrezic-Renault, N.; Dejous, C.; Othmane, A.; Duncan, A. *Escherichia coli*-functionalized magnetic nanobeads as an ultrasensitive biosensor for heavy metals. *Procedia Chem.* **2009**, *1*(1), 1027–1030.
151. Su, X.; Sutarlie, L.; Loh, X. J. Sensors and analytical technologies for air quality: particulate matters and bioaerosols. *Chem. Asian J.* **2020**, *15*(24), 4241–4255.
152. Kumar, A.; Maiti, P. Paper based sustainable biosensors. *Mater. Adv.* **2024**, *5*, 3563–3586.
153. Oujji, N. B.; Bakas, I.; Istamboulié, G.; Ait-Ichou, I.; Ait-Addi, E.; Rouillon, R.; Noguer, T. A Simple Colorimetric Enzymatic-Assay, based on immobilization of acetylcholinesterase by adsorption, for sensitive detection of organophosphorus insecticides in olive oil. *Food Control* **2014**, *46*, 75–80.
154. Durrieu, C.; Chouteau, C.; Barthet, L.; Chovelon, J. M.; Tran-Minh, C. A bi-enzymatic whole-cell algal biosensor for monitoring waste water pollutants. *Anal. Lett.* **2004**, *37*(8), 1589–1599.
155. Al Mamun, A.; Nafiujjaman, M.; Ahammad, A. S. Carbon nanomaterial-based sensors in air pollution remediation. In Jamballi G. Manjunatha and Chaudhery Mustansar Hussain (eds.), *Carbon Nanomaterials-Based Sensors*, **2022**, 105–123.
156. González, E.; Casanova-Chafer, J.; Romero, A.; Vilanova, X.; Mitrovics, J.; Llobet, E. LoRa sensor network development for air quality monitoring or detecting gas leakage events. *Sensors* **2020**, *20*(21), 6225.
157. Siontorou, C. G.; Georgopoulos, K. N. A biosensor platform for soil management: the case of nitrites. *J. Clean. Prod.* **2016**, *111*, 133–142.
158. Qiu, G.; Yue, Y.; Tang, J.; Zhao, Y.-B.; Wang, J. Total bioaerosol detection by a Succinimidyl-Ester-functionalized Plasmonic biosensor to reveal different characteristics at three locations in Switzerland. *Environ. Sci.Technol.* **2020**, *54*(3), 1353–1362.
159. Suo, Z.; Liang, R.; Liu, R.; Wei, M.; He, B.; Jiang, L.; Sun, X.; Jin, H. A convenient paper-based fluorescent aptasensor for high-throughput detection of Pb2+ in multiple real samples (water-soil-food). *Anal. Chim. Acta.* **2023**, *1239*, 340714.
160. Zielinski, O.; Busch, J. A.; Cembella, A. D.; Daly, K. L.; Engelbrektsson, J.; Hannides, A. K.; Schmidt, H. Detecting marine hazardous substances and organisms: sensors for pollutants, toxins, and pathogens. *Ocean Sci.* **2009**, *5*(3), 329–349.
161. Laneve, G.; Bruno, M.; Mukherjee, A.; Messineo, V.; Giuseppetti, R.; De Pace, R.; Magurano, F.; D'Ugo, E. Remote sensing detection of algal blooms in a lake impacted by petroleum hydrocarbons. *Remote Sens.* **2021**, *14*(1), 121.
162. Saravanakumar, K.; SivaSantosh, S.; Sathiyaseelan, A.; Naveen, K. V.; AfaanAhamed, M. A.; Zhang, X.; Priya, V. V.; MubarakAli, D.; Wang, M.-H. Unraveling the hazardous impact of diverse contaminants in the marine environment: Detection and remedial approach through nanomaterials and nano-biosensors. *J. Hazard. Mater.* **2022**, *433*, 128720.

13 Biosensors for Food Safety and Quality Assurance

Sumitha M. S. and Xavier T. S.

13.1 INTRODUCTION

Food safety and quality assurance must be included in the worldwide food chain as fundamental components to ensure that the food we eat is real, tasty, safe, and healthy. Food safety raises several different issues. The public health issues of contamination, adulteration, and foodborne diseases are major; the described activities try to lower these hazards. Ensuring that final products follow set criteria depends on quality assurance in food production, processing, and distribution. Rising consumer expectations and laws have created an unmatched market for consistent and efficient food safety control systems

In this field, the biosensor represents a noteworthy success since it combines sensor technology with biological systems to allow the identification of particular chemical or biological substances (Bhatia et al., 2024). Particularly for the identification of food allergies, poisons, pollutants, and various other food-related diseases, the use of biosensors in the food sector is an increasing trend. Superior selectivity, improved sensitivity, and fast response times define these sensor systems (Hierlemann & Gutierrez-Osuna, 2008). This performance is a model of their accuracy, efficiency, and financial feasibility. Their potential exceeds those of more traditional approaches. Rapid identification of chemical pollutants including pesticides and detection of dangerous organisms such as *Salmonella* and *Escherichia coli* are made possible by biosensors. This guarantees that food is safe and of better quality for farmers, authorities, and consumers alike.

The development and improvement of biosensors used in the food industry have been greatly impacted by recent scientific breakthroughs in nanotechnology, molecular biology, microelectronics, and artificial intelligence (AI)-assisted point-of-care devices (Wasilewski et al., 2024). When it started in the 1960s, the field was mostly used for medical testing. Since then, it has grown to include uses in food safety. Modern technologies have made it possible to create sophisticated biosensors able to detect minute amounts of pollutants in food. The rapid development of biosensors has made them indispensable instruments for food testing laboratories, regulatory

 DOI: 10.1201/9781003536239-13

authorities, and manufacturers who have to ensure the safety of their goods and follow ever-stricter worldwide rules.

The application of biosensors in improving food safety and quality control will be under focus in this chapter. The chapter intends to give a comprehensive overview of biosensors, together including their several forms, uses in the food sector, and historical evolution. With particular attention to their impact on food safety procedures, this will investigate the relevance of biosensors in identifying different pollutants, diseases, poisons, and allergies. Together with expected future advancements in the field, it will examine the underlying limits and constraints of biosensor technology. Also, the ability of technology to change food safety monitoring is investigated. Its integration with other developing technologies and the Internet of Things (IoT) will offer constant, instantaneous control.

13.2 FUNDAMENTALS OF BIOSENSORS

A biosensor is an analytical device that combines a biological element with a sensor to detect and quantify specific chemical or biological substances (Purohit et al., 2020). The biological component, which could be an enzyme, antibody, nucleic acid, or microorganism, interacts with the target substance (analyte), producing a measurable signal. The components of a biosensor typically include the biological recognition element, a transducer, and a signal processing system. The biological recognition element is responsible for the specific interaction with the analyte, while the transducer converts the biochemical interaction into an electrical signal. This signal is then processed and analyzed to provide information about the analyte's concentration or presence.

Biosensors can be classified based on the type of biological recognition element used. Some common types include enzyme-based biosensors, which use enzymes to catalyze reactions with the target substance; immunosensors, which rely on antibodies for antigen detection; DNA-based biosensors, which utilize nucleic acids for the identification of specific genetic sequences; and microbial biosensors, which employ microorganisms as biological elements to detect contaminants or toxins. Other variations include optical biosensors, which measure changes in light; and electrochemical biosensors, which detect changes in electrical properties such as current or potential (Sumitha & Xavier, 2023). Figure 13.1 schematically illustrates the working of a biosensor (Suvarnaphaet & Pechprasarn, 2017).

The working principle of a biosensor involves a two-step process: the biological recognition of the target analyte followed by a transduction of this interaction into a detectable signal. When the analyte binds to the biological recognition element, a biochemical reaction occurs, such as enzyme–substrate interaction, antigen–antibody binding, or nucleic acid hybridization. This interaction causes a change in the properties of the sensor, such as a shift in pH, temperature, or light absorption (ManoharanNair Sudha Kumari & Thankappan Suryabai, 2024). The transducer captures this change and converts it into an electrical, optical, or other measurable signal, which is then processed and displayed as data representing the concentration or presence of the analyte. This rapid conversion of biological events into a

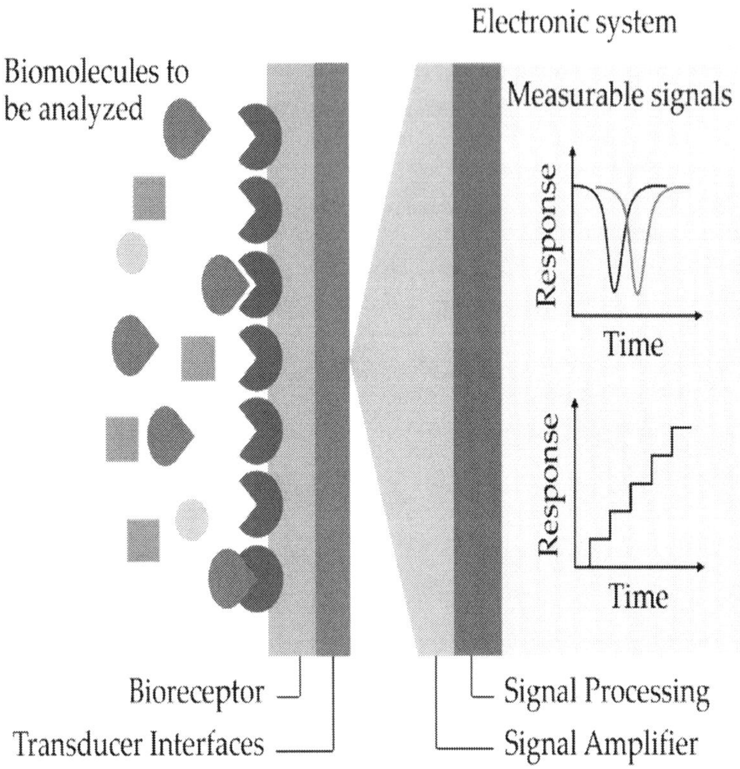

FIGURE 13.1 Schematic illustration of the working of a typical biosensor (Suvarnaphaet & Pechprasarn, 2017).

quantifiable signal makes biosensors powerful tools for real-time monitoring in various applications, including food safety.

13.2.1 Advantages and Limitations of Biosensors in Food Safety

Food safety regulators and the food industry use biosensors because of their versatility. Biosensors can detect germs, poisons, and other things in real time, preventing foodborne infections. Biosensors save time and money compared to typical lab methods. Biosensors can detect trace levels of hazardous substances. Biosensors are the best real-time surveillance method for warehouses and grocery stores because of their portability. However, restrictions apply. Compounds in complex food matrices can reduce biosensor sensitivity. Reliable results require precise calibration and constant maintenance. Biosensors that require government permission are confined to certain demographics. The chance that biosensors miss specific molecules may exacerbate the issue. Conventional methods may work. Biosensors accurately detect analytes. Food safety is monitored via biosensors, overcoming industry difficulties (Curulli, 2021; Lv et al., 2018; Silva et al., 2018).

13.3 BIOSENSOR TECHNOLOGIES IN FOOD SAFETY

13.3.1 Electrochemical Biosensors

A biological recognition element, such as an enzyme or antibody, engages with a target analyte to generate an electrical signal in electrochemical biosensors. The biological element either facilitates catalysis or specifically binds to the analyte, thereby inducing a measurable variation in the sensor's current, voltage, or impedance (Manjunatha et al., 2021; Monnappa et al., 2022; Monnappa et al., 2020; Raril et al., 2020). Signals are converted into analyte concentrations or their existence. Electron transfer between the recognition element and the electrode surface generates a signal directly proportional to the analyte concentration (Manjunatha et al., 2011; Manjunatha et al., 2021). Electrochemical biosensors are commonly utilized in food safety applications to detect *Salmonella*, *E. coli*, mycotoxins, and heavy metals. Pathogen detection in food manufacturing is defined by speed, cost-efficiency, and portability due to the utilization of these sensors, thereby guaranteeing food safety. Electrochemical biosensors are employed for the real-time surveillance of pesticide residues and food additives, enabling the identification of potential health hazards (Bobrinetskiy et al., 2021; Castle et al., 2021; Mei et al., 2022; Viswanathan et al., 2009).

13.3.2 Optical Biosensors

Optical biosensors detect changes in light absorption, fluorescence, and reflectance resulting from interactions between an analyte and a biological identifying element. The modifications help one to better understand the nature of the partnership. Analytes touching the sensor produce changes in its optical characteristics that result in a signal being generated. To identify particular compounds, optical biosensors combine changes in absorbance with surface plasmon resonance (SPR) and fluorescence resonance energy transfer (FRET). Several hazardous compounds and other substances can be found in food using an optical biosensor. Low levels of heavy met als, organophosphate pesticides, and pesticide leftovers have been discovered. The method is sensitive and non-intrusive, as using optical biosensors to check the safety and quality of food does not need much sample preparation (Balbinot et al., 2021; Mahmood Khan et al., 2023; Meira et al., 2024; Singh et al., 2023).

13.3.3 Mass-Based Biosensors

When analytes interact with one another, biological sensors that conduct their functions based on mass are able to detect changes in surface mass. QCM and SAW sensors are considered to be the two most important components of the approach. When the sensing element comes into contact with the analyte, the mass of the element undergoes a change, which makes it easier to identify modifications in either frequency or velocity. The mass of the element that is being detected is altered as a result of the analyte. Because of their extraordinary sensitivity and ability to evaluate changes in mass, these sensors are able to detect analytes even at low concentrations.

Biosensors that are based on mass are capable of accurately recognizing mycotoxins and allergens found in food. Additionally, in addition to peanut proteins, they may also be able to identify fungal toxins, such as aflatoxins, which have the potential to cause irritation or even pose a mortal threat to individuals. Because of their speed and accuracy, these biosensors provide regulatory bodies with assistance in protecting customers who may be more susceptible to illness (Kuchmenko & Lvova, 2019; Länge, 2022; Liang et al., 2022; Zeng et al., 2024).

13.3.4 Thermal Biosensors

Thermal biosensors monitor changes in temperature or heat production caused by biological processes. When a process's biological recognition element heats up or cools down, a thermal transducer can detect the temperature difference. This allows for sensitive evaluation of biological interactions by establishing a one-to-one relationship between the concentration or presence of analytes and the subsequent temperature change. Thermal biosensors can detect metabolites produced by microbes, including compounds from *Listeria* and *Salmonella* species (Ivnitski et al., 1999). Thermodynamic measurements reveal the presence of pollutants in food. If we are concerned about the safety of our food while it is in transit or storage, we absolutely need these sensors (Dharshni et al., 2019; Luong et al., 1991).

13.3.5 Nanomaterial-Based Biosensors

Nanotechnology has the potential to greatly improve biosensors' sensitivity, selectivity, and overall performance. Nanomaterials, such as nanoparticles, nanowires, and nanotubes, can be incorporated into biosensors in order to improve their surface area, conductivity, and reaction times (Eivazzadeh-Keihan et al., 2022; Hareesha & Manjunatha, 2021; Pushpanjali et al., 2021). Nanomaterials are superior to other materials when it comes to biological identifying components because of the unique properties that they possess. It is now possible to achieve both real-time monitoring and increased detection levels, respectively.

The identification of illnesses, toxins, and herbicides in food products is being accomplished through the use of a growing number of biosensors that make use of nanomaterials in the context of food safety applications (Fernández et al., 2017; Lata et al., 2024). By utilizing sensors that are able to accurately identify contaminants, even at low concentrations, this technology makes a substantial contribution to the improvement of food safety management. It has been demonstrated that the use of gold nanoparticles can facilitate the rapid detection of *Salmonella* and *E. coli* (Du et al., 2020). The development of sophisticated monitors that can detect the adulteration of food and verify the genuineness of food products is currently underway. This will ensure that food safety is maintained and that consumer health is protected.

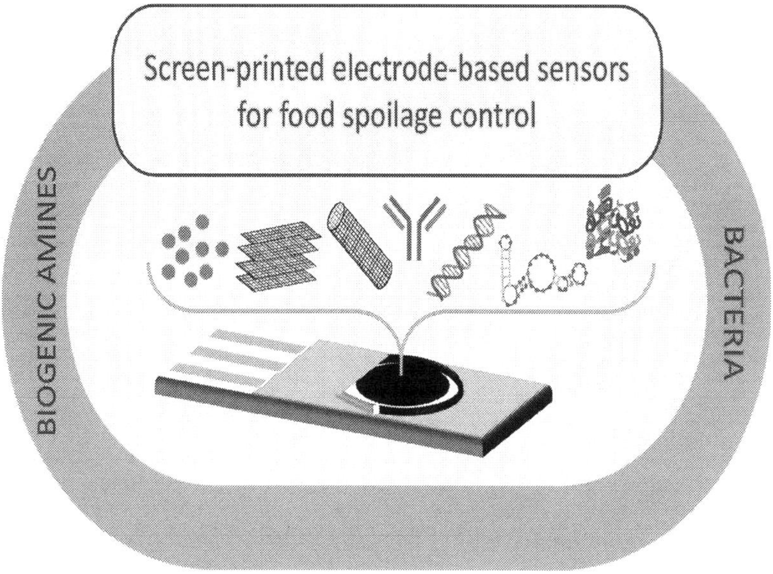

FIGURE 13.2 Schematic illustration of biosensor working for food spoilage control (Torre et al., 2020).

13.4 APPLICATIONS OF BIOSENSORS IN FOOD SAFETY

Figure 13.2 schematically illustrates the biosensor working in food spoilage control (Torre et al., 2020).

A brief summary of each application is given in the table formats (Tables 13.1–13.5)

13.5 ADVANCEMENTS IN BIOSENSOR DEVELOPMENT

Recent advancements in biosensor technology have resulted in a novel food safety methodology. As a result of the breakthroughs, numerous other disciplines have progressed swiftly. Incorporating additional biorecognition elements improves the sensitivity and specificity of disease detection. Despite the challenges posed by these components, specific projects might optimize their potential. This category encompasses genetically engineered proteins and polymers exhibiting atypical molecular structures. Mobile devices and wireless biosensors provide real-time surveillance and extensive data transmission. Recent advancements in Internet of Things (IoT) technology have enhanced food safety assessment procedures. The advancements in mobility and miniaturization tools have enhanced the efficiency and accuracy of in-person testing. This commitment can now be realized due to lab-on-a-chip technology and portable biosensors. Recent technological advancements have underscored this purpose. Thorough assessments of food production necessitate high-throughput screening techniques. Multiplexed biosensors identify multiple analytes during the

TABLE 13.1
Detection of Microbial Contaminants

Analytes	Remarks	References
Bacteria (e.g., *Salmonella*, *E. coli*)	With better efficiency and less cost than traditional microbiological techniques, this review emphasizes the major developments in biosensor technology for quick, selective, sensitive detection of foodborne pathogens like *Listeria*, *Campylobacter*, *Salmonella*, and *E. coli*	Cossettini et al. (2022)
	This work shows the creation of a multiplex PCR assay as a fast, sensitive, and dependable method for the simultaneous detection of *Salmonella* spp. and *Escherichia coli* O157:H7, therefore illustrating its possible for effective pathogen monitoring in many food and environmental samples	Fratamico and Strobaugh (1998)
	With great advantages for food safety and public health, this work shows the efficiency of the novel SELA-BARDOT detection platform as a fast, label-free, non-invasive method for simultaneous detection and differentiation of *Salmonella*, *Escherichia coli*, and *Listeria monocytogenes*	Abdelhaseib et al. (2019)
Viruses (e.g., norovirus, hepatitis A)	This study highlights molluscs as the most contaminated category and underlines the need for better sampling strategies and microbiological criteria in European food safety rules. It also offers important new insights on the frequency of hepatitis A virus and norovirus contamination in different food types throughout Italy.	Pavoni et al. (2022)
	This work emphasizes the important function of process control viruses in validating viral detection techniques in food virology and shows that the food matrix and targeted pathogenic virus determine the choice between murine norovirus (MNV-1) and Mengovirus, so guaranteeing consistent RT-qPCR results for HAV and noroviruses.	Hennechart-Collette et al. (2015)
	This systematic review highlights the need for more accurate proxies to evaluate pathogen contamination in fresh produce and related samples, since routinely used microbial indicators show weak to moderate predictive relationships with the presence of norovirus and hepatitis A virus in agricultural environments.	Victor et al. (2021)
Fungi and mycotoxins (e.g., aflatoxins)	Offering fast, precise, and stable solutions for early detection of spoilage fungi and mycotoxins in the food sector, this review highlights current developments in biosensor technology using different transducers and biological or bioinspired features.	Santana Oliveira et al. (2019)
	This work emphasizes the integration of microsystem technology tools—DNA arrays, biosensors, and electrical sensing systems—as part of a multidisciplinary strategy to enable fast, on-site detection of toxigenic fungus and mycotoxin contamination, therefore ensuring the safety and quality of foodstuffs	Logrieco et al. (2005)
	While investigating new technologies for enhancing detection efficiency and food safety, this review thoroughly reviews developments in aflatoxin analysis, stressing the important function of extraction and cleanup techniques alongside analytical methods including chromatography and mass spectrometry	Zhang and Banerjee (2020)

TABLE 13.2
Detection of Chemical Contaminants

Analytes	Remarks	References
Pesticides and herbicides	This study addresses important food safety issues by stressing contemporary electrochemical techniques and sophisticated sensors as effective, sensitive, and dependable tools for identifying pesticide residue in food.	Bakirhan et al. (2018)
	This review opens the path for increased food safety by stressing the possibilities of nanoscale substance-integrated optical sensing platforms as creative and effective instruments for identifying pesticide and herbicide residues in food and agriculture.	Lee et al. (2024)
	Emphasizing developments in surface-enhanced Raman spectroscopy (SERS), a highly sensitive and simple approach for pesticide residue detection, this study presents a viable substitute for current chromatographic methods for food safety.	Xu et al. (2017)
Heavy metals	Emphasizing their potential to enhance diagnostics in resource-limited environments, this review notes the developments in smartphone-based point-of-care tests (POCTs) as affordable, flexible tools for identifying food contamination and metal ions.	Sivakumar and Lee (2021)
	Emphasizing creative optical and electrochemical techniques as well as new device applications and commercialization obstacles, this review covers developments in biosensors for heavy metal ions in food.	Aihaiti et al. (2024)
	This paper addresses the important necessity to monitor heavy metal contamination using methods like atomic absorption spectroscopy to guarantee food safety, and stresses the nutritional value of milk.	Aggarwal et al. (2022)
Residual antibodies	Emphasizing their relevance for maintaining food safety, this article reviews the most recent developments in fast and sensitive detection techniques for chemical and veterinary drug residues in animal-derived foods	Toldrá and Reig (2006)
	The great importance of nanotechnology-enhanced optical aptasensors in developing multiplex screening techniques for the sensitive and specific detection of antimicrobial residues in food products originating from animals is underlined in this review	Gaudin (2020)
	This work introduces a very sensitive sandwich ELISA including nanobodies and monoclonal antibodies, therefore providing a strong methodological progress for the correct detection and separation of β-lactoglobulin allergens in thermal-processed dairy and dairy-free goods	Li et al. (2023)
Food additives and preservatives	Emphasizing their safety consequences, regulatory frameworks, and sophisticated analytical techniques for guaranteeing compliance and quality in meat products, this chapter offers a thorough summary of present and new meat preservatives and additives	Surendran Nair et al. (2020)
	To minimize negative health effects, guarantee quality, and solve the growing concern over food adulteration and its influence on consumer safety, this article underlines the vital need for strict control of synthetic additions and color preservatives in food	Sen (2021)
	The publication emphasizes the transforming power of nanomaterials in improving the sensitivity, selectivity, and efficiency of food additive detection techniques, therefore addressing both normal and illicit additives with great future uses in food safety	Zhang et al. (2020)

TABLE 13.3
Detection of Physical Contaminants

Analytes	Remarks	References
Foreign particles (plastics, glass, etc.)	Reviewing mechanical, optical, and metal detection methods as well as new sensor technologies for improved quality assurance, this article updates developments in detection systems for foreign bodies in food	Graves et al. (1998)
	This work shows that grating-based multimodal X-ray imaging is superior for novelty detection of foreign objects in meals, thereby greatly improving detection of organic components over standard X-ray techniques	Einarsdóttir et al. (2016)
	This innovative work highlights their bioavailability and offers the first proof of plastic particles in human bloodstream, so stressing the need for more investigation into their possible effects on health and exposure concerns	Leslie et al. (2022)

procedure. Public health is anticipated to be enhanced due to these technological advancements. This situation occurs as these advancements are anticipated to enhance the reliability and effectiveness of food safety assurance systems.

13.5.1 Genetically Engineered Proteins

Rapid agricultural expansion requires genetically engineered proteins for biosensor development. Agricultural methods drive the spread. These proteins may improve biosensors for diseases, contaminants, and food quality due to their increased sensitivity, specificity, and adaptability. Molecularly engineered enzymes, receptors, and antibodies detect chemicals precisely. Many foodborne illnesses are linked to *Salmonella* and *Listeria*. Synthetic proteins can selectively attach to analytes to detect tiny pollution quantities in complex food matrices. This allows biosensors to identify contaminated substances. One benefit of biosensors is their capacity to detect trace amounts of toxins. Genetically modified protein biosensors may improve food safety and manufacturing quality. These sensors cost less and work faster than lab detection methods. This enhances real-time monitoring and testing (Arugula et al., 2014; Mishra et al., 2018; Scognamiglio et al., 2014; Warriner et al., 2014).

13.5.2 Synthetic Receptors (e.g., Molecularly Imprinted Polymers)

Molecularly imprinted polymers (MIPs) and synthetic receptors can detect diseases, poisons, and environmental pollutants in food. Like antibodies, molecularly imprinted polymers (MIPs) feature niche binding sites that recognize and attach target molecules. Synthetic receptors improve food preparation stability, reusability, and temperature and pH resistance. Food companies are increasingly using molecularly imprinted polymer (MIPs) biosensors to detect pesticides, foodborne illnesses,

TABLE 13.4
Monitoring Food Quality Parameters

Quality Metrics	Remarks	References
Freshness indicators (pH, ripeness, etc.)	This review emphasizes developments in intelligent pH indicator packaging and stresses its ability to improve food safety by means of creative materials and pH-sensitive dyes, therefore indicating quality changes	Balbinot-Alfaro et al. (2019)
	This work presents novel biodegradable intelligent packaging systems with a bioinspired pH-sensitive dye that show outstanding halochromic qualities, stability, and environmental sustainability for advanced food packaging uses	Păușescu et al. (2022)
	This work efficiently classifies ripening phases and origin with great precision, therefore showing the potential of the patented FLUTE and BIONOTE systems for non-destructive, IoT-integrated monitoring of dry-cured gammon quality	Zompanti et al. (2020)
Nutrient contents	Through multidisciplinary cooperation, this review highlights the transforming possibilities of wearable sensor technologies for real-time, non-invasive nutrient monitoring, so opening the path for tailored precision diets	Shi et al. (2022)
	Emphasizing their accuracy, dual-signal processing ability, and future potential for innovation, this study examines the developments, design techniques, and uses of ratiometric electrochemical sensors in food analysis	Hu et al. (2024)
	The significance of nutrient-sensing pathways in metabolic control and consequences in human metabolic illnesses are highlighted in this work by means of cellular and organismal levels of integration	Efeyan et al. (2015)
Spoilage detection (volatile organic compounds)	This article stresses the possibility of biosensors and loss models to improve the preservation of unique fragrance components and shows the processes of VOC degradation in fresh and processed foods	Gong et al. (2023)
	By means of the classification of bacterial and fungal volatiles, this review emphasizes the possibilities of electronic nose technologies for fast and accurate microbiological detection, therefore illustrating uses in medical diagnostics and food quality control	Casalinuovo et al. (2006)
	This review emphasizes the need to recognize seafood spoilage bacteria, volatile organic compounds, and the effects of storage and packing, thereby supporting culture-independent approaches to improving shelf-life management and quality control	Odeyemi et al. (2018)

TABLE 13.5
Biosensors for Packaging and Shelf-Life Extension

Application	Remarks	References
Intelligent packaging systems	Leading the way for a new generation of "smart sensors," which will mold the future commercial markets for gas sensors, conducting polymer composite, inherently conducting polymer and metal oxide conductivity gas sensors offer remarkable discrimination	Matindoust et al. (2016)
	Smart packaging integrates chemical and biosensors to monitor food quality and safety in real time, enhancing control from production to consumption	Kuswandi et al. (2011)
	Smart packaging uses advanced sensors to ensure food safety and quality throughout the supply chain	Kuswandi et al. (2011)
Real-time monitoring of food conditions	This work verifies a sophisticated RFID smart tag system for real-time food traceability and cold-chain monitoring.	Abad et al. (2009)
	Using litmus paper for real-time monitoring of beef freshness, this study shows the development of a basic and affordable on-package sticker sensor that precisely detects rotting by successfully linking sensor response with pH, microbial development, and sensory analysis	Kuswandi et al. (2015)
	By raising genetic resolution, boosting cluster detection, and integrating with epidemiological data, whole-genome sequencing (WGS) has transformed listeriosis epidemic surveillance	Jackson et al. (2016)

and other toxins. They are easy to use, precise, and fast, allowing for on-site testing and quality assurance in food safety monitoring systems. MIP-based biosensors improve food safety and quality by detecting contamination across the supply chain quickly and accurately (Arreguin-Campos et al., 2021; Cao et al., 2019; Geng et al., 2024; Villa et al., 2021; Wang et al., 2016).

13.5.3 Wireless and Smartphone-Based Biosensors

Food testing is being changed by mobile and instantaneous technologies. One such technology is wireless biosensors linked to cellphones. These biosensors precisely and rapidly detect food quality, pollutants, and viruses. Bluetooth and cameras on smartphones with biosensors allow them to examine biosensor data in real time and can be used for field testing without the need for costly lab equipment. This benefits consumers, processors, and growers. Data can flow from wireless networking to central databases or cloud services easily. This simplifies remote monitoring, data distribution, and automated decision-making. Wireless biosensors made available through cell phones could improve food safety. Among the examples are customer assessment, supply chain management, and manufacturing quality control. The portability and simplicity of biosensors help to enable many uses. These systems improve traceability, food safety, and consumer confidence (Chen et al., 2024; Jha et al., 2021; Lu et al., 2019; Yang et al., 2022).

13.5.4 Internet of Things (IoT) in Food Safety

The Internet of Things (IoT) and biosensors are revolutionizing food safety. This is accomplished via constant, real-time food supply chain monitoring. Food temperature, humidity, contamination, and microbial proliferation are monitored, transmitted, and evaluated by biosensors with IoT capability. Placed throughout the food supply chain, these sensors evaluate food safety at manufacturing, transportation, and storage stages. Food producers, distributors, and stores can all help (Chen et al., 2024; Jha et al., 2021; Upadhyay et al., 2024) by using IoT systems with biosensors to avoid foodborne diseases. This is accomplished by rapidly spotting contamination, remotely tracking product conditions, and taking corrective action. The IoT gathers and sends biosensor data to improve traceability and supply chain visibility. Exchange of information helps companies satisfy food safety standards. The IoT and biosensors enhance consumer food safety by maximizing manufacturing, waste reduction, and improvement of food safety management systems (Ben-Daya et al., 2020; Guo & Jayan, 2023; Liu et al., 2016; Popa et al., 2019; Witjaksono et al., 2018).

13.5.5 Lab-on-a-Chip Devices

Laboratory-on-a-chip (LOC) technologies improve customer safety and food quality. These minuscule devices incorporate several scientific techniques to accelerate the food assessment procedure. This technology offers services in chemical analysis, detection, and sample preparation. LOC devices facilitate the identification of pollutants, infectious pathogens, and adherence to quality standards via microfluidics, biochemistry, and miniature biosensors. The employment of this particular technology facilitates this. It functions through the utilization of biosensors. Because of LOC, food prodcuers and merchants may now perform testing that is both swift and portable. These technologies are far more cost-effective, more sensitive, and quicker than laboratory procedures. Furthermore, they are more efficient than other options. Professionals in food processing and quality control can use them to swiftly ascertain the suitability of a certain food item for human ingestion. Low-overhead (LOC) devices are characterized by their relatively small sample size, as well as their rapidity, simplicity, and sensitivity. This renders them appropriate for the food logistics chain, as they excel in maintaining product quality and compliance. Consequently, they are considered outstanding alternatives (Burns et al., 2016; Kumar & Panda, 2022; Park et al., 2024; Rani et al., 2024; Sobhan et al., 2021).

13.5.6 Handheld Biosensors for On-Site Testing

Handheld biosensors are revolutionizing on-site testing in the food industry by providing portable, user-friendly solutions for detecting contaminants, pathogens, and quality indicators in food products (Boodhoo et al., 2024; Kapoor et al., 2024; Srinivasan & Tung, 2015; Umapathi et al., 2022). These compact devices enable quick, real-time analysis without the need for an extensive laboratory equipment or expertise, making them particularly valuable for food safety monitoring in production facilities, retail environments, and even during transport. Handheld biosensors often incorporate advanced sensing technologies, such as electrochemical, optical, or enzymatic detection, to identify specific contaminants like *Salmonella*, *Listeria*, pesticides, or spoilage organisms in food matrices. Their ease of use and rapid response times—often delivering results within minutes—allow food producers, processors, and inspectors to take immediate corrective actions, reducing the risk of contamination and ensuring food quality. Additionally, many handheld biosensors are designed to be rugged and adaptable to various environmental conditions, making them ideal for diverse applications within the food supply chain. By providing quick, accurate, and on-site testing, handheld biosensors enhance food safety management and improve operational efficiency.

13.6 CHALLENGES AND FUTURE PROSPECTS

Generally speaking, biosensors have great trouble determining if food is safe. The intrinsic complexity of food matrices makes creating biosensors displaying sensitivity and selectivity challenging. Given the complexity of the situation, accurate identification of even the slightest level of contaminants is vital. Furthermore, many biosensors degrade with time and require expensive replacement parts, which increases the difficulties with stability and reusability. The design must be strong and flexible to withstand the demanding conditions of an industrial environment and also fit for food preparation tools. Before they may be used widely and certified, biosensors must go through standardization and regulatory validation. Protracted and complex government licensing procedures make overcoming the usual obstacles of entering new markets much more challenging. Only if ethical issues of biosensor technology are sufficiently resolved will they be able to help rebuild confidence and equity. Examining the problems resulting from unequitable access to this technology and the possibility of data leaks from biosensors is necessary.

Overcoming these obstacles and remaining sensitive to these trends would help biosensors in food safety to be developed more easily. Modern materials and technologies that could soon be included in biosensors to increase their sensitivity, robustness, and flexibility are nanostructures and advanced synthetic biology. The above-described developments could improve customized nutrition by customizing food safety assessments to meet different dietary needs. This could follow from changing the food safety evaluations. Global food access is crucial as well as helping to enable the use of customized diet plans. This goal can be achieved by the quick and continuous identification of contaminants in several food systems.

Developing more scalable and reasonably priced manufacturing techniques helps to build momentum in hitherto untapped fields. These markets must quicken if anything is to change. The degree of commercial success of a company determines its chances of continuous economic survival. Validation of the efficacy and lifetime of the given technology is crucial to reaching the goals of acquiring new customers and increasing market share. Major developments in biosensor technology could lead to a radical transformation in the validation of food safety and the building of a stronger worldwide food chain.

13.7 CASE STUDIES AND INDUSTRY EXAMPLES

13.7.1 Case Study 1: Implementation of Biosensors in the Dairy Industry

The study titled "Detection of *Escherichia coli* in unpasteurized raw milk," published in the *International Journal of Agriculture and Food Science*, examines the incidence of *E. coli* contamination in raw milk from the Sfax governorate in Tunisia, revealing that 32.5% of the samples tested positive for the pathogen, with contamination levels varying across localities (Samet-Bali et al., 2013). The study emphasizes the need for stringent hygienic practices and pasteurization to mitigate public health risks.

13.7.2 Case Study 2: Use of Biosensors in Meat Processing and Safety

A case study stressed the need of biosensors in the meat processing industry. This effort primarily aimed to enable quick identification of possibly harmful bacteria, such as *Salmonella* and *Campylobacter*, usually linked to contaminated meat products. The development of the instrument started from an aim to improve safety precautions. Using nanoparticles, the research team basically raised the general capacity, sensitivity, and efficiency in pathogen detection of the electrochemical biosystem. Chicken flesh samples were taken from a fully running processing plant, and in 30 min, the biosensor displayed good performance in *Salmonella* and *Campylobacter* detection. LOD was 90 pM(Morant-Miñana & Elizalde, 2015). This obviously shows a change from conventional cultural methods, which might help to identify these diseases within a three-day maximum period. A portable biosensor connected to the processing line can be used for on-site constant testing. Another benefit was the fast response capacity in the case of found pollution. The device seems to be very effective in guarding cattle given its detection accuracy above 95%. Especially in terms of improving disease detection methods defined by higher accuracy, efficiency, and cost-effectiveness, the case study aimed to show the transforming potential of biosensor technology within the cattle sector. This development helps the meat processing sector to create beef products satisfying strict food safety standards, therefore lowering consumer risk.

13.7.3 Case Study 3: Application of Biosensors in Monitoring Fresh Produce

With a focus on pesticide residues and microbiological contamination, this case study examined the degree to which biosensors might evaluate the quality and safety of fresh fruit (Guo et al., 2015). Scientists designed a portable biosensor to evaluate vegetable pesticide levels. This biosensor makes advantage of enzyme-based identifying elements, SnO_2 and carbon nanotubes. The biosensor found pesticide levels in fresh food samples from several retail stores to be below the legal limit. The detection limit was 100ng/L. This approach generates data in a mere 15 min, and hence it is a significant improvement over standard testing techniques requiring extensive chemical investigations. Farmers, companies, and consumers would all be better off choosing the amazing, extremely accurate biosensor since it is quite dependable. By tracking pesticide residues and contamination in real time, this case study shows how biosensors may improve food safety and guarantee the quality of fresh produce. Health risks are reduced and safety standards are satisfied.

13.7.4 Case Study 4: Role of Biosensors in Beverage Quality Control

This case study looked at the ways biosensors evaluate beverage quality with specific regard to the identification of harmful organisms and the assurance of the safety of bottled beverages. Designed to find Brettanomyces, a prevalent yeast causing beverage degradation including wine and beer, an immunosensor was developed. Twenty minutes after testing, the antibody-based, electrochemical detection biosensor showed a quite low concentration of Brettanomyces contamination (Borisova et al., 2017). This was a big step forward from standard microbiological techniques, which produced findings over several days. Designed into a quality control system for a brewery, the biosensor tracked bacterial contamination in real time during beer production and bottling. The method's remarkable precision and sensitivity allowed it to find decaying yeast in even the most complex drink mixtures. This case study shows how quickly and dependably cheaply priced biosensors can ensure beverage freshness and safety, hence reducing spoilage and offering quality control.

13.8 CONCLUSION

By enabling fast, on-site identification of pollutants, pathogens, and quality indicators in many different food products, biosensors can change food safety and quality assurance. The ability of the food sector to offer real-time results with enhanced sensitivity and specificity is changing its attitude toward safety and quality surveillance. Integration of biosensors into food manufacturing processes is claimed to improve food safety measures, save costs, and increase consumer confidence by ensuring that products satisfy quality criteria and safety regulations. Food industry biosensors enable automatic, more exact, efficient systems of food safety. The continuous supervision of food items from conception to consumption is now feasible due to developments in digital connectivity, miniaturization, biorecognition properties, and biosensors. This guarantees the following of safety criteria and helps to lower the expected pollutant potential. Being part of IoT networks helps biosensors to become progressively more critical in large-scale food production and distribution systems and hence a required instrument for ensuring world food security. Comparing biosensor-based technology with more traditional approaches, there are a range of advantages. Conventional techniques including culture-based testing could demand a lot of time, effort, and knowledge; the results require days to determine. Small sample preparation and fast, real-time data allow biosensors to provide swift decision-making and remedial steps to be quickly implemented. Furthermore, portable biosensors enable on-site testing, which is not always possible with traditional laboratory methods. While conventional techniques have historical constancy, biosensors are a major development in the modern food industry, offering competitively priced solutions for food safety and quality control. Including biosensors in the food sector marks significant progress in food quality and safety control. Biosensors will challenge existing food safety rules since they can offer quick, accurate, reasonably priced solutions for the identification of contaminants and pathogens. By reducing human error, improving productivity, and offering real-time monitoring—which thereby helps to limit human error—they increase consumer confidence in food safety and improve food manufacturing operations. Should biosensor technology be fully applied, the food industry will find it quite advantageous, and thus preservation of public health and enhancement of the sustainability of world food systems will follow. The special properties of biosensors will anchor food safety and ensure food quality, bringing in a new era of accuracy, simplicity, and dependability.

REFERENCES

Abad, E., Palacio, F., Nuin, M., Zárate, A. G. de, Juarros, A., Gómez, J. M. & Marco, S. (2009). RFID smart tag for traceability and cold chain monitoring of foods: Demonstration in an intercontinental fresh fish logistic chain. *Journal of Food Engineering*, *93*(4), 394–399. https://doi.org/https://doi.org/10.1016/j.jfoodeng.2009.02.004

Abdelhaseib, M. U., Singh, A. K. & Bhunia, A. K. (2019). Simultaneous detection of *Salmonella enterica*, *Escherichia coli* and *Listeria monocytogenes* in food using a light scattering sensor. *Journal of Applied Microbiology*, *126*(5), 1496–1507. https://doi.org/10.1111/jam.14225

Aggarwal, A., Verma, T. & Ghosh, S. (2022). Heavy Metal Residues in Milk and Milk Products and Their Detection Method. In Y. El-Samragy (Ed.), *Trends and innovations in food science* (p. Ch. 7). IntechOpen. https://doi.org/10.5772/intechopen.105004

Aihaiti, A., Wang, J., Zhang, W., Shen, M., Meng, F., Li, Z., Zhang, Y., Ren, M. & Zhang, M. (2024). Recent advances and trends in innovative biosensor-based devices for heavy metal ion detection in food. *Comprehensive Reviews in Food Science and Food Safety*, *23*(4), e13358. https://doi.org/https://doi.org/10.1111/1541-4337.13358

Arreguin-Campos, R., Jiménez-Monroy, K. L., Diliën, H., Cleij, T. J., van Grinsven, B. & Eersels, K. (2021). Imprinted polymers as synthetic receptors in sensors for food safety. *Biosensors*, *11*(2). https://doi.org/10.3390/bios11020046

Arugula, M. A., Zhang, Y. & Simonian, A. L. (2014). Biosensors as 21st century technology for detecting genetically modified organisms in food and feed. *Analytical Chemistry*, *86*(1), 119–129. https://doi.org/10.1021/ac402898j

Bakirhan, N. K., Uslu, B. & Ozkan, S. A. (2018). Chapter 5 - The detection of pesticide in foods using electrochemical sensors. In A. M. Grumezescu & A. M. Holban (Eds.), *Food safety and preservation* (pp. 91–141). Academic Press. https://doi.org/https://doi.org/10.1016/B978-0-12-814956-0.00005-6

Balbinot, S., Srivastav, A. M., Vidic, J., Abdulhalim, I. & Manzano, M. (2021). Plasmonic biosensors for food control. *Trends in Food Science & Technology*, *111*, 128–140. https://doi.org/https://doi.org/10.1016/j.tifs.2021.02.057

Balbinot-Alfaro, E., Craveiro, D. V., Lima, K. O., Costa, H. L. G., Lopes, D. R. & Prentice, C. (2019). Intelligent packaging with pH indicator potential. *Food Engineering Reviews*, *11*(4), 235–244. https://doi.org/10.1007/s12393-019-09198-9

Ben-Daya, M., Hassini, E., Bahroun, Z. & Banimfreg, B. H. (2020). The role of internet of things in food supply chain quality management: A review. *Quality Management Journal*, *28*(1), 17–40. https://doi.org/10.1080/10686967.2020.1838978

Bhatia, D., Paul, S., Acharjee, T. & Ramachairy, S. S. (2024). Biosensors and their widespread impact on human health. *Sensors International*, *5*, 100257. https://doi.org/https://doi.org/10.1016/j.sintl.2023.100257

Bobrinetskiy, I., Radovic, M., Rizzotto, F., Vizzini, P., Jaric, S., Pavlovic, Z., Radonic, V., Nikolic, M. V. & Vidic, J. (2021). Advances in nanomaterials-based electrochemical biosensors for foodborne pathogen detection. *Nanomaterials*, *11*(10). https://doi.org/10.3390/nano11102700

Boodhoo, N., Shoja Doost, J. & Sharif, S. (2024). Biosensors for monitoring, detecting, and tracking dissemination of poultry-borne bacterial pathogens along the poultry value chain: A review. *Animals*, *14*(21). https://doi.org/10.3390/ani14213138

Borisova, B., Villalonga, M. L., Arévalo-Villena, M., Boujakhrout, A., Sánchez, A., Parrado, C., Pingarrón, J. M., Briones-Pérez, A. & Villalonga, R. (2017). Disposable electrochemical immunosensor for *Brettanomyces bruxellensis* based on nanogold-reduced graphene oxide hybrid nanomaterial. *Analytical and Bioanalytical Chemistry*, *409*(24), 5667–5674. https://doi.org/10.1007/s00216-017-0505-5

Burns, M., Wiseman, G., Knight, A., Bramley, P., Foster, L., Rollinson, S., Damant, A. & Primrose, S. (2016). Measurement issues associated with quantitative molecular biology analysis of complex food matrices for the detection of food fraud. *Analyst*, *141*(1), 45–61. https://doi.org/10.1039/C5AN01392E

Cao, Y., Feng, T., Xu, J. & Xue, C. (2019). Recent advances of molecularly imprinted polymer-based sensors in the detection of food safety hazard factors. *Biosensors and Bioelectronics*, *141*, 111447. https://doi.org/https://doi.org/10.1016/j.bios.2019.111447

Casalinuovo, I. A., Di Pierro, D., Coletta, M. & Di Francesco, P. (2006). Application of electronic noses for disease diagnosis and food spoilage detection. *Sensors*, *6*(11), 1428–1439. https://doi.org/10.3390/s6111428

Castle, L. M., Schuh, D. A., Reynolds, E. E. & Furst, A. L. (2021). Electrochemical sensors to detect bacterial foodborne pathogens. *ACS Sensors*, *6*(5), 1717–1730. https://doi.org/10.1021/acssensors.1c00481

Chen, Y., Wang, Y., Zhang, Y., Wang, X., Zhang, C. & Cheng, N. (2024). Intelligent biosensors promise smarter solutions in food safety 4.0. *Foods*, *13*(2). https://doi.org/10.3390/foods13020235

Cossettini, A., Vidic, J., Maifreni, M., Marino, M., Pinamonti, D. & Manzano, M. (2022). Rapid detection of *Listeria monocytogenes*, *Salmonella*, *Campylobacter* spp., and *Escherichia coli* in food using biosensors. *Food Control*, *137*, 108962. https://doi.org/https://doi.org/10.1016/j.foodcont.2022.108962

Curulli, A. (2021). Electrochemical biosensors in food safety: Challenges and perspectives. *Molecules*, *26*(10). https://doi.org/10.3390/molecules26102940

Dharshni, Ra., Dharshini, Rd. & Sivarajasekar, N. (2019). Thermal biosensors and their applications. *American International Journal of Research in Science*, 50. http://www.iasir.net

Du, J., Wu, S., Niu, L., Li, J., Zhao, D. & Bai, Y. (2020). A gold nanoparticles-assisted multiplex PCR assay for simultaneous detection of *Salmonella typhimurium*, *Listeria monocytogenes* and *Escherichia coli* O157:H7. *Analytical Methods*, *12*(2), 212–217. https://doi.org/10.1039/C9AY02282A

Efeyan, A., Comb, W. C. & Sabatini, D. M. (2015). Nutrient-sensing mechanisms and pathways. *Nature*, *517*(7534), 302–310. https://doi.org/10.1038/nature14190

Einarsdóttir, H., Emerson, M. J., Clemmensen, L. H., Scherer, K., Willer, K., Bech, M., Larsen, R., Ersbøll, B. K. & Pfeiffer, F. (2016). Novelty detection of foreign objects in food using multi-modal X-ray imaging. *Food Control*, *67*, 39–47. https://doi.org/https://doi.org/10.1016/j.foodcont.2016.02.023

Eivazzadeh-Keihan, R., Bahojb Noruzi, E., Chidar, E., Jafari, M., Davoodi, F., Kashtiaray, A., Ghafori Gorab, M., Masoud Hashemi, S., Javanshir, S., Ahangari Cohan, R., Maleki, A. & Mahdavi, M. (2022). Applications of carbon-based conductive nanomaterials in biosensors. *Chemical Engineering Journal*, *442*, 136183. https://doi.org/https://doi.org/10.1016/j.cej.2022.136183

Fernández, H., Arévalo, F. J., Granero, A. M., Robledo, S. N., Nieto, C. H. D., Riberi, W. I. & Zon, M. A. (2017). Electrochemical biosensors for the determination of toxic substances related to food safety developed in South America: Mycotoxins and herbicides. *Chemosensors*, *5*(3). https://doi.org/10.3390/chemosensors5030023

Fratamico, P. M. & Strobaugh, T. P. (1998). Simultaneous detection of *Salmonella* spp and *Escherichia coli* O157:H7 by multiplex PCR. *Journal of Industrial Microbiology and Biotechnology*, *21*(3), 92–98. https://doi.org/10.1038/sj.jim.2900520

Gaudin, V. (2020). The growing interest in development of innovative optical aptasensors for the detection of antimicrobial residues in food products. *Biosensors*, *10*(3). https://doi.org/10.3390/bios10030021

Geng, L., Huang, J., Fang, M., Wang, H., Liu, J., Wang, G., Hu, M., Sun, J., Guo, Y. & Sun, X. (2024). Recent progress of the research of metal-organic frameworks-molecularly imprinted polymers (MOFs-MIPs) in food safety detection field. *Food Chemistry*, *458*, 140330. https://doi.org/https://doi.org/10.1016/j.foodchem.2024.140330

Gong, X., Huang, J., Xu, Y., Li, Z., Li, L., Li, D., Belwal, T., Jeandet, P., Luo, Z. & Xu, Y. (2023). Deterioration of plant volatile organic compounds in food: Consequence, mechanism, detection, and control. *Trends in Food Science & Technology*, *131*, 61–76. https://doi.org/https://doi.org/10.1016/j.tifs.2022.11.022

Graves, M., Smith, A. & Batchelor, B. (1998). Approaches to foreign body detection in foods. *Trends in Food Science & Technology*, *9*(1), 21–27. https://doi.org/https://doi.org/10.1016/S0924-2244(97)00003-4

Guo, Y., Sun, X., Liu, X., Sun, X., Zhao, G., Chen, D. & Wang, X. (2015). A miniaturized portable instrument for rapid determination pesticides residues in vegetables and fruits. *IEEE Sensors Journal*, *15*(7), 4046–4052. https://doi.org/10.1109/JSEN.2015.2410532

Guo, Z. & Jayan, H. (2023). Fast nondestructive detection technology and equipment for food quality and safety. *Foods*, *12*(20). https://doi.org/10.3390/foods12203744

Hareesha, N. & Manjunatha, J. G. (2021). Electro-oxidation of formoterol fumarate on the surface of novel poly(thiazole yellow-G) layered multi-walled carbon nanotube paste electrode. *Scientific Reports*, *11*(1), 12797. https://doi.org/10.1038/s41598-021-92099-x

Hennechart-Collette, C., Martin-Latil, S., Guillier, L. & Perelle, S. (2015). Determination of which virus to use as a process control when testing for the presence of hepatitis A virus and norovirus in food and water. *International Journal of Food Microbiology*, *202*, 57–65. https://doi.org/https://doi.org/10.1016/j.ijfoodmicro.2015.02.029

Hierlemann, A. & Gutierrez-Osuna, R. (2008). Higher-order chemical sensing. *Chemical Reviews*, *108*(2), 563–613. https://doi.org/10.1021/cr068116m

Hu, X., Wei, W., Li, X., Yang, Y. & Zhou, B. (2024). Recent advances in ratiometric electrochemical sensors for food analysis. *Food Chemistry: X*, *23*, 101681. https://doi.org/https://doi.org/10.1016/j.fochx.2024.101681

Ivnitski, D., Abdel-Hamid, I., Atanasov, P. & Wilkins, E. (1999). Biosensors for detection of pathogenic bacteria. *Biosensors and Bioelectronics*, *14*(7), 599–624. https://doi.org/https://doi.org/10.1016/S0956-5663(99)00039-1

Jackson, B. R., Tarr, C., Strain, E., Jackson, K. A., Conrad, A., Carleton, H., Katz, L. S., Stroika, S., Gould, L. H., Mody, R. K., Silk, B. J., Beal, J., Chen, Y., Timme, R., Doyle, M., Fields, A., Wise, M., Tillman, G., Defibaugh-Chavez, S., … Gerner-Smidt, P. (2016). Implementation of nationwide real-time whole-genome sequencing to enhance listeriosis outbreak detection and investigation. *Clinical Infectious Diseases*, *63*(3), 380–386. https://doi.org/10.1093/cid/ciw242

Jha, A., Moses, J. A. & Anandharamakrishnan, C. (2021). 13 - Smartphone-based detection devices for the agri-food industry. In C. Hussain (Ed.), *Smartphone-based detection devices* (pp. 269–310). Elsevier. https://doi.org/https://doi.org/10.1016/B978-0-12-823696-3.00001-5

Kapoor, A., Ramamoorthy, S., Sundaramurthy, A., Vaishampayan, V., Sridhar, A., Balasubramanian, S. & Ponnuchamy, M. (2024). Paper-based lab-on-a-chip devices for detection of agri-food contamination. *Trends in Food Science & Technology*, *147*, 104476. https://doi.org/https://doi.org/10.1016/j.tifs.2024.104476

Kuchmenko, T. A. & Lvova, L. B. (2019). A perspective on recent advances in piezoelectric chemical sensors for environmental monitoring and foodstuffs analysis. *Chemosensors*, *7*(3). https://doi.org/10.3390/chemosensors7030039

Kumar, A. & Panda, U. (2022). Chapter 12 - Microfluidics-based devices and their role on point-of-care testing. In R. Khan, A. Parihar & S. K. Sanghi (Eds.), *Biosensor based advanced cancer diagnostics* (pp. 197–224). Academic Press. https://doi.org/https://doi.org/10.1016/B978-0-12-823424-2.00011-9

Kuswandi, B., Damayanti, F., Jayus, Abdullah, A. & Heng, L. Y. (2015). Simple and low-cost on-package sticker sensor based on litmus paper for real-time monitoring of beef freshness. *Journal of Mathematical and Fundamental Sciences*, *47*(3), 236–251. https://doi.org/10.5614/j.math.fund.sci.2015.47.3.2

Kuswandi, B., Wicaksono, Y., Jayus, Abdullah, A., Heng, L. Y. & Ahmad, M. (2011). Smart packaging: Sensors for monitoring of food quality and safety. *Sensing and Instrumentation for Food Quality and Safety*, *5*(3), 137–146. https://doi.org/10.1007/s11694-011-9120-x

Länge, K. (2022). Bulk and surface acoustic wave biosensors for milk analysis. *Biosensors*, *12*(8). https://doi.org/10.3390/bios12080602

Lata, S., Bhardwaj, S. & Garg, R. (2024). Nanomaterials for sensing and biosensing: Applications in agri-food diagnostics. *International Journal of Environmental Analytical Chemistry*, *104*(16), 4868–4885. https://doi.org/10.1080/03067319.2022.2115895

Lee, D. H., Ahmed, M. W. & Kamruzzaman, M. (2024). Nanoscale substance–integrated optical sensing platform for pesticide detection in perishable foods. *Current Opinion in Food Science*, *60*, 101227. https://doi.org/https://doi.org/10.1016/j.cofs.2024.101227

Leslie, H. A., van Velzen, M. J. M., Brandsma, S. H., Vethaak, A. D., Garcia-Vallejo, J. J. & Lamoree, M. H. (2022). Discovery and quantification of plastic particle pollution in human blood. *Environment International*, *163*, 107199. https://doi.org/https://doi.org/10.1016/j.envint.2022.107199

Li, S., Nie, L., Wang, Y., Wang, Y., Fan, D., Wang, J., Hu, Y., Dong, L., Zhang, Y. & Wang, S. (2023). Detection of β-lactoglobulin under different thermal-processing conditions by immunoassay based on nanobody and monoclonal antibody. *Food Chemistry, 424*, 136337. https://doi.org/https://doi.org/10.1016/j.foodchem.2023.136337

Liang, S., Sutham, P., Wu, K., Mallikarjunan, K. & Wang, J.-P. (2022). Giant magnetoresistance biosensors for food safety applications. *Sensors, 22*(15). https://doi.org/10.3390/s22155663

Liu, Y., Han, W., Zhang, Y., Li, L., Wang, J. & Zheng, L. (2016). An Internet-of-Things solution for food safety and quality control: A pilot project in China. *Journal of Industrial Information Integration, 3*, 1–7. https://doi.org/https://doi.org/10.1016/j.jii.2016.06.001

Logrieco, A., Arrigan, D. W. M., Brengel-Pesce, K., Siciliano, P. & Tothill, I. (2005). DNA arrays, electronic noses and tongues, biosensors and receptors for rapid detection of toxigenic fungi and mycotoxins: A review. *Food Additives & Contaminants, 22*(4), 335–344. https://doi.org/10.1080/02652030500070176

Lu, Y., Shi, Z. & Liu, Q. (2019). Smartphone-based biosensors for portable food evaluation. *Current Opinion in Food Science, 28*, 74–81. https://doi.org/https://doi.org/10.1016/j.cofs.2019.09.003

Luong, J. H. T., Groom, C. A. & Male, K. B. (1991). The potential role of biosensors in the food and drink industries. *Biosensors and Bioelectronics, 6*(7), 547–554. https://doi.org/https://doi.org/10.1016/0956-5663(91)80018-S

Lv, M., Liu, Y., Geng, J., Kou, X., Xin, Z. & Yang, D. (2018). Engineering nanomaterials-based biosensors for food safety detection. *Biosensors and Bioelectronics, 106*, 122–128. https://doi.org/https://doi.org/10.1016/j.bios.2018.01.049

Mahmood Khan, I., Niazi, S., Akhtar, W., Yue, L., Pasha, I., Khan, M. K. I., Mohsin, A., Waheed Iqbal, M., Zhang, Y. & Wang, Z. (2023). Surface functionalized AuNCs optical biosensor as an emerging food safety indicator: Fundamental mechanism to future prospects. *Coordination Chemistry Reviews, 474*, 214842. https://doi.org/https://doi.org/10.1016/j.ccr.2022.214842

Manjunatha, J. G. G., Pushpanjali, P. A. & Nagarajappa, H. (2021). An overview of recent developments of carbon-based sensors for the analysis of drug molecules. *Journal of Electrochemical Science and Engineering*. https://api.semanticscholar.org/CorpusID:237723445

Manjunatha, J. G., Swamy, B. E. K., Ganjeenahalli, P., Mamatha, Gilbert, O. & Sherigara, B. S. (2011). *Poly (maleic acid) modified carbon paste electrode for simultaneous detection of dopamine in the presence of uric acid: A cyclic voltammetric study*. https://api.semanticscholar.org/CorpusID:201800234

Manoharan Nair Sudha Kumari, S. & Thankappan Suryabai, X. (2024). Sensing the future—frontiers in biosensors: Exploring classifications, principles, and recent advances. *ACS Omega*. https://doi.org/10.1021/acsomega.4c07991

Matindoust, S., Baghaei-Nejad, M., Zou, Z. & Zheng, L.-R. (2016). Food quality and safety monitoring using gas sensor array in intelligent packaging. *Sensor Review, 36*(2), 169–183. https://doi.org/10.1108/SR-07-2015-0115

Mei, Y., He, C., Zeng, W., Luo, Y., Liu, C., Yang, M., Kuang, Y., Lin, X. & Huang, Q. (2022). Electrochemical biosensors for foodborne pathogens detection based on carbon nanomaterials: Recent advances and challenges. *Food and Bioprocess Technology, 15*(3), 498–513. https://doi.org/10.1007/s11947-022-02759-7

Meira, D. I., Barbosa, A. I., Borges, J., Reis, R. L., Correlo, V. M. & Vaz, F. (2024). Recent advances in nanomaterial-based optical biosensors for food safety applications: Ochratoxin-A detection, as case study. *Critical Reviews in Food Science and Nutrition, 64*(18), 6318–6360. https://doi.org/10.1080/10408398.2023.2168248

Mishra, G. K., Barfidokht, A., Tehrani, F. & Mishra, R. K. (2018). Food safety analysis using electrochemical biosensors. *Foods*, *7*(9). https://doi.org/10.3390/foods7090141

Monnappa, A. B., Manjunatha, J. G., Bhatt, A. S., Nemumoolya, A., Medappa, S. B. & Karthammaiah, G. N. (2022). Methyl orange electropolymerized composite carbon paste electrode as a sensitive and selective sensor for the electrochemical determination of riboflavin. *Bentham Science, 17,* 117–133. http://doi.org/10.2174/97898150397 19122010006.

Monnappa, A. B., Manjunatha, J. G. & Bhatt, A. S. (2020). Design of a sensitive and selective voltammetric sensor based on a cationic surfactant-modified carbon paste electrode for the determination of alloxan. *ACS Omega*, *5*(36), 23481–23490. https://doi.org/10.1021/acsomega.0c03517

Morant-Miñana, M. C. & Elizalde, J. (2015). Microscale electrodes integrated on COP for real sample Campylobacter spp. detection. *Biosensors and Bioelectronics*, *70*, 491–497. https://doi.org/https://doi.org/10.1016/j.bios.2015.03.063

Odeyemi, O. A., Burke, C. M., Bolch, C. C. J. & Stanley, R. (2018). Seafood spoilage microbiota and associated volatile organic compounds at different storage temperatures and packaging conditions. *International Journal of Food Microbiology*, *280*, 87–99. https://doi.org/https://doi.org/10.1016/j.ijfoodmicro.2017.12.029

Park, J., Kim, Y. W. & Jeon, H.-J. (2024). Machine learning-driven innovations in microfluidics. *Biosensors*, *14*(12). https://doi.org/10.3390/bios14120613

Păuşescu, I., Dreavă, D.-M., Bîtcan, I., Argetoianu, R., Dăescu, D. & Medeleanu, M. (2022). Bio-based pH indicator films for intelligent food packaging applications. *Polymers*, *14*(17). https://doi.org/10.3390/polym14173622

Pavoni, E., Bertasi, B., Galuppini, E., Mangeri, L., Meletti, F., Tilola, M., Carta, V., Todeschi, S. & Losio, M.-N. (2022). Detection of hepatitis a virus and norovirus in different food categories: A 6-year survey in Italy. *Food and Environmental Virology*, *14*(1), 69–76. https://doi.org/10.1007/s12560-021-09503-y

Popa, A., Hnatiuc, M., Paun, M., Geman, O., Hemanth, D. J., Dorcea, D., Son, L. H. & Ghita, S. (2019). An intelligent IoT-based food quality monitoring approach using low-cost sensors. *Symmetry*, *11*(3). https://doi.org/10.3390/sym11030374

Purohit, B., Vernekar, P. R., Shetti, N. P. & Chandra, P. (2020). Biosensor nanoengineering: Design, operation, and implementation for biomolecular analysis. *Sensors International*, *1*, 100040. https://doi.org/https://doi.org/10.1016/j.sintl.2020.100040

Pushpanjali, P. A., Manjunatha, J. G., Hareesha, N., Souza, E. S. D', Charithra, M. M. & Prinith, N. S. (2021). Voltammetric analysis of antihistamine drug cetirizine and paracetamol at poly(L-Leucine) layered carbon nanotube paste electrode. *Surfaces and Interfaces*, *24*, 101154. https://doi.org/https://doi.org/10.1016/j.surfin.2021.101154

Rani, N., Yadav, S., Saini, K. & Maity, D. (2024). Effect of biosensors on food dyes: A review. In J. G. Manjunatha (Ed.), *Biosensing technology for human health: Eco-friendly materials and real-world applications* (Vol. 27). Royal Society of Chemistry. https://doi.org/10.1039/9781837676323-00316

Raril, C., Manjunatha, J. G., Ravishankar, D. K., Fattepur, S., Siddaraju, G. & Nanjundaswamy, L. (2020). Validated electrochemical method for simultaneous resolution of tyrosine, uric acid, and ascorbic acid at polymer modified nano-composite paste electrode. *Surface Engineering and Applied Electrochemistry*, *56*, 415–426. https://api.semanticscholar.org/CorpusID:225527755

Samet-Bali, O., Lajnaf, R., Felfoul, I., Attia, H. & Ayadi, M. (2013). Detection of Escherichia coli in unpasteurized raw Milk. *International Journal of Agricultural and Food Science, 3*(2): 53–55.

Santana Oliveira, I., da Silva Junior, A. G., de Andrade, C. A. S. & Lima Oliveira, M. D. (2019). Biosensors for early detection of fungi spoilage and toxigenic and mycotoxins in food. *Current Opinion in Food Science*, *29*, 64–79. https://doi.org/https://doi.org/10.1016/j.cofs.2019.08.004

Scognamiglio, V., Arduini, F., Palleschi, G. & Rea, G. (2014). Biosensing technology for sustainable food safety. *TrAC Trends in Analytical Chemistry*, *62*, 1–10. https://doi.org/https://doi.org/10.1016/j.trac.2014.07.007

Sen, M. (2021). Food chemistry: Role of additives, preservatives, and adulteration. In *Food chemistry* (pp. 1–42). https://doi.org/https://doi.org/10.1002/9781119792130.ch1

Shi, Z., Li, X., Shuai, Y., Lu, Y. & Liu, Q. (2022). The development of wearable technologies and their potential for measuring nutrient intake: Towards precision nutrition. *Nutrition Bulletin*, *47*(4), 388–406. https://doi.org/https://doi.org/10.1111/nbu.12581

Silva, N. F. D., Magalhães, J. M. C. S., Freire, C. & Delerue-Matos, C. (2018). Electrochemical biosensors for Salmonella: State of the art and challenges in food safety assessment. *Biosensors and Bioelectronics*, *99*, 667–682. https://doi.org/https://doi.org/10.1016/j.bios.2017.08.019

Singh, A. K., Mittal, S., Das, M., Saharia, A. & Tiwari, M. (2023). Optical biosensors: A decade in review. *Alexandria Engineering Journal*, *67*, 673–691. https://doi.org/https://doi.org/10.1016/j.aej.2022.12.040

Sivakumar, R. & Lee, N. Y. (2021). Recent progress in smartphone-based techniques for food safety and the detection of heavy metal ions in environmental water. *Chemosphere*, *275*, 130096. https://doi.org/https://doi.org/10.1016/j.chemosphere.2021.130096

Sobhan, A., Muthukumarappan, K. & Wei, L. (2021). Biosensors and biopolymer-based nanocomposites for smart food packaging: Challenges and opportunities. *Food Packaging and Shelf Life*, *30*, 100745. https://doi.org/https://doi.org/10.1016/j.fpsl.2021.100745

Srinivasan, B. & Tung, S. (2015). Development and applications of portable biosensors. *Journal of Laboratory Automation*, *20*(4), 365–389. https://doi.org/10.1177/2211068215581349

Sumitha, M. S. & Xavier, T. S. (2023). Recent advances in electrochemical biosensors – A brief review. *Hybrid Advances*, *2*, 100023. https://doi.org/https://doi.org/10.1016/j.hybadv.2023.100023

Surendran Nair, M., Nair, D. V. T., Kollanoor Johny, A. & Venkitanarayanan, K. (2020). Chapter 12 - Use of food preservatives and additives in meat and their detection techniques. In A. K. Biswas & P. K. Mandal (Eds.), *Meat quality analysis* (pp. 187–213). Academic Press. https://doi.org/https://doi.org/10.1016/B978-0-12-819233-7.00012-4

Suvarnaphaet, P. & Pechprasarn, S. (2017). Graphene-based materials for biosensors: A review. *Sensors*, *17*(10). https://doi.org/10.3390/s17102161

Toldrá, F. & Reig, M. (2006). Methods for rapid detection of chemical and veterinary drug residues in animal foods. *Trends in Food Science & Technology*, *17*(9), 482–489. https://doi.org/https://doi.org/10.1016/j.tifs.2006.02.002

Torre, R., Costa-Rama, E., Nouws, H. P. A. & Delerue-Matos, C. (2020). Screen-printed electrode-based sensors for food spoilage control: Bacteria and biogenic amines detection. *Biosensors*, *10*(10). https://doi.org/10.3390/bios10100139

Umapathi, R., Ghoreishian, S. M., Rani, G. M., Cho, Y. & Huh, Y. S. (2022). Review—emerging trends in the development of electrochemical devices for the on-site detection of food contaminants. In *ECS sensors plus* (Vol. 1, No. 4). Institute of Physics. https://doi.org/10.1149/2754-2726/ac9d4a

Upadhyay, S., Kumar, A., Srivastava, M., Srivastava, A., Dwivedi, A., Singh, R. K. & Srivastava, S. K. (2024). Recent advancements of smartphone-based sensing technology for diagnosis, food safety analysis, and environmental monitoring. *Talanta*, *275*, 126080. https://doi.org/https://doi.org/10.1016/j.talanta.2024.126080

Victor, C. P., Ellis, K., Lamar, F. & Leon, J. S. (2021). Agricultural detection of norovirus and hepatitis a using fecal indicators: A systematic review. *International Journal of Microbiology, 2021*(1), 6631920. https://doi.org/https://doi.org/10.1155/2021/6631920

Villa, C. C., Sánchez, L. T., Valencia, G. A., Ahmed, S. & Gutiérrez, T. J. (2021). Molecularly imprinted polymers for food applications: A review. *Trends in Food Science & Technology, 111*, 642–669. https://doi.org/https://doi.org/10.1016/j.tifs.2021.03.003

Viswanathan, S., Radecka, H. & Radecki, J. (2009). Electrochemical biosensors for food analysis. *Monatshefte Für Chemie - Chemical Monthly, 140*(8), 891–899. https://doi.org/10.1007/s00706-009-0143-5

Wang, P., Sun, X., Su, X. & Wang, T. (2016). Advancements of molecularly imprinted polymers in the food safety field. *Analyst, 141*(12), 3540–3553. https://doi.org/10.1039/C5AN01993A

Warriner, K., Reddy, S. M., Namvar, A. & Neethirajan, S. (2014). Developments in nanoparticles for use in biosensors to assess food safety and quality. *Trends in Food Science & Technology, 40*(2), 183–199. https://doi.org/https://doi.org/10.1016/j.tifs.2014.07.008

Wasilewski, T., Kamysz, W. & Gębicki, J. (2024). AI-assisted detection of biomarkers by sensors and biosensors for early diagnosis and monitoring. In *Biosensors* (Vol. 14, Issue 7). Multidisciplinary Digital Publishing Institute (MDPI). https://doi.org/10.3390/bios14070356

Witjaksono, G., Saeed Rabih, A. A., Yahya, N. B. & Alva, S. (2018). IOT for agriculture: Food quality and safety. *IOP Conference Series: Materials Science and Engineering, 343*(1). https://doi.org/10.1088/1757-899X/343/1/012023

Xu, M.-L., Gao, Y., Han, X. X. & Zhao, B. (2017). Detection of pesticide residues in food using surface-enhanced raman spectroscopy: A review. *Journal of Agricultural and Food Chemistry, 65*(32), 6719–6726. https://doi.org/10.1021/acs.jafc.7b02504

Yang, T., Luo, Z., Bewal, T., Li, L., Xu, Y., Mahdi Jafari, S. & Lin, X. (2022). When smartphone enters food safety: A review in on-site analysis for foodborne pathogens using smartphone-assisted biosensors. *Food Chemistry, 394*, 133534. https://doi.org/https://doi.org/10.1016/j.foodchem.2022.133534

Zeng, Y., Yuan, R., Fu, H., Xu, Z. & Wei, S. (2024). Foodborne pathogen detection using surface acoustic wave biosensors: A review. In *RSC advances* (Vol. 14, No. 50, pp. 37087–37103). Royal Society of Chemistry. https://doi.org/10.1039/d4ra06697a

Zhang, J., Hu, S., Du, Y., Cao, D., Wang, G. & Yuan, Z. (2020). Improved food additive analysis by ever-increasing nanotechnology. In *Journal of food and drug analysis* (Vol. 28, Issue 4, pp. 622–640). Taiwan Food and Drug Administration. https://doi.org/10.38212/2224-6614.1152

Zhang, K. & Banerjee, K. (2020). A review: Sample preparation and chromatographic technologies for detection of aflatoxins in foods. *Toxins, 12*(9). https://doi.org/10.3390/toxins12090539

Zompanti, A., Grasso, S., Santonico, M. & Pennazza, G. (2020). *A sensor system for non-destructive monitoring of food ripening processes.* 2020 IEEE International Workshop on Metrology for Industry 4.0 & IoT, 80–83. https://doi.org/10.1109/MetroInd4.0IoT48571.2020.9138308

14 Perspectives on the Role of Biosensors in Biomedical Diagnostics

Amar K. Durgannavar, Geetha B. Heggannavar and Ramesha Hanumanthappa

14.1 INTRODUCTION

A biosensor is a subclass of sensor and is a diagnostic device that has an electronic component (sensing element) associated with a transducer to detect a measurement. The advantage of biosensors over other detection methods lies in their ability to selectively recognize biological chemicals and respond to relevant chemical assays in the body (Metkar & Girigoswami, 2018). The advent of biosensors has revolutionized the landscape of biomedical applications, enhancing our capability to monitor and diagnose health conditions with unprecedented precision and efficiency. These analytical devices, which fuse biological recognition elements with transducer technology, facilitate the real-time detection of biochemical substances, rendering them indispensable in various clinical settings. As the global burden of chronic diseases escalates, the integration of biosensors into early detection and continuous monitoring protocols not only promises to improve patient outcomes but also reduces pressures on healthcare systems by enabling decentralized diagnostics. As the world population ages and grows, lifelong diseases, biomedical diagnostics, and biosensor challenges become increasingly important. One of the most important benefits of biomarker testing is its ability to aid in accurate and targeted drug discovery and early diagnosis. It offers a way to make better diagnoses based on the patient's molecular information, rather than external medical treatment (Pourasl et al., 2014; Baird et al., 2001). With the development of compact and portable microsensors, different biomarkers of human health can be continuously measured and monitored. They identify molecular markers using easily accessible variables such as temperature, heart rate amplitude, electrical concentration, dehydration, and glucose. Biosensors eliminate the need for repeated testing by allowing the collected data to be saved for comparison in the future and added to the patient's medical file (Chandra et al., 2012; Choudhary et al., 2016). The development of nanomaterials, particularly graphene quantum

DOI: 10.1201/9781003536239-14

dots (GQDs), has significantly enhanced the functionality of biosensors by improving their sensitivity, selectivity, and reproducibility, which is crucial for detecting ultralow concentrations of biomarkers in biological samples. Recent studies emphasize that GQDs not only serve as effective electro-catalysts and fluorophores but also facilitate label-free immunoassays, thereby democratizing access to rapid diagnostics, especially in resource-limited settings (Mansuriya & Altintas, 2020). Consequently, the evolution of biosensors signifies a pivotal advancement towards innovative, cost-effective healthcare solutions, underscoring their indispensable role in modern biomedicine. Nowadays, a large number of biosensors that are used in the medical field act as a tool to detect and analyze tissues, bacteria, viruses, cell receptors, enzymes, antibodies, and nucleic acids. Their development as biosensors is of great importance and will lead to significant advances in the field of medicine and allow the discovery of new, accurate and useful analytical sensors. Clinical laboratories use biosensors to monitor a variety of processes, including the interaction of antibodies, glucose levels, catalytic enzymes, proliferation of tumors, microbial infections, pathogen and toxin detection, and so on (Haleem et al., 2021). This chapter aims to elucidate the multifaceted roles biosensors occupy within the biomedical arena, encompassing the technological advancements that underpin their functionality, their applications across diverse health domains, and the ethical considerations that accompany their deployment. In doing so, it aspires to provide a holistic perspective on the transformative potential of biosensors in contemporary medicine as diagnostic tools.

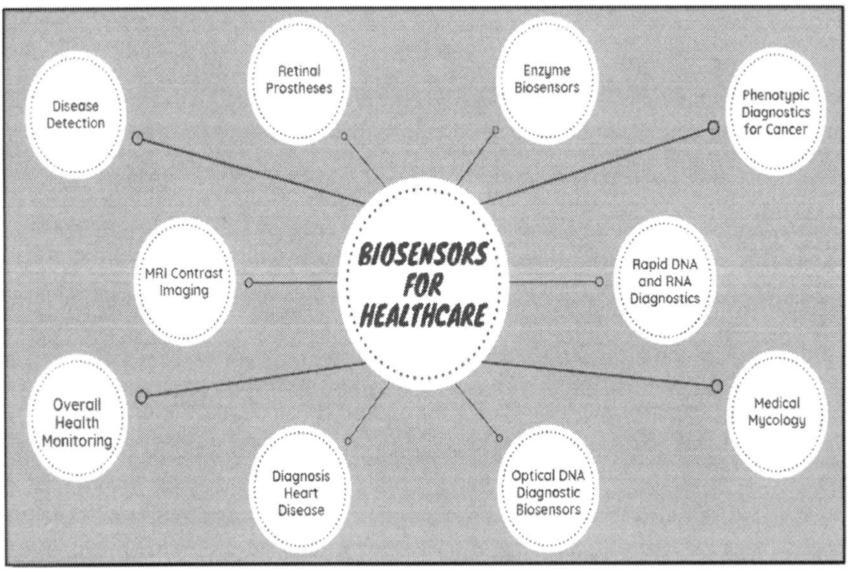

FIGURE 14.1 Biosensors offering diagnostics in various areas of healthcare (Haleem et al., 2021).

14.2 OVERVIEW OF BIOSENSORS AND THEIR APPLICATIONS IN BIOMEDICAL DIAGNOSTICS

The integration of biosensors in biomedical fields represents a paradigm shift in the early detection and management of various health conditions. These analytical devices leverage biorecognition elements to identify specific biomolecules, enabling the quantification of markers associated with diseases such as cancer and cardiovascular disorders. Consequently, the evolution of biosensors signifies a pivotal advance toward innovative, cost-effective healthcare solutions, underscoring their indispensable role in modern biomedicine. Key perspectives on the role of biosensors in biomedical diagnostics include the following:

14.2.1 Early Disease Detection

Early detection of diseases using biosensors is a transformative advancement in modern healthcare. Biosensors are analytical tools that identify and measure biomarkers in biological samples through the integration of a transducer with a biological recognition element, such as enzymes, antibodies, or nucleic acids. By identifying disease-specific molecules at low concentrations, biosensors enable the rapid and accurate diagnosis of conditions like diabetes, cardiovascular diseases, and cancer, often before symptoms manifest. These devices are highly sensitive, cost-effective, and capable of delivering real-time results. Early disease detection through biosensors could improve patient outcomes by facilitating timely treatment, but biosensors can also reduce healthcare costs by preventing disease progression.

i. *Cardiovascular diseases diagnosis*: Heart conditions are thought to be the world's biggest cause of death. For such a diagnosis, prompt and early diagnosis is crucial. Numerous biomarkers, including myoglobin, B-type natriuretic peptide (BNP), cardiac troponin I (cTnI), C-reactive protein (CRP), interleukin, interferon, and others, have been employed in conjunction with the developments in nanotechnology and medicine. These cardiovascular signatures use various biosensor technologies such as optical, acoustic, electrochemical and magnetic-based biosensors. An example of a cardiac biomarker is the troponin biomarker used for diagnosis and risk stratification in patients with suspected heart disease. These types of biosensors can be used in various cardiovascular and other applications. Classification box, cardiac labeling system, stratus, alpha, and so on are several biosensors used for early detection and diagnosis (Nam et al., 2021; Girigoswami & Akhtar, 2019; Choi, 2020; Zadran et al., 2012).
ii. *Cancer diagnosis*: In Western nations, cancer ranks as the second leading cause of death and morbidity. Cancer diagnosis and treatment are a significant unmet need worldwide. However, there is a strong link between early diagnosis and positive patient outcomes; early diagnosis offers the best chance of improving patient survival and prognosis. It can reduce pain in all cancers and contribute to cancer prevention. The most important

markers for cancer diagnosis today are tumor morphological and histological features or specific biomarkers (Mascini & Tombelli, 2008). Using biosensors, a variety of biomarkers, including peptides, proteins, antibodies, DNA, RNA, and microRNA, can be employed as predictive and prognostic tools for quick cancer diagnosis. In cancer research, various biosensor technologies have been used to investigate the effects of anticancer/cytotoxic drugs and to study various markers of metastatic cancer (Arafa et al., 2022). Analysis of specific chemical compounds (e.g., extracellular metabolites, nucleotides, expressed protein biomarkers, or whole cancer cells) is the driving force behind the operation of electrochemical biosensors. There are many types of electrochemical biosensors used in cancer diagnosis, including microfluidic analysis, gene sensors, impedance-based biosensors, electrochemiluminescence biosensors, and immunosensors. A combination of cell chemical methods with bioelectrochemical and biophysical methods has been proposed in the literature to better understand cancer and apoptosis (Arafa et al., 2022; Hassan, 2022).

Prostate cancer has become a major health problem, being the third most common cancer in men worldwide (Jemal et al., 2006). In recent years, various biosensors have been proposed for prostate-specific antigen (PSA) detection based on different technologies, ranging from electrochemical (Meyerhoff et al., 1995; Fernandez-Sanchez et al., 2004; Sarkar, 2002) to piezoelectric (Wu, 2001; Wee et al., 2005) and optical methods (Besselink et al., 2004; Huang et al., 2005). It has recently been reported that surface plasmon resonance (SPR) can be used to identify different PSA isoforms, even at concentrations as low as 1 ng/ml (Cao et al., 2006). In summary, electrochemical biosensors are a promising tool for early target detection due to their role in the detection of cancer and apoptosis, their straightforward design, and their capacity to translate biological parameters (such as cell viability, protein binding affinity, and DNA or RNA) to particular viruses. In addition, cancer cells may enable the discovery of anti-cancer drugs by determining the immediate and sustained response of cancer or normal cells.

iii. *Infectious disease diagnosis*: Biosensors have garnered significant attention for monitoring infectious diseases, particularly those caused by microbial pathogens such as fungi, bacteria, and viruses. Molecularly imprinted polymers (MIPs) have emerged as promising tools for biomedical applications due to their ability to interact with specific proteins and biomaterials. Research has explored MIPs for precise disease detection (Gast et al., 2019). For instance, Piletska et al. (2020) developed MIPs for virus identification and analyzing different epitopes of adeno-associated virus (AAV) immobilized on glass beads using acrylate and methacrylate polymer nanoparticles. An anti-MIP sensor for early dengue virus detection in blood samples was created using dopamine self-polymerization in the presence of the target non-structural protein 1 (NS1), achieving a detection limit of 0.3 ng/ml (Arshad et al., 2020).

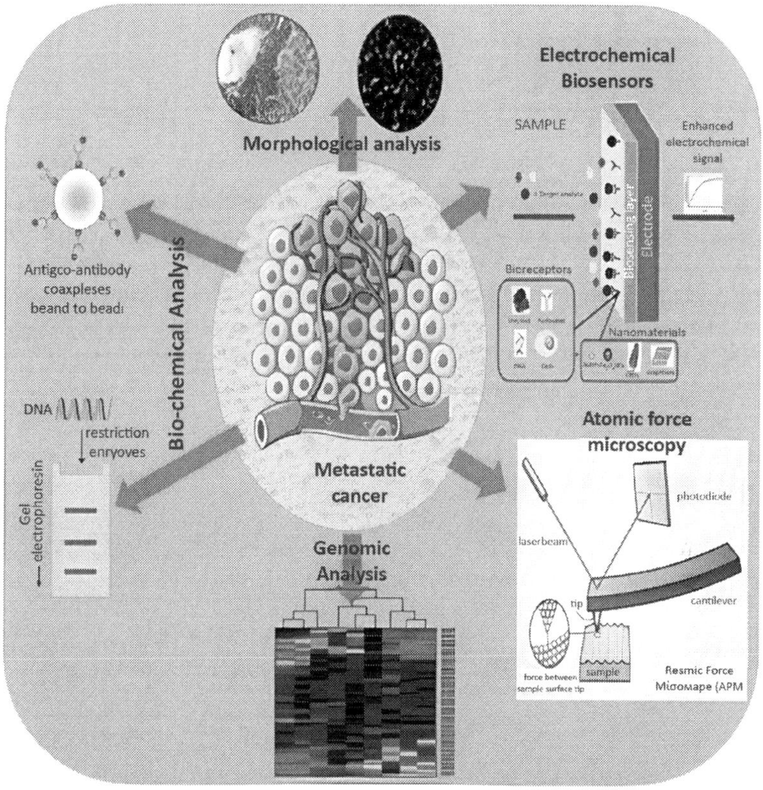

FIGURE 14.2 Cancer investigation through combinatorial biochemical and biophysical techniques using biosensors (Hassan, 2022). This figure is used from an article which is an open access article distributed under the terms and conditions of the Creative Commons Attribution (CC BY) license (https://creativecommons.org/licenses/by/4.0/).

Chowdhury et al. (2019) designed a sensor incorporating antibody-conjugated polyaniline nanowires, graphene quantum dots, and nano-gold to detect hepatitis E virus (HEV) in clinical samples. Sensitivity was enhanced by external radiation during HEV accumulation, which expanded the sensor surface and lengthened the antibody-conjugated polyaniline chain. This sensor successfully detected various HEV genotypes, including G1, G3, G7, and ferret HEV from cell culture supernatants and fecal samples of G7 HEV-infected monkeys. Rapid detection of the Zika virus (ZIKV) was achieved using electrochemical biosensors based on surface-imprinted polymers and graphene oxide composites. Quantitative virus measurements were made by monitoring electrical signal changes corresponding to virus concentration (Tancharoen et al., 2019). In all these studies, virus biosensors represent effective diagnostic methods for rapid response and are highly sensitive, allowing accurate selectivity and the quantitation of viruses in real samples without labeling and without the need for the extraction or purification of genetic material (DNA and RNA).

14.2.2 Point-of-Care Testing (POCT)

POCT entails "patient specimens assayed at or near the patient with the assumption that test results will be available instantly or in a very short timeframe to assist caregivers with immediate diagnosis and/or clinical intervention" (Ehrmeyer et al.,2007). A portable and user-friendly biosensor for biomedical applications refers to a small, easy-to-operate device that can detect specific biological molecules (like proteins, DNA, or glucose) directly on-site, often using technologies like electrochemical sensing, microfluidics, or optical detection, allowing for rapid and convenient point-of-care diagnostics without requiring complex laboratory setups. POCT using biosensors has revolutionized healthcare by enabling rapid, accurate, and decentralized diagnostic capabilities. These compact devices provide immediate results at or near the patient's location, eliminating the need for time-consuming laboratory procedures. Timely results are crucial for test outcomes to impact clinical decisions, including triage, referral, and discharge planning. "Rapid" typically refers to a timeframe ranging from seconds to minutes or a few hours, ensuring the patient remains at the site during testing. Figure 14.3 illustrates the rapid turnaround of results with POCT when compared to a conventional testing procedure (Srinivasan, 2015). By integrating biosensors, POCT offers high sensitivity and specificity in detecting biomarkers, pathogens, or disease states, facilitating timely decision-making and treatment. This technology is particularly vital in resource-limited settings, emergency care, and remote areas, where access to traditional laboratory infrastructure is limited. Moreover, POCT biosensors enhance patient outcomes by reducing

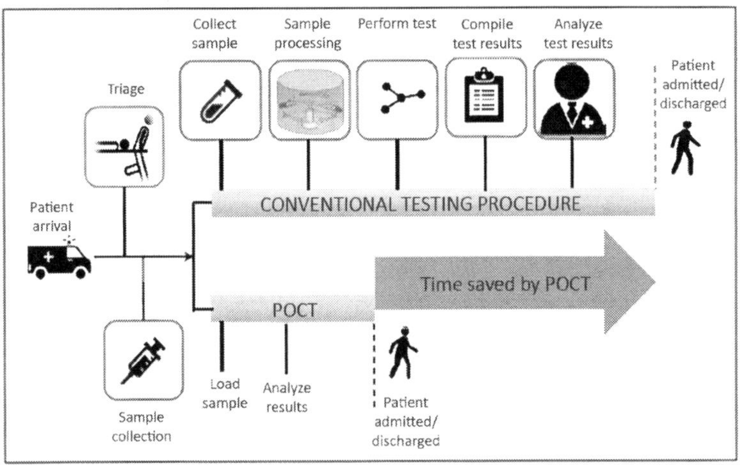

FIGURE 14.3 Schematic representation comparing conventional testing procedure with POCT (Srinivasan, 2015).

diagnostic delays, lowering healthcare costs, and empowering individuals to manage their health proactively.

In healthcare settings (Cohen-Bacrie et al., 2011), POCT can include testing for HIV, tuberculosis, malaria, hepatitis B, syphilis, dengue fever, and other diseases. These tools include not only rapid diagnostic tests (RDTs) but also molecular tests, enzyme-linked immunosorbent assays (ELISAs), and microscopic tests. POCT can also be performed in a hospital setting for diagnosis, treatment, and monitoring purposes. Since the tests are performed by hospital staff in the emergency room, operating room, and intensive care unit, no laboratory testing is required. An impedance biosensor has been developed to detect H5N1 influenza virus (Wang et al., 2009) at titers greater than 10^3 EID 50/ml (EID50: 50% egg infectivity) within 2 h and 32 s. A novel potentiometric immunosensor (Tang et al., 2004) has been developed to detect hepatitis B surface antigen (HBsAg) by immobilizing it on a gold nanoparticle-based platinum disk electrode via self-assembly. The dynamic concentration range of the assay was 4–800 ng/ml, the detection limit was 1.3 ng/ml, and the results were performed according to the standard ELISA method.

Bioelectric recognition analysis (Kintzios et al., 2004) (BERA) is a cell-based biosensor method that measures the electrical response (membrane potential) of cultures suspended in a gel matrix to various ligands that bind and interact with cells. BERA has been used to detect hepatitis C virus (HCV) in blood samples (Kintzios et al., 2001), and can rapidly (detection time 3–5 min) and specifically detect serum levels below 100 pg/ml. The use of nanomaterials, such as carbon nanotube replacement electrodes (Deo & Wang, 2004), has significantly enhanced the detection capabilities of biosensors for various biomolecules. Recently, screen-printed electrodes (Susmel et al., 2003) have gained considerable attention for use in DNA, antibody, and enzyme-based biosensors. These electrodes offer a commercially viable solution for the large-scale production of disposable electronic devices. As biosensor technology continues to evolve, POC testing is expected to play a significant role in precision medicine and global health.

14.2.3 Continuous Health Monitoring

Continuous health monitoring through biosensor devices is revolutionizing personalized healthcare by facilitating real-time tracking of vital signs, physiological metrics, and biomarkers. These implantable devices provide continuous, non-invasive monitoring, offering valuable insights into an individual's health trends and early detection of potential issues. By delivering data instantaneously, they empower patients and healthcare providers to make informed decisions, improving chronic disease management, early intervention, and preventive care. This technology is particularly critical for managing conditions like diabetes, cardiovascular diseases, and sleep disorders, where ongoing monitoring is essential. Additionally, continuous biosensors enhance patient engagement by fostering a proactive approach to health and reducing hospital visits, ultimately improving overall quality of life.

The rapid development of electronic and digital health care has promoted the participation of doctors and patients in biomedical processes and modern health with the help of wearable health monitors (Carreiro et al., 2015). Smart wearable body sensors can be used to monitor the patient's activity. These wearable devices provide important information about chronic and acute diseases, helping doctors to effectively treat diseases with the help of modern medical techniques. Wearable sensing devices can measure variables and send data to online data storage systems or personal devices. These sensors can be effective diagnostic tools in the diagnosis and treatment of various diseases. The adaptability of wearable body sensors offers patients both comfort and convenience, leading to rapid growth in the wearable sensor market in recent years. Wearable sensors are commonly designed to be attached to clothing or worn on various body parts, including the chest, wrist, arm, ankle, leg, and hip. These sensors enable continuous monitoring of parameters such as body temperature, heart rate, blood pressure, galvanic skin response, electrocardiogram (ECG), calorie consumption, step count, distance traveled, and speed. Some wearable sensors can be used as accessories on body which can be attached on gloves, belt buckles, necklaces, earrings, rings, and brooches (Appelboom et al., 2014).

Wearable body sensors are applicable in medical fields such as vascular and cardiopulmonary monitoring, blood glucose tracking, and neurological assessment. In 2016, Stephanie Carreiro et al. (2016) reported a biosensor device to detect opioid use. Use of this measurement device may help reduce deaths from opioid overdose. In 2018, Y. Kim et al. (2018) developed wearable sweat biosensors which could measure cation concentrations in human sweat in real time. Current advancements in Internet of Things (IoT) technology, biomedical engineering, and wireless communication have enabled comprehensive monitoring of the human body and mind. Flexible, stretchable, and biocompatible electronic devices are needed for this purpose (J. Kim et al., 2018). In 2018, Marín-Morales et al. (2018) reported that wearable devices can be used to identify emotions based on heart rate and brain wave abnormalities. Kweon et al. (2018) illustrated high-performance wearable devices using 3D electrospun conductive nanofibers. This wearable pressure sensor can function as a wireless blood pressure measurement belt. Hallfors et al. (2018) designed a wearable electrocardiogram biosensor using reduced graphene oxide and nylon. Additionally, flexible wearable body sensors are suitable for monitoring human movement, vital signs, body temperature, heart rate, respiratory rate, blood pressure, blood glucose (Saha et al., 2017) and pulse oximetry (Gao et al., 2019).

In the real world, wearable sensors (such as smartphones or smartwatches) can also be used to passively collect health data (Bender & Cortés-Ciriano, 2021). Traditionally, signals are received through gel electrodes applied to the body. Along with the use of modern devices like smartwatches and wearables, recent advancements in manufacturing and electronics have enabled the integration of biosensor electrodes into other devices, such as eyeglasses (Vourvopoulos et al., 2019), virtual reality head-mounted displays (Bernal et al., 2018), and textiles (Nittala et al., 2020). A new category of computational devices, known as epidermal devices, enables non-invasive interaction with body tissues through interactive tattoos (Figure 14.4) (Nittala & Steimle, 2022; Wang et al., 2020). These epidermal devices can measure electrophysiological and electrochemical signals in the body (Bandodkar et al., 2019). As innovations in biosensor technology advance, these devices are expected

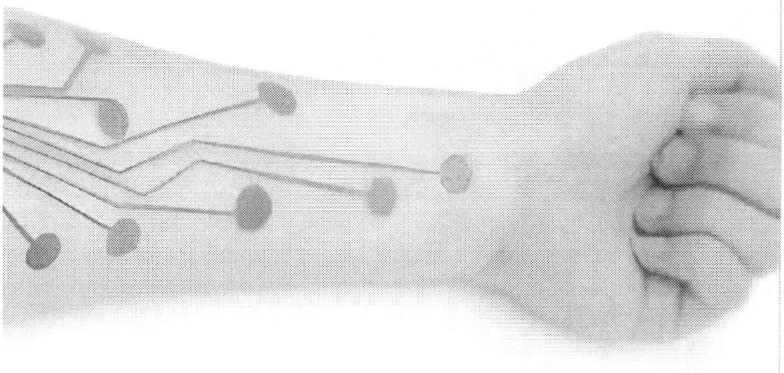

FIGURE 14.4 A wearable biosensor in the form of a multi-modal physiological sensing tattoo, capable of detecting ECG, EDA, and EMG signals on the forearm (Nittala et al., 2020).

to become even more integral to modern healthcare, promoting better outcomes and more efficient healthcare delivery.

14.2.4 Drug Discovery and Development

Biosensors play a significant role in drug discovery and development by enabling highly sensitive and specific detection of potential drug candidates, allowing for rapid screening of large compound libraries, real-time monitoring of drug–target interactions, and evaluation of drug efficacy and toxicity, often with high throughput capabilities, thus accelerating the drug development process considerably, essentially acting as a critical tool for identifying promising lead compounds and optimizing their properties early on in the research pipeline. Additionally, biosensors reduce reliance on animal testing by offering alternative *in-vitro* platforms for toxicity screening and mechanistic studies. By enhancing throughput and precision, biosensors streamline the drug development pipeline, saving time and resources while ensuring more targeted and effective therapeutics.

Surface plasmon resonance (SPR) biosensors have established themselves as the gold standard for investigating drug interactions, offering a crucial platform for drug screening. SPR has two major challenges. First, it is sustainable. After covalent attachment to the biosensor surface, the protein must retain its ligand binding ability. Another is the expertise required to develop efficient SPR bioassays. The number of drug screening studies using SPR biosensors is continuously increasing and covers a wide range of drug targets, including metallo-β-lactamase VIM-2 (Christopeit et al., 2015), Clostridial collagenase (Schönauer et al., 2017), tumor necrosis factor receptor (Cao et al., 2016), signal transduction and transcription activator 3 (Chen et al., 2018), C-X-C chemokine receptor type 4 (Chen et al., 2020b), α-glucosidase

(Kong et al., 2018), FK506 binding protein (Zhou et al., 2018), R15B-PP1c holoenzyme (Krzyzosiak et al., 2018), PI3Kγ (Gangadhara et al., 2019), xanthine oxidase (Tao et al., 2021), DNA MTase (Xia et al., 2019), and others. SPR technology has been expanded to target species that are difficult to attack, such as membrane-bound proteins. Optical waveguide grating (OWG) biosensors utilize resonant waveguide gratings or nanostructured gratings to detect substances that are bound to proteins (Geschwindner et al., 2012). Biolayer interferometry (BLI) biosensors do not have microfluidics. They analyze the interference patterns of white light reflected by the drug target layer immobilized on the biosensor tip using built-in sensors (Kaminski et al., 2017). Drug targets include glutamate carboxypeptidase II (Choy & Berkman, 2016), glutaminase (Zhu et al., 2017), double-stranded DNA (Overacker et al., 2021), and protein–protein interactions such as tissue fibronectin and transglutaminase interactions (Sima et al., 2019) and SARS-CoV-2 RBD and hACE2 interaction (Yang et al., 2021). It provides rapid and accurate measurement of kinetic rates and determination of affinity constants for interacting ligands.

Among other methods, electroanalytical techniques also offer several advantages, including easy operation, suitability for routine analysis, minimal sensitivity to interfering molecules, affordability, rapid analysis, and an efficient alternative for studying medicinal compounds (Beitollahi et al., 2016; Jahani & Beitollahi, 2016; Ibrahim et al., 2018). Pushpanjali et al. (2021) developed an electrochemical sensor using a carbon nanotube paste electrode modified with a thin layer of electrochemically polymerized L-leucine to detect cetirizine (CTZ) in the presence of paracetamol. The method achieved a detection limit of 0.17 μM for CTZ and was successfully applied to analyze CTZ in pharmaceutical samples, blood serum, and urine, yielding a recovery rate of 96.49%–99.85%. Similarly, Chenthattil et al. (2020) designed an electrochemical sensor using a poly-threonine-modified graphite-carbon nanotube paste electrode to analyze tyrosine. The electrode demonstrated high selectivity by effectively detecting tyrosine in the presence of ascorbic acid and uric acid. It exhibited a linear response within concentration ranges of 2×10^{-6}–2.5×10^{-5} M and 3×10^{-5}–1.2×10^{-4} M, with detection and quantification limits of 2.9×10^{-7} M and 9.6×10^{-7} M, respectively. The modified electrodes' low detection limit and high selectivity suggest their strong potential as sensor candidates (Chenthattil et al., 2020).

Foster resonance energy transfer (FRET)-based biosensors have been created to detect biomolecules or changes in the microenvironment. FRET involves the non-radiative transfer of energy from an excited donor fluorophore molecule to a nearby acceptor fluorophore molecule. Fluorescence detection is widely used in HIV high-throughput testing (HTS) services due to its high demand and commercialization. Typical FRET fluorophores are coated with UV dyes and near-infrared (NIR) dyes. The former includes coumarin, naphthalene, and pyrene-based analogs, while the latter include cyanine (e.g., Cy3, Cy5, Cy5.5, Cy7, and Cy7.5) and rhodamine (Zhang et al., 2019), which are drug targets that can block the binding of donor or acceptor fluorophores. The use of FRET biosensors in drug development is increasing year by year. A variety of drug targets have been targeted, including acetylcholinesterase (Sun & Yang, 2015), tyrosinase (Liu et al., 2016; Chen et al., 2020a), HIV-1 reverse transcriptase (Sharma et al., 2016), α-glucosidase (Li et al., 2016), and many more

(Tao et al., 2021). The impact of biosensors on the pharmaceutical industry will play a significant role in the delivery of innovative and patient-centered medicines.

14.2.5 Environmental Health Monitoring

Biosensors are essential tools in environmental health monitoring, providing rapid, accurate, and cost-effective solutions for detecting pollutants, toxins, and pathogens in air, water, and soil. These devices enable real-time tracking of environmental contaminants, such as heavy metals, pesticides, and industrial waste, helping to identify potential health hazards before they escalate. Biosensors are particularly valuable in addressing challenges like water quality monitoring, air pollution control, and ecosystem management by offering high sensitivity and specificity. Their portable and user-friendly design allows for on-site applications, reducing reliance on traditional laboratory testing. By facilitating early detection and continuous surveillance, biosensors contribute to safeguarding public health, preserving biodiversity, and promoting sustainable environmental practices.

i. *Heavy Metal Measurement*: Heavy metals are a primary cause of today's most severe pollution issues. Even in trace amounts, they pose significant risks to both the environment and human health because they do not break down biologically. People are continuously exposed to heavy metals in their surroundings. The hazards associated with heavy metals arise from their widespread presence in the biosphere, their high bioavailability from both natural and human-made sources, and their extreme toxicity. As a result, numerous studies have documented serious health damage, including fatalities, caused by exposure to these pollutants. Many bacterial biosensors designed to detect heavy metals in environmental samples utilize specific genes as bioreceptors, which are responsible for the bacteria's sensitivity to these elements. Bacterial cells resistant to various metals (including mercury, zinc, copper, silver, cobalt, and tin) have been isolated as bioreceptors (Nasipude et al., 2021).
ii. *Nitrogen compound measurement*: Nitrite is commonly used in food preservation and soil fertilization. However, the prolonged presence of these ions in the body can pose significant health risks, particularly due to their harmful interaction with heme. High nitrate levels in groundwater and surface waters are a concern because they can harm aquatic life. A biosensor for measuring nitrite was designed by immobilizing cytochrome c nitrite reductase (ccNiR) from Desulfovibrio in redox-active [ZnCr-AQS] double-layer hydroxide and electrically connecting it to a glassy carbon electrode and encapsulating it in anthraquinone-2-sulfonic acid (AQS). This device responds very quickly (5 s) to nitrite, with a wide linear range between nitrite concentrations of 0.015 and 2.35 μM and a detection limit of 4 nM. It is a highly rapid, sensitive and stable conductometric enzyme biosensor for the determination of nitrite in water (Nasipude et al., 2021).

iii. *Polychlorinated biphenyls (PCBs) measurement*: The amount of PCBs in the environment depends on the matrix from which they are derived. There are 209 PCB congeners present in the global environment and food chain. Based on the arrangement of the chlorine atoms, these congeners are classified into three groups: coplanar, mono-ortho-coplanar, and non-coplanar. The traditional method of analyzing PCBs is mainly based on gas chromatography mass spectrometry (GCMS). Among the different immunoassay techniques, enzyme-linked immunosorbent assay (ELISA) with colorimetric endpoint detection is the most widely used. Another promising method is the application of immunosensor devices. Immunosensors are a type of biosensor that uses chemical reagents, antibodies, or antibodies to detect PCBs. Pribyl et al. (2006) created an innovative piezoelectric immunosensor designed to detect PCB congeners within the concentration range commonly found in real matrices such as soil. As the use of the technology increases, biosensors will play an important role in protecting the global environment.

14.2.6 Tissue Engineering and Regenerative Medicine

Biosensors are revolutionizing tissue engineering and regenerative medicine by providing real-time monitoring and precise control over critical biological processes. These devices enable the detection of cellular behavior, tissue growth, and biochemical markers during tissue fabrication and repair. By offering insights into parameters like pH, oxygen levels, glucose concentration, and mechanical stress, biosensors ensure optimal conditions for tissue regeneration and scaffold development. They also facilitate the evaluation of tissue functionality and integration post-implantation, accelerating the development of effective therapeutic solutions. Moreover, biosensors enhance the precision of biofabrication technologies, such as 3D bioprinting, by providing feedback for process optimization.

Biosensors are especially valuable in tissue engineering applications, such as monitoring three-dimensional (3D) printed cell cultures (Wheeldon et al., 2011) and developing "organ-on-a-chip" models, where the levels of biomolecules like glucose, adenosine, and hydrogen peroxide are crucial. These biosensors are essential in influencing the behavior of cells and tissues. Physical and chemical signals, including changes in oxygen consumption, pH, membrane potential, ion concentrations, and the release of various metabolic chemicals and proteins, are well-known indicators transmitted by living organisms, particularly in the brain (Wang et al., 2010). Real-time monitoring of these assays can provide insight into cell function. The process of depositing bioink, which includes living cells and biomaterials, onto a printing surface is called bioprinting; it is a way to control the process of various biomaterials (such as biomolecules and new biocells). It has many features that can be explored in biosensor applications, such as patterning and rapid release of proteins and other biomolecules (Salimi et al., 2011). There are many bioprinting technologies used in biosensor fabrication. For high extrusion bioprinting, such as coaxial or triaxial, it is important to optimize the bioink viscosity to prevent clogging. Other properties such as pore size and cell behavior can also affect the biosensor (Dias et al., 2014). Some printing processes, such as electroplating, can use electricity to transfer thin films of metal nanoparticles (Salimi et al., 2011) or nanowires (Xiang et al., 2008) onto a substrate. By printing thin metal sheets (Peng & Zhang, 2012; Loget et al., 2013), it is possible to create circuits that are an important part of biosensors, including some immunoassays or microarrays. Thin films of biomaterials like polysaccharides, enzymes, nucleic acids, proteins, and cell membranes can be printed using electrodeposition techniques (Novak et al., 2009; Ammam & Fransaer, 2010; Poortinga et al., 2000). Further research is needed on bioprinting technology, which can be used to precisely position a variety of biological products and mammalian cells for biosensor applications for a variety of purposes, including biomanufacturing, organ-on-a-chip technology, and performance measurement devices. Biomanufacturing is a new business concept that uses biological systems to produce biological products, such as human tissue. Bioproducts on a commercial scale are being produced through additive manufacturing methods like 3D printing and other biofabrication technologies. Biomanufacturing facilities can use cells designed to make chemical or molecular products, as well as more cells to produce organs. Controlling the quality and nature of biological processes is essential for the production of reliable products, and biosensing technology can help achieve this. For example, electrochemical enzyme-based biosensors have been used in the past to monitor metabolites in cell cultures (Boero et al., 2014). Organ-on-a-chip technology has revolutionized biomedical research by leveraging microfluidic technology and organoids, combined with biosensors, to enable organ-on-a-chip integration. Organoids are small groups of cells composed of specialized tissues that outperform tissues and organs in their behavior. Organ-on-a-chip technology has a wide range of applications, including monitoring the response of organoids to drugs and other stimuli (Shafiee & Atala, 2017). The use of biosensors to monitor the behavior of microtissues and organoids in real time has advanced this technology. Figure 14.5 shows the development of an organ-on-a-chip system using biosensors (Zhu et al., 2021).

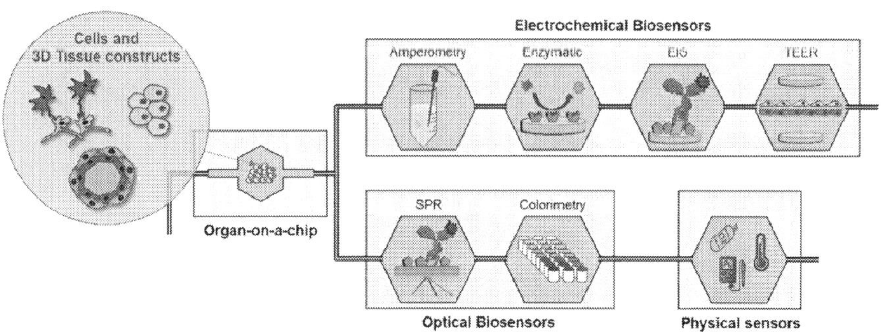

FIGURE 14.5 Representation of organ-on-a-chip integration using biosensors (Zhu et al., 2021).

14.2.7 Neuroscience Applications

Biosensors enable the detection of neurotransmitters and other neurochemicals, advancing our understanding of brain function and neurological disorders. In neuroscience research, 2D electrochemical biosensors are often used to detect neurochemicals in bodily fluids and cellular environments. Many neurotransmitters such as monoamines (Baharfar et al., 2021), choline (Gu et al., 2021), and amino acids (Wang et al., 2022) and other neurochemicals like glucose (Hassan et al., 2021), lactic acid (Wang et al., 2021), ascorbic acid (Singh et al., 2022), nucleic acid (Wang et al., 2024; Yang et al., 2023), and so on are commonly detected by 2D biosensors. Researchers can culture nerve cells on the sensing interface and observe the effect of external stimulation on the release of chemicals from cells *in-situ* (Zeng et al., 2023; Liu et al., 2017; Zhou et al., 2020). Electrochemical biosensors offer a simple and reliable platform for neuroscience experiments. Zeng et al. (2023) designed a biosensor inspired by the ion exchange mechanism involved in coral formation to detect hydrogen peroxide released by tumor cells in real time. The authors examined the chemical behavior of the cells *in-situ* by planting them on the surface of the electrode and then applying stimulation. The results revealed that some cancer cells exhibit a dynamic imbalance of REDOX under the action of dehydroascorbic acid, which is significantly different from what occurs in normal cells. Moreover, due to the stretchability advantage of flexible electrochemical sensing substrates, researchers can fabricate deformations to study mechanobiological problems. Liu et al. (2017) designed a hybrid CNT-AuNT (carbon nanotube nanotube) nanocomposite with good electrochemical performance and stretchability. Due to the electrochemical properties of CNTs and AuNTs to some neurochemicals this strategy can provide a reference for studying the mechanotransduction responses of mechanosensitive cells. Neural microelectrode arrays (MEAs) are among the most commonly used brain–computer interfaces (BCIs). Neural MEAs are composite sensors composed of multiple independent sensing interfaces. MEAs can accept data from multiple detection sites at the same time and record selectively (Zhou et al., 2013). Therefore, spatial resolution is one of the main advantages of MEAs. When a high-density MEA is used, electrical signals emitted by brain tissue can be acquired with greater precision, allowing researchers to obtain high-quality, rapidly changing neuronal action potentials (Bakkum et al., 2013; Dragas et al., 2017). Eickenscheidt et al. (2022) designed a miniaturized optoelectronic device that allows optical and electrical activation at the same point and simultaneous monitoring of the electrical activity of retinal neurons. In this device, LED light passes through polyimide and indium tin oxide to the tissue surface, providing light stimulation and electrical stimulation at the same time. The results of this study indicated that the action potentials of the two stimulation modes during simultaneous stimulation and the direct evoked action potentials were similar. Due to the high spatial resolution and high spatial accessibility of MEAs, they are suitable for more accurately analyzing neural activity at a microscopic scale.

14.2.8 Implantable Devices

Biosensors integrated into implantable medical devices can provide continuous monitoring of vital parameters and drug delivery, improving patient care and treatment efficacy. Implantable and semi-implantable biosensors are capable of providing localized, particularly comprehensive data without environmental interference. Over the past few decades, these biosensors have been developed to perform a variety of functions, including the measurement of chemical, physical, and biological data. Glucose is one of the most important energy sources for humans. However, consuming sugar or storing it in cells will make people with diabetes worse. People with diabetes are at risk of life-threatening complications (Klonoff 2007). The World Health Organization reports that in 2019, around 1.5 million people died from diabetes, with half of those fatalities occurring before the age of 70. Due to the high reliability of glucose oxidase (GOx) in glucose oxidation, most implantable glucose sensors are enzyme-based electrochemical sensors. The redox reaction occurring in these glucose sensors leads to the production of hydrogen peroxide (H_2O_2) and generates electrical energy, transforming glucose concentration into an electrical signal. GOx can be directly applied to working electrodes made of materials like zinc (Li et al., 2023) and carbon (Molinnus et al., 2022). Molinnus et al. developed screen-printed carbon electrodes on biodegradable and biocompatible fibroin substrates integrated with a GOx membrane, allowing glucose monitoring within a range of 0.5–4 mM. Their device remained effective for up to one year with minimal interference from ascorbic acid, epinephrine, and adrenaline. In addition, some non-enzymatic glucose sensing methods have been developed to convert glucose information into optical or fluorescent signals. Bai et al. developed single crystal silicon (m-Si) encapsulated in PLGA, which could provide continuous blood glucose monitoring (Bai et al., 2018). Silicon (Si) has a high refractive index, which results in a tight optical mode limitation at the submicron scale, allowing light to be precisely directed to a target as small as a single cell. Near-infrared light absorbed by tissues provides information about chemical composition by creating spectral signatures due to harmonic overtones and combinations of molecular vibrations of specific chemical groups. These sensors effectively detected glucose concentrations based on optical return loss influenced by chemical groups, particularly at wavelengths of 1160 and 1330 nm. The detection range is 120–180 mg/dL, and the sensitivity is 15 mg/dL. pH deficiency can cause serious problems such as ischemia, kidney disease, brain disease, and so on (Kahlert 2008; Andrews et al., 1994). It can also be an indicator of health; for example, postoperative infection often creates an acidic environment that lowers the pH from 7.5 to 6.7 (Wijayaratna et al., 2021). A low pH value can also indicate cancer (Swietach et al., 2014) or heart disease (Ghoneim et al., 2019). Incomplete gastric bypass surgery can lead to anastomotic leaks that can alter gastric pH and affect other organs (Girard et al., 2014). In this context, continuous point-of-care pH sensors have gained popularity in the field of implantable biomedical devices. As we all know, pH depends on the concentration of hydrogen and hydroxide ions. Various techniques have been developed to measure ion concentrations, one of which involves converting these concentrations into electrical signals. These sensor designs often incorporate nanomaterials. For instance, Chou et al. created a basic polypropylene micromembrane (PPMM) infused with iridium(IV) oxide (IrO_2) nanoparticles (Chou et al., 2022). The introduction of IrO_2 leads to a pH response due to redox interactions, and pH affects the redox potential. It was reported that their sensor can operate within a pH range of 2–12, with a sensitivity of -74.45 mV/pH. Additionally, the sensor demonstrates resistance to fatigue over 5,000 bending cycles and exhibits good biocompatibility. The importance of biosensors in implantable devices cannot be overstated. As technology continues to advance, biosensors are expected to become even more sophisticated, enabling breakthroughs in diagnostics, monitoring, and therapeutic interventions, ultimately transforming healthcare delivery and enhancing the quality of life for patients (Xu et al., 2024).

14.3 FUTURE DIRECTIONS AND POTENTIAL IMPACT OF BIOSENSORS IN BIOMEDICAL DIAGNOSTICS

The future of biosensors in biomedical applications is poised for transformative advancements driven by innovations in nanotechnology, materials science, and data analytics. Biosensor-integrated biomarker technologies will explore options for developing precision medicine, devices, and diagnostics. Implantable biosensors, which could play a key role in accelerating personalized medicine, could also be used for continuous monitoring, allowing researchers to observe the effects of competing drugs on the body and determine if the drugs could be used as treatments. Biosensor chip technology might be implanted in the body to detect changes in blood DNA in the early stages of disease, before symptoms appear. Emerging trends focus on the miniaturization of biosensors, leading to more portable and user-friendly devices that can facilitate real-time monitoring of physiological parameters outside clinical settings. Additionally, the integration of biosensors with wearable technology and the Internet of Things (IoT) is expected to foster a new era of personalized medicine, enabling continuous health surveillance and early detection of diseases.

Smart devices of the future will be able to track multiple health conditions simultaneously and provide data collection through mobile applications. The advancement of this technology is expected to drive significant growth in the medical sector, including patient satisfaction. Biosensors, together with IoT, artificial intelligence, and 5G, are making the field more reliable, sensitive, and treatable. The market for biosensors is growing due to their widespread use in medicine. Biosensors have also been developed across various domains, including medical diagnostics, patient health monitoring, disease detection, and human health management, and they are expected to drive further advancements in these areas. As these technologies continue to evolve, their potential impact on preventive healthcare, therapeutic decision-making, and overall patient outcomes underscores the necessity for interdisciplinary collaboration in biosensor research and application development.

14.4 CONCLUSION

In conclusion, the integration of biosensors within biomedical applications represents a pivotal advancement in diagnostic and therapeutic strategies. These innovative devices enhance the detection of biochemical markers, offering improved sensitivity and specificity, which are critical in clinical settings. Detection and diagnosis of many human diseases at an early stage of development is essential to effectively treat patients. Hence, the development of simple, sensitive, and cost-effective diagnostic tools like biosensors is crucial for the efficient detection of pathogens. Biosensors have a wide range of applications in healthcare and can help doctors and patients manage diseases, provide treatment, prevent diseases, provide patient health information, conduct disease testing, and so on. It is widely believed that biosensors can help personalize medicine in today's world and provide new avenues for medical treatment. This approach has had a profound impact on healthcare and has led to the development of many different therapeutic and diagnostic approaches. In this context, collaboration and technological development in biosensors can lead to a broad knowledge base and multiple competitive areas and products. Biosensors will support the development of personalized cancer treatments by measuring changes in tumor tissue rapidly, locally, and quantitatively during radiation and chemotherapy. There will be many opportunities in the design and development of biosensors that address various medical problems using nanomaterial-based mechanisms. Collectively, these advancements not only facilitate early disease detection but also contribute to personalized medicine, highlighting the transformative role of biosensors in enhancing patient outcomes and guiding therapeutic interventions.

REFERENCES

Ammam, M., & Fransaer, J. (2010) Two enzyme lactose biosensor based on β-galactosidase and glucose oxidase deposited by AC-electrophoresis: Characteristics and performance for lactose determination in milk. *Sensors & Actuators, B: Chemical*, 148, 583–589. https://doi.org/10.1016/j.snb.2010.05.027

Andrews, R. J., Bringas, J. R., & Alonzo, G. (1994). Cerebrospinal fluid pH and PCO2 rapidly follow arterial blood pH and PCO2 with changes in ventilation. *Neurosurgery*, 34(3), 466–470. https://doi.org/10.1227/00006123-199403000-00012

Appelboom, G., Camacho, E., Abraham, M. E., Bruce, S. S., Dumont, E. L., Zacharia, B. E., D'Amico, R., Slomian, J., Reginster, J. Y., Bruyère, O., & Connolly, E. S., Jr (2014). Smart wearable body sensors for patient self-assessment and monitoring. *Archives of Public Health = Archives Belges de Sante Publique*, 72(1), 28. https://doi.org/10.1186/2049-3258-72-28

Arafa, K. K., Ibrahim, A., Mergawy, R., El-Sherbiny, I. M., Febbraio, F., & Hassan, R. Y. A. (2022). Advances in cancer diagnosis: Bio-electrochemical and biophysical characterizations of cancer cells. *Micromachines*, 13(9), 1401. https://doi.org/10.3390/mi13091401

Arshad, R., Rhouati, A., Hayat, A., Nawaz, M. H., Yameen, M. A., Mujahid, A., & Latif, U. (2020). MIP-based impedimetric sensor for detecting dengue fever biomarker. *Applied Biochemistry and Biotechnology*, 191(4), 1384–1394. https://doi.org/10.1007/s12010-020-03285-y

Baharfar, M., Mayyas, M., Rahbar, M., Allioux, F. M., Tang, J., Wang, Y., Cao, Z., Centurion, F., Jalili, R., Liu, G., & Kalantar-Zadeh, K. (2021). Exploring interfacial graphene oxide reduction by liquid metals: Application in selective biosensing. *ACS Nano*, 15(12), 19661–19671. https://doi.org/10.1021/acsnano.1c06973

Bai, W., Yang, H., Ma, Y., Chen, H., Shin, J., Liu, Y., Yang, Q., Kandela, I., Liu, Z., Kang, S. K., Wei, C., Haney, C. R., Brikha, A., Ge, X., Feng, X., Braun, P. V., Huang, Y., Zhou, W., & Rogers, J. A. (2018). Flexible transient optical waveguides and surface-wave biosensors constructed from monocrystalline silicon. *Advanced Materials* (Deerfield Beach, Fla.), 30(32), e1801584. https://doi.org/10.1002/adma.201801584

Baird, C. L., & Myszka, D. G. (2001). Current and emerging commercial optical biosensors. *Journal of Molecular Recognition: JMR*, 14(5), 261–268. https://doi.org/10.1002/jmr.544

Bakkum, D. J., Frey, U., Radivojevic, M., Russell, T. L., Müller, J., Fiscella, M., Takahashi, H., & Hierlemann, A. (2013). Tracking axonal action potential propagation on a high-density microelectrode array across hundreds of sites. *Nature Communications*, 4, 2181. https://doi.org/10.1038/ncomms3181

Bandodkar, A. J., Gutruf, P., Choi, J., Lee, K., Sekine, Y., Reeder, J. T., Jeang, W. J., Aranyosi, A. J., Lee, S. P., Model, J. B., Ghaffari, R., Su, C. J., Leshock, J. P., Ray, T., Verrillo, A., Thomas, K., Krishnamurthi, V., Han, S., Kim, J., Krishnan, S., & Rogers, J. A. (2019). Battery-free, skin-interfaced microfluidic/electronic systems for simultaneous electrochemical, colorimetric, and volumetric analysis of sweat. *Science Advances*, 5(1), eaav3294. https://doi.org/10.1126/sciadv.aav3294

Beitollahi, H., Ebadinejad, F., Shojaie, F., & Torkzadeh-Mahani, M. (2016). A magnetic core–shell Fe_3O_4@SiO_2/MWCNT nanocomposite modified carbon paste electrode for amplified electrochemical sensing of amlodipine and hydrochlorothiazide. *Analytical Methods*, 8, 6185–6193, https://doi.org/10.1039/c6ay01438k

Bender, A., & Cortés-Ciriano, I. (2021). Artificial intelligence in drug discovery: what is realistic, what are illusions? Part 1: Ways to make an impact, and why we are not there yet. *Drug Discovery Today*, 26(2), 511–524. https://doi.org/10.1016/j.drudis.2020.12.009

Bernal, G., Yang, T., Jain, A., & Maes, P. (2018). PhysioHMD: a conformable, modular toolkit for collecting physiological data from head-mounted displays. *Proceedings of the 2018 ACM International Symposium on Wearable Computers*, 160–167. https://doi.org/10.1145/3267242.3267268

Besselink, G. A., Kooyman, R. P., van Os, P. J., Engbers, G. H., & Schasfoort, R. B. (2004). Signal amplification on planar and gel-type sensor surfaces in surface plasmon resonance-based detection of prostate-specific antigen. *Analytical Biochemistry*, 333(1), 165–173. https://doi.org/10.1016/j.ab.2004.05.009

Boero, C., Casulli, M. A., Olivo, J., Foglia, L., Orso, E., Mazza, M., Carrara, S., & De Micheli, G. (2014). Design, development, and validation of an in-situ biosensor array for metabolite monitoring of cell cultures. *Biosensors & Bioelectronics*, 61, 251–259. https://doi.org/10.1016/j.bios.2014.05.030

Cao, C., Kim, J. P., Kim, B. W., Chae, H., Yoon, H. C., Yang, S. S., & Sim, S. J. (2006). A strategy for sensitivity and specificity enhancements in prostate specific antigen-alpha1-antichymotrypsin detection based on surface plasmon resonance. *Biosensors & Bioelectronics*, 21(11), 2106–2113. https://doi.org/10.1016/j.bios.2005.10.014

Cao, Y., Li, Y. H., Lv, D. Y., Chen, X. F., Chen, L. D., Zhu, Z. Y., Chai, Y. F., & Zhang, J. P. (2016). Identification of a ligand for tumor necrosis factor receptor from Chinese herbs by combination of surface plasmon resonance biosensor and UPLC-MS. *Analytical and Bioanalytical Chemistry*, 408(19), 5359–5367. https://doi.org/10.1007/s00216-016-9633-6

Carreiro, S., Smelson, D., Ranney, M., Horvath, K. J., Picard, R. W., Boudreaux, E. D., Hayes, R., & Boyer, E. W. (2015). Real-time mobile detection of drug use with wearable biosensors: A pilot study. *Journal of Medical Toxicology: Official Journal of the American College of Medical Toxicology*, 11(1), 73–79. https://doi.org/10.1007/s13181-014-0439-7

Carreiro, S., Wittbold, K., Indic, P., Fang, H., Zhang, J., & Boyer, E. W. (2016). Wearable Biosensors to Detect Physiologic Change During Opioid Use. *Journal of Medical Toxicology: Official Journal of the American College of Medical Toxicology*, 12(3), 255–262. https://doi.org/10.1007/s13181-016-0557-5

Chandra, P., Koh, W. C., Noh, H. B., & Shim, Y. B. (2012). In vitro monitoring of i-NOS concentrations with an immunosensor: the inhibitory effect of endocrine disruptors on i-NOS release. *Biosensors & Bioelectronics*, 32(1), 278–282. https://doi.org/10.1016/j.bios.2011.11.027

Chen, J., Wei, X., Tang, H., Ali, M. C., Han, Y., Li, Z., & Qiu, H. (2020a). Highly discriminative fluorometric sensor based on luminescent covalent organic nanospheres for tyrosinase activity monitoring and inhibitor screening. *Sensors and Actuators B: Chemical*, 305, 127386. https://doi.org/10.1016/j.snb.2019.127386

Chen, L., Lv, D., Chen, X., Liu, M., Wang, D., Liu, Y., Hong, Z., Zhu, Z., Hu, X., Cao, Y., Yang, J., & Chai, Y. (2018). Biosensor-based active ingredients recognition system for screening STAT3 ligands from medical herbs. *Analytical Chemistry*, 90(15), 8936–8945. https://doi.org/10.1021/acs.analchem.8b01103

Chen, L., Lv, D., Wang, S., Wang, D., Chen, X., Liu, Y., Hong, Z., Zhu, Z., Cao, Y., Chai, Y. (2020b). Surface plasmon resonance-based membrane protein-targeted active ingredients recognition strategy: Construction and implementation in ligand screening from herbal medicines. *Analytical Chemistry*, 92, 3972–3980. https://doi.org/10.1021/acs.analchem.9b05479

Chenthattil, R., Manjunatha, J. G., Ravishankar, D. K., et al. (2020). Validated electrochemical method for simultaneous resolution of tyrosine, uric acid, and ascorbic acid at polymer modified nano-composite paste electrode. *Surface Engineering and Applied Electrochemistry*, 56, 415–426. https://doi.org/10.3103/S1068375520040134

Choi, J. R. (2020). Development of point-of-care biosensors for COVID-19. *Frontiers in Chemistry*, 8, 517. https://doi.org/10.3389/fchem.2020.00517

Chou, S. C., Hsieh, Y. C., Cheang, W. H., Sun, B. Y., Chu, C. Y., Chen, S. Y., Chiao, J. C., & Wu, P. W. (2022). A flexible IrO2 membrane for pH sensing. *Scientific Reports*, 12(1), 11712. https://doi.org/10.1038/s41598-022-15961-6

Choudhary, M., Yadav, P., Singh, A., Kaur, S., Ramirez-Vick, J., Chandra, P., Arora, K., Singh, S. P. (2016). CD 59 targeted ultrasensitive electrochemical immunosensor for fast and noninvasive diagnosis of oral cancer. *Electroanalysis*, 28(10), 2565–2574. https://doi.org/10.1002/elan.201600238

Chowdhury, A. D., Takemura, K., Li, T. C., Suzuki, T., & Park, E. Y. (2019). Electrical pulse-induced electrochemical biosensor for hepatitis E virus detection. *Nature Communications*, 10(1), 3737. https://doi.org/10.1038/s41467-019-11644-5

Choy, C. J., & Berkman, C. E. (2016). A method to determine the mode of binding for GCPII inhibitors using bio-layer interferometry. *Journal of Enzyme Inhibition and Medicinal Chemistry*, 31(6), 1690–1693. https://doi.org/10.3109/14756366.2015.1132208

Christopeit, T., Carlsen, T. J., Helland, R., & Leiros, H. K. (2015). Discovery of novel inhibitor scaffolds against the metallo-β-lactamase VIM-2 by Surface Plasmon Resonance (SPR) based fragment screening. *Journal of Medicinal Chemistry*, 58(21), 8671–8682. https://doi.org/10.1021/acs.jmedchem.5b01289

Cohen-Bacrie, S., Ninove, L., Nougairède, A., Charrel, R., Richet, H., Minodier, P., Badiaga, S., Noël, G., La Scola, B., de Lamballerie, X., Drancourt, M., & Raoult, D. (2011). Revolutionizing clinical microbiology laboratory organization in hospitals with in situ point-of-care. *PLoS One*, 6(7), e22403. https://doi.org/10.1371/journal.pone.0022403

Deo, R. P., & Wang, J. (2004). Electrochemical detection of carbohydrates at carbon-nanotube modified glassy-carbon electrodes. Electrochemistry Communications, 6 (3), 284–287. https://doi.org/10.1016/j.elecom.2004.01.003

Dias, A. D., Kingsley, D. M., & Corr, D. T. (2014). Recent advances in bioprinting and applications for biosensing. *Biosensors*, 4(2), 111–136. https://doi.org/10.3390/bios4020111

Dragas, J., Viswam, V., Shadmani, A., et al. (2017) A multi-functional microelectrode array featuring 59,760 electrodes, 2048 electrophysiology channels, stimulation, impedance measurement and neurotransmitter detection channels. IEEE Journal of Solid-State Circuits, 52(6), 1576–1590. https://doi.org/10.1109/jssc.2017.2686580

Ehrmeyer, S. S., & Laessig, R. H. (2007). Point-of-care testing, medical error, and patient safety: A 2007 assessment. *Clinical Chemistry and Laboratory Medicine*, 45(6), 766–773. https://doi.org/10.1515/CCLM.2007.164

Eickenscheidt, M., Herrmann, T., Weisshap, M., Mittnacht, A., Rudmann, L., Zeck, G., & Stieglitz, T. (2022). An optoelectronic neural interface approach for precise superposition of optical and electrical stimulation in flexible array structures. *Biosensors & Bioelectronics*, 205, 114090. https://doi.org/10.1016/j.bios.2022.114090

Fernández-Sánchez, C., McNeil, C. J., Rawson, K., & Nilsson, O. (2004). Disposable noncompetitive immunosensor for free and total prostate-specific antigen based on capacitance measurement. *Analytical -Chemistry*, 76(19), 5649–5656. https://doi.org/10.1021/ac0494937

Gangadhara, G., Dahl, G., Bohnacker, T., Rae, R., Gunnarsson, J., Blaho, S., Öster, L., Lindmark, H., Karabelas, K., Pemberton, N., Tyrchan, C., Mogemark, M., Wymann, M. P., Williams, R. L., Perry, M. W. D., Papavoine, T., & Petersen, J. (2019). A class of highly selective inhibitors bind to an active state of PI3Kγ. *Nature Chemical Biology*, 15(4), 348–357. https://doi.org/10.1038/s41589-018-0215-0

Gao, Q., Zhang, J., Xie, Z., Omisore, O.M., Zhang, J., Wang, L., & Li, H. (2019). Highly stretchable sensors for wearable biomedical applications. *Journal of Materials Science*, 54, 5187-5223. https://doi.org/10.1007/s10853-018-3171-x

Gast, M., Sobek, H., & Mizaikoff, B. (2019). Advances in imprinting stretchable sensors for wearable biomedical applications. Journal of Materials Science, 114, 218–232. https://doi.org/10.1016/j.trac.2019.03.010

Geschwindner, S., Carlsson, J. F., & Knecht, W. (2012). Application of optical biosensors in small-molecule screening activities. *Sensors*, 12(4), 4311–4323. https://doi.org/10.3390/s120404311

Ghoneim, M. T., Nguyen, A., Dereje, N., Huang, J., Moore, G. C., Murzynowski, P. J., & Dagdeviren, C. (2019). Recent progress in electrochemical pH-sensing materials and configurations for biomedical applications. *Chemical Reviews*, 119(8), 5248–5297. https://doi.org/10.1021/acs.chemrev.8b00655

Girard, E., Messager, M., Sauvanet, A., Benoist, S., Piessen, G., Mabrut, J. Y., & Mariette, C. (2014). Anastomotic leakage after gastrointestinal surgery: Diagnosis and management. *Journal of Visceral Surgery*, 151(6), 441–450. https://doi.org/10.1016/j.jviscsurg.2014.10.004

Girigoswami, K., & Akhtar, N. (2019). Nanobiosensors and fluorescence based biosensors: An overview. *International Journal of Nano Dimension*, 10(1), 1-17. https://dorl.net/dor/20.1001.1.20088868.2019.10.1.1.2

Gu, X., & Wang, X. (2021). An overview of recent analysis and detection of acetylcholine. *Analytical Biochemistry*, 632, 114381. https://doi.org/10.1016/j.ab.2021.114381

Haleem, A., Javaid, M., Singh R. P., Suman, R., & Rab, S. (2021). Biosensors applications in medical field: A brief review. *Sensors International*, 2, 100100. https://doi.org/10.1016/j.sintl.2021.100100

Hallfors, N., Alhawari, M., Jaoude, M. A., Kifle, Y., Saleh, H.H., Liao, K., Ismail, M., & Isakovic, A. F. (2018). Graphene oxide: Nylon ECG sensors for wearable IoT healthcare—nanomaterial and SoC interface. *Analog Integrated Circuits and Signal Processing*, 96, 253–260. https://doi.org/10.1007/s10470-018-1116-6

Hassan, M. H., Vyas, C., Grieve, B., & Bartolo, P. (2021). Recent advances in enzymatic and non-enzymatic electrochemical glucose sensing. *Sensors*, 21(14), 4672. https://doi.org/10.3390/s21144672

Hassan, R. Y. A. (2022). Advances in electrochemical nano-biosensors for biomedical and environmental applications: From current work to future perspectives. *Sensors*, 22(19), 7539. https://doi.org/10.3390/s22197539

Huang, L., Reekmans, G., Saerens, D., Friedt, J. M., Frederix, F., Francis, L., Muyldermans, S., Campitelli, A., & Van Hoof, C. (2005). Prostate-specific antigen immunosensing based on mixed self-assembled monolayers, camel antibodies and colloidal gold enhanced sandwich assays. *Biosensors & Bioelectronics*, 21(3), 483–490. https://doi.org/10.1016/j.bios.2004.11.016

Ibrahim, M., Ibrahim, H., Almandil, N., Kawde, A. N. (2018). Gold nanoparticles/f-MWCNT nanocomposites modified glassy carbon paste electrode as a novel voltammetric sensor for the determination of cyproterone acetate in pharmaceutical and human body fluids. *Sensors and Actuators B: Chemical*, 274, 123–132. https://doi.org/10.1016/j.snb.2018.07.105

Jahani, S., & Beitollahi, H. (2016). Selective detection of dopamine in the presence of uric acid using NiO nanoparticles decorated on graphene nanosheets modified screenprinted electrodes. *Electroanalysis*, 28, 2022–2028, https://doi.org/10.1002/elan.201501136

Jemal, A., Siegel, R., Ward, E., Murray, T., Xu, J., Smigal, C., & Thun, M. J. (2006). Cancer statistics, 2006. *CA: A Cancer Journal for Clinicians*, 56(2), 106–130. https://doi.org/10.3322/canjclin.56.2.106

Kahlert, H. (2008). Functionalized carbon electrodes for pH determination. *Journal of Solid State Electrochemistry*, 12, 1255–1266. https://doi.org/10.1007/s10008-008-0566-7

Kaminski, T., Gunnarsson, A., & Geschwindner, S. (2017). Harnessing the versatility of optical biosensors for target-based small-molecule drug discovery. *ACS Sensors*, 2(1), 10–15. https://doi.org/10.1021/acssensors.6b00735

Kim, J. H., Hwang, J. Y., Hwang, H. R., Kim, H. S., Lee, J. H., Seo, J. W., Shin, U. S., & Lee, S. H. (2018). Simple and cost-effective method of highly conductive and elastic carbon nanotube/polydimethylsiloxane composite for wearable electronics. *Scientific Reports*, 8(1), 1375. https://doi.org/10.1038/s41598-017-18209-w

Kim, Y., Lim, T., Kim, C. H., et al. (2018). Organic electrochemical transistor-based channel dimension-independent single-strand wearable sweat sensors. *NPG Asia Materials*, 10, 1086–1095. https://doi.org/10.1038/s41427-018-0097-3

Kintzios, S., Bem, F., Mangana, O., Nomikou, K., Markoulatos, P., Alexandropoulos, N., Fasseas, C., Arakelyan, V., Petrou, A. L., Soukouli, K., Moschopoulou, G., Yialouris, C., & Simonian, A. (2004). Study on the mechanism of bioelectric recognition assay: Evidence for immobilized cell membrane interactions with viral fragments. *Biosensors & Bioelectronics*, 20(4), 907–916. https://doi.org/10.1016/j.bios.2004.04.009

Kintzios, S., Pistola, E., Panagiotopoulos, P., Bomsel, M., Alexandropoulos, N., Bem, F., Ekonomou, G., Biselis, J., & Levin, R. (2001). Bioelectric recognition assay (BERA). *Biosensors & Bioelectronics*, 16(4-5), 325–336. https://doi.org/10.1016/s0956-5663(01)00127-0

Klonoff, D. C. (2007). The benefits of implanted glucose sensors. *Journal of Diabetes Science and Technology*, 1(6), 797–800. https://doi.org/10.1177/193229680700100601

Kong, W., Wu, D., Hu, N., Li, N., Dai, C., Chen, X., Suo, Y., Li, G., & Wu, Y. (2018). Robust hybrid enzyme nanoreactor mediated plasmonic sensing strategy for ultrasensitive screening of anti-diabetic drug. *Biosensors & Bioelectronics*, 99, 653–659. https://doi.org/10.1016/j.bios.2017.08.009

Krzyzosiak, A., Sigurdardottir, A., Luh, L., Carrara, M., Das, I., Schneider, K., & Bertolotti, A. (2018). Target-based discovery of an inhibitor of the regulatory phosphatase PPP1R15B. *Cell*, 174(5), 1216–1228.e19. https://doi.org/10.1016/j.cell.2018.06.030

Kweon, O. Y., Lee, S. J., & Oh, J. H. (2018). Wearable high-performance pressure sensors based on three-dimensional electrospun conductive nanofibers. *NPG Asia Materials*, 10, 540–551. https://doi.org/10.1038/s41427-018-0041-6

Li, G., Kong, W., Zhao, M., Lu, S., Gong, P., Chen, G., Xia, L., Wang, H., You, J., & Wu, Y. (2016). A fluorescence resonance energy transfer (FRET) based "Turn-On" nanofluorescence sensor using a nitrogen-doped carbon dot-hexagonal cobalt oxyhydroxide nanosheet architecture and application to α-glucosidase inhibitor screening. *Biosensors & Bioelectronics*, 79, 728–735. https://doi.org/10.1016/j.bios.2015.12.094

Li, J., Liu, J., Wu, Z., Shang, X., Li, Y., Huo, W., & Huang, X. (2023). Fully printed and self-compensated bioresorbable electrochemical devices based on galvanic coupling for continuous glucose monitoring. *Science Advances*, 9(29), eadi3839. https://doi.org/10.1126/sciadv.adi3839

Liu, J. W., Wang, Y. M., Xu, L., Duan, L. Y., Tang, H., Yu, R. Q., & Jiang, J. H. (2016). Melanin-like nanoquencher on graphitic carbon nitride nanosheets for tyrosinase activity and inhibitor assay. *Analytical Chemistry*, 88(17), 8355–8358. https://doi.org/10.1021/acs.analchem.6b01667

Liu, Y. L., Qin, Y., Jin, Z. H., Hu, X. B., Chen, M. M., Liu, R., Amatore, C., & Huang, W. H. (2017). A stretchable electrochemical sensor for inducing and monitoring cell mechanotransduction in real time. *Angewandte Chemie* (International ed. in English), 56(32), 9454–9458. https://doi.org/10.1002/anie.201705215

Loget, G., Wood, J. B., Cho, K., Halpern, A. R., & Corn, R. M. (2013). Electrodeposition of polydopamine thin films for DNA patterning and microarrays. *Analytical Chemistry*, 85(21), 9991–9995. https://doi.org/10.1021/ac4022743

Mansuriya, B. D., & Altintas, Z. (2020). Graphene quantum dot-based electrochemical immunosensors for biomedical applications. *Materials*, 13(1), 96. https://doi.org/10.3390/ma13010096

Marín-Morales, J., Higuera-Trujillo, J. L., Greco, A., Guixeres, J., Llinares, C., Scilingo, E. P., Alcañiz, M., & Valenza, G. (2018). Affective computing in virtual reality: Emotion recognition from brain and heartbeat dynamics using wearable sensors. *Scientific Reports*, 8(1), 13657. https://doi.org/10.1038/s41598-018-32063-4

Mascini, M., & Tombelli, S. (2008). Biosensors for biomarkers in medical diagnostics. *Biomarkers: Biochemical Indicators of Exposure, Response, and Susceptibility to Chemicals*, 13(7), 637–657. https://doi.org/10.1080/13547500802645905

Metkar, S. K., & Girigoswami, K. (2018). Diagnostic biosensors in medicine- a review. *Biocatalysis and Agricultural Biotechnology*, 17, 271–283. https://doi.org/10.1016/j.bcab.2018.11.029

Meyerhoff, M. E., Duan, C., & Meusel, M. (1995). Novel nonseparation sandwich-type electrochemical enzyme immunoassay system for detecting marker proteins in undiluted blood. *Clinical Chemistry*, 41(9), 1378–1384. https://pubmed.ncbi.nlm.nih.gov/7544708/

Molinnus, D., Janus, K. A., Fang, A. C., Drinic, A., Achtsnicht, S., Köpf, M., Keusgen, M., & Schöning, M. J. (2022). Thick-film carbon electrode deposited onto a biodegradable fibroin substrate for biosensing applications. *Physica Status Solidi (A)*, 219: 2200100. https://doi.org/10.1002/pssa.202200100

Nam, D., Cha, J. M., & Park, K. (2021). Next-generation wearable biosensors developed with flexible bio-chips. *Micromachines*, 12(1), 64. https://doi.org/10.3390/mi12010064

Nasipude, S., Bagwan, S., & Ranveer, A. (2021). Biosensor for environmental monitoring. *International Journal of Innovations in Engineering Research and Technology*, 2(4), 1–8. https://repo.ijiert.org/index.php/ijiert/article/view/439

Nittala, A. S., & Steimle, J. (2022). *Next steps in epidermal computing: Opportunities and challenges for soft on-skin devices.* CHI Conference on Human Factors in Computing Systems (CHI '22), April 29-May 5, 2022, New Orleans, LA. ACM, New York, NY, USA 22 Pages. https://doi.org/10.1145/3491102.3517668

Nittala, A. S., Khan, A., Kruttwig, K., Kraus, T., & Steimle, J. (2020). *PhysioSkin: Rapid fabrication of skin-conformal physiological interfaces.* Proceedings of the 2020 CHI Conference on Human Factors in Computing Systems, 1–10. https://doi.org/10.1145/3313831.3376366

Novak, S., Maver, U., Peternel, Š., Venturini, P., Bele, M., Gaberšček, M. (2009) Electrophoretic deposition as a tool for separation of protein inclusion bodies from host bacteria in suspension. *Colloids and Surfaces A: Physicochemical and Engineering Aspects*, 340, 155–160. https://doi.org/10.1016/j.colsurfa.2009.03.023

Overacker, R. D., Plitzko, B., & Loesgen, S. (2021). Biolayer interferometry provides a robust method for detecting DNA binding small molecules in microbial extracts. *Analytical and Bioanalytical Chemistry*, 413(4), 1159–1171. https://doi.org/10.1007/s00216-020-03079-5

Peng, S., & Zhang, X. (2012). Electrodeposition of CdSe quantum dots and its application to an electrochemiluminescence immunoassay for α-fetoprotein. *Microchim Acta*, 178, 323–330. https://doi.org/10.1007/s00604-012-0844-z

Piletska, E. V., Mirkes, E., Piletsky, S. S., Abosoglu, H., Cassim, A., Chu, E., Doughty, S., Eganda, S.-J., Fuchigami, H., Hussein, A., et al. (2020). Combinatorial screening of polymer nanoparticles for their ability to recognize epitopes of AAV-neutralizing antibodies. *Journal of Molecular Recognition*, 33, e2824. https://doi.org/10.1002/jmr.2824

Poortinga, A. T., Bos, R., & Busscher, H. J. (2000). Controlled electrophoretic deposition of bacteria to surfaces for the design of biofilms. *Biotechnology and Bioengineering*, 67(1), 117–120. https://doi.org/10.1002/(sici)1097-0290(20000105)67:1%3C117::aid-bit14%3E3.0.co;2-6

Pourasl, A. H., Ahmadi, M. T., Rahmani, M., et al. (2014). Analytical modeling of glucose biosensors based on carbon nanotubes. *Nanoscale Research Letters*, 9, 33. https://doi.org/10.1186/1556-276X-9-33

Přibyl, J., Hepel, M., & Skládal, P. (2006). Piezoelectric immunosensors for polychlorinated biphenyls operating in aqueous and organic phases. *Sensors and Actuators B: Chemical*, 113(2), 900–910. https://doi.org/10.1016/j.snb.2005.03.077

Pushpanjali, P. A., Manjunatha, J. G., Hareesha, N., et al. (2021). Voltammetric analysis of antihistamine drug cetirizine and paracetamol at poly(L-Leucine) layered carbon nanotube paste electrode. *Surfaces and Interfaces*, 24, 101154. https://doi.org/10.1016/j.surfin.2021.101154

Saha, S., Cano-Garcia, H., Sotiriou, I., Lipscombe, O., Gouzouasis, I., Koutsoupidou, M., Palikaras, G., Mackenzie, R., Reeve, T., Kosmas, P., & Kallos, E. (2017). A glucose sensing system based on transmission measurements at millimetre waves using micro strip patch antennas. *Scientific Reports*, 7(1), 6855. https://doi.org/10.1038/s41598-017-06926-1

Salimi, A., Noorbakhsh, A., Rafiee-Pour, H.-A., & Ghourchian, H. (2011). Direct voltammetry of copper, zinc-superoxide dismutase immobilized onto electrodeposited nickel oxide nanoparticles: Fabrication of amperometric superoxide biosensor. *Electroanalysis*, 23, 683–691. https://doi.org/10.1002/elan.201000519

Sarkar, P., Pal, P. S., Ghosh, D., Setford, S. J., & Tothill, I. E. (2002). Amperometric biosensors for detection of the prostate cancer marker (PSA). *International Journal of Pharmaceutics*, 238(1–2), 1–9. https://doi.org/10.1016/s0378-5173(02)00015-7

Schönauer, E., Kany, A. M., Haupenthal, J., Hüsecken, K., Hoppe, I. J., Voos, K., Yahiaoui, S., Elsässer, B., Ducho, C., Brandstetter, H., & Hartmann, R. W. (2017). Discovery of a potent inhibitor class with high selectivity toward clostridial collagenases. *Journal of the American Chemical Society*, 139(36), 12696–12703. https://doi.org/10.1021/jacs.7b06935

Shafiee, A., & Atala, A. (2017). Tissue engineering: Toward a new era of medicine. *Annual Review of Medicine*, 68, 29–40. https://doi.org/10.1146/annurev-med-102715-092331

Sharma, K. K., Przybilla, F., Restle, T., Godet, J., & Mély, Y. (2016). FRET-based assay to screen inhibitors of HIV-1 reverse transcriptase and nucleocapsid protein. *Nucleic Acids Research*, 44(8), e74. https://doi.org/10.1093/nar/gkv1532

Sima, L. E., Yakubov, B., Zhang, S., Condello, S., Grigorescu, A. A., Nwani, N. G., Chen, L., Schiltz, G. E., Arvanitis, C., Zhang, Z. Y., & Matei, D. (2019). Small molecules target the interaction between tissue transglutaminase and fibronectin. *Molecular Cancer Therapeutics*, 18(6), 1057–1068. https://doi.org/10.1158/1535-7163.MCT-18-1148

Singh, A., Sharma, A., Ahmed, A. et al. (2022). Highly selective and efficient electrochemical sensing of ascorbic acid via CuO/rGO nanocomposites deposited on conductive fabric. *Applied Physics A*, 128, 262. https://doi.org/10.1007/s00339-022-05436-w

Srinivasan, B., & Tung, S. (2015). Development and applications of portable biosensors. *Journal of Laboratory Automation*, 20(4), 365–389. https://doi.org/10.1177/2211068215581349

Sun, J., & Yang, X. (2015). Gold nanoclusters-Cu($^{2+}$) ensemble-based fluorescence turn-on and real-time assay for acetylcholinesterase activity and inhibitor screening. *Biosensors & Bioelectronics*, 74, 177–182. https://doi.org/10.1016/j.bios.2015.06.013

Susmel, S., Guilbault, G. G., & O'Sullivan, C. K. (2003). Demonstration of labeless detection of food pathogens using electrochemical redox probe and screen printed gold electrodes. *Biosensors & Bioelectronics*, 18(7), 881–889. https://doi.org/10.1016/s0956-5663(02)00214-2

Swietach, P., Vaughan-Jones, R. D., Harris, A. L., & Hulikova, A. (2014). The chemistry, physiology and pathology of pH in cancer. *Philosophical Transactions of the Royal Society of London. Series B, Biological Sciences*, 369(1638), 20130099. https://doi.org/10.1098/rstb.2013.0099

Tancharoen, C., Sukjee, W., Thepparit, C., Jaimipuk, T., Auewarakul, P., Thitithanyanont, A., & Sangma, C. (2019). Electrochemical biosensor based on surface imprinting for zika virus detection in serum. *ACS Sensors*, 4(1), 69–75. https://doi.org/10.1021/acssensors.8b00885

Tang, D. P., Yuan, R., Chai, Y. Q., Zhong, X., Liu, Y., Dai, J. Y., & Zhang, L. Y. (2004). Novel potentiometric immunosensor for hepatitis B surface antigen using a gold nanoparticle-based biomolecular immobilization method. *Analytical Biochemistry*, 333(2), 345–350. https://doi.org/10.1016/j.ab.2004.06.035

Tao, Y., Chen, L., & Jiang, E. (2021). Layer-by-layer assembly strategy for fabrication of polydopamine-polyethyleneimine hybrid modified fibers and their application to solid-phase microextraction of bioactive molecules from medicinal plant samples followed by surface plasmon resonance biosensor validation. *Analytica Chimica Acta*, 1146, 155–165. https://doi.org/10.1016/j.aca.2020.11.016

Tao, Y., Chen, L., Pan, M., Zhu, F., & Zhu, D. (2021). Tailored biosensors for drug screening, efficacy assessment, and toxicity evaluation. *ACS Sensors*, 6(9), 3146–3162. https://doi.org/10.1021/acssensors.1c01600

Vourvopoulos, A., Niforatos, E., & Giannakos, M. (2019). *EEGlass: An EEG-eyeware prototype for ubiquitous brain-computer interaction.* Adjunct Proceedings of the 2019 ACM International Joint Conference on Pervasive and Ubiquitous Computing and Proceedings of the 2019 ACM International Symposium on Wearable Computers (UbiComp/ISWC '19 Adjunct). Association for Computing Machinery, New York, NY, 647–652. https://doi.org/10.1145/3341162.3348383

Wang, F., Zhang, C., Deng, S., Jiang, Y., Zhang, P., Yang, H., Xiang, L., Lyu, Y., Cai, R., & Tan, W. (2024). Dual-responsive 3D DNA nanomachines cascaded hybridization chain reactions for novel self-powered flexible microRNA-detecting platform. *Biosensors and Bioelectronics*, 252, 116149. https://doi.org/10.1016/j.bios.2024.116149

Wang, M., Yang, Y., Min, J., Song, Y., Tu, J., Mukasa, D., Ye, C., Xu, C., Heflin, N., McCune, J. S., Hsiai, T. K., Li, Z., & Gao, W. (2022). A wearable electrochemical biosensor for the monitoring of metabolites and nutrients. *Nature Biomedical Engineering*, 6(11), 1225–1235. https://doi.org/10.1038/s41551-022-00916-z

Wang, R., Wang, Y., Lassiter, K., Li, Y., Hargis, B., Tung, S., Berghman, L., & Bottje, W. (2009). Interdigitated array microelectrode based impedance immunosensor for detection of avian influenza virus H5N1. *Talanta*, 79(2), 159–164. https://doi.org/10.1016/j.talanta.2009.03.017

Wang, R., Zhai, Q., An, T., Gong, S., & Cheng, W. (2021). Stretchable gold fiber-based wearable textile electrochemical biosensor for lactate monitoring in sweat. *Talanta*, 222, 121484. https://doi.org/10.1016/j.talanta.2020.121484

Wang, Y., Chen, Q., & Zeng, X. (2010). Potentiometric biosensor for studying hydroquinone cytotoxicity in vitro. *Biosensors & Bioelectronics*, 25(6), 1356–1362. https://doi.org/10.1016/j.bios.2009.10.027

Wang, Y., Yin, L., Bai, Y., Liu, S., Wang, L., Zhou, Y., Hou, C., Yang, Z., Wu, H., Ma, J., Shen, Y., Deng, P., Zhang, S., Duan, T., Li, Z., Ren, J., Xiao, L., Yin, Z., Lu, N., & Huang, Y. (2020). Electrically compensated, tattoo-like electrodes for epidermal electrophysiology at scale. *Science Advances*, 6(43), eabd0996. https://doi.org/10.1126/sciadv.abd0996

Wee, K. W., Kang, G. Y., Park, J., Kang, J. Y., Yoon, D. S., Park, J. H., & Kim, T. S. (2005). Novel electrical detection of label-free disease marker proteins using piezoresistive self-sensing micro-cantilevers. *Biosensors & Bioelectronics*, 20(10), 1932–1938. https://doi.org/10.1016/j.bios.2004.09.023

Wheeldon, I., Farhadi, A., Bick, A. G., Jabbari, E., & Khademhosseini, A. (2011). Nanoscale tissue engineering: spatial control over cell-materials interactions. *Nanotechnology*, 22(21), 212001. https://doi.org/10.1088/0957-4484/22/21/212001

Wijayaratna, U., Kiridena, S., Adams, J. D., Behrend, C. J., & Anker, J. N. (2021). Synovial fluid pH sensor for early detection of prosthetic hip infections. *Advanced Functional Materials*, 31(37), 2104124. https://doi.org/10.1002/adfm.202104124

Wu, G., Datar, R. H., Hansen, K. M., Thundat, T., Cote, R. J., & Majumdar, A. (2001). Bioassay of prostate-specific antigen (PSA) using microcantilevers. *Nature Biotechnology*, 19(9), 856–860. https://doi.org/10.1038/nbt0901-856

Xia, Y., Wu, L., Hu, Y., He, Y., Cao, Z., Zhu, X., Yi, X., & Wang, J. (2019). Sensitive surface plasmon resonance detection of methyltransferase activity and screening of its inhibitors amplified by p53 protein bound to methylation-specific ds-DNA consensus sites. *Biosensors & Bioelectronics*, 126, 269–274. https://doi.org/10.1016/j.bios.2018.10.054

Xiang, C., Kung, S. C., Taggart, D. K., Yang, F., Thompson, M. A., Güell, A. G., Yang, Y., & Penner, R. M. (2008). Lithographically patterned nanowire electrodeposition: A method for patterning electrically continuous metal nanowires on dielectrics. *ACS Nano*, 2(9), 1939–1949. https://doi.org/10.1021/nn800394k

Xu, Y., Zhang, J., Ray, W. Z., & MacEwan, M. R. (2024). Implantable and semi-implantable biosensors for minimally invasive disease diagnosis. *Processes*, 12(7), 1535. https://doi.org/10.3390/pr12071535

Yang, L. J., Chen, R. H., Hamdoun, S., Coghi, P., Ng, J. P. L., Zhang, D. W., Guo, X., Xia, C., Law, B. Y. K., & Wong, V. K. W. (2021). Corilagin prevents SARS-CoV-2 infection by targeting RBD-ACE2 binding. *Phytomedicine: International Journal of Phytotherapy and Phytopharmacology*, 87, 153591. https://doi.org/10.1016/j.phymed.2021.153591

Yang, Y., Wang, F., Li, J., He, S., Lyu, Y., Yang, H., Cai, R., & Tan, W. (2023). Self-powered biosensor based on DNA walkers for ultrasensitive MicroRNA detection. *Analytical Chemistry*, 95(40), 15042–15048. https://doi.org/10.1021/acs.analchem.3c03087

Zadran, S., Standley, S., Wong, K., Otiniano, E., Amighi, A., & Baudry, M. (2012). Fluorescence resonance energy transfer (FRET)-based biosensors: Visualizing cellular dynamics and bioenergetics. *Applied Microbiology and Biotechnology*, 96(4), 895–902. https://doi.org/10.1007/s00253-012-4449-6

Zeng, Z., Wang, J., Zhao, S., Zhang, Y., Fan, J., Wu, H., Chen, J., Zhang, Z., Meng, Z., Yang, L., Wang, R., Zhang, B., Wang, G., Li, C.-Z., & Zang, G. (2023). A bioinspired flexible sensor for electrochemical probing of dynamic redox disequilibrium in cancer cells. *Advanced Science*, 10, 2304079. https://doi.org/10.1002/advs.202304079

Zhang, X., Hu, Y., Yang, X., Tang, Y., Han, S., Kang, A., Deng, H., Chi, Y., Zhu, D., & Lu, Y. (2019). FÖrster resonance energy transfer (FRET)-based biosensors for biological applications. *Biosensors & Bioelectronics*, 138, 111314. https://doi.org/10.1016/j.bios.2019.05.019

Zhou, M., Jiang, Y., Wang, G., Wu, W., Chen, W., Yu, P., Lin, Y., Mao, J., & Mao, L. (2020). Single-atom Ni-N4 provides a robust cellular NO sensor. *Nature Communications*, 11(1), 3188. https://doi.org/10.1038/s41467-020-17018-6

Zhou, S., Liu, C., Song, Y., & Cai, X. (2013). Highly selective and sensitive determination of dopamine using nafion coated microelectrode arrays. *Journal of Nanoscience and Nanotechnology*, 13, 1598. https://doi.org/10.1166/jnn.2013.6077

Zhou, W., Yang, M., Li, S., & Zhu, J. (2018). Surface plasmon resonance imaging validation of small molecule drugs binding on target protein microarrays. *Applied Surface Science*, 450, 328–335. https://doi.org/10.1016/j.apsusc.2018.04.072

Zhu, M., Fang, J., Zhang, J., Zhang, Z., Xie, J., Yu, Y., Ruan, J. J., Chen, Z., Hou, W., Yang, G., Su, W., & Ruan, B. H. (2017). Biomolecular interaction assays identified dual inhibitors of glutaminase and glutamate dehydrogenase that disrupt mitochondrial function and prevent growth of cancer cells. *Analytical Chemistry*, 89(3), 1689–1696. https://doi.org/10.1021/acs.analchem.6b03849

Zhu, Y., Mandal, K., Hernandez, A. L., Kawakita, S., Huang, W., Bandaru, P., Ahadian, S., Kim, H. J., Jucaud, V., Dokmeci, M. R., & Khademhosseini, A. (2021). State of the art in integrated biosensors for organ-on-a-chip applications. *Current Opinion in Biomedical Engineering*, 19, 100309. https://doi.org/10.1016/j.cobme.2021.100309

Index

A

B

C

D

Q

R

S

X

Y

Z

Made in the USA
Las Vegas, NV
19 September 2026

4d5d09ee-ccfb-421a-8efa-e8f2140e101fR01